Functional Electrodes for Enzymatic and Microbial Electrochemical Systems

Functional Electrodes for Enzymatic and Microbial Electrochemical Systems

Editors

Nicolas Brun
CNRS, France

Victoria Flexer
CONICET, Argentina

World Scientific

NEW JERSEY · LONDON · SINGAPORE · BEIJING · SHANGHAI · HONG KONG · TAIPEI · CHENNAI · TOKYO

Published by

World Scientific Publishing Europe Ltd.

57 Shelton Street, Covent Garden, London WC2H 9HE

Head office: 5 Toh Tuck Link, Singapore 596224

USA office: 27 Warren Street, Suite 401-402, Hackensack, NJ 07601

Library of Congress Cataloging-in-Publication Data
Names: Brun, Nicolas, 1984– editor. | Flexer, Victoria, editor.
Title: Functional electrodes for enzymatic and microbial electrochemical systems /
edited by Nicolas Brun (CNRS, France), Victoria Flexer (CONICET, Argentina).
Description: New Jersey : World Scientific, 2017. | Includes bibliographical references.
Identifiers: LCCN 2017005406 | ISBN 9781786343536 (hc : alk. paper)
Subjects: LCSH: Electric power production from chemical action. | Electrodes, Enzyme. |
Biomass energy. | Renewable energy sources.
Classification: LCC TK2901 .F86 2017 | DDC 621.31/2424--dc23
LC record available at https://lccn.loc.gov/2017005406

British Library Cataloguing-in-Publication Data
A catalogue record for this book is available from the British Library.

Desk Editors: Ram Mohan/Shi Ying Koe

Typeset by Stallion Press
Email: enquiries@stallionpress.com

Printed in Singapore

Preface

Purified enzymes and whole microbial cells are inexpensive or even free catalysts that researchers have mastered to mass grow and purify, and that can be produced almost in unlimited amounts. In addition, they are interesting from the perspective that they catalyze an extremely large variety of chemical reactions, ranging from the oxidation of dozens of biofuels, and waste materials, to oxygen reduction, CO_2 fixation, and the synthesis of multi-carbon compounds. Moreover, they operate at close to ambient temperature and pressure and, usually show optimal activity at circumneutral pH. The capacity of whole microbial cells to harvest electrons from organic matter was realized over 100 years ago. But it was not until the 1960s that researchers started to alternatively propose the use of purified enzymes or bacteria in combination with electrodes to convert chemical energy into electrical energy, either as a means of electricity generation, or as a way of transducing a biological signal in a biosensor. Aside from the pioneering work from Potter[1] and Cohen,[2] enlightening but of clear fundamental nature, electrochemical devices based on the catalytic activity of bacteria[3,4] and enzymes[5] saw the light almost simultaneously, a most interesting coincidence. Since the 1960s, the field of bioelectrochemistry evolved steadily but slowly, until a real boom aroused in the 1990s. Starting from the more than attractive characteristics that these biocatalysts present, lots of hope was put in the prospective use of both enzymes and whole microbial cells as catalysts to be immobilized on electrode surfaces. Unfortunately, these also included some unreasonable suggestions, including that they might one day replace noble metal catalysts, or that biofuel cells might become a new green energy source replacing, at least partially, fossil resources.

Since those early claims, bioelectrochemical systems have largely matured, and some very smart real-life devices have been launched into the market, with enzymatic strips for diabetes management being the flagship in the field. Reasonable and yet great expectations have replaced unreasonable claims, including resource recovery in microbial systems, and subcutaneous implantable enzymatic fuel cells. Research in the field has evolved substantially, from a crazy and rather blind search for optimized performance, to the realization that there is a thermodynamic limit to performance and that some specialty applications are now the driving force to move the field forward. Moreover, researchers and engineers now acknowledge that we will only approach that thermodynamic limit if we improve our fundamental understanding of bioelectrochemical systems and we perform a rational design of bioelectrodes, based on all the previous knowledge that has been gathered in the last decades in enzyme and microbial kinetics, electron transfer mechanisms, diffusional process in bioelectrodes, mass transfer phenomena, etc.

In a field where so many scientific reports have been written, we have decided to tackle this overwhelming amount of information by attempting to establish general trends and highlighting similarities among allegedly different techniques and reports. The cited literature is extensive, although it mainly includes seminal papers and reviews. It would have been pointless to try to cover the huge body of primary literature in this field, and therefore we have made no attempt to do so. We have aimed at focusing on the design of functional bioelectrodes, *i.e.* materials synthesis, structuration, surface functionalization and modification. We have compiled an overview of the most suitable electrode materials and the most efficient strategies proposed up to date to deposit enzyme layers and to grow microbial biofilms that have shown good electron transfer capabilities, *i.e.* that efficiently communicate with the electrode surface. We have asked authors to contribute to a critical discussion of the different electrode configurations reported so far.

This book covers both processes of technological interest with scale-up possibilities, as well as advanced approaches aiming at more in-depth understanding of the processes involved in electron transfer or proof-of-concept of processes and mechanisms. Most importantly, we have aimed at establishing a parallelism between functional enzymatic and microbial electrodes, pointing at the several similarities and also complementarities and differences between these related systems. We are happy to have embarked on board of this project several authors who have an extensive personal expertise on both microbial and enzymatic systems, and thus have contributed their vision to this book trying to bridge the two worlds. We have to concede, however, that beyond the many similarities, there are some important differences between the two systems and therefore two sections make almost exclusive focus on either enzymatic or microbial systems. Conversely, we open and close the book with introduction and characterization sections that englobe both systems. Because of our interest in stablishing analogies both between microbial and enzymatic systems, and between different architectures some topics will be approached in more than one chapter, analyzed every time from different perspectives.

Victoria Flexer
Nicolas Brun

References

1. M.C. Potter (1911). Electrical effects accompanying the decomposition of organic compounds. *Proceedings of the Royal Society of London. Series B* **84**(571), 260–276.
2. B. Cohen (1931). The bacteria culture as an electrical half-cell. *Journal of Bacteriology* **21**, 18–19.
3. R.S. Berk and J.H. Canfield (1964). Bioelectrochemical energy conversion. *Applied Microbiology* **12**, 10–12.
4. W. van Hees (1965). A bacterial methane fuel cell. *Journal of the Electrochemical Society* **112**(3), 258–262.
5. L.C. Clark, Jr. and C. Lyons (1962). Electrode systems for continuous monitoring in cardiovascular surgery. *Annals of the New York Academy of Sciences* 29–45.

About the Editors

Victoria Flexer is a Professor in Electrochemical Processes and a Research Fellow from CONICET, the Argentinean National Scientific and Technical Research Council. She obtained her Ph.D. from the University of Buenos Aires, and later held positions in the CNRS, France, The University of Queensland, Australia, and Ghent University, Belgium. She has worked with enzymatic and microbial electrochemical systems in both fundamental understanding and modeling and in applied topics, including new materials and surface functionalization. Her research interests also include the world of inorganic electrochemistry, and she has recently started work on electrochemical technologies applied to mining and element recovery from unusable deposits.

Nicolas Brun is a CNRS associate scientist at the Charles Gerhardt Institute (Montpellier, France). He obtained his Ph.D. in 2010 at the University of Bordeaux (France) on the design and synthesis of hierarchical porous materials via integrative chemistry. Then, he spent two years as a post-doctoral fellow at the Max Planck Institute of Colloids and Interfaces (Potsdam, Germany), working on the hydrothermal synthesis of porous carbonaceous materials. In 2013, he was awarded a one-year postdoctoral fellowship at Kyoto University from the Japan Society for the Promotion of Science (JSPS). His research interests include porous carbon materials for environmental remediation, energy storage, electrocatalysis, and bioelectrocatalysis.

Contents

Preface v

About the Editors ix

Section I: Fundamentals **1**

1. Fundamentals of Enzymatic Electrochemical Systems 3
Victoria Flexer and Nicolas Brun

2. Fundamentals of Microbial Electrochemical Systems 51
Stefano Freguia, Kun Guo, and Pablo Ledezma

3. Continuum in Enzymatic and Microbial Bioelectrocatalysis 77
Frédéric Barrière

4. Electron Transfer Between Bacteria and Electrodes 93
Lucie Semenec, Sanja Aracic, Elizabeth R. Mathews, and Ashley E. Franks

Section II: Electrodes for Enzymatic Electrochemical Systems **171**

5. Architectures of Enzyme Electrodes Using Redox Mediators 173
Victoria Flexer, Antonin Prévoteau, and Nicolas Brun

6. Functional Electrodes for Enzymatic Electrosynthesis 215
Lin Zhang, Mathieu Etienne, Neus Vilà, and Alain Walcarius

7. Redox Hydrogels as an Efficient Strategy for Immobilization of Enzymes at Electrode Interfaces 273
Joshua W. Gallaway and Scott Calabrese Barton

8. Conducting Polymer Hydrogels and Their Applications as Electrode Materials 291
Yu Zhao, Lanlan Li, Lijia Pan, Guihua Yu, and Yi Shi

9. Nanocarbon-Based Enzymatic Electrodes 341
Nicolas Brun, Mohamed Baccour, and Victoria Flexer

10. Carbonaceous Electrodes Featuring Tunable Mesopores for Use as Enzyme Electrodes 381
Seiya Tsujimura

Section III: Electrodes for Microbial Electrochemical Systems **401**

11. Materials and Their Surface Modification for Use as Anode in Microbial Bioelectrochemical Systems 403
Kun Guo, Antonin Prévoteau, Sunil A. Patil, and Korneel Rabaey

12. Electrodes for Cathodic Microbial Electrosynthesis Processes: Key Developments and Criteria for Effective Research and Implementation 429

Ludovic Jourdin and David Strik

13. Non-Carbonaceous Electrodes for Microbial Electrochemical Systems 475

Hernán Romeo, Diego Massazza, Rodrigo Parra, and Juan Pablo Busalmen

Section IV: Imaging and Characterization of Bioelectrodes 523

14. Imaging and Characterization of Microbial Electrodes 525

Yang Lu and Bogdan C. Donose

15. Spectroscopic Methods for Characterizing Redox Chemistry at Metalloprotein-Modified Electrodes 545

Philip A. Ash and Kylie A. Vincent

16. Spectroelectrochemistry of Microbial Biofilms 563

Diego Millo and Bernardino Virdis

17. Scanning Electrochemical Microscopy: A New Tool for Studying Enzymatic Reactions 599

Dodzi Zigah and Olivier Fontaine

Outlook 627

Index 633

Section I

Fundamentals

Chapter 1

Fundamentals of Enzymatic Electrochemical Systems

Victoria Flexer*,†,§ and Nicolas Brun‡

Centro de Investigación y Desarrollo en Materiales Avanzados y Almacenamiento de Energía de Jujuy-CIDMEJu (CONICET-Universidad Nacional de Jujuy), Argentina

†*Centro de Investigación y Transferencia Jujuy, CIT-JUJUY, Argentina*

‡*Institut Charles Gerhardt de Montpellier, UMR5253 CNRS, ENSCM, Université de Montpellier, France*

§*vflexer@unju.edu.ar*

1.1 Enzymatic Electrochemical Systems

Enzymes are widely used as the key component in the construction of amperometric, coulometric and potentiometric electrodes. Enzymes are inexpensive and widely available biological catalysts, which operate under conditions of mild temperature, pressure and pH. They are ideal candidates for the construction of biosensors for the specific detection of multiple type of analytes in complex media, because of their intrinsic specificity and their incredible diversity. Therefore, oxydoreductases have been extensively incorporated into electrochemical biosensors due to their ability to quantitatively transduce the presence of their substrate into a measurable electric signal. More recently, the concept of enzymatic fuel cells (EFC) was introduced, and similar enzymatic electrodes were incorporated into devices aimed at energy production.

Although the use of enzymes as building blocks in bioelectrodes is not a recent trend, real applications of protein-based electrochemical systems are often circumscribed by limitations arising from the enzyme itself. Consequently, a lot of work has been carried out for over three decades in understanding the intrinsic characteristics and improving the performance of enzymatic electrodes.

1.1.1 Enzymes, why?

Enzymes are macromolecules that catalyze chemical reactions, most of the time inside cells. The great majority of enzymes are proteins, with an active center or active site, most often buried or hidden inside a complex 3D aminoacidic structure that guarantees that only one specific substrate (or at the most a very limited number of substrates) can reach that active center and undergo a chemical reaction. This is called the "lock and key" model. Therefore, enzymes have extremely high specificity and are able to very selectively react with the target molecule even in a very complex media, such as inside the cell. In order to achieve this extremely high specificity, the 3D structure creates restricted spaces, usually referred to as binding pockets, where only one, or at the most very few, specific substrates will fit. This is possible because the binding pockets have shape complementary, and/or charge and hydrophilic/hydrophobic characteristics for the specific substrate of every enzyme.

Coming closer to our topic of interest, because of their very high specificity, enzymes were thought of as the main component of biosensors to quantitatively determine the composition of a very specific analyte in a complex media without need for sample pretreatment: blood, urine and natural waters are just a few examples. Since that preliminary concept, the challenge has been to incorporate enzymes into devices, while preserving their biological activity and specificity, and transduce the biological response into a quantitative signal. Electrochemistry has proved to be one of the most successful ways of transducing a biological response.

Aside from very rare exceptions of rather small enzymes that can be chemically synthesized in the lab,[1,2] enzymes are synthesized by living organisms. In nature, most enzymes are meant to work inside the cell, or, in some cases, be excreted in a very specific media.[3] Scientists have managed to extract and purify enzymes, and even to mass produce a large variety of enzymes. However, because of their biological nature, and because their complex biochemical structure that has been developed over long years of adaptation, not all enzymes are able to keep their biological activity once outside the cell, *i.e.* not all enzymes are able to work *in vitro.* This is the first limitation to bear in mind when attempting to build devices with these biological entities. Even those enzymes that manage to keep biological activity in a media other than their natural environments, are fragile entities that have to be kept with care, under quite specific stocking and working conditions. Otherwise, they will irreversibly lose their activity. Most interestingly, every enzyme is unique, and researchers will have to ascertain which are the specific conditions that apply to each particular enzyme. Moreover, some enzymes are *tougher* than others and will withstand exposure to non-natural or non-optimal conditions for longer periods of time, while others will only survive very briefly outside the cell, or when exposed to pH, temperature and ionic strength that deviate from the optimal conditions. Scientists are struggling hard to find mimicking conditions that will allow more enzymes to work *in vitro,* and to withstand *in vitro* conditions for longer periods of time. This is certainly work carried out by biochemists; however, when the final aim is to build devices, the work is also carried out by materials scientists, who devote a lot of time to find materials and functionalize surfaces that will mimic the natural environment of the enzymes' and hence might keep the enzymes' functionality and specificity when immobilized or trapped on those surfaces.

In summary, in order to build devices, researchers and engineers have to evaluate very carefully the intrinsic characteristics of different enzymes before choosing a suitable candidate. Evidently, the first characteristic to look at will be the selectivity towards a certain substrate/analyte. A suitable dynamic range and a high activity are obviously other issues to pay attention to. Parameters such as shelf-life and capacity to withstand temperature variations and changes in media composition will be very important assets, and even a must, when attempting to build real-life devices.

1.1.2 Oxidoreductases

Oxidoreductases are enzymes that catalyze redox reactions, *i.e.* reactions where the substrate either loses or gains electrons (oxidation and reduction reactions, respectively). They are extremely important in nature because of the wide range of reactions of biological importance that they catalyze.

Oxidoreductases follow a particular mechanism usually referred to as a ping–pong mechanism. In a ping–pong mechanism, the enzyme is not left unchanged upon completion of the reaction, but it is left with either an excess or a deficit of electrons (respectively, in an oxidation or a reduction), and often a few extra atoms as well. Therefore, a second substrate is yet needed to bring back the enzyme to its original condition, so that the enzyme can be considered a catalyst, in the sense that it is left unchanged after accelerating a given chemical reaction.

A cofactor is a non-protein chemical compound that is a must for the biological activity of the enzyme to take place. A cofactor can be an organic molecule, or an inorganic ion or group of ions. Some enzymes require several cofactors. If the cofactor is tightly and permanently bound to the enzyme, this is usually referred to as a prosthetic group (*e.g.* FAD, PQQ, heme, transition metals, etc.). Conversely, when it is only attached for short-time intervals to perform the biological task, it is referred to as a cosubstrate (*e.g.* FMN, NAD^+, $NADP^+$, etc.). For oxidoreductases, either a prosthetic group or a cosubstrate is directly involved in the oxidation/reduction reaction. They are the chemical species that either gain or lose electrons when the reaction takes place. For reactions catalyzed by oxidoreductases, when an oxidation or reduction reaction takes place, the excess (or deficit) of electrons from the analyte is temporarily stored in either the prosthetic group or the enzyme cosubstrate.[4,5]

For example, glucose oxidase (GOx, EC 1.1.3.4) is an enzyme capable of oxidizing glucose ($C_6H_{12}O_6$) to gluconolactone ($C_6 H_{12}O_7$).[6]

$$C_6H_{12}O_6 + \text{GOx(FAD)} \overset{K_{\text{MS}}}{\longleftrightarrow} C_6H_{12}O_6 + \text{GOx(FAD)} \xrightarrow{k_{\text{cat}}} \text{GOx(FADH}_2) + C_6H_{12}O_7 \quad (1)$$

Upon conversion of glucose into gluconolactone, the prosthetic group of GOx, flavin adenine dinucleotide (FAD) is left with an excess of two protons and two electrons ($FADH_2$), as a reduced FAD, which is completely unable to oxidase another glucose molecule. A second substrate, O_2, is needed to bring back the catalyst to its original state.[6] The second substrate is often named the natural redox mediator.

$$O_2 + \text{GOx(FADH}_2) \xrightarrow{k} \text{GOx(FAD)} + H_2O_2 \quad (2)$$

k, K_{MS}, k_{cat} are kinetic constants that will be further discussed in Section 1.4. Because the enzyme is jumping back and forth between two different oxidation states, reacting with two substrates, this mechanism has been nicknamed ping–pong mechanism. Reactions (1) and (2) are classical examples of such a mechanism. Oxidoreductases are also known as two-substrate enzymes.

It is very interesting to note that while enzymes are highly specific towards their first substrate, they are quite unspecific towards the second substrate. This unspecificity has been crucial in developing enzyme electrodes,[7] and we will come back to this point later in this chapter.

Coming back to the example of GOx, it is sometimes considered that it can only oxidize β-D-glucose. It can, however, also oxidize maltose and galactose, among other sugars, although the turnover number for other sugars is below 1% of the rate of oxidation of glucose.[6,8] However, reduced [GOx($FADH_2$)] will be reoxidized by many different molecules, other than O_2, the natural second substrate, acting *in vivo*. These molecules are usually called artificial mediators, or artificial redox mediators. A wide selection of artificial mediators has been reported for GOx, including substituted ferrocenes, ferrocyanide, quinones, Ru and Os complexes with many different chemical structures, viologens and tetrathiafulvalene, among others.[6,8–10] A full discussion of the biochemistry of enzymes is well beyond the scope of the present chapter. For further information, the reader should consult some of the more specialized texts.[3,11]

1.1.3 Enzyme electrodes: What for and why?

So why should we immobilize enzymes on electrode surfaces? The answer is in the envisioned applications. Enzyme electrodes were first designed as biosensors,[12–19] and almost two decades later gave rise to a fast-developing application aiming at energy generation: enzymatic fuel cells, EFCs.[20–24] Biosensors and biofuel cells are by far the most common envisioned applications of enzyme electrodes and a very large proportion of academic work on such electrodes is aimed at advancing those fields. Very recently, work on enzymatic electrosynthesis has started to appear.[25,26] Last, but not least, to immobilize enzymes and to interrogate them by means of an electrical signal is also a powerful way to study many fundamental aspects related to the biochemistry of enzymes, particularly its kinetics and reaction mechanisms.[27,28]

Enzyme electrodes for all the above-mentioned applications are built in extremely similar ways, and several characteristics and requirements for enzyme immobilization are shared.[20] The field of EFCs was clearly pushed forward because of the promise of a green, new energy generation source. However, the field would not have developed so incredibly fast if it had not been for the very solid foundations that had been laid by previous studies in the related biosensor field.

Dozens of techniques for enzyme immobilization have been reported to date. While attempting to build an enzyme electrode, the absolute must of any immobilization technique is that the enzyme keeps its activity upon immobilization. Enzyme inactivation should be avoided at all costs, or at least, reduced to minimal proportions.

This is a shared requirement, regardless of the immobilization technique, and even whether we would like to build a mediated or a mediatorless system (see below). Electrodes with inactivated enzymes are nothing but a waste of time and money.

1.2 Electrochemical Biosensors

So, why would we use electrochemistry to transduce an enzymatic signal?

Electrochemistry is a simple and economical way to transduce an enzymatic signal. Despite the low price of the equipment involved, electrochemical detection of enzymatic signals offers high accuracy and precision, together with low detection limits.[20,29] Moreover, electrochemistry allows for portability and miniaturization. Therefore, electrochemical biosensors have evolved from the experimental lab bench in academia to be present in dozens of industrial and daily life applications. There is no doubt that electrochemical enzymatic sensors are much more popular than optical,[30,31] piezoelectric[32,33] or other kinds of enzymatic sensors.[34]

Immobilizing enzymes on an electrode surface allows the confinement of the reaction of interest to a small volume. This is equivalent to considerably reducing the amount to reactants needed, starting by reduced amount of enzyme, in addition to a reduced amount of sample to be analyzed. Finally, an entrapped enzyme is often, though not always, a better way to control the electrochemical reaction.

Electrochemical biosensors are based on the catalyzed reaction of a target analyte that is either oxidized or reduced by the immobilized enzyme. If the reaction produces a faradaic reaction at the electrode surface, *i.e.* the transfer of electrons across the electrode surface, the sensor will either be classified as an "amperometric" or a "coulometric" biosensor. On the contrary, if no electrons are transferred across the electrode surface, but the reaction contributes to a change in double layer potential, and therefore produces a measurable voltage change, the biosensor is called a "potentiometric" biosensor. In amperometric biosensors, the working electrode is poised at a defined and constant potential, and the electric current at the steady state is measured. In a coulometric sensor, the total charge generated by either the complete oxidation or reduction of the analyte is measured. To date, amperometric and coulometric biosensors gave rise to a much larger field of work than potentiometric biosensors. Thus, the remainder of this book will be devoted *solely to amperometric and coulometric* enzyme electrodes.

How to build an enzymatic biosensor? First, we clearly need a conducting solid-state electrode and choose the right enzyme that will catalyze the reaction of interest. Second, we have to think of a way of immobilizing the enzyme on the electrode surface, while keeping its enzymatic activity. Next, for amperometric and coulometric sensors, we need to think of how to translate the biological response into an electrical current. This point leads us to the question of how the enzyme exchanges electrons with the electrode surface, *i.e.* how to electrically connect the enzyme to the electrode, or which is the electron transfer pathway between the enzyme and the solid-state electrode. The performance of the biosensor will clearly be tied to the kinetics of the electron transfer process. The underlying principle of an efficient (fast response and high sensitivity) amperometric/coulometric enzyme electrode is the establishment of a fast electron transfer from the transfer from the enzyme to the electrode. This very indeed holds true as well for both microbial and organelle electrodes: fast electron transfer from the biological component to the solid-state electrode is the key to the existence of the field.

Interestingly, mother nature, by a long history of evolution, has designed enzymes so as to protect the primary redox site, and avoid unwanted redox reactions with the very many free-diffusing species present in complex natural environments, be it the cellular or extracellular environments. Again, this holds true for organelles and microbial cells. Therefore, establishing an electrical communication between biological entities and solid-state electrodes is not a trivial matter. Enzymes can either undergo direct or mediated electron transfer with the electrode, and these two different mechanisms are discussed in the next sections.

1.2.1 Direct electron transfer

Figure 1 is a schematic representation of the two classical electron transfer mechanisms in amperometric or coulometric enzyme electrodes. An efficient electron transfer mechanism holds the key for an efficient biosensor. An electrochemical signal will only be obtained if the electron transfer between enzyme and solid-state electrode takes place. In Figure 1, the active site of the enzyme, containing either a tightly bound prosthetic group or a temporarily attached cosubstrate, is depicted in black. An electron transfer pathway has to be found to electrically connect this buried and small redox moiety to the electrode surface.

It is usually very difficult, though not impossible, to get an enzyme to directly exchange electrons with an electrode surface.[5] Very intuitively, the capability of an enzyme to directly exchange electrons with an electrode surface is called direct electron transfer (DET). In DET, the prosthetic group, or the enzyme's cosubstrate is directly oxidized or reduced by the electrode when this is poised at the right potential (Figure 1(a)).

Surprisingly, cosubstrates that are able to detach from the active site of the enzyme will usually not react directly at the electrode

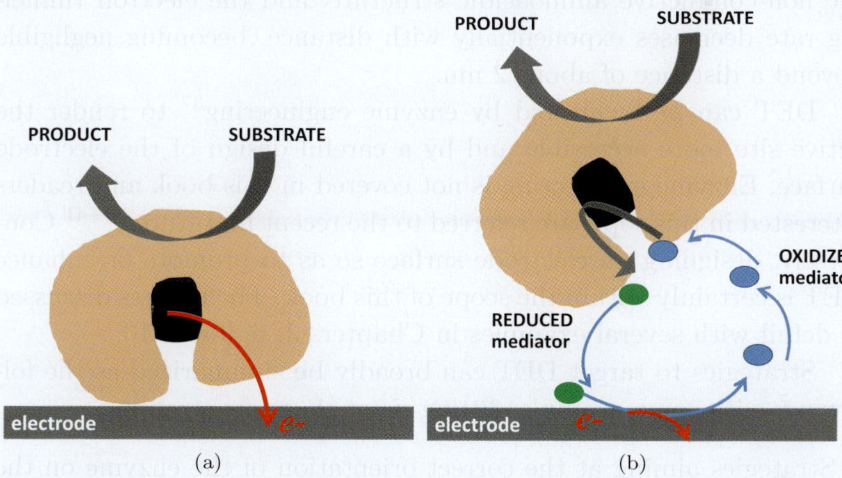

Figure 1. Schematic representation of electron transfer mechanisms between enzyme and electrode. (a) DET. (b) Mediated electron transfer (MET).

surface, or will only do so at redox potentials that are either too high or too low, to allow construction of real devices. This is clearly the case of NAD^+, and NAP^+, which only directly react at the electrode surface at extremely high potentials.

Many of the enzymes of interest in our field possess a tightly bound prosthetic group, and this is the redox center that we aim at connecting with the electrode surface. The electron transfer kinetics between two redox moieties are ruled by the potential difference or driving force for the reaction; the reorganization energy, associated with the flexibility in the structure of the redox species (the enzyme, in our case of interest); and the separation/distance between the two redox centers, as it has been largely discussed by Marcus.[35,36] The capability of an enzyme to undergo DET depends primarily on the enzyme itself. For some enzymes, it is relatively easy, for others very difficult and yet for others DET to an electrode surface has never been reported. Following Marcus's theory, it will mostly depend on how exposed the enzyme's active site is within the 3D protein structure, and how flexible the protein structure is, so as to temporarily expose the prosthetic group to allow for reaction. Usually, the deeper the active site is buried, the more difficult it will be to achieve DET. This is because the electrons have to tunnel across the non-conductive aminoacidic structure, and the electron tunneling rate decreases exponentially with distance, becoming negligible beyond a distance of about 2 nm.

DET can be facilitated by enzyme engineering[37] to render the active site more accessible and by a careful design of the electrode surface. Enzyme engineering is not covered in this book and readers interested in this topic are referred to the recent literature.[38–41] Conversely, designing an electrode surface so as to promote or enhance DET is certainly within the scope of this book. The topic is discussed in detail with several examples in Chapters 3, 6, 9 and 10.

Strategies to target DET can broadly be summarized as the following, with some examples fitting more than one strategy:

- Strategies aiming at the correct orientation of the enzyme on the electrode surface, so that the prosthetic group will directly face the electrode surface[42,43];

- Strategies aiming at surfaces with rugosity in the Angstrom or nanometer scale that will produce protuberances that will somehow penetrate the aminoacidic structure, or approach the prosthetic group[44,45];
- Strategies aiming at the entrapment of enzymes into pores with the right size for a certain enzyme, which will force the enzyme to be fully surrounded by a conducting electrode surface, facilitating the interaction.[45–49]

In general, DET will only work when the enzyme is in direct contact with the electrode surface, *i.e.* it will only work for monolayer or submonolayer enzyme coverage.

Reductases that have repeatedly been reported to undergo DET fairly easy are blue multi-copper oxidases (BMCOs),[50–52] *i.e.* ascorbate oxidase, laccase, bilirubin oxidase and ceruloplasmin. The catalytic site of most of BMCOs consists of four copper atoms (sites). The T1 site is quite exposed and it is through this atom that the enzyme couples directly to electrode surfaces. Laccases and bilirubin oxidases reduce O_2 to water. The ping–pong mechanism is completed by oxidizing a large number of organic molecules. Oxidases with DET capabilities include several dehydrogenases (cellobiose dehydrogenase,[53] glucose dehydrogenase,[54,55] fructose dehydrogenase[49,56]); and several peroxidases (cytochrome *c* peroxidase,[57] horseradish peroxidase,[58] lactoperoxidase,[59] etc.), among others.

1.2.2 Mediated electron transfer

If attempts to connect the enzyme directly to the electrode surface are unsuccessful, or if the electrical communication is not efficient enough, what can we do? We should recall that in nature, redox enzymes always resort to a second substrate, the natural redox mediator, that is responsible for bringing back the enzyme to its original redox state.[3] Many of these natural redox mediators can readily be reduced or oxidized at the electron surface; therefore, they could actually be used as *electron shuttles*, or transducers of the enzymatic signal.[5,7] Alternatively, we could also take advantage from the fact that the coproduct is often also electroactive at the electrode surface, and hence could be used as an electron shuttle. The coproduct is the product from the second reaction in the ping–pong mechanism. Electron shuttles are indeed very efficient in delivering electrons both from the enzyme to the electrode surface, and in the opposite direction, from the electrode, towards the enzyme.[13,16,60–62] In the field of enzymatic electrodes, electron shuttles are also widely known as *redox mediators*, *molecular wires* or simply *mediators*. Mediated electron transfer is schematically shown in Figure 1(b); the redox mediator in the scheme can alternatively be a natural or an artificial redox mediator (see below).

Let us go back to the classical example of GOx. We have just seen that molecular oxygen is the natural second substrate or natural redox mediator of GOx. As shown in reaction (2), O_2 is reduced to H_2O_2 by accepting the two electrons and the two protons in excess in reduced GOx after the enzyme oxidizes glucose. Both O_2 and H_2O_2 have been repeatedly used to fabricate GOx-based electrodes, since both species are electroactive and can readily react on an electrode surface when posed at the right potential. Indeed, already back in 1962 the first enzymatic electrochemical biosensor was described by Clark and Lyons.[63] In that enlightening work, the authors fixed GOx on top of an oxygen-sensitive electrode, a Clark electrode. This is an oxygen sensor, *i.e.* the depletion in O_2 concentration is measured, as a function of glucose concentration, on the platinum electrode surface at a potential sufficiently low for O_2 reduction.[63] Since then, many biosensors have been built that measure either the consumption of dissolved O_2, or the production of H_2O_2, as a way of actually assessing the glucose concentration in solution,[5] based on the following overall reaction:

$$\text{glucose} + O_2 \xrightarrow{\text{GOx}} \text{gluconolactone} + H_2O_2 \qquad (3)$$

Although this methodology seems to be a rather straightforward way of analysis, there is a major problem in assessing glucose concentrations by this method in real samples. The analyst is looking for the glucose concentration, but what we are actually measuring are variations in the O_2 concentration, due to the presence of glucose. Therefore, we must precisely know the O_2 concentration in the absence of glucose, which is not always possible, since O_2 has a varying concentration in most samples (particularly in physiological samples).

Therefore, some authors have resorted to a new strategy, in their attempt to assess the glucose concentration. The first option would be to measure the oxidation of H_2O_2. However, H_2O_2 is oxidized at relatively high potentials (roughly +0.6 V *vs.* Ag/AgCl), where other chemical species, often present in physiological samples at varying concentrations, are directly oxidized at the electrode surface. This is the case, for example, for ascorbate (vitamin C), urate, or acetaminophen (paracetamol).[64] Thinking of the other common application, this high potential would make it impossible to build a biofuel cell using H_2O_2 as mediator for GOx. In addition, H_2O_2 at relatively high concentrations is toxic for GOx.[65] Finally, H_2O_2 could also be reduced at the electrode surface. However, H_2O_2 is only reduced at potentials around −0.2 to 0 V, where O_2 will also be reduced. Therefore, we come back again to the same puzzle: the reduction current is caused by O_2 or H_2O_2 and what is the departing concentration of the former.

1.2.2.1 *Artificial redox mediators*

We have discussed that while enzymes are highly specific for their substrates, they are quite unspecific for their second substrates. This characteristic is most interesting, and it has opened a whole new world of possibilities to MET reactions. Indeed, for most enzymes, it is possible to replace the natural mediator by other redox species that are also efficient in shuttling electrons from the enzyme to the electrode surface or *vice versa.* Species with these capabilities are called *artificial redox mediators*, and to date, they have largely replaced natural redox mediators. Why? Researchers have resorted to artificial mediators for several reasons.

- The natural cosubstrate and/or the natural coproduct might have an oxidation/reduction potential that is not suitable to build devices. In the case of biofuel cells, we would like the mediators to have potentials relatively close to the enzymes, so as to maximize the voltage output. In the case of biosensors, we would usually like to work at potentials as low as possible (in the case of oxidation reactions) so as to avoid the direct oxidation on the electrode surface of other species that might be present in the matrix. By carefully playing with the chemical structure of artificial redox mediators, a whole library of compounds can be synthesized that will yield a mediator with the desired redox potential, with a precision of just a few mV. Particularly for organometallic compounds, this is a relatively easy task to perform, if the researcher is skilled in organic/organometallic chemistry.[66,67] Os and Ru compounds are particularly interesting to play around with, and several dozen Os compounds have been reported in the enzyme electrode literature, encompassing a potential range of over 1 V.[16,64,66–74] For those who are not lucky enough to have a skilled synthetic chemist in their team, quite a large library of soluble compounds are commercially available, with fairly fast electron transfer kinetics for many enzymes encompassing an interesting potential window, such as ferrocene derivatives.

- The natural cosubstrate and/or the natural coproduct might be a gaseous species, and therefore, its initial concentration will be very difficult to control. This is the case of O_2.
- If accumulated, the coproduct might be toxic for the enzyme.

As exemplified in the previous section, once again, GOx is a great example for all of the above. Even though the first GOx bioelectrodes used either O_2 or H_2O_2 as electron shuttles, for more than two decades, these have largely been replaced by artificial mediators.

Moreover, in many bioelectrode architectures, the artificial mediator might have a double role, as an electron shuttle, and as *cement* or *glue* to attach the enzyme to the electrode surface,[13,14,16,61,70,72,74,75] as it will be largely discussed in Chapter 5. Finally, we cannot help but remind the reader that the most important characteristic of an artificial redox mediator is that it shows fast electron transfer kinetics. Therefore, for the particular examples of enzymes that use oxygen as a natural redox mediator, the kinetics of electron transfer for the artificial redox mediator should be even faster than the kinetics of the enzyme/O_2 reaction. Otherwise, two competing redox mediators will simultaneously act, and it will be difficult to assess which proportion of the amperometric current corresponds to the reaction of interest.

1.2.3 What is more advantageous — Direct electron transfer or mediated electron transfer?

There is no unique answer to this question, and the researcher should analyze the particular enzymatic system keeping in mind the desired application. The first obvious advantage of DET is that it usually makes the construction of the bioelectrode simpler and less expensive, since no mediator is needed. Most importantly, DET allows working at potentials closer to the redox potential of the prosthetic group of the enzyme. This is a clear, undisputable advantage regardless of the envisioned application. In the field of biosensors, it will allow operation of the biosensor at a potential range that will be helpful to avoid interferences (usually produced by direct oxidation of organic molecules on the electrode surface at high potentials). In the field of biofuel cells, it will allow for a larger potential difference between anode and cathode, *i.e.* a higher potential output, and hence a higher power output. Whereas fundamental studies on the electrochemistry and biochemistry of redox enzymes are almost exclusively performed on DET configuration. On one hand, this is to keep the experimental setup simple and to ensure that all electrochemical signals are exclusively generated by the enzyme and not by other compounds. Moreover, the overall kinetics of a mediated system are much more complex than those of a DET configuration, usually without an analytical formula defining the current output of the system (see Section 1.4).[76–78] Therefore, it is usually highly non-trivial to correlate the electrical signals of mediated systems with pure enzymatic system.

On a second analysis, while enzymes are very good catalysts in terms of substrate specificity, and relatively high turnover numbers, reaching values of up to 5,000 s^{-1}, they are, however, very large molecules as compared to non-biological catalysts, be it inorganic/organic compounds or small nanoparticles. The large size of enzymes is a drawback when attempting to build up electrodes with high activity (mostly desired for biofuel cell applications, see below). Calabrese-Barton *et al.* calculated that for an average enzyme with a 100 nm^2 footprint and a turnover number of 500 s^{-1}, fully covering a flat surface would give a theoretical current limit of about 80 $\mu A\,cm^{-2}$,[20] well below the desired values to drive the field forward (in the order of mA cm^{-2}). Therefore, if higher currents are desired, enzyme multi-layers should be deposited on electrode surfaces. However, we have just seen that in order to get DET from enzyme to electrode, enzymes need to be in direct contact with the electrode surface. Therefore, it would be rather useless to build enzyme multilayers if no redox mediators are used. This is because only the first layer, in direct contact with the electrode surface, would be effectively connected to the electrode.

There is, however, a way to accommodate more than an enzyme monolayer and still get DET from a large proportion of those enzymes. This can be accomplished by using porous electrodes, especially electrodes with nanometric features that will provide an increased surface area, which will accommodate larger enzyme loadings. Dozens of electrodes with nanometric features have been reported to successfully trigger DET,[46,47,49,51,55] and several prominent examples are discussed in Chapters 6, 9 and 10.

1.2.4 Amperometric biosensors

We have just presented the different mechanisms of electron transfer from enzyme to the electrode, but we have not yet discussed in detail how exactly a biosensor works, *i.e.* how the presence of the analyte is quantitatively translated into a catalytic current value. In an amperometric and in a coulometric biosensor, different molecules can produce the faradaic reaction of interest:

- In DET mode, the enzyme itself is oxidized/reduced at the electrode surface, in which case, the electrode surface is acting as a second substrate.

- The product of the enzymatic reaction can react at the electrode surface.
- If a redox mediator is used, be it the natural mediator acting *in vivo*, or an artificial mediator, then this redox mediator can react at the electrode surface, either in its oxidized or reduced form.

The historical terminology to classify enzymatic biosensors is based on the mechanism of electron transfer. Even though we did discuss DET before MET, this choice was not based on historical development of the field. We will narrow our discussion to amperometric sensors. The field of coulometric sensors is very tightly linked to amperometric sensors, with most characteristics in common (most used enzymes, mediators, operating conditions, etc.). Readers interested in the few particularities of coulometric sensors are referred to the literature.[69,70,79]

First-generation biosensors made use of either the natural electron acceptor or the natural coproduct of the enzymatic reaction (for example, O_2 and H_2O_2, respectively, in the case of GOx). Second generation biosensors use artificial redox mediators. Finally, third generation biosensors do not use redox mediators and rely exclusively on DET.

Figure 2 shows the schematic representation of a second generation amperometric enzymatic biosensor, where the enzyme, E_{OX} is an oxidase. First, there is a molecular recognition step, where the analyte forms an enzyme–substrate complex, ES, and the biochemical reaction takes place with product formation. The two catalytic constants of the Michaelis–Menten mechanism[80] (see Section 1.4) govern each of these two reactions. Following reaction with the analyte, the enzyme is left with an excess of electrons, a state that is called E_{RED}. This reduced enzyme reacts in turn with the oxidized mediator, MED_{OX}, with second order kinetics represented by the kinetic constant k. The enzyme is brought back to state E_{OX} and is ready to react with another substrate molecule. If the electrode is poised at a sufficiently high redox potential, MED_{RED} will react at the electrode surface, regenerating MED_{OX}. Thus, an amperometric current is generated, and the cycle is maintained by the auto-regeneration of both enzyme and redox mediator.

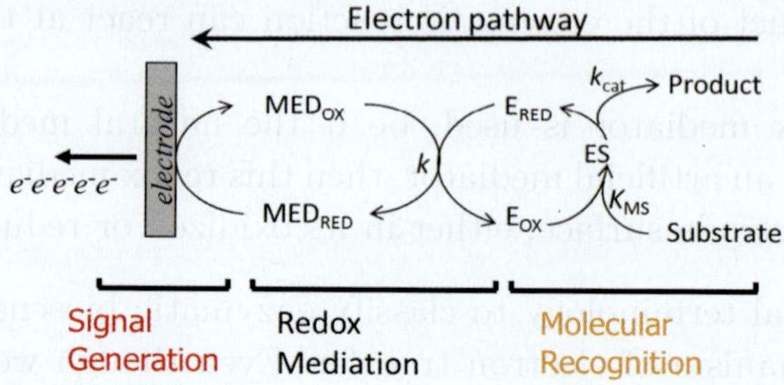

Figure 2. From analyte to electricity. Schematic representation of the different steps in an amperometric biosensor (second generation). In the figure, E is an oxidase, E_{OX} represents the enzyme in its oxidized state, ES is the enzyme–substrate complex, E_{RED} is the enzyme in its reduced state, MED_{OX} is the redox mediator in its oxidized state, and MED_{RED} is the redox mediator in its reduced state.

Figure 2 highlights the electron pathway from the substrate to the electrode surface. In the particular example of an oxidase, the electrons pass from the analyte to the enzyme, then to the artificial mediator, and finally to the electrode surface and the external circuit. In order to make this electron transfer possible, thermodynamic forces us to choose enzyme and mediator in such a way that the redox potentials follow the sequence as given below:

$$E'_{\mathrm{SUBSTRATE/PRODUCT}} < E'_{E_{\mathrm{OX}}/E_{\mathrm{RED}}} < E'_{\mathrm{MED_{OX}/MED_{RED}}} < E'_{\mathrm{electrode}} \quad (4)$$

The prime in the super index in the above schematic of electrode potentials is meant to exemplify that we are referring to those hemi-reactions in the given conditions at which the electrodes are meant to work (pH, temperature, ionic strength, etc.). The potentials should also consider the concentrations of analyte and products; although these are supposed to be variable, they should still hold for the whole range of concentrations at which the enzyme electrode is meant to be used. In addition, one should not work with potentials that are too close to each other; otherwise the driving force and the kinetics will end up being too slow.

For a biosensor to achieve all the desired properties we enumerated above, an efficient electron transfer from analyte to the solid-state electrode is needed. Evidently, for an efficient electron transfer, the choice of redox mediator is crucial. An ideal artificial redox mediator will show the following characteristics[19,69,81–83]:

- Good chemical stability both in the reduced and oxidized state.
- Its oxidized form should react with the reduced form of the enzyme with fast kinetics (or the reduced form of the mediator should react with the oxidized form of the enzyme).
- For an oxidase, the mediator redox potential must be higher than that of the prosthetic group of the enzyme, but low enough to avoid direct oxidation of other species on the electrode surface (conversely, for a reductase, the mediator redox potential must be lower than that of the prosthetic group of the enzyme).
- The mediator redox potential should be pH independent.
- The heterogeneous reaction of the mediator with the solid-state electrode must be fast and not imply high overpotentials.
- It should react only with the enzyme, and not with any other molecule potentially present in the matrix (*e.g.* O_2, amongst others).[84,85]
- It should be non-toxic (if *in vivo* implantation is envisioned).
- For real applications/device fabrication, low cost and abundance will be an asset.

Other characteristics/requirements might be necessary depending on the desired application, and whether the mediator is going to be coimmobilized with the enzyme, or free to diffuse in solution.

Non-mediated enzymatic systems have also thrived in the field of biosensors.[55,86–90] The electron transfer pathway is simplified in the absence of a redox mediator. Taking again the example of an oxidase, electrons are transferred from the analyte to the enzyme, and from the enzyme directly to the solid-state electrode. In this scenario, the requirement for electrode potentials is reduced to

$$E'_{\text{SUBSTRATE/PRODUCT}} < E'_{E_{\text{OX}}/E_{\text{RED}}} < E'_{\text{electrode}}$$

The most interesting results in terms of potential real-life devices and applications have been obtained with fully integrated systems, *i.e.* systems where the enzyme and the mediator, if present, are both immobilized on the electrode surface. We will not discuss in detail in this chapter the extremely large variety of electrode configurations that have been proposed to-date. This will be the matter of discussion in Chapters 3 and 5 through 10, both for mediated and DET systems. Even if we will not enter into a detailed discussion of the matter at this point, the reader should still bear in mind that a large proportion of work reported in the last 20 years has been produced with fully integrated electrodes. Therefore, in that light, an extra requirement for the mediator will be the possibility to integrate the redox moiety to an immobilized layer at the electrode surface. Moreover, the reader should not forget that fully integrated systems are at the core of the ultra-fast development of the field of EFCs, including, of course, the promise of implantable miniaturized devices that we will discuss in the following section.

Last but not least, if we want the sensor to work as a quantitative sensor, and not just qualitative, we are forced to work at an analyte concentration range where the enzyme will not be substrate saturated.[12,17–19,82,91] This requirement forces us to carefully choose the enzyme not only according to the selectivity towards a certain analyte, but also according to its affinity for such an analyte, which is given by the Michaelis–Menten constant, K_{MS}. As it will be outlined in Section 1.4, both the immobilization strategy and the electron transfer kinetics with the redox mediator, if present, might modify the K_{MS} with respect to the value measured in solution.

1.3 Enzymatic Fuel Cells

1.3.1 Generalities

Originally named enzymatic biofuel cells, emphasizing the biological nature of enzymes, in the last couple of years, this terminology has been updated, and most publications today refer to these devices as enzymatic fuel cells,[20,22,23,92] to differentiate them from microbial fuel cells. The general terminology of biofuel cells is used alternatively to jointly refer to enzymatic and microbial devices, and, as the attentive reader might have already notice we do in this chapter, to shortly refer to either enzymatic or microbial devices. Readers new to the field should be warned that most of the old literature, roughly until 2010, still used the emphatic terminology of enzymatic biofuel cells.[20] A few new reports still use the old terminology.

EFCs are electrochemical devices that rely on the catalytic activity of enzymes to transform chemical energy into electrical energy. Like classical fuel cells, they run while there is a continuous feed of fuel and oxidant, *i.e.* they should not be mistaken with batteries.

In its most simple form, EFCs consist of an enzymatic fuel oxidizing anode, an enzymatic reducing cathode, separated by electrolyte and an external circuit with a load. The solid-state electrode acts only as a conducting support for the enzymes and the mediator, if present. Its chemical structure and morphology, however, will certainly be determinant of the rate of electron transfer from enzyme/mediator to the said electrode. Enzymes are very attractive catalysts. They are inexpensive and relatively easy to produce in unlimited and large quantities, which is in marked contrast with classical precious metal catalysts and other inorganic compounds which are rare and hence very costly. Second, they work under very mild conditions, very close to ambient temperature and pressure, and physiological conditions, at neutral or close to neutral pH, *i.e.* they are safer and easier to work with than classical inorganic catalysts. In addition, by choosing the right enzyme, it is possible to gain electrical work by oxidizing an extremely large selection of fuels, including, of course, many different biofuels, such as sugars, alcohols, aldehydes, organic acids and many others. Oxygen is in most cases the oxidant of choice. Most interestingly, because of its high substrate specificity, enzymes do not usually suffer from inactivation by other molecules present in the fluid carrying the fuel and oxidant. Once again, another great benefit when compared to classical fuel cells, where these fuels need to be highly pure to avoid poisoning/inactivation of inorganic catalysts. These biofuels are inexpensive, and the possibility of feeding a very complex media containing only relatively dilute quantities of fuel allows us to further decrease the operational costs, since impure substances, and even waste streams can be used as fuels. Indeed, EFCs have gained the highest level of popularity because of their use in implantable devices in living organisms, harvesting glucose and O_2 from blood, either in humans or animals.[12,15,93–96] In addition, no membrane or casing is needed. Therefore, EFCs consisting of only enzymatic anode and cathode are ideal for miniaturized devices with less than a 1 mm^2 fingerprint.

Because there is no need to apply high temperatures and pressures in sealed electrochemical reactors, EFCs are much safer to operate than their classical analog counterparts. Moreover, both enzymes and their reaction products can be considered as relatively safe, and therefore, the implantation of such devices into the human body, without any type of casing has been proposed.[69,70,97]

Despite the great advantages that we have just enumerated, EFCs have certainly not managed to replace their classical abiotic counterparts. They have not even taken a certain share of the market, and most likely they never will. Indeed, the electrical power input required by many everyday life applications, such as household electricity needs or vehicle propulsion, cannot be met by EFCs. Performance of EFCs is limited, with current and power density values several orders of magnitude below classical fuel cells. Moreover, they suffer from the added disadvantage of limited lifetimes due to enzyme instability and denaturation. Following a boom at the turn of the century, when high expectations were placed on them, thermodynamic and kinetic limitations have put a realistic limit to what should be expected from them. Conversely, EFCs have managed to find interesting and vital niche applications, such as powering remotely located electronics, toys for infants needing ultra-safe power sources and implantable biomedical devices (not including cardiac pace makers).

The glucose/oxygen EFC has become the undisputable flagship of EFCs, and implantable medical devices in particular. For this particular application, an EFC displays undisputable advantages *vs.* either a classical fuel cell or a battery[7,17,20,22–24,94,96,98]:

(i) The possibility for extreme miniaturization, since only two electrodes and their external electrical connections are needed. Both the fuel and the oxidant are naturally present in the organisms, *i.e.* it is not necessary to include a reservoir for them. Similarly, the products of the reaction are also naturally present in blood;

therefore, there is no need to recur to a mechanism or a reservoir to get rid of them. Blood or any fluid in the body can act as the electrolyte separating anode and cathode, while the natural selectivity and specificity of enzymes guarantees that the system will work perfectly fine in the absence of a membrane. Finally, no casing or sealing is needed. Therefore, the whole EFC can be readily miniaturized, its construction and implantation should be relatively simple and mass production would decrease its costs well below the dollar per unit barrier.

(ii) Both reactants and products are present in large quantities in the body, and therefore, the system can be considered harmless and safe. Conversely, enzymes and eventually mediators are immobilized on the inert electrodes, and while they can denaturate, their leaching should not be important if good immobilization mechanisms are chosen.

(iii) They can virtually provide infinite specific energy, considering that both oxygen and glucose are constantly renewed. The lifetime of the fuel cell is given by the durability of the catalysts, and not by fuel exhaustion.

Implantable biofuel cells have been envisioned to power different types of biosensors: measuring local temperature (an indication of infection and inflammation), local pressure (veins or artery partial blockage) abnormal variations of glucose concentration (glycaemia), among others.[7,15,69,94,97] These biosensors could transmit data to an external device that would allow the patient to take action.[97] In a more futuristic scenario, the biosensor could be coupled to a second device acting upon the local need, *e.g.* delivering sugar or insulin.[97] Both the sensor and the second device need energy to work. Miniaturized biosensors are relatively easy to fabricate.[69] The point, however, is how to miniaturize the energy source powering the whole system. And EFCs are a most interesting answer.[97]

A summarized comparison between classical fuel cells and EFCs is given in Table 1.

Table 1. Comparison of classical and EFCs.

	Classical fuel cells H_2/O_2 or methanol/O_2	Enzymatic fuel cells
Catalyst	Noble metal or Noble metal particles	Enzymes
Case/seal	Essential	Not needed
Membrane	Essential	Not needed
Operating pH	pH $\approx$ 0	Neutral or close to neutral
Operating temperature	$\approx$70–120°C	$\approx$20–40°C
Average operating life	>1 year	$\approx$1 month (or less)[a]
Miniaturization/ fingerprint	>10 mm^2	>0.1 mm^2
Current density	0.1–1 A cm^{-2}	<25 mA cm^{-2}
Power density	1 W cm^{-2}	<5 mW cm^{-2}
Cost/cell	$\gg$\$100	~10 ¢[b]

Notes: [a]Several examples with much longer Operating life, but average values usually do not exceed 1 month.
[b]Estimated if mass produced.

1.3.2 Working principle

The working principle of an EFC will be exemplified by the mediated glucose/oxygen EFC, which is depicted in Figure 3. The key issue on these systems is that, in order to avoid the need to use a membrane, both enzymes and mediators, if present, are coimmobilized on their respective electrode surfaces. Regardless of the immobilization strategy, the active components, enzymes and mediators, are never free in solution to diffuse. The anode works pretty much in the same way as biosensors using artificial redox mediators (either second or third generation, explained in Section 1.2.4). In the example depicted in Figure 3, at the anode, the enzyme GOx is coimmobilized with redox mediator I. At the cathode, the enzyme bilirubin oxidase (BOD) is coimmobilized with redox mediator II. GOx reduces glucose to gluconolactone. GOx is reoxidized by redox mediator I, which in turn passes the excess of electrons to the inert electrode. The electrons travel via the external electrical circuit to the cathode, where they are transferred from the inert electrode to redox mediator II. BOD reduces oxygen to water. BOD is reduced back to its original state via the electrons supplied by redox mediator II. Anode and cathode are

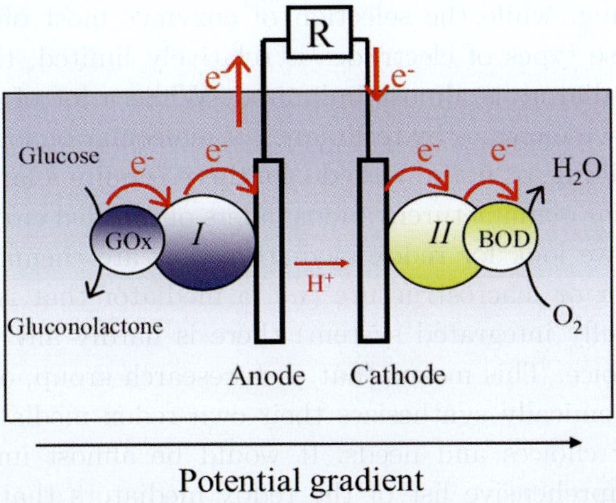

Figure 3. General example of a membraneless glucose/oxygen fuel cell. GOx and BOD represent the enzymes GOx and BOD, respectively. I and II represent two artificial redox mediators specially designed to mediate the electron transfer from the said enzymes.

connected via the external electrical circuit, and via the electrolyte solution in which both of them are immersed. The chemical reactions fueling the system are:

Anode: $2C_6H_{12}O_6 \rightarrow 2C_6H_{10}O_7 + 4H^+ + 4e^-$
Cathode: $O_2 + 4H^+ + 4e^- \rightarrow 2H_2O$
Biofuel cell: $2C_6H_{12}O_6 + O_2 \rightarrow 2C_6H_{10}O_7 + 2H_2O$

1.3.3 About enzymes and redox mediators

There are other enzymes capable of oxidizing glucose, most notably, PQQ-glucose dehydrogenase. Other than BOD, there are many different types of laccase enzymes, which are also able to reduce O_2 to H_2O. BOD and laccases are not unique enzymes, but the nomenclature encompasses a rather large family of enzymes, originating in different microorganisms, all of them reducing O_2 to H_2O. Laccases usually show higher voltage values than BOD. However, BOD has optimum activity in physiological pH, while, laccases have optimum pH values around pH 5, and its activity largely decreases at physiological pH.

Interesting, while the selection of enzymes most often used to prepare these types of electrodes is relatively limited, the selection of redox mediators is almost unlimited. While a lot of groups produce their own enzymes by techniques of molecular biology,[99–103] for those not willing, or not able to do so, there is quite a large selection of commercial manufacturers and retailers of purified enzymes. However, when we look for redox mediators that are chemically bound to a polymer or macrostructure (*i.e.* a mediator that is most suitable for a fully integrated system) there is hardly any commercial available choice. This means that each research group, or industrial company chemically synthesizes their own redox mediators according to their choices and needs. It would be almost impossible to build a comprehensive list of the redox mediators that have been used to wire enzymes to electrodes.[5,7,13,16,17,20–22,24,104] Compiling such a list, even for GOx electrodes, would be tedious and extremely long. Most artificial redox mediators, both for oxidases, dehydrogenases and reductases belong to one of the following families of compounds: substituted ferrocenes, quinones, viologens, tetrathiafulvalene and organometallic compounds of Ru and Os.[6,8–10] Because of its extremely high chemical stability, fast reaction kinetics to interconvert between the reduced and oxidized form, and the versatility for organometallic synthesis, the library of Os complexes that has been reported to date is larger than for any other species. Often, the most complex part of the chemical synthesis is the synthesis of the organometallic complex, with the right selection of ligands to match a desired redox potential. Later on, the organometallic complex is chemically grafted to a commercially available non-redox polymer, or to a non-redox polymer made out of two or more monomers, *i.e.* a copolymer.[5,16,68,74,104]

Interestingly, most researchers who have been in the field since the early times would argue that even if they were available they would not purchase commercial mediators. This is because chemically synthesizing their own mediators allows them to specifically design them to the right choice of redox potential, and chemical functionalities needed to best suit the immobilization method both to the electrode material and to the enzyme.[66–68,74,104,105] This unicity of mediators allows for specific chemical functionalities, and for companies manufacturing commercial disposable glucose biosensors, it is a way to keep their technological secrets.[71,73] One issue with these complicated chemical synthesis is that while the redox complexes yield very reproducible organometallic structures, the ratio of organometallic complex to monomer is not very reproducible, even between different synthetic batches in the same research and even industrial laboratories. Others, especially newcomers to the field, and those specializing in either materials chemistry or molecular biology, but not in organometallic synthesis, will argue that it is not straightforward to test their newly synthesized materials or enzymes in the absence of standardized mediators. Moreover, it is difficult to compare results with published reports when changing either the enzyme or the electrode material, if the same mediators are not available for comparison. Specific examples of Os-polymers are presented in Chapter 5, Section 5.5.1.

1.3.4 Performance

Optimizing performance has been the driving force behind most of the research carried out in the field. The power density output of an EFC is the product of the cell voltage and the current normalized by the surface area of the electrodes.

It is not trivial to compare results from different reports, since there is no agreement in the standardized way in which results should be reported. Therefore, some authors report results normalized only by the surface area of one electrode, while others take into account both anode and cathode. Hardly anyone takes into account the whole volume of the electrode assembly. EFC voltages depend above all on the fuel and oxidant selected, the enzymes catalyzing the reaction, and the redox potential of the mediators, if present. Moreover, the rate of electron transfer, the flow of current, resistances within the whole assembly (Ohmic losses) and mass transport processes will also affect the voltage output.

Most of the work to date has focused on electrodes containing only one enzyme. This means that most electrodes only release a few electrons from the many possible in a given organic fuel molecule. For example, only two electrons are extracted from glucose, out of the 24 possible ones if it would be fully oxidized to CO_2. This is in clear contrast to the optimized fuel efficiency that microbial fuel cells can afford, a topic that will be discussed in more detail in Chapters 2 and 3. Whilst some big effort has been devoted on the assembly of multiple enzymes, or enzyme cascades, at the anode of an EFC, these are mostly isolated examples within an otherwise very vast field.[106–110]

When no mediators are used, *i.e.* if both enzymes directly transfer electrons to the solid-state electrode, the maximum cell voltage that can be extracted from an EFC is given by the difference between the formal redox potentials (E') of the redox enzyme cofactors. Conversely, when mediators are used, the maximum voltage output that can be extracted is that given by the formal redox potentials of the redox mediators. This is schematically presented in Figure 4. The redox potentials of enzymes and redox mediators must be chosen so that the kinetics of electron transfer are facilitated, while the overall voltage of the fuel cell is maximized. At the risk of being repetitive,

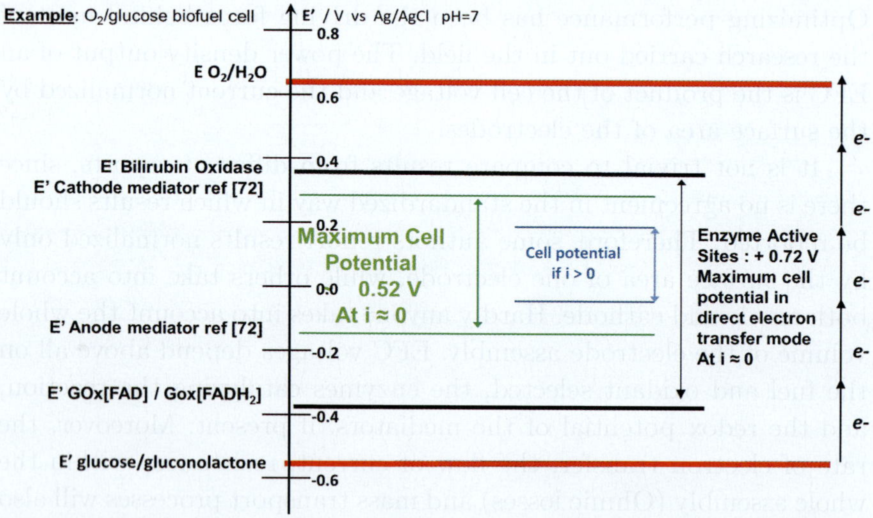

Figure 4. Example of redox potentials for an EFC. Potentials of the uncatalyzed reactions, potentials of the redox centers of the enzymes, potentials for the redox mediators and potential at which a real cell can work. The potentials of the redox mediators correspond to those given in Ref. 72.

we must emphasize again that the prime in the super index in the electrode potentials in Figure 4 is meant to exemplify that we are referring to those hemi-reactions in the given conditions at which the electrodes are meant to work (pH, temperature, ionic strength, and the corresponding concentration ranges for substrate/product/enzyme/mediator).

The best performing EFCs to date produce power densities between 1 and 3 mW cm^{-2}. If we look backwards, for many years, it used to be mediated systems that outperformed DET systems in terms of power density.[111–113] This was because mediated systems were able to produce much higher current densities than their DET analogues. However, with the optimization of different nanomaterials that are able to simultaneously host large amounts of enzymes, while at the same time efficiently deliver substrate and electrically connect a large proportion of those enzymes, DET-based EFCs have caught up their delay in performance. And in recent years, most examples of high-performance EFCs are mediatorless.[45,114–116] DET-based systems have the added advantage of operating at higher voltages, since the open circuit potential is established by the redox centers of the enzymes, and not the mediators, as it was shown in Figure 4.

The last 20 years have seen outstanding advancements in the field of EFCs.[98] Multi-disciplinary work has led to the increase in open circuit potentials from around 150 mV to almost 1 V. Concomitantly, current densities have increased from a few nA cm^{-2} to up to 30 mA cm^{-2},[47] and electrodes lasting from a few hours to a month and longer. Advancements have come from several fields, from selection of new enzymes and mediators, enzyme engineering and new materials. In the following chapters, we will concentrate on advancements in electrode design that have pushed the field forward.

1.4 Some Thoughts on Enzyme Kinetics

The simplest and yet most of the time accurate description of enzyme kinetics is based on work by Michaelis and Menten.[80] It assumes a two-step reaction, enzyme E and substrate S form an enzyme–substrate complex, ES, in a reversible step, that is at equilibrium under steady-state conditions. The irreversible rupture of ES yields product P, as depicted in Reaction (5). Two kinetic constants characterize these reactions, K_{MS} and k_{cat}. Because we are working with two substrate enzymes (recall the ping–pong mechanism), in particular with an oxidase, we further need reaction (6), where the reduced enzyme will react, with a kinetic constant k, with the oxidized mediator to be fully regenerated and be able to react with another substrate molecule. ζ is a stoichiometric coefficient. Finally, the reduced mediator will react at the electrode surface to leave the excess of electrons (Reaction 7).

$$S + E_{\mathrm{OX}} \overset{K_{\mathrm{MS}}}{\longleftrightarrow} \mathrm{ES} \xrightarrow{k_{\mathrm{cat}}} P + E_{\mathrm{RED}} \tag{5}$$

$$\zeta\mathrm{MED}_{\mathrm{OX}} + E_{\mathrm{RED}} \xrightarrow{k} \zeta\mathrm{MED}_{\mathrm{RED}} + E_{\mathrm{OX}} \tag{6}$$

$$\mathrm{MED}_{\mathrm{RED}} \rightarrow \mathrm{MED}_{\mathrm{OX}} + ne^{-} \tag{7}$$

k_{cat}, also called turnover number is a measure of how active an enzyme is towards a given substrate; the larger k_{cat}, the faster the reaction will be. K_{MS} is a measure of the affinity of an enzyme for the substrate. The smaller K_{MS}, the lower the concentration at which the enzyme will be saturated in substrate, and will hence reach its maximum possible reaction rate. For enzymes with affinity to more than one substrate, each substrate will have characteristic k_{cat} and K_{MS} values. The general expression for the reaction rate of the first enzymatic reaction is given by

$$v = \frac{\mathrm{d}[S]}{\mathrm{d}t} = \frac{k_{\mathrm{cat}}[E][S]}{K_{\mathrm{MS}} + [S]} = k_{\mathrm{E}}[E][S] \tag{8}$$

where

$$k_{\mathrm{E}} = \frac{k_{\mathrm{cat}}}{K_{\mathrm{MS}} + [S]} \tag{9}$$

This shows that, even in the simplest possible scenario, all the components in solution, enzyme kinetics are nonlinear. From (9), when ($[S] \ll K_{\mathrm{MS}}$) we obtain a linear expression for the reaction rate at low substrate concentrations:

$$v = \frac{k_{\text{cat}}[E][S]}{K_{\text{MS}}} \tag{10}$$

The reaction with the mediator is usually characterized by a simple second order kinetic expression:

$$v = k[E_{\text{RED}}][\text{MED}_{\text{OX}}] \tag{11}$$

where for a given enzyme, k, the reoxidation rate constant, is a function of the specific redox mediator (be it artificial or natural). Every enzyme–mediator couple will have a distinct k value. The concepts of k_{cat}, K_{MS} and k are of paramount importance to understand how to choose a certain enzyme to build an enzyme electrode for a given purpose. However, all three kinetic constants might show large degrees of variability depending on whether the enzyme is free in solution or whether it is immobilized on an electrode surface. In what follows, the terminology "enzyme in solution" will exclusively refer to enzyme activity tests performed by classical biochemistry techniques (usually UV–Vis). In this homogeneous reaction media, the enzyme, its substrate and the redox mediator are all in solution free to diffuse, the reaction takes place in the whole reaction volume and no concentration profiles are developed.

An attentive reader can already wonder if all of the above will still hold, when the enzyme is immobilized at the electrode surface. And the answer is a clear no. First, the kinetic equations will be coupled to diffusion equations, and together, they will give rise to complex nonlinear differential equations without analytical solution that are way beyond the scope of this introductory chapter. Second, as already suggested, the catalytic constants themselves might change values when the enzyme and/or the mediator are no longer free in solution to diffuse. Even if we do not dig into the detailed treatment of the kinetics of immobilized enzyme electrodes, it is important to stress some key aspects of the problem that will help to better understand how the researcher should rationally attack the problem of designing a new immobilized enzyme electrode. In a field that is already over 30 years old, and where very comprehensive theoretical treatments of the coupled reaction–diffusion problem have already been presented,[76–78,91,117,118] it would be silly to attempt the optimization of enzyme electrodes by a blind trial and error approach.

Three different phenomena might change the enzyme response with respect to solution experiments. First, if the enzyme is immobilized on the electrode surface, many of its properties might change. It is not unreasonable to think that the adsorbed enzyme will partially change its conformation. This might in turn change, either for better or for worse (most likely), the reactivity towards substrate and/or mediator, stability and tolerance towards other chemical species. In the bioelectrochemistry literature, countless reports can be found that give evidence of this phenomenon.[76–78] Interestingly, different electrodes might produce opposite effects on the same enzyme. Second, one or more reaction steps will occur in heterogeneous media, *i.e.* at the electrode surface, and this might create concentration gradients in the direction perpendicular to the electrode surface. For instance, in many experimental setups, the redox mediator in the active form is produced at the electrode surface, and therefore it will only be present in a rather thin layer in the vicinity of the electrode. If there is not enough mediator, the global reaction rate, measured by the generated amperometric current, might seem slower than what the experiment in solution had predicted. Third and finally, interaction of mediator and enzyme when both of them are immobilized should not be disregarded. Free mediators in solution will usually approach the enzyme at an optimum contact point, whereas immobilized mediators might be conformationally hindered from doing so. The fact that the mediator often plays a double role, as both electron relay and as the *cement* or *glue*, keeping the enzyme attached to the electrode surface (see Chapter 5), should not be forgotten. In view of these three important differences, we are particularly interested in comparing the properties of the enzyme in solution to the properties of the same enzyme when incorporated into an electrochemical system.

A first simple consideration is to analyze the amperometric response of an enzyme electrode as a function of substrate concentration. In an analogous way to kinetics in solution, for a given enzyme-modified electrode, we can define the parameter $K_{\mathrm{MS,app}}$, the apparent Michaelis-Menten constant, as the substrate concentration for which half the maximum amperometric output is achieved. It is important to notice that $K_{MS,app}$ is unique for each system, and it will change if we change the enzyme immobilization method, the redox mediator, the amount of enzyme and/or mediator on the electrode, etc. If $K_{MS,app}$ obtained from kinetic analyses of the bioelectrode is equal to K_{MS} for the enzyme, then one can conclude that the amperometric enzyme electrode response is limited by the enzyme kinetics.

For applications such as biofuel cells, the ideal option is an enzyme that allows for maximum current output, *i.e.* with maximum turnover (maximize k_{cat}). For particular applications, such as biofuel cells working in real media with rather fixed and low substrate concentrations, we will also be interested in finding an enzyme with a maximum turnover at the concentration of interest (*i.e.* low K_{MS}).[119] This is the case of implantable biofuel cells, working in physiological conditions where both glucose and O_2 concentrations are rather low (around 5 and 0.2 mM, respectively). It is also the case of biofuel cells where diluted wastewater is used as fuel.

On the contrary, for biosensor applications, as wide dynamic range as possible is desired, *i.e.* large K_{MS}. This is needed to avoid sample dilutions, which would allow for fast analysis and, ideally, to be used by non-trained personnel or even the general public, such as in the case of commercial glucose biosensors. A too narrow dynamic range also reduces the applicability of the biosensors for *in situ* sample analysis. In addition, to discriminate between close values, good sensitivity is required, *i.e.* high k_{cat}/K_{MS}. Eventually, for certain biosensor applications, as low as possible detection limit is required, again a combination of K_{MS} and k_{cat}.

From the examples discussed above, the rational design of an enzyme electrode should evidently include a careful study of the reported enzymatic parameters for the free enzyme in solution. Choosing an enzyme and mediator system with promising k_{cat}, K_{MS}, and k parameters in solution is certainly a great starting point when one wishes to design a new enzyme electrode. However, it is not trivial to predict to which extent these parameters will be affected upon immobilization, although some general trends usually hold true.

Immobilization approaches that allow for relatively large degrees of mobility within the enzymatic film are less likely to negatively affect the solution kinetic parameters. This is the case of redox hydrogels, and particularly, of redox mediators that are tethered via a long arm to a polymeric backbone, allowing for a very large mobility.[74] Conversely, immobilization strategies that restrict enzyme and mediator mobility are more likely to produce considerable changes in the kinetic parameters previously measured in solution. These examples will be further discussed in Chapter 5.

Finally, thick enzyme films (thicker than three to four times the enzyme diameter) are more likely to produce concentration gradients, both within the film, and in the solution adjacent to the enzymatic film. Concentration gradients might affect the ratio of reduced and oxidized mediator, with variations with respect to the Nernst-established ratio at the electrode surface. Sometimes, concentration gradients will also affect substrate and product concentrations with variations *vs.* the bulk solution concentrations. Mediator and substrate gradients will seemingly produce a lower current output than predicted. Product gradients might, however, severely affect the enzyme, particularly if pH gradients are being formed. Formation of substrate and product concentration gradients are also often generated when working with porous electrode materials, whether mediatorless or not. Within some restricted geometries, or when the pore sizes are not optimized, such gradients will be formed since mass transfer phenomena are not efficient within the porous structure; see Chapters 9 and 10 for some detailed examples.

In summary, the kinetics of immobilized enzyme electrodes are non-trivial, and there is no single analytical equation that can predict the amperometric output of every enzyme electrode. This is because the kinetics are coupled to diffusion equations and the resulting differential equations are highly nonlinear and therefore do not have analytical solutions. While the need for device calibration is still a must, several approximate equations have been derived for different limiting cases. Moreover, many numerical resolutions of the coupled differential equations have been proposed.[76–78] Together, approximate analytical solutions and numerical simulations are an invaluable tool for the rational design of enzyme electrodes. More could be written on the kinetics of modified enzyme electrodes. The special case of redox hydrogel electrodes is addressed in Chapter 7. Readers further interested in the general treatment of enzyme electrode kinetics are referred to our book chapter published in 2008.[91]

1.5 Advanced Electrode Materials: Is it Simply Just More Surface Area?

Many examples of advanced electrode materials will be discussed in the following chapters in this book. Advanced electrode materials encompass mostly porous materials. Evidently, one of the key advantages of porous electrodes is that they present much more surface area, and hence they allow for the immobilization of a larger amount of enzyme (and mediator, if present). As such, for most practical applications and technological developments, it is more than obvious that porous electrodes will be used almost exclusively. Naturally, the same can be said for porous electrodes in the field of microbial electrodes; more surface area is a synonym of more available surface area for biofilm development. A debate has been ongoing for many years between researchers working on materials engineering, and their colleagues working on more fundamental aspects of bioelectrodes. The latter sometimes accusing the former of merely increasing the surface area of new materials, which is certainly an oversimplification of the whole debate.

Of course, porous electrodes have larger surface areas. Consequently, they greatly serve to get closer to the holy grail of bioelectrodes, both enzymatic and microbial, of increased current (and power) densities. However, advanced (porous) electrodes are designed to improve yet many other characteristics of bioelectrodes. Advanced electrodes with features in the nanometric or Angstrom scale can very often trigger and/or facilitate DET from enzyme to the solid-state electrode or *vice versa*. Once again, exactly the same can be said of electrodes with such features used for the development of microbial biofilms.[120–122] In the following chapters, many examples will be discussed where nanofeatures approach the prosthetic group of the enzyme and therefore facilitate DET by decreasing the tunneling distance.

The third goal in the rational design of advanced electrode materials is enzyme stabilization. Enzymes trapped in purposely designed materials have shown much longer stability times and such bioelectrodes have proven to keep catalytic activity for weeks and even months, when the same enzymes usually lose their activity on conventional electrodes, or in homogenous systems, much earlier. In addition, some of these materials might also favor the stability of the enzyme in conditions where it would otherwise denaturate straightaway, such as outside the normal pH range, at higher temperatures, or salinity. Finally, many of these advanced electrodes serve the double function of both stabilizing the enzyme and favoring DET. Detailed examples of all the above-mentioned electrodes will be discussed in Chapters 6–10.

It will also be discussed in detail in the following chapters how trying to get all of the above characteristics together often renders the design of electrodes highly non-trivial. Increasing the surface area and the amount of immobilized enzymes (or bacteria) calls for larger pores to allow for efficient substrate delivery to, and product diffusion away from the enzymes/bacteria. Moreover, it will be useless to create all those pores, if it is later impossible to fill them with enzyme and/or mediator molecules, because they are partially inaccessible. There will also be a delicate equilibrium between increasing porosity, keeping conductivity and avoiding fragility of the electrodes.

It is usually very difficult to precisely measure the amount of enzyme that has been deposited or adsorbed within a porous electrode. Consequently, it is also difficult to assess what percentage of these enzymes are contributing to the measured catalytic current. Experimental data suggests that most of the time, there is a very delicate equilibrium when trying to find the optimum porosity. While decreasing the pore sizes will allow for an increase in surface area, and, provided the pores are accessible, an increase in enzyme loading, this strategy might lead to pores that are too small and will not afford for efficient substrate delivery. In such a scenario, the substrate concentration in the vicinity of a large proportion of enzymes will be significantly lower than in the bulk solution. And eventually, the current will be far from that expected from multiplying the turnover number by the enzyme concentration.

Eventually, the ultimate goal to allow for the largest possible amount of immobilized enzyme is to work with 3D electrodes, *i.e.* electrodes with a true volume, with dimensions in x, y and z, in the same order of magnitude, as opposed to merely rough electrodes. This has been common practice in classical fuel cells, since the dawn of the field.[46,47,123,124] In order to get efficient mass transport towards and out of the whole volume of these electrodes, a hierarchical porous structure will be needed. Mesopores in the 10th of nanometer scale are necessary to entrap the enzymes. In turn, macropores are needed to deliver substrate and take away the products. Finally, microporosity in the nano or Angstrom scale might help to increase the rate of electron transfer from enzyme and/or mediator towards the electrode.

1.6 Reporting Results and Comparisons

In order to rightly compare results between different reports, for EFCs, it is very important to clearly specify how current density values are being reported. The first discussion to settle is whether results should be normalized by the area/volume of one or the two electrodes. Beyond that discussion, current densities can be either normalized by *projected surface area*, *total surface area* or *volume* of the electrodes. Normalization by projected surface area is one of the most common ways of reporting data. The projected surface area usually refers to the fingerprint area of the electrode, regardless of its thickness or total volume. From an engineering perspective, this is indeed quite appropriate, since what really matters is the total current value that a certain electrode (device) can produce. Lately, some reports have started to provide current values by electrode volume, which is again quite appropriate from an engineering perspective. This is indeed slightly more accurate than projected surface area, and somehow it takes into account the fact that thicker electrodes (usually scaffolds) will accommodate more enzyme/mediator.

Most often, porous or rough electrodes have displayed higher current than flat electrodes, when the current is normalized by the projected surface area. However, it would be over-simplistic to say that porous electrodes work better than their flat counterparts. Normalization by total surface area is calculated by dividing the output current by the electrode surface area that has been determined by either mercury intrusion or nitrogen sorption porosimetry. Often, when current normalized by these values is compared to current density values on flat electrodes, porous electrodes usually underperform. Although less useful from an engineering perspective, these current density values are useful to compare the intrinsic efficiency of different electrode materials. There are several reasons why 3D and rough electrodes usually show a lower current by total surface area than flat electrodes. First, in most cases, mass transport in thick electrodes is not 100% optimized, *i.e.* not all enzymes are exposed to the bulk fuel concentration or are quickly cleared from potentially damaging products. Second, because rough and 3D electrodes are sometimes not fully loaded with enzyme (*i.e.* partial surface coverage only). Finally, because the enzyme and the mediator are poorly mixed, *i.e.* partially segregated, within the pores. These limitations are clearly a driving force to keep research on porous and advanced electrode materials ongoing.

Finally, it should be noted that often, most of these reports only take into account the active electrode itself, but not the complex structures that are sometimes needed to hold the electrodes, and/or electrically connect them to the outside circuitry; *i.e.* they do not take into account the whole biosensor/biofuel cell setup, but only the electrodes themselves.

1.7 Conclusions

An efficient enzymatic electrode will show good/optimized

(1) kinetics, including high enzymatic activity, optimal substrate dynamic range and sensitivity to a given substrate;
(2) stability;

(3) selectivity;
(4) electron transfer to the electrode; and
(5) redox potential.

The remainder of this book will discuss selected examples from recent years, where one or more of the above characteristics have been studied and/or improved. Naturally, the different properties are interconnected and changing one characteristic with a particular aim might influence other properties, and not always in the desired direction.

List of Abbreviations

BMCOs	blue multi-copper oxidases
BOD	bilirubin oxidase
DET	direct electron transfer
E	enzyme
EFC	enzymatic fuel cell
E_{OX}	oxidized enzyme
E_{RED}	reduced enzyme
ES	enzyme-substrate complex
FAD	flavin adenine dinucleotide
$FADH_2$	flavin mononucleotide (reduced form)
FMN	flavin mononucleotide (oxidized form)
GOx	enzyme glucose oxidase
MED_{OX}	oxidized mediator
MED_{RED}	reduced mediator
MET	mediated electron transfer
NAD^+	nicotinamide adenine dinucleotide (oxidized form)
$NADP^+$	nicotinamide adenine dinucleotide phosphate (oxidized form)
P	product
PQQ	pyrroloquinoline quinone
S	substrate
UV–Vis	ultraviolet–visible spectroscopy

References

1. B.L. Nilsson, M.B. Soellner and R.T. Raines (2005). Chemical synthesis of proteins. *Annual Review of Biophysics and Biomolecular Structure* **34**, 91–118.
2. S. Kent (2003). Total chemical synthesis of enzymes. *Journal of Peptide Science* **9**(9), 574–593.
3. D.L. Nelson, M.M. Cox and A.L. Lehninger (2013). *Lehninger Principles of Biochemistry,* New York: W.H. Freeman.
4. W. Schuhmann (2002). Amperometric enzyme biosensors based on optimised electron-transfer pathways and non-manual immobilisation procedures. *Reviews in Molecular Biotechnology* **82**(4), 425–441.
5. K. Habermüller, M. Mosbach and W. Schuhmann (2000). Electron-transfer mechanisms in amperometric biosensors. Fresenius' *Journal of Analytical Chemistry* **366**(6–7), 560–568.
6. R. Wilson and A.P.F. Turner (1992). Glucose oxidase: An ideal enzyme. *Biosensors and Bioelectronics* **7**(3), 165–185.
7. P.N. Bartlett (2008). *Bioelectrochemistry: Fundamentals, Experimental Techniques and Applications.* John Wiley & Sons, Ltd.
8. S.B. Bankar, M.V. Bule, R.S. Singhal and L. Ananthanarayan (2009). Glucose oxidase — An overview. *Biotechnology Advances* **27**(4), 489–501.
9. C. Bourdillon, C. Hervagault and D. Thomas (1985). Increase in operational stability of immobilized glucose oxidase by the use of an artifical cosubstrate. *Biotechnology and Bioengineering* **27**(11), 1619–1622.
10. P.N. Bartlett, P. Tebbutt and R.G. Whitaker (1991). Kinetic aspects of the use of modified electrodes and mediators in bioelectrochemistry. *Progress in Reaction Kinetics* **16**(2), 55–155.
11. J.M. Berg, J.L. Tymoczko and L. Stryer (2002). *Biochemistry,* 5th edition. New York: W H Freeman.
12. A.P.F. Turner (2013), Biosensors: Sense and sensibility. *Chemical Society Reviews* **42**(8), 3184–3196.
13. A. Heller (1992). Electrical connection of enzyme redox centers to electrodes. *Journal of Physical Chemistry* **96**(9), 3579–3587.
14. B.A. Gregg and A. Heller (1990). Cross-linked redox gels containing glucose oxidase for amperometric biosensor applications. *Analytical Chemistry* **62**(3), 258–263.
15. A. Heller (1999). Implanted electrochemical glucose sensors for the management of diabetes, in *Annual Review of Biomedical Engineering,* pp. 153–175.
16. A. Heller (2006). Electron-conducting redox hydrogels: Design, characteristics and synthesis. *Current Opinion in Chemical Biology* **10**(6), 664–672.
17. C.R. Ispas, G. Crivat and S. Andreescu (2012). Review: Recent developments in enzyme-based biosensors for biomedical analysis. *Analytical Letters* **45**(2–3), 168–186.
18. N.J. Ronkainen, H.B. Halsall and W.R. Heineman, Electrochemical biosensors. *Chemical Society Reviews* **39**(5), 1747–1763.

19. J. Wang (2008). Electrochemical glucose biosensors. *Chemical Reviews* **108**(2), 814–825.
20. S. Calabrese Barton, J. Gallaway and P. Atanassov (2004). Enzymatic biofuel cells for implantable and microscale devices. *Chemical Reviews* **104**(10), 4867–4886.
21. J.A. Cracknell, K.A. Vincent and F.A. Armstrong (2008). Enzymes as working or inspirational electrocatalysts for fuel cells and electrolysis. *Chemical Reviews* **108**(7), 2439–2461.
22. D. Leech, P. Kavanagh and W. Schuhmann (2012). Enzymatic fuel cells: Recent progress. *Electrochimica Acta* **84**, 223–234.
23. M.H. Osman, A.A. Shah and F.C. Walsh (2011). Recent progress and continuing challenges in bio-fuel cells. Part I: Enzymatic cells. *Biosensors and Bioelectronics* **26**(7), 3087–3102.
24. R.A. Bullen, T.C. Arnot, J.B. Lakeman and F.C. Walsh (2006). Biofuel cells and their development. *Biosensors and Bioelectronics* **21**(11), 2015–2045.
25. X. Dominguez-Benetton, S. Srikanth, Y. Satyawali, K. Vanbroekhoven and D. Pant (2013). Enzymatic electrosynthesis: An overview on the progress in enzyme-electrodes for the production of electricity, fuels and chemicals. *Journal of Microbial and Biochemical Technology* **5**(2).
26. I. Mazurenko, W. Ghach, G.W. Kohring, C. Despas, A. Walcarius and M. Etienne (2015). Immobilization of membrane-bounded (S)-mandelate dehydrogenase in sol-gel matrix for electroenzymatic synthesis. *Bioelectrochemistry* **104**, 65–70.
27. F.A. Armstrong (2005). Recent developments in dynamic electrochemical studies of adsorbed enzymes and their active sites. *Current Opinion in Chemical Biology* **9**(2), 110–117.
28. C. Léger, S.J. Elliott, K.R. Hoke, L.J.C. Jeuken, A.K. Jones and F.A. Armstrong (2003). Enzyme electrokinetics: Using protein film voltammetry to investigate redox enzymes and their mechanisms. *Biochemistry* **42**(29), 8653–8662.
29. H. Yang (2012). Enzyme-based ultrasensitive electrochemical biosensors. *Current Opinion in Chemical Biology* **16**(3–4), 422–428.
30. X.M. Chen, B.Y. Su, X.H. Song, Q.A. Chen, X. Chen and X.R. Wang (2011). Recent advances in electrochemiluminescent enzyme biosensors. *TrAC — Trends in Analytical Chemistry* **30**(5), 665–676.
31. M.M.F. Choi (2004). Progress in enzyme-based biosensors using optical transducers. *Microchimica Acta* **148**(3–4), 107–132.
32. Y. Ding, J. Liu, H. Wang, G. Shen and R. Yu (2007). A piezoelectric immunosensor for the detection of α-fetoprotein using an interface of gold/hydroxyapatite hybrid nanomaterial. *Biomaterials* **28**(12), 2147–2154.
33. Z. Yang, S. Si, H. Dai and C. Zhang (2007). Piezoelectric urea biosensor based on immobilization of urease onto nanoporous alumina membranes. *Biosensors and Bioelectronics* **22**(12), 3283–3287.
34. T.A.P. Rocha-Santos (2014). Sensors and biosensors based on magnetic nanoparticles. *TrAC — Trends in Analytical Chemistry* **62**, 28–36.

35. R.A. Marcus (1993). Electron transfer reactions in chemistry: Theory and experiment (Nobel Lecture). *Angewandte Chemie International Edition in English* **32**(8), 1111–1121.
36. R.A. Marcus and N. Sutin (1985). Electron transfers in chemistry and biology. *BBA Reviews on Bioenergetics* **811**(3), 265–322.
37. O. Courjean, F. Gao and N. Mano (2009). Deglycosylation of glucose oxidase for direct and efficient glucose electrooxidation on a glassy carbon electrode. *Angewandte Chemie International Edition* **48**(32), 5897–5899.
38. D.J. Caruana and S. Howorka (2010). Biosensors and biofuel cells with engineered proteins. *Molecular BioSystems* **6**(9), 1548–1556.
39. T.S. Wong and U. Schwaneberg (2003). Protein engineering in bioelectrocatalysis. *Current Opinion in Biotechnology* **14**(6), 590–596.
40. J.N. Renner and S.D. Minteer (2016). The use of engineered protein materials in electrochemical devices. *Experimental Biology and Medicine* **241**(9), 980–985.
41. M. Campàs, B. Prieto-Simón and J.L. Marty (2009). A review of the use of genetically engineered enzymes in electrochemical biosensors. *Seminars in Cell and Developmental Biology* **20**(1), 3–9.
42. C.F. Blanford, R.S. Heath and F.A. Armstrong (2007). A stable electrode for high-potential, electrocatalytic O2 reduction based on rational attachment of a blue copper oxidase to a graphite surface. *Chemical Communications* **17**, 1710–1712.
43. M. Zayats, B. Willner and I. Willner (2008). Design of amperometric biosensors and biofuel cells by the reconstitution of electrically contacted enzyme electrodes. *Electroanalysis* **20**(6), 583–601.
44. H. Li, S. Liu, Z. Dai, J. Bao and X. Yang (2009). Applications of nanomaterials in electrochemical enzyme biosensors. *Sensors* **9**(11), 8547–8561.
45. D. Wen and A. Eychmüller (2016). Enzymatic biofuel cells on porous nanostructures. *Small* **12**(34), 4649–4661.
46. V. Flexer, N. Brun, O. Courjean, R. Backov and N. Mano (2011). Porous mediator-free enzyme carbonaceous electrodes obtained through Integrative Chemistry for biofuel cells. *Energy & Environmental Science* **4**(6), 2097–2106.
47. V. Flexer, N. Brun, M. Destribats, R. Backov and N. Mano (2013). A novel three-dimensional macrocellular carbonaceous biofuel cell. *Physical Chemistry Chemical Physics* **15**(17), 6437–6445.
48. H. Funabashi, K. Murata and S. Tsujimura (2015). Effect of pore size of MgO-templated carbon on the direct electrochemistry of D-fructose dehydrogenase. *Electrochemistry* **83**(5), 372–375.
49. S. Tsujimura, A. Nishina, Y. Hamano, K. Kano and S. Shiraishi (2010). Electrochemical reaction of fructose dehydrogenase on carbon cryogel electrodes with controlled pore sizes. *Electrochemistry Communications* **12**(3), 446–449.
50. S. Shleev, J. Tkac, A. Christenson, T. Ruzgas, A.I. Yaropolov, J.W. Whittaker and L. Gorton (2005). Direct electron transfer between

copper-containing proteins and electrodes. *Biosensors and Bioelectronics* **20**(12), 2517–2554.

51. S. Tsujimura, Y. Kamitaka and K. Kano (2007). Diffusion-controlled oxygen reduction on multi-copper oxidase-adsorbed carbon aerogel electrodes without mediator. *Fuel Cells* **7**(6), 463–469.
52. C. Vaz-Dominguez, S. Campuzano, O. Rüdiger, M. Pita, M. Gorbacheva, S. Shleev, V.M. Fernandez and A.L. De Lacey (2008). Laccase electrode for direct electrocatalytic reduction of O2 to H2O with high-operational stability and resistance to chloride inhibition. *Biosensors and Bioelectronics* **24**(4), 531–537.
53. R. Ludwig, W. Harreither, F. Tasca and L. Gorton (2010). Cellobiose dehydrogenase: A versatile catalyst for electrochemical applications. *ChemPhysChem* **11**(13), 2674–2697.
54. D. Ivnitski, P. Atanassov and C. Apblett (2007). Direct bioelectrocatalysis of PQQ-dependent glucose dehydrogenase. *Electroanalysis* **19**(15), 1562–1568.
55. V. Flexer, F. Durand, S. Tsujimura and N. Mano (2011). Efficient direct electron transfer of PQQ-glucose dehydrogenase on carbon cryogel electrodes at neutral pH. *Analytical Chemistry* **83**(14), 5721–5727.
56. Y. Kamitaka, S. Tsujimura, N. Setoyama, T. Kajino and K. Kano (2007). Fructose/dioxygen biofuel cell based on direct electron transfer-type bioelectrocatalysis. *Physical Chemistry Chemical Physics* **9**(15), 1793–1801.
57. F.A. Armstrong and A.M. Lannon (1987). Fast interfacial electron transfer between cytochrome c peroxidase and graphite electrodes promoted by aminoglycosides: Novel electroenzymic catalysis of H2O2 reduction. *Journal of the American Chemical Society* **109**(23), 7211–7212.
58. U. Wollenberger, J. Wang, M. Ozsoz, E. Gonzalez-Romero and F. Scheller (1991). Bulk modified enzyme electrodes for reagentless detection of peroxides. *Bioelectrochemistry and Bioenergetics* **26**(2), 287–296.
59. E. Csöregi, G. Jönsson-Pettersson and L. Gorton (1993). Mediatorless electrocatalytic reduction of hydrogen peroxide at graphite electrodes chemically modified with peroxidases. *Journal of Biotechnology* **30**(3), 315–337.
60. Y. Degani and A. Heller (1989). Electrical communication between redox centers of glucose oxidase and electrodes via electrostatically and covalently bound redox polymers. *Journal of the American Chemical Society* **111**(6), 2357–2358.
61. A. Heller (1990). Electrical wiring of redox enzymes. *Accounts of Chemical Research* **23**(5), 128–134.
62. A.E.G. Cass, G. Davis, G.D. Francis, H. Allen O Hill, W.J. Aston, I. John Higgins, E.V. Plotkin, L.D.L. Scott and A.P.F. Turner (1984). Ferrocene-mediated enzyme electrode for amperometric determination of glucose. *Analytical Chemistry* **56**(4), 667–671.
63. L.C. Clark, Jr. and C. Lyons (1962). Electrode systems for continuous monitoring in cardiovascular surgery, in *Annals of the New York Academy of Sciences*, pp. 29–45.

64. T.J. Ohara, R. Rajagopalan and A. Heller (1994). "Wired" enzyme electrodes for amperometric determination of glucose or lactate in the presence of interfering substances. *Analytical Chemistry* **66**(15), 2451–2454.
65. K. Kleppe (1966). The effect of hydrogen peroxide on glucose oxidase from aspergillus niger*. *Biochemistry* **5**(1), 139–143.
66. J.W. Gallaway and S.A.C. Barton (2008). Kinetics of redox polymer-mediated enzyme electrodes. *Journal of the American Chemical Society* **130**(26), 8527–8536.
67. J.W. Gallaway and S.A. Calabrese Barton (2009). Effect of redox polymer synthesis on the performance of a mediated laccase oxygen cathode. *Journal of Electroanalytical Chemistry* **626**(1–2), 149–155.
68. C. Danilowicz, E. Cortón, F. Battaglini and E.J. Calvo (1998). An Os(byp)2ClPyCH2NHPoly(allylamine) hydrogel mediator for enzyme wiring at electrodes. *Electrochimica Acta* **43**(23), 3525–3531.
69. A. Heller and B. Feldman (2008). Electrochemical glucose sensors and their applications in diabetes management. *Chemical Reviews* **108**(7), 2482–2505.
70. A. Heller and B. Feldman (2010). Electrochemistry in diabetes management. *Accounts of Chemical Research* **43**(7), 963–973.
71. Z. Liu, B.J. Feldman, F. Mao and A. Heller (2013). *Redox Polymers for Use in Analyte Monitoring*. Google Patents.
72. N. Mano, F. Mao and A. Heller (2002). A miniature biofuel cell operating in a physiological buffer. *Journal of the American Chemical Society* **124**(44), 12962–12963.
73. F. Mao and A. Heller (2013). *Redox Polymers*. Google Patents.
74. F. Mao, N. Mano and A. Heller (2003). Long tethers binding redox centers to polymer backbones enhance electron transport in enzyme "wiring" hydrogels. *Journal of the American Chemical Society* **125**(16), 4951–4957.
75. J. Hodak, R. Etchenique, E.J. Calvo, K. Singhal and P.N. Bartlett (1997). Layer-by-layer self-assembly of glucose oxidase with a poly(allylamine) ferrocene redox mediator. *Langmuir* **13**(10), 2708–2716.
76. V. Flexer, E.J. Calvo and P.N. Bartlett (2010). The application of the relaxation and simplex method to the analysis of data for glucose electrodes based on glucose oxidase immobilised in an osmium redox polymer. *Journal of Electroanalytical Chemistry* **646**(1–2), 24–32.
77. V. Flexer, M.V. Ielmini, E.J. Calvo and P.N. Bartlett (2008). Extracting kinetic parameters for homogeneous [Os(bpy)2ClPyCOOH]+ mediated enzyme reactions from cyclic voltammetry and simulations. *Bioelectrochemistry* **74**(1), 201–209.
78. V. Flexer, K.F.E. Pratt, F. Garay, P.N. Bartlett and E.J. Calvo (2008). Relaxation and Simplex mathematical algorithms applied to the study of steady-state electrochemical responses of immobilized enzyme biosensors: Comparison with experiments. *Journal of Electroanalytical Chemistry* **616**(1–2), 87–98.
79. S. Tsujimura, A. Nishina, Y. Kamitaka and K. Kano (2009). Coulometric D-fructose biosensor based on direct electron transfer using D-fructose dehydrogenase. *Analytical Chemistry* **81**(22), 9383–9387.

80. L. Michaelis and M.L. Menten (1913). Die kinetik der invertinwirkung. *Biochem. Z.* **49**, 333–369.
81. S.A. Jaffari and A.P.F. Turner (1995). Recent advances in amperometric glucose biosensors for in vivo monitoring. *Physiological Measurement* **16**(1), 1–15.
82. E.H. Yoo and S.Y. Lee (2010). Glucose biosensors: An overview of use in clinical practice. *Sensors* **10**(5), 4558–4576.
83. A. Chaubey and B.D. Malhotra (2002). Mediated biosensors. *Biosensors and Bioelectronics* **17**(6–7), 441–456.
84. A. Prévoteau and N. Mano (2012). Oxygen reduction on redox mediators may affect glucose biosensors based on "wired" enzymes. *Electrochimica Acta* **68**, 128–133.
85. A. Prévoteau and N. Mano (2013). How the reduction of O2 on enzymes and/or redox mediators affects the calibration curve of "wired" glucose oxidase and glucose dehydrogenase biosensors. *Electrochimica Acta* **112**, 318–326.
86. M. Zhao, Y. Gao, J. Sun and F. Gao (2015). Mediatorless glucose biosensor and direct electron transfer type glucose/air biofuel cell enabled with carbon nanodots. *Analytical Chemistry* **87**(5), 2615–2622.
87. Y. Wu and S. Hu (2007). Biosensors based on direct electron transfer in redox proteins. *Microchimica Acta* **159**(1), 1–17.
88. L. Gorton, A. Lindgren, T. Larsson, F.D. Munteanu, T. Ruzgas and I. Gazaryan (1999). Direct electron transfer between heme-containing enzymes and electrodes as basis for third generation biosensors. *Analytica Chimica Acta* **400**(1–3), 91–108.
89. Q. Xu, S.-X. Gu, L. Jin, Y.-E. Zhou, Z. Yang, W. Wang and X. Hu (2014). Graphene/polyaniline/gold nanoparticles nanocomposite for the direct electron transfer of glucose oxidase and glucose biosensing. *Sensors and Actuators B: Chemical* **190**, 562–569.
90. M.M. Barsan, K.P. Prathish, X. Sun and C.M.A. Brett (2014). Nitrogen doped graphene and its derivatives as sensors and efficient direct electron transfer platform for enzyme biosensors. *Sensors and Actuators B: Chemical* **203**, 579–587.
91. P.N. Bartlett, C.S. Toh, E.J. Calvo and V. Flexer (2008). Modelling biosensor responses, in *Bioelectrochemistry: Fundamentals, Experimental Techniques and Applications*, pp. 267–325.
92. U. Schröder, F. Harnisch and L.T. Angenent (2015). Microbial electrochemistry and technology: Terminology and classification. *Energy and Environmental Science* **8**(2), 513–519.
93. P. Cinquin, C. Gondran, F. Giroud, S. Mazabrard, A. Pellissier, F. Boucher, J.P. Alcaraz, K. Gorgy, F. Lenouvel, S. Mathé, P. Porcu and S. Cosnier (2010). A glucose BioFuel cell implanted in Rats. *PLoS One* **5**(5).
94. S. Cosnier, A. Le Goff and M. Holzinger (2014). Towards glucose biofuel cells implanted in human body for powering artificial organs: Review. *Electrochemistry Communications* **38**, 19–23.

95. A. Pothukuchy, N. Mano, G. Georgiou and A. Heller (2006). A potentially insect-implantable trehalose electrooxidizing anode. *Biosensors and Bioelectronics* **22**(5), 678–684.
96. S. Cosnier, A.J. Gross, A. Le Goff and M. Holzinger (2016). Recent advances on enzymatic glucose/oxygen and hydrogen/oxygen biofuel cells: Achievements and limitations. *Journal of Power Sources* **325**, 252–263.
97. A. Heller (2005). Integrated medical feedback systems for drug delivery. *AIChE Journal* **51**(4), 1054–1066.
98. M. Rasmussen, S. Abdellaoui and S.D. Minteer (2016). Enzymatic biofuel cells: 30 years of critical advancements. *Biosensors and Bioelectronics* **76**, 91–102.
99. J. Gallaway, I. Wheeldon, R. Rincon, P. Atanassov, S. Banta and S.C. Barton (2008). Oxygen-reducing enzyme cathodes produced from SLAC, a small laccase from Streptomyces coelicolor. *Biosensors and Bioelectronics* **23**(8), 1229–1235.
100. F. Durand, S. Gounel, C.H. Kjaergaard, E.I. Solomon and N. Mano (2012). Bilirubin oxidase from Magnaporthe oryzae: An attractive new enzyme for biotechnological applications. *Applied Microbiology and Biotechnology* **96**(6), 1489–1498.
101. F. Durand, C.H. Kjaergaard, E. Suraniti, S. Gounel, R.G. Hadt, E.I. Solomon and N. Mano (2012). Bilirubin oxidase from Bacillus pumilus: A promising enzyme for the elaboration of efficient cathodes in biofuel cells. *Biosensors and Bioelectronics* **35**(1), 140–146.
102. F. Durand, C. Stines-Chaumeil, V. Flexer, I. André and N. Mano (2010). Designing a highly active soluble PQQ-glucose dehydrogenase for efficient glucose biosensors and biofuel cells. *Biochemical and Biophysical Research Communications* **402**(4), 750–754.
103. R. Ludwig, R. Ortiz, C. Schulz, W. Harreither, C. Sygmund and L. Gorton (2013). Cellobiose dehydrogenase modified electrodes: Advances by materials science and biochemical engineering. *Analytical and Bioanalytical Chemistry* **405**(11), 3637–3658.
104. P. Kavanagh and D. Leech (2013). Mediated electron transfer in glucose oxidising enzyme electrodes for application to biofuel cells: Recent progress and perspectives. *Physical Chemistry Chemical Physics* **15**(14), 4859–4869.
105. C. Danilowicz, E. Cortón and F. Battaglini (1998). Osmium complexes bearing functional groups: Building blocks for integrated chemical systems. *Journal of Electroanalytical Chemistry* **445**(1–2), 89–94.
106. P.K. Addo, R.L. Arechederra and S.D. Minteer (2010). Evaluating enzyme cascades for methanol/air biofuel cells based on NAD+-dependent enzymes. *Electroanalysis* **22**(7–8), 807–812.
107. D. Sokic-Lazic, R.L. Arechederra, B.L. Treu and S.D. Minteer (2010). Oxidation of biofuels: Fuel diversity and effectiveness of fuel oxidation through multiple enzyme cascades. *Electroanalysis* **22**(7–8), 757–764.
108. R.L. Arechederra and S.D. Minteer (2009). Complete oxidation of glycerol in an enzymatic biofuel cell. *Fuel Cells* **9**(1), 63–69.

109. S. Xu and S.D. Minteer (2012). Enzymatic biofuel cell for oxidation of glucose to CO2. *ACS Catalysis* **2**(1), 91–94.
110. G.T.R. Palmore, H. Bertschy, S.H. Bergens and G.M. Whitesides (1998). A methanol/dioxygen biofuel cell that uses NAD+-dependent dehydrogenases as catalysts: application of an electro-enzymatic method to regenerate nicotinamide adenine dinucleotide at low overpotentials. *Journal of Electroanalytical Chemistry* **443**(1), 155–161.
111. Z. Zhu, T. Kin Tam, F. Sun, C. You and Y.H. Percival Zhang (2014). A high-energy-density sugar biobattery based on a synthetic enzymatic pathway. *Nature Communications* **5**.
112. B. Reuillard, A. Le Goff, C. Agnès, M. Holzinger, A. Zebda, C. Gondran, K. Elouarzaki and S. Cosnier (2013). High power enzymatic biofuel cell based on naphthoquinone-mediated oxidation of glucose by glucose oxidase in a carbon nanotube 3D matrix. *Physical Chemistry Chemical Physics* **15**(14), 4892–4896.
113. H. Sakai, T. Nakagawa, Y. Tokita, T. Hatazawa, T. Ikeda, S. Tsujimura and K. Kano (2009). A high-power glucose/oxygen biofuel cell operating under quiescent conditions. *Energy and Environmental Science* **2**(1), 133–138.
114. A. Zebda, C. Gondran, A. Le Goff, M. Holzinger, P. Cinquin and S. Cosnier (2011). Mediatorless high-power glucose biofuel cells based on compressed carbon nanotube-enzyme electrodes. *Nature Communications* **2**, 370.
115. T. Miyake, S. Yoshino, T. Yamada, K. Hata and M. Nishizawa (2011). Self-regulating enzyme-nanotube ensemble films and their application as flexible electrodes for biofuel cells. *Journal of the American Chemical Society* **133**(13), 5129–5134.
116. A. De Poulpiquet, A. Ciaccafava, R. Gadiou, S. Gounel, M.T. Giudici-Orticoni, N. Mano and E. Lojou (2014). Design of a H2/O2 biofuel cell based on thermostable enzymes. *Electrochemistry Communications* **42**, 72–74.
117. P.N. Bartlett and K.F.E. Pratt (1993). Modeling of processes in enzyme electrodes. *Biosensors and Bioelectronics* **8**(9–10), 451–462.
118. P.N. Bartlett and K.F.E. Pratt (1995). Theoretical treatment of diffusion and kinetics in amperometric immobilized enzyme electrodes Part I: Redox mediator entrapped within the film. *Journal of Electroanalytical Chemistry* **397**(1–2), 61–78.
119. N. Mano (2008). A 280 μW cm-2 biofuel cell operating at low glucose concentration. *Chemical Communications* **19**, 2221–2223.
120. V. Flexer, J. Chen, B.C. Donose, P. Sherrell, G.G. Wallace and J. Keller (2013). The nanostructure of three-dimensional scaffolds enhances the current density of microbial bioelectrochemical systems. *Energy and Environmental Science* **6**(4), 1291–1298.
121. L. Jourdin, S. Freguia, B.C. Donose, J. Chen, G.G. Wallace, J. Keller and V. Flexer (2014). A novel carbon nanotube modified scaffold as an efficient biocathode material for improved microbial electrosynthesis. *Journal of Materials Chemistry A* **2**(32), 13093–13102.

122. L. Jourdin, T. Grieger, J. Monetti, V. Flexer, S. Freguia, Y. Lu, J. Chen, M. Romano, G.G. Wallace and J. Keller (2015). High acetic acid production rate obtained by microbial electrosynthesis from carbon dioxide. *Environmental Science and Technology* **49**(22), 13566–13574.
123. S.D. Minteer, B.Y. Liaw and M.J. Cooney (2007). Enzyme-based biofuel cells. *Current Opinion in Biotechnology* **18**(3), 228–234.
124. V. Flexer, N. Brun, R. Backov and N. Mano (2010). Designing highly efficient enzyme-based carbonaceous foams electrodes for biofuel cells. *Energy and Environmental Science* **3**(9), 1302–1306.

Chapter 2

Fundamentals of Microbial Electrochemical Systems

Stefano Freguia[*,‡], Kun Guo[†], and Pablo Ledezma[*]

[*] *Advanced Water Management Centre, The University of Queensland, Brisbane, QLD 4072, Australia*
[†] *Center for Microbial Ecology and Technology, Ghent University, Coupure Links 653, B-9000 Ghent, Belgium*
[‡] *s.freguia@awmc.uq.edu.au*

2.1 Introduction to Bioelectrochemical Systems: The Possibility to Make Living Microorganisms Interact Directly With Solid Electrodes

Microbial bioelectrochemical systems (BESs), including microbial fuel cells (MFCs), microbial electrolysis cells (MECs), and microbial electrosynthesis (MES), are bioreactors that use electrochemically active microorganisms (EAMs) to catalyze oxidation and/or reduction reactions at the electrode surface.[1] At the anode, the bacteria catalyze oxidation of a number of organic compounds (most commonly acetate and glucose) to carbon dioxide; at the cathode an array of reduction reactions may occur, as shown in Figure 1, including oxygen reduction to water,[2] sulfate reduction to sulfide,[3] proton reduction to hydrogen,[4] CO_2 reduction to methane or acetate,[5] and nitrate reduction to nitrogen gas.[6]

EAMs are defined as microorganisms that exchange electrons with conductive surfaces (*e.g.* electrode) and the biofilms they form on electrode surfaces are referred to as electrochemically active

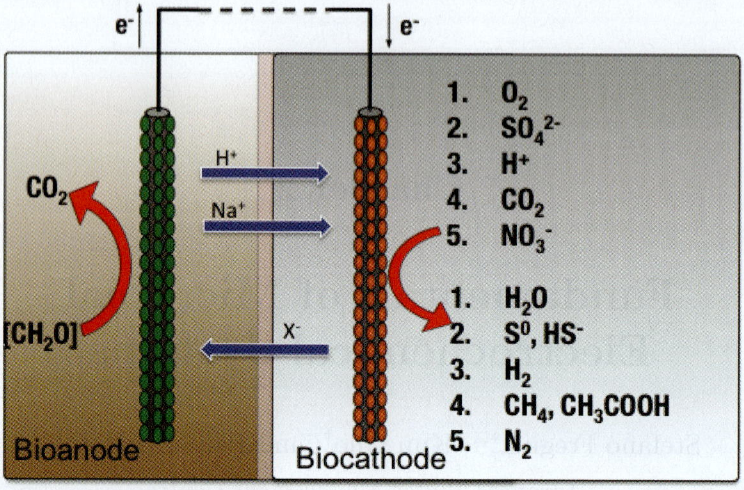

Figure 1. A BES which can achieve multiple microbially catalyzed anodic oxidations and cathodic reductions. Electrons move from anode to cathode via electrical circuitry, ions close the circuit migrating through the electrolyte and optional membrane separator.

biofilms (EABs). EABs that deliver electrons to the electrode are called anodic biofilms, while those that accept electrons are called cathodic biofilms.[7]

2.1.1 EABs formation process

In nature, most microorganisms live within biofilms on biotic or abiotic material surfaces, because biofilms provide them with reliable access to food, protect them from predators and enhance their resistance to harsh environments.[8] However, the main driving force for EAMs to form biofilms is to dump electrons to, or uptake electrons from, the solid surfaces (*e.g.* electrodes). In other words, exchanging of electrons with the electrodes is crucial for the living of EABs. Similar to other biofilms, the formation process of EABs can be divided into four steps, namely bacteria transport onto the surface, bacteria adhesion, biofilm matrix synthesis and maturation and detachment (Figure 2).[9]

The planktonic bacteria can be passively transported to electrode surfaces by a series of long-range physical forces including hydrodynamic forces, gravitational forces, and Brownian movement.[10] In BESs

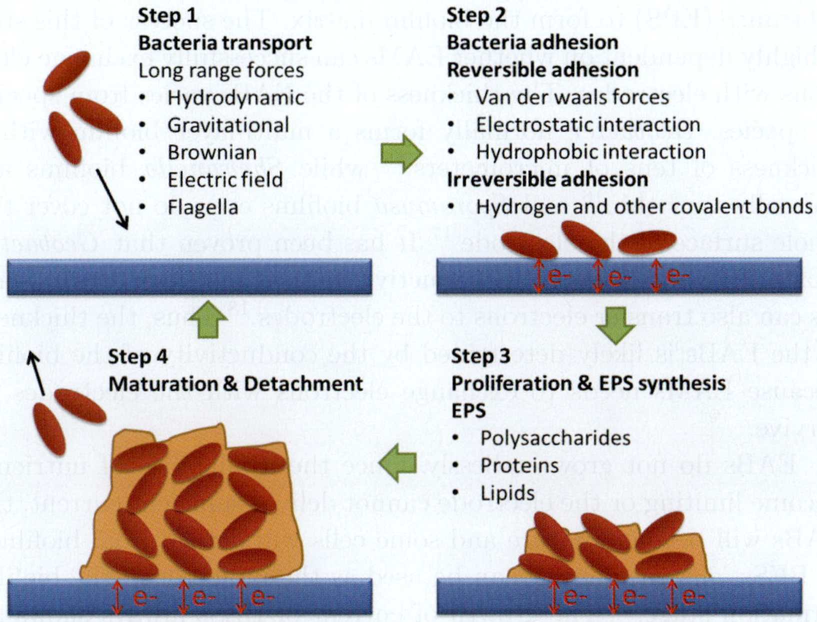

Figure 2. Schematic diagram of the electroactive biofilm formation process.

that start with an applied voltage (*e.g.* MEC), the electric field force may also play a role in bacteria transport because an additional electrical field is added and bacteria normally carry a negative charge. The most commonly known electroactive bacteria such as *Geobacter, Shewanella,* and *Sporomusa* are all flagella-containing bacteria,[11–13] so these EAMs can also actively swim to the surfaces using their flagella.

Bacteria adhesion can be further divided into two stages: reversible adhesion and irreversible adhesion.[14] The forces involved in the reversible bacteria adhesion are Van der Waals forces, electronic interactions and hydrophobic interactions. In the irreversible stage, the bacterial anchorage on the surface is strengthened by short-range forces such as hydrogen bonding and other covalent bonds. For EAMs, it is believed that the electron transfer units located on the outer cell membrane such as cytochrome *c* or conductive pili are involved in the cell adhesion.

After adhesion on the electrode surface, bacteria start to proliferate to form bacterial colonies and synthesize extracellular polymeric substance (EPS) to form the biofilm matrix. The success of this step is highly dependent on whether EAMs can successfully exchange electrons with electrodes. The thickness of the EABs varies from species to species. *Geobacter* normally forms a multi-layer biofilm with a thickness of tens of micrometers,[15] while *Shewanella* biofilms are normally very thin[16] and *Sporomusa* biofilms even do not cover the whole surface of the electrode.[17] It has been proven that *Geobacter* biofilms are highly electron conductive, so the bacteria on the top layers can also transfer electrons to the electrodes.[18] Thus, the thickness of the EABs is likely determined by the conductivity of the biofilm because EAMs needs to exchange electrons with the electrodes to survive.

EABs do not grow endlessly. Once the availability of nutrients become limiting or the electrode cannot deliver sufficient current, the EABs will become mature and some cells will detach from biofilms. In BESs, current density can be used as the indicator of the biofilm formation stage.[19] The growth of current of BESs always shows an S-shaped curve. The lag phase corresponds to steps 1 and 2 of biofilm formation process. The log phase can be attributed to the step 3 of biofilm formation. When the current density of the BESs reached the stationary phase, the EAB can be considered mature at the current environmental conditions.

2.1.2 EAB formation impacting factors

The biofilm formation process is an extremely variable process, which is affected by a series of factors, including bacteria's cell wall properties, material surface properties, and the environmental conditions (Figure 3).[10] In order to improve the EAB formation, a lot of research has been done on electrode surface modifications, genetic modifications of EAMs, and optimization of the environmental conditions.

2.1.3 Electrode surface properties

In terms of surface chemistry, it has been demonstrated that positively charged surfaces and hydrophilic surface are more attractive for both anodic biofilm[19] and cathodic biofilm[20] formation. This can

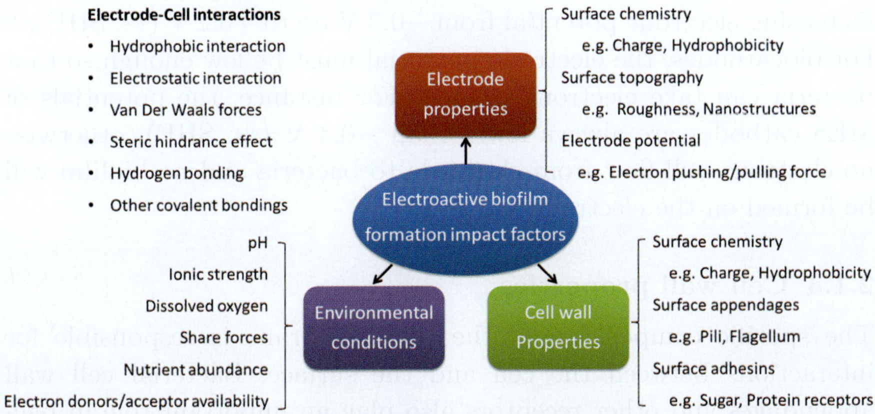

Figure 3. Factors impacting the EAB formation and electrode–cell interactions.

be explained by the fact that bacteria surface are normally negatively charged and hydrophilic. The surface chemical composition also determines the biocompatibility of the materials. For instance, introducing iron oxide, which is the natural electron acceptor of *Geobacter*, dramatically improved the anodic biofilm formation on stainless steel (SS) felt.[21] The topography of the electrode surface influences bacterial adhesion significantly. It has been found that bacteria prefer to colonize porous, grooved and braided surfaces.[10] For example, coating a smooth material such as reticulated vitreous carbon electrodes (RVC) with carbon nanotubes (CNTs) not only improved a mixed culture anodic biofilm formation[22] but also a mixed culture acetate-producing cathodic biofilm formation.[23]

In BESs, electrodes are not only the biofilm carriers but also the electron acceptors or donors. Thus, the electrode potential dramatically impacts the EAB biofilm formation. For bioanodes, the anode potential must be set higher than the standard biological redox potential of the substrate so that electrons can flow from the bacteria to electrode. It has been revealed that the increase of the potential from −0.25 to +0.21 V *vs.* standard hydrogen electrode (SHE) increased the amount of biomass (*Geobacter* dominated biofilm) on the electrode and the current production.[24] It is also shown that in *Shewanella putrefaciens*, the growth of the biofilm is increasing with increasing electrode potential from −0.3 V up to +0.2 V (*vs.* SHE).[25] For biocathodes, the electrode potential must be low enough so that bacteria can take electrons from it. For instance, the potentials of MES cathodes are always lower than −0.4 V (*vs.* SHE); otherwise no electrons will flow from electrode to bacteria and no biofilm will be formed on the electrode surfaces.[26]

2.1.4 Cell wall properties

The specific composition of the cell membrane is responsible for interactions between the cell and the surface. Bacterial cell wall appendages and other receptors also play an important role in bacteria adhesion.[10] It is demonstrated the pili of *G. sulfurreducens* permit increased stacking of cells on the anode surface[15] and it is well known that pili and flagella are required for the biofilm formation of *P. aeruginosa*.[27] Moreover, it is found that disruption of the putative cell surface polysaccharide biosynthesis gene SO3177 in *Shewanella oneidensis* MR-1 enhances cell adhesion to graphite felt surface.[28] Recently, it was reported that heterologously overexpressed a c-di-GMP biosynthesis gene *ydeH* in *S. oneidensis* MR-1 had significantly enhanced biofilm formation and bioelectricity generation.

2.1.5 Environmental conditions

The environmental conditions impact the biofilm formation by changing the physicochemical properties of the bacterial and electrode surface. Therefore, the temperature, pH, ion strength, nutrient abundance, dissolved oxygen and flow conditions greatly influence the bacterial adhesion and biofilm development.[14] It has been reported that applying high shear rates in an MFC can result in a specific EAB that is thicker and denser and attaches better.[29] Patil *et al.* demonstrated that only a narrow pH-window, ranging from 6 to 9, was suitable for the growth of anodic biofilms derived from pH-neutral wastewater.[30] They also reported that anodic biofilms are active in a temperature range between 5°C and 45°C, and elevated temperatures during initial biofilm growth greatly accelerate the biofilm formation process.[31] It is well known that dissolved oxygen (DO) lowers Coulombic efficiency of the anodic biofilm,[32] but there is no specific study that investigated the impact of the DO on the anodic biofilm formation. However, it was found in unpublished experiments that exposing the anode to air dramatically increases the anodic current start-up time, implying that DO inhibits the anodic biofilm formation. This is very likely caused by the fact DO is a competitive electron acceptor to the electrode. Cathodic biofilms for MES are all strict anaerobes; therefore DO is extremely detrimental to their growth and biofilm formation. However, oxygen-reducing microbial cathodes have been described, which utilize electrons from the electrode to reduce oxygen to water.[2]

2.2 Principal Components in BESs

This section discusses the essential materials that are commonly employed in the research and development of BES. Although the exponential growth of the field in the past 20 years has seen the list of materials expand accordingly, there are some fundamental principles that can guide the selection of BES components. These include, but are not limited to the following

Biocompatibility

As mentioned earlier in the chapter, BESs depend fundamentally on the interaction between living microorganisms and solid conductive materials. Accordingly, any material that may interact and/or be exposed to the electroactive cultures should be biocompatible, that is, it should not generate any physicochemical conditions that may have deleterious effects to the metabolic processes, affect the viability of the bacteria nor be subject to bacterial degradation for the duration of the experiment.[33]

Stability

Further to resistance of bacterial degradation, the components that form a BES reactor should suffer no major physical or chemical alterations when exposed to normal experimental conditions. Special attention should be paid to maintaining all components within intended application ranges. Failure to monitor the latter may, for example, result in considerable changes in electrolyte pH or working electrode potentials, which can easily cause membrane fouling[34] or electrode corrosion,[35] respectively.

2.2.1 Electrode materials

Electrodes are solid conductive materials that are fundamental to all BESs, acting as the terminal electron acceptor/donor for the EAB that drive the redox reactions of interest. They can also undertake abiotic electrochemical reactions such as hydrogen or oxygen evolution, depending on their working potential, which is the principle for MECs and commonly utilized in MES.

2.2.1.1 *Carbon-based electrodes*

Based on the aforementioned criteria of biocompatibility and stability, the absolute majority of investigations and implementations of BES to date utilize carbon-based electrodes.[36,37] The overall tendency over the last 15 years has been to move from conductive planar materials with little surface roughness, such as carbon paper (Figure 4(a)) and graphite plates (Figure 4(b)) to filamentous (Figures 4(c), 4(f) and 4(i)) and granular carbonaceous materials (see Figures 4(e) and 4(f)) that significantly increase the surface area/volume ratios. These larger ratios allow for the attachment of greater amounts of EAB, thus increasing the catalytic capacity of the BES per unit of volume.[38] Further enhancements have resulted from the utilization of '3D' electrodes (Figures 4(g) and 4(h)), whereby the juxtaposition of numerous conductive filamentous planes combine to form larger pores that allow for fast reactant perfusion, significantly reducing mass-transfer limitations and further enhancing the reaction rates in BES.[22,39] A major tendency in the field since 2012 is to furthermore decorate these 3D porous electrodes with nanoparticles that alter the electrode surfaces at the micro and nanoscales.[40] A significant number of recent investigations have revealed that these additions significantly enhance overall performance,[20] indeed resulting in the highest current densities of carbon-based electrodes

Figure 4. Photograph of various carbon-based materials utilized for BES: (a) carbon paper, (b) graphite plates, (c) carbon cloth, (d) graphite mesh. (e) granular graphite, (f) granular activated graphite, (g) carbon felt, (h) reticulated vitreous carbon, (i) carbon brush and (j) SS mesh. Adapted from Ref. 37.

reported to date for anodic MFC[22,41] and cathodic MES[23,42,43] processes.

2.2.1.2 *Metallic electrodes*

As discussed in the previous sections, iron in its oxidized forms is one of the electron acceptors for species such as *Geobacter* in the natural environment; given this possibility, direct electron transfer (DET) between EAB and solid metallic electrodes is also possible and has indeed been demonstrated for Ag, Au, Cu, Ni, Ti, W and SS mesh (see Figure 4(j)) electrodes.[44,45] Research on bioanodic processes with this type of electrodes remains nevertheless limited to date (when compared to carbonaceous materials) but has been utilized on a broad spectrum, from fundamental studies into EAB electron transfer rates,[46] to applied investigations that have produced the highest current densities observed to date for bioanodes.[21,35,47,48]

DET with metals has been found to occur even for non-EAB species — such as sulfate-reducing bacteria[49] — and is now believed to be one of the fundamental mechanisms behind microbially induced corrosion.[50,51] This inherently limits the use of metallic electrodes in biocathodic applications to working potentials below each material's corrosion onset.[35]

2.2.2 Membranes and separators

Membranes are layered materials that are typically placed between two (or more) chambers in a BES cell, acting both as an electrolyte for ion-exchange and as a partition to prevent the mixing of reactive species from one chamber to the other.[52] Separators typically cover only the latter function, but can also have additional roles in BES (see Section 2.2.2.2).

The importance of membranes and separators is better understood when revisiting the basic hydrogen fuel cell (HFC). This electrochemical cell (see Figure 5) utilizes the overall redox reaction below (see Eq. (1)), whereby hydrogen and oxygen gases combine, to produce electricity via oxidation and reduction half-reactions.[53] First and foremost, it should be noted that mixing these two gases

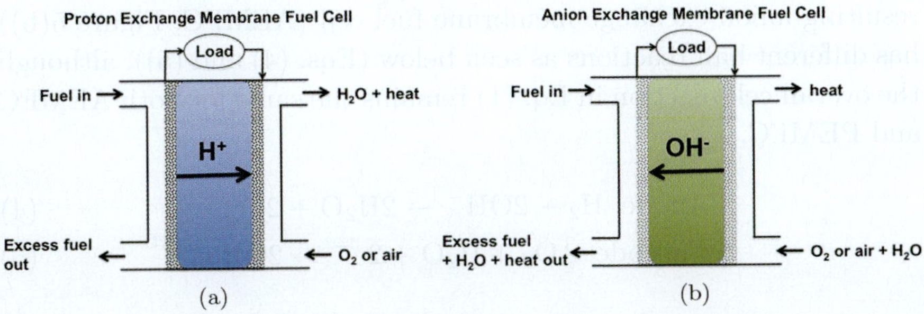

Figure 5. (a) PEMFC. (b) AEMFC. Schematics by Hickner *et al.*[54]

directly poses an explosion risk, hence the need for a partition that safely confines each reactant to its half-cell.[52]

$$\text{Overall cell reaction: } H_2 + \tfrac{1}{2}\,O_2 \rightarrow H_2O. \tag{1}$$

Furthermore, the production of electricity necessitates that electrons flow from the oxidation half-cell electrode (*i.e.* the anode) to the reduction half-cell electrode (cathode). However, for each electron that is displaced, one negative charge also moves with it. To maintain electroneutrality in the system, either (a) a positive charge needs to move in the same direction or (b) a negative charge should move from cathode to anode effectively replacing the negative charge of the displaced electron. The ion-exchange properties of membranes are therefore designed to facilitate this selective charge migration. For a fuel cell with option (a), positive charge migration is facilitated by a proton exchange membrane (PEM), which selectively allows the flow of protons (H^+) only; such a cell is termed proton exchange membrane fuel cell (PEMFC; see Figure 5(a)). The pairing of H^+ and electrons can be seen in the PEMFC anodic (Eq. (2)) and cathodic (Eq. (3)) half-reactions below:

$$\text{Anode: } H_2 \rightarrow 2H^+ + 2e^-, \tag{2}$$

$$\text{Cathode: } \tfrac{1}{2}O_2 + 2e^- + 2H^+ \rightarrow H_2O. \tag{3}$$

A fuel cell with option (b) however, necessitates an anion-exchange membrane (AEM), which has a different chemistry and only allows negatively charged anionic species (mainly OH^-) through. The resulting anion exchange membrane fuel cell (AEMFC; Figure 5(b)) has different half reactions as seen below (Eqs. (4) and (5)), although the overall cell reaction in Eq. (1) remains the same for both AEMFC and PEMFC.

$$\text{Anode: } H_2 + 2OH^- \rightarrow 2H_2O + 2e^-, \tag{4}$$

$$\text{Cathode: } \tfrac{1}{2}O_2 + H_2O + 2e^- \rightarrow 2OH^-. \tag{5}$$

2.2.2.1 *Ion exchange membranes for BES*

BESs also necessitate sufficient separation to avoid the mixing of reactants for the anodic and cathodic half-reactions (which can also have deleterious effects, though perhaps not as significant as in HFCs). Ion exchange membranes have historically been the preferred option for these purposes, with two basic functional types.

(i) Proton exchange membranes and cation exchange membranes

PEMs from the HFC industry were without doubt the preferred membrane material for BES in the first half of the last 25 years of BES research.[55] Nafion® is one of the most widely utilized PEMs and consists of sulfonated perfluoro polymers linked to a polytetrafluoroethylene (PTFE) backbone (see structure in Figure 6), which gives it excellent proton conductivity and good stability over broad ranges of pH and temperature. However, observations of biofouling (which may require constant reactivation)[34] and, particularly, its high cost

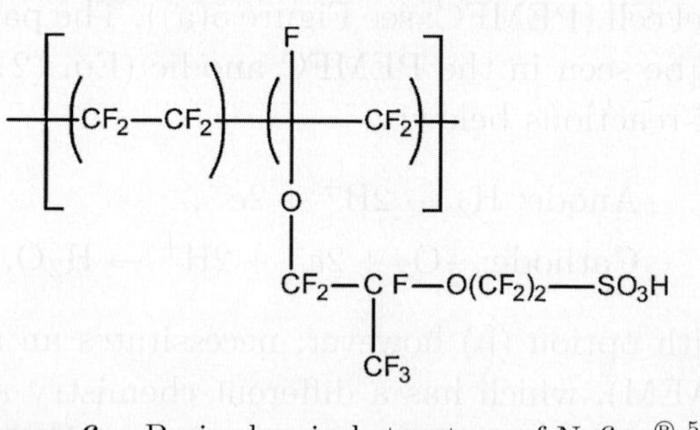

Figure 6. Basic chemical structure of Nafion®.[56]

due to the complex fluorochemistry in its synthesis are a strong limitation for any practical implementations in BES.[55,56]

To alleviate these issues and also because specific proton conduction is generally negligible in BESs (due to physiological pH, proton concentrations are negligible), research — particularly on the bioanode/MFC space — has moved from PEMs to cation exchange membranes (CEMs), many of which have been recently commercialized for a growing demand in the general fuel cell industry. CEMs have demonstrated good resistance to biofouling and ever-decreasing costs, which makes them ideal for BES, although the specific chemistry of each electrochemical system should be considered when selecting a particular material. Compositions differ significantly between brands, and whilst fluorinated CEMs are still widely manufactured, several other chemistries have emerged, many specifically to outperform Nafion®. Some of the most promising materials are non-fluorinated hydrocarbon polymers such as polyalyrene ether sulfone (PAES), polyalyrene ether ketone (PAEK), and polyvinyl chloride (PVC).[56]

(ii) Anion exchange membranes

Anion exchange membranes (AEMs) are typically fabricated via chloromethylation of an aromatic polymer (such as poly(sulfone) or poly(styrene)) and quaternization of benzylchloride with a tertiary amine to form quaternary ammonium cationic groups that repel other cationic species, therefore allowing for the selective permeation of anions.[54] AEMs could be a promising alternative in some particular BES applications due to the different half-reactions that they imply (see Section 2.2.2), but there are only handful investigations to date in the field (*e.g.*[57–59]) with further research needed to demonstrate whether AEMs can be a viable alternative to CEMs.

2.2.2.2 *Separators*

The development of membrane-less MFCs over 10 years ago demonstrated that in some situations, membranes may not be necessary for successful operation of BES[60] and may even be beneficial, because removing the membrane also removes part of the system's internal resistance, which can result in higher catalytic rates. Ion-exchange membranes will lead to the formation of pH gradients between chambers;[61] for example, migration of cationic species other than H^+ through a CEM during MFC operation leads to acidification of the anodic chamber, which can become sufficient to inhibit the EAB activity.[62] Simple porous separators (*e.g.* *j*-cloth, glass fibers or porous pots) can prevent the formation of these gradients while concomitantly allowing for very short distances between anode and cathode, thus increasing the catalytic rate density on a volumetric basis.[63] More importantly, separators have costs that can be orders-of-magnitude lower than ion-exchange membranes, thus making are very strong argument towards the economic viability of BES.

However, excessive non-selective porosity can also have deleterious effects, the most notable example being the permeation of DO from cathodic to anodic chamber in MFCs, which severely decreases the Coulombic efficiency of the system.[64] Moreover, membrane-less MEC reactors normally suffer from serious hydrogen losses through reoxidation by anodic EAMs or methanogens in the reactors. Accordingly, the utilization of separators should probably be paired with specific reactor designs that can capitalize on the benefits of membrane-less systems and avoid the problems caused by non-selective permeation.

2.2.3 Current collectors

As previously highlighted, current collectors are essential to sustain electron flow in scaled up systems, due to the limited electron conductivity of typical electrode materials in BESs.[65] While cm^2-sized electrodes do not suffer from large Ohmic losses, moving to m^2-sized electrodes would lead to unaffordable voltage drops as the electrons travel through the electrode to the external circuits. Therefore, scaled up BESs need to be fitted with highly conductive materials such as steel plates or meshes, to minimize the distances travelled by electrons in the electrode itself. Rozendal *et al.*[64] forecast that current collectors will become the largest cost in future BESs.[64]

2.2.4 Reactor configurations

A broad variety of reactors for BESs have been described in the literature, with numerous shapes, sizes, compartments and volumes ranging from the microliter scale to pilot systems of up to 1 m^3.[66] The most fundamental design is known as an H-shaped fuel cell, whereby two bottles are linked by a salt bridge or separated by an ion-exchange membrane, forming two chambers — one for the anode and one for the cathode. This type of reactor can be used for fundamental biofilm studies or for educational purposes, but its application value is limited because it has a massive internal resistance (due to very large distances between anode and cathode).[39] Flat-plate reactors bring the chambers closely together by tightening two or more flat acrylic chambers together with fasteners. Membranes and/or separators and the electrolytes are held in place with the use of gaskets, which are also tightened between the various plates. Such designs have low internal resistances due to narrow electrode spacing, can be operated in continuous mode with liquid and gaseous reactants and can easily be disassembled for experimental/prototyping purposes; they are extensively utilized in contemporary BES investigations.[66] A final notable example concerns tubular reactors, which have also been explored extensively over the last 10 years. Their geometrical shape allows for a more efficient utilization of electrode surface area *vs.* total reactor volume,[67] particularly if fluidized or packed-bed electrodes are employed.[66] In certain fluidic configurations, this type of reactors can be the most suitable for membrane-less operation — with liquid/gaseous species being constrained to specific regions of the reactor, resulting in ultra-low reactor costs.[68]

2.3 Who Is Doing It?

Most bacteria can exchange electrons extracellularly when supplemented with suitable synthetic redox mediators, however most BES applications rely on special bacteria which are capable of self-expressing the ability to transfer electrons to and from electrodes. These bacteria are termed *electroactive.* Despite much achievement in understanding the mechanisms and applications, there is still incomplete knowledge about how broad the range of electroactive bacteria is in nature. There seems to be consensus on the fact that electroactive bacteria are ubiquitous,[69] as they are found in most natural and engineered environments. However, ubiquity does not automatically imply that there is a broad diversity of these microorganisms. Indeed, it appears that most anodic EABs are dominated by *Geobacter* sp., which is known to accept only short chain fatty acids or hydrogen as electron donors. This suggests that anodic electroactive bacteria diversity is in fact limited to a few ubiquitous species, which may work in syntrophy with fermenters and other organisms to extract energy out of a variety of organic substrates.

Another electroactive anodic bacterium that has been thoroughly described is *Shewanella* sp., which is known to oxidize lactate to acetate in the presence of electrodes,[70] but is less frequently found in mixed culture anodic communities. Other anodic electroactive species are niche species, which are not usually encountered in mixed anodic biofilms for electron extraction from the oxidation of organic matter. These include *Pseudomonas* sp.[71] and *Lactococcus* sp.,[72] which owe their electron transfer ability to the excretion of redox active compounds, phenazines and quinones, respectively.

Cathodic electroactive bacteria are even less understood. It is known that biofilms enhance cathodic current in oxygen-reducing cathode,[73] however no single study has been able to univocally link the observed enhanced currents to microbial metabolism and extracellular electron transfer: non-biological side effects are still a possible cause.

Cathodic biofilms for MES of acetate from CO_2 have been extensively studied in the past five years. There is general consensus that high rate systems are not based on direct electron transfer (DET), but on abiotic hydrogen production at the electrode, followed by hydrogen consumption for acetate production by homoacetogenic populations.[42] However, one species, namely *Sporomusa ovata*, has been shown to be directly electroactive at a cathode surface, converting CO_2 to acetate without hydrogen as intermediate.[17]

Beese-Vasbender *et al.*[74] showed that a similar process led to the production of methane from CO_2 at a biocathode, using a pure archaeal strain.

The above discussion summarizes 15 years of intense research in a booming field. The question whether extracellular electron transfer (EET) is widespread can therefore at the stage be answered with a cautious statement: EET-capable bacteria are ubiquitous in natural and engineered environments; however, they are not diverse, being limited to a small number of genera.

2.4 Prospective Applications

After nearly 20 years of research, BESs have not breached into the real world. Researchers worldwide are still busy improving rates and efficiencies and reducing the cost of materials, in order to make some of the associated prospective technologies economically viable. The following is a non-exhaustive list of some of the key prospective applications, their working principles and the reasons why they have not yet been taken up at full scale.

2.4.1 Microbial fuel cells

MFCs couple the oxidation of organics at the anode with the reduction of oxygen at the cathode (option 1 in Figure 1). The exothermic nature of the reaction allows for generation of electricity by the BES. When wastewater is used at the anode, the system is capable of removing wastewater organics while producing electricity, simultaneously saving aeration costs for conventional treatment. As attractive as this may seem, the problem of insufficient salinity and alkalinity of most wastewaters[64] limits the achievable current densities due to Ohmic losses and acidification in the anodic EAB, respectively. Thus far, a maximal current density (normalized to membrane area) of 2 A/m^2 has been attained with domestic wastewater.[75] At the current cost of electrodes, separators and electrochemical cells MFCs for wastewater treatment cannot pay themselves back.

2.4.2 Microbial electrolysis cells

MECs replace oxygen reduction with hydrogen production at the cathode (option 3 in Figure 1). They consume electricity to deliver a value-added product, in this case hydrogen gas. When wastewater is used at the anode, they are affected by the same issues as MFCs. As the value of hydrogen gas is low, MECs are also not economically viable at current cost of materials.[64]

2.4.3 Nutrient recovery systems

Several researchers have recognized the value of recovering nutrients in wastewater. This comes mostly from the reduction in energy consumption if the nutrients (especially nitrogen) are recovered rather than destroyed: nitrogen removal from domestic wastewater accounts for approximately 25% of the operating cost of conventional sewage treatment plants. The technology is essentially a MEC, where ammonium (NH_4^+) is the primary ion migrating from anode to cathode. The cathode is operated at high pH, so that ammonia is formed and stripped by hydrogen and finally recovered by absorption into an acid.[76] The economic benefit of energy savings and the value of the recovered nutrient may make this technology economically attractive even at current material costs.[62]

2.4.4 Microbial electrosynthesis

Cathodic MES is the bacterial production of value-added compounds using cathode-supplied electrons. The technology is a modification of MECs, whereby hydrogen production is replaced (or in most cases, followed) by a reduction reaction, leading to a desired product. In MES, the anode reaction often is not biological, but mere abiotic oxygen evolution. This is for two reasons: (i) the product should be valuable enough to justify the application of extra voltage; (ii) to prevent rising cathode pH, the anolyte pH is drive to very low values so that protons are the main migrating ion. As mentioned above, electrosynthesis of acetate is the only successful example of MES from CO_2. This has been shown to occur via EET through pure cultures, or via hydrogen through mixed cultures of homoacetogens,[5] the latter achieving high rate/high titer acetate production. Unfortunately, acetic acid has limited economic value and its production by this method is short of matching the market price. Ongoing work is aimed at the attainment of higher volatile fatty acids (such as butyrate and caproate), which have higher value per electron.

Cathodic electrosynthesis from organic is also limited to a few examples, notably 1,3-propanediol production from glycerol.[77] This process is also thought to proceed via hydrogen as intermediate.

2.4.5 Biogas polishing systems

While production of methane from CO_2 by cathodic microbial reduction (option 4 in Figure 1) has only marginal economics, it may be more attractive if the source of CO_2 is biogas, which normally requires polishing before it can be used as energy source. In this case, the biogas would be fed to the cathode, which would increase its methane content, by both converting CO_2 to CH_4 and by migration of bicarbonate ions from cathode to anode through an AEM.[78]

2.4.6 Bioremediation

Bioremediation is the removal of target pollutants from contaminated natural and engineered environments. Several contaminants can be removed via reductive cathodic processes using bioelectrochemical systems. This includes removal of nitrate from groundwater via reduction to N_2 (option 5, Figure 1[79]), sulfate removal from acid mine drainage (option 2, Figure 1[3]), as well as reduction of harmful compounds such as nitrobenzene (reduced to aniline[80]), and reductive dechlorination of organic compounds.[81] With the exception of nitrate reduction, all these reactions proceed via hydrogen as abiotically produced intermediate. The economic potential for future applications derives from the fact that most of these reactions are not readily achieved in any other ways, or they require large sources of organic matter, which are usually not readily available at the contaminated sites.

2.4.7 Biosensors

Biosensing is the new frontier of BESs. If the applicability of BES to full-scale systems for bioproduction or nutrient recovery is still debatable, small devices which selectively detect certain target compounds or bacteria are not affected as much by the costs of electrodes and membranes. A number of biosensing devices have been proposed that rely on the microbial electrical interaction with electrodes. These are still limited to sensing of toxicity,[82] but an array of applications of bioelectrochemical sensing is envisaged.

List of Abbreviations

AEM	anion exchange membrane
AEMFC	anion exchange membrane fuel cell
BESs	microbial bioelectrochemical systems
CEM	cation exchange membrane
DET	direct electron transfer
EAB	electrochemically active biofilms
EAM	electrochemically active microorganisms
EET	extracellular electron transfer
EPS	extracellular polymeric substance
HFCs	hydrogen fuel cells
MFCs	microbial fuel cells
MECs	microbial electrolysis cells
MES	microbial electrosynthesis
PEM	proton exchange membrane
PEMFC	proton exchange membrane fuel cell
RVC	reticulated vitreous carbon
SHE	standard hydrogen electrode

References

1. K. Rabaey (2010). Bioelectrochemical Systems: From Extracellular Electron Transfer to Biotechnological Application, London, IWA Publishing.
2. K. Rabaey *et al.* (2008). Cathodic oxygen reduction catalyzed by bacteria in microbial fuel cells. *Isme Journal* **2**(5), 519–527.
3. G. Pozo *et al.* (2015). Methanobacterium enables high rate electricity-driven autotrophic sulfate reduction. *Rsc Advances* **5**(111), 91821–91821.

4. R.A. Rozendal *et al.* (2006). Principle and perspectives of hydrogen production through biocatalyzed electrolysis. *International Journal of Hydrogen Energy* **31**(12), 1632–1640.
5. K.P. Nevin *et al.* (2011). Electrosynthesis of organic compounds from carbon dioxide is catalyzed by a diversity of acetogenic microorganisms. *Applied and Environmental Microbiology* **77**(9), 2882–2886.
6. B. Virdis *et al.* (2008). Microbial fuel cells for simultaneous carbon and nitrogen removal. *Water Research* **42**(12), 3013–3024.
7. J. Babauta *et al.* (2012). Electrochemically active biofilms: Facts and fiction. A review. *Biofouling* **28**(8), 789–812.
8. K.K. Jefferson (2004). What drives bacteria to produce a biofilm? *Fems Microbiology Letters* **236**(2), 163–173.
9. G. O'Toole, H.B. Kaplan and R. Kolter (2000). Biofilm formation as microbial development. *Annual Review of Microbiology* **54**, 49–79.
10. L. Ploux, A. Ponche and K. Anselme (2010). Bacteria/material interfaces: Role of the material and cell wall properties. *Journal of Adhesion Science and Technology* **24**(13–14), 2165–2201.
11. S.E. Childers, S. Ciufo and D.R. Lovley (2002). Geobacter metallireducens accesses insoluble Fe(III) oxide by chemotaxis. *Nature* **416**(6882), 767–769.
12. L. Wu *et al.* (2011). Genetic and molecular characterization of flagellar assembly in shewanella oneidensis. *Plos One* **6**(6).
13. B. Moller *et al.* (1984). Sporomusa, a new genus of gram-negative anaerobic-bacteria including sporomusa-sphaeroides spec-nov and sporomusa-ovata spec-nov. *Archives of Microbiology* **139**(4), 388–396.
14. M. Katsikogianni and Y.F. Missirlis (2004). Concise review of mechanisms of bacterial adhesion to biomaterials and of techniques used in estimating bacteria-material interactions. *European Cells & Materials* **8**, 37–57.
15. G. Reguera *et al.* (2006). Biofilm and nanowire production leads to increased current in Geobacter sulfurreducens fuel cells. *Applied and Environmental Microbiology* **72**(11), 7345–7348.
16. J.A. Cornejo *et al.* (2015). Surface modification for enhanced biofilm formation and electron transport in shewanella anodes. *Journal of the Electrochemical Society* **162**(9), H597–H603.
17. K.P. Nevin *et al.* (2010). Microbial electrosynthesis: feeding microbes electricity to convert carbon dioxide and water to multicarbon extracellular organic compounds. *Mbio* **1**(2).
18. N.S. Malvankar, M.T. Tuominen and D.R. Lovley (2012). Biofilm conductivity is a decisive variable for high-current-density Geobacter sulfurreducens microbial fuel cells. *Energy & Environmental Science* **5**(2), 5790–5797.
19. K. Guo *et al.* (2013). Effects of surface charge and hydrophobicity on anodic biofilm formation, community composition and current generation in bioelectrochemical systems. *Environ Sci Technol* **47**(13), 7563–7570.
20. T. Zhang *et al.* (2013). Improved cathode materials for microbial electrosynthesis. *Energy & Environmental Science* **6**(1), 217–224.

21. K. Guo *et al.* (2014). Flame oxidation of stainless steel felt enhances anodic biofilm formation and current output in bioelectrochemical systems. *Environmental Science & Technology* **48**(12), 7151–7156.
22. V. Flexer *et al.* (2013). The nanostructure of three-dimensional scaffolds enhances the current density of microbial bioelectrochemical systems. *Energy & Environmental Science* **6**(4), 1291–1298.
23. L. Jourdin *et al.* (2014). A novel carbon nanotube modified scaffold as an efficient biocathode material for improved microbial electrosynthesis. *Journal of Materials Chemistry A* **2**(32), 13093–13102.
24. X.P. Zhu *et al.* (2014). Microbial community composition is unaffected by anode potential. *Environmental Science & Technology* **48**(2), 1352–1358.
25. A.A. Carmona-Martinez *et al.* (2013). Electron transfer and biofilm formation of Shewanella putrefaciens as function of anode potential. *Bioelectrochemistry* **93**, 23–29.
26. E. Blanchet *et al.* (2015). Importance of the hydrogen route in up-scaling electrosynthesis for microbial CO_2 reduction. *Energy & Environmental Science.*
27. P. Chiang and L.L. Burrows (2003). Biofilm formation by hyperpiliated mutants of Pseudomonas aeruginosa. *Journal of Bacteriology* **185**(7), 2374–2378.
28. A. Kouzuma *et al.* (2010). Disruption of the putative cell surface polysaccharide biosynthesis gene SO3177 in Shewanella oneidensis MR-1 enhances adhesion to electrodes and current generation in microbial fuel cells. *Applied and Environmental Microbiology* **76**(13), 4151–4157.
29. H.T. Pham *et al.* (2008). High shear enrichment improves the performance of the anodophilic microbial consortium in a microbial fuel cell. *Microbial Biotechnology* **1**(6), 487–496.
30. S.A. Patil *et al.* (2011). Electroactive mixed culture derived biofilms in microbial bioelectrochemical systems: The role of pH on biofilm formation, performance and composition. *Bioresource Technology* **102**(20), 9683–9690.
31. S.A. Patil *et al.* (2010). Electroactive mixed culture biofilms in microbial bioelectrochemical systems: The role of temperature for biofilm formation and performance. *Biosensors & Bioelectronics* **26**(2), 803–808.
32. S.E. Oh *et al.* (2009). Effects of applied voltages and dissolved oxygen on sustained power generation by microbial fuel cells. *Water Science and Technology* **60**(5), 1311–1317.
33. X. Xie, C. Criddle and Y. Cui (2015). Design and fabrication of bioelectrodes for microbial bioelectrochemical systems. *Energy & Environmental Science* **8**(12), 3418–3441.
34. M.-J. Choi *et al.* (2011). Effects of biofouling on ion transport through cation exchange membranes and microbial fuel cell performance. *Bioresource Technology* **102**(1), 298–303.
35. P. Ledezma *et al.* (2015). Oxidised stainless steel: A very effective electrode material for microbial fuel cell bioanodes but at high risk of corrosion. *Electrochimica Acta* **158**(0), 356–360.

36. Z. Du, H. Li and T. Gu (2007). A state of the art review on microbial fuel cells: A promising technology for wastewater treatment and bioenergy. *Biotechnology Advances* **25**(5), 464–482.
37. J. Wei, P. Liang and X. Huang (2011). Recent progress in electrodes for microbial fuel cells. *Bioresource Technology* **102**(20), 9335–9344.
38. U. Schroder (2007). Anodic electron transfer mechanisms in microbial fuel cells and their energy efficiency. *Physical Chemistry Chemical Physics* **9**(21), 2619–2629.
39. P. Ledezma, J. Greenman and I. Ieropoulos (2012). Maximising electricity production by controlling the biofilm specific growth rate in microbial fuel cells. *Bioresource Technology* **118**, 615–618.
40. K. Guo *et al.* (2015). Engineering electrodes for microbial electrocatalysis. *Current Opinion in Biotechnology* **33**, 149–156.
41. S. Chen *et al.* (2012). Layered corrugated electrode macrostructures boost microbial bioelectrocatalysis. *Energy & Environmental Science* **5**(12), 9769–9772.
42. L. Jourdin *et al.* (2015). High acetic acid production rate obtained by microbial electrosynthesis from carbon dioxide. *Environmental Science & Technology.*
43. L. Jourdin *et al.* (2016). Bringing high-rate, CO_2-based microbial electrosynthesis closer to practical implementation through improved electrode design and operating conditions. *Environmental Science & Technology.*
44. D. Pocaznoi *et al.* (2012). Stainless steel is a promising electrode material for anodes of microbial fuel cells. *Energy & Environmental Science* **5**(11), 9645–9652.
45. A. Baudler *et al.* (2015). Does it have to be carbon? Metal anodes in microbial fuel cells and related bioelectrochemical systems. *Energy & Environmental Science* **8**(7), 2048–2055.
46. H. Richter *et al.* (2008). Electricity generation by geobacter sulfurreducens attached to gold electrodes. *Langmuir* **24**(8), 4376–4379.
47. S.F. Ketep *et al.* (2014). Stainless steel foam increases the current produced by microbial bioanodes in bioelectrochemical systems. *Energy & Environmental Science* **7**(5), 1633–1637.
48. D. Massazza *et al.* (2015). New ceramic electrodes allow reaching the target current density in bioelectrochemical systems. *Energy & Environmental Science* **8**(9), 2707–2712.
49. H.T. Dinh *et al.* (2004). Iron corrosion by novel anaerobic microorganisms. *Nature* **427**(6977), 829–832.
50. H. Venzlaff *et al.* (2013). Accelerated cathodic reaction in microbial corrosion of iron due to direct electron uptake by sulfate-reducing bacteria. *Corrosion Science* **66**(0), 88–96.
51. B.J. Little and J.S. Lee (2014). Microbiologically influenced corrosion: An update. *International Materials Reviews* **59**(7), 384–393.
52. J.L.A.A. Dicks (2003). *Fuel Cell Systems Explained.* 2nd ed. New York: John Wiley & Sons.

53. W. Vielstich, A. Lamm and H.A. Gasteiger (2003). Handbook of Fuel Cells: Fundamentals, Technology and Applications. Chichester, England; Hoboken, N.J.: Wiley.
54. M.A. Hickner, A.M. Herring and E.B. Coughlin (2013). Anion exchange membranes: Current status and moving forward. *Journal of Polymer Science Part B: Polymer Physics* **51**(24), 1727–1735.
55. B.E. Logan *et al.* (2006). Microbial fuel cells: Methodology and technology. *Environmental Science & Technology* **40**(17), 5181–5192.
56. R.S.L. Yee *et al.* (2012). Cost effective cation exchange membranes: A review. *Chemical Engineering Research and Design* **90**(7), 950–959.
57. S. Pandit *et al.* (2011). Performance of electron acceptors in catholyte of a two-chambered microbial fuel cell using anion exchange membrane. *Bioresource Technology* **102**(3), 2736–2744.
58. E. Ji *et al.* (2011). Interface resistances of anion exchange membranes in microbial fuel cells with low ionic strength. *Biosensors and Bioelectronics* **26**(7), 3266–3271.
59. X. Zhang *et al.* (2010). Improved performance of single-chamber microbial fuel cells through control of membrane deformation. *Biosensors and Bioelectronics* **25**(7), 1825–1828.
60. J.K. Jang *et al.* (2004). Construction and operation of a novel mediator- and membrane-less microbial fuel cell. *Process Biochemistry* **39**(8), 1007–1012.
61. R.A. Rozendal, H.V.M. Hamelers and C.J.N. Buisman (2006). Effects of membrane cation transport on pH and microbial fuel cell performance. *Environmental Science & Technology* **40**(17), 5206–5211.
62. P. Ledezma *et al.* (2015). Source-separated urine opens golden opportunities for microbial electrochemical technologies. *Trends in Biotechnology.*
63. B. Logan (2010). Scaling up microbial fuel cells and other bioelectrochemical systems. *Applied Microbiology and Biotechnology* **85**(6), 1665–1671.
64. R.A. Rozendal *et al.* (2008). Towards practical implementation of bioelectrochemical wastewater treatment. *Trends in Biotechnology* **26**(8), 450–459.
65. F. Zhang *et al.* (2011). Mesh optimization for microbial fuel cell cathodes constructed around stainless steel mesh current collectors. *Journal of Power Sources* **196**(3), 1097–1102.
66. T. Krieg *et al.* (2014). Reactor concepts for bioelectrochemical syntheses and energy conversion. *Trends in Biotechnology* **32**(12), 645–655.
67. K. Rabaey *et al.* (2005). Tubular microbial fuel cells for efficient electricity generation. *Environmental Science & Technology* **39**(20), 8077–8082.
68. I.A. Ieropoulos *et al.* (2013). Waste to real energy: The first MFC powered mobile phone. *Physical Chemistry Chemical Physics* **15**(37), 15312–15316.
69. L.E. Doyle and E. Marsili (2015). Methods for enrichment of novel electrochemically-active microorganisms. *Bioresource Technology* **195**, 273–282.
70. H.J. Kim *et al.* (2002). A mediator-less microbial fuel cell using a metal reducing bacterium, Shewanella putrefaciense. *Enzyme and Microbial Technology* **30**(2), 145–152.

71. K. Rabaey *et al.* (2005). Microbial phenazine production enhances electron transfer in biofuel cells. *Environmental Science & Technology* **39**(9), 3401–3408.
72. S. Freguia *et al.* (2009). Lactococcus lactis catalyses electricity generation at microbial fuel cell anodes via excretion of a soluble quinone. *Bioelectrochemistry* **76**(1–2), 14–18.
73. S. Freguia, S. Tsujimura and K. Kano (2010). Electron transfer pathways in microbial oxygen biocathodes. *Electrochimica Acta* **55**(3), 813–818.
74. P.F. Beese-Vasbender *et al.* (2015). Selective microbial electrosynthesis of methane by a pure culture of a marine lithoautotrophic archaeon. *Bioelectrochemistry* **102**, 50–55.
75. X.Y. Zhang *et al.* (2015). COD removal characteristics in air-cathode microbial fuel cells. *Bioresource Technology* **176**, 23–31.
76. P. Kuntke *et al.* (2012). Ammonium recovery and energy production from urine by a microbial fuel cell. *Water Research* **46**(8), 2627–2636.
77. M. Zhou *et al.* (2013). Carbon and electron fluxes during the electricity driven 1,3-propanediol biosynthesis from glycerol. *Environmental Science & Technology* **47**(19), 11199–11205.
78. H. Xu, K.J. Wang and D.E. Holmes (2014). Bioelectrochemical removal of carbon dioxide (CO_2): An innovative method for biogas upgrading. *Bioresource Technology* **173**, 392–398.
79. N. Pous *et al.* (2013). Bioremediation of nitrate-polluted groundwater in a microbial fuel cell. *Journal of Chemical Technology and Biotechnology* **88**(9), 1690–1696.
80. A.J. Wang *et al.* (2011). Efficient reduction of nitrobenzene to aniline with a biocatalyzed cathode. *Environmental Science & Technology* **45**(23), 10186–10193.
81. P. Leitao *et al.* (2015). Bioelectrochemically-assisted reductive dechlorination of 1,2-dichloroethane by a Dehalococcoides-enriched microbial culture. *Bioresource Technology* **195**, 78–82.
82. N.E. Stein *et al.* (2011). Kinetic models for detection of toxicity in a microbial fuel cell based biosensor. *Biosensors & Bioelectronics* **26**(7), 3115–3120.

Chapter 3

Continuum in Enzymatic and Microbial Bioelectrocatalysis

Frédéric Barrière
Université de Rennes 1,
Institut des Sciences Chimiques de Rennes, France
frederic.barriere@univ-rennes1.fr

3.1 Introduction

Redox reactions at electrodes may be catalyzed by biological macromolecules such as (inert) oxidoreductase proteins or by living and self-replicating microorganisms like bacteria. In this short concept introductory chapter, the continuum between the inert (enzymes) and living (microbes) biocatalysis at electrodes is delineated. The motivation for catalysis is first discussed with simple case-study redox reactions at electrodes. Then, enzymatic biocatalysis of these illustrative redox reactions is addressed. The need for redox mediation is explained together with the possibility of direct enzymatic bioelectrocatalysis. In this context, the effect and value of modifying the electrode surface is introduced. Cellular bioelectrocatalysis with living microorganisms is then considered and illustrated mainly for electroactive bacteria at anodes and cathodes. The specificity of microbial bioelectrocatalysis is stressed. Similar to the case of enzymes, several electron transfer mechanisms to/from the electrode are discussed and the possible effect of modified electrodes emphasized. Finally, the pros and cons of the two types of catalysts are addressed with a discussion of possible applications.

3.2 Electron Transfer and Redox Reactions at Electrodes

Virtually all molecules may transfer or accept electrons to/from an electrode. The equilibrium potential at which electron transfer may thermodynamically occur for a given redox couple is defined in the Nernst equation as the standard potential E^0.[1] Many molecules, however, when studied with a potentiostat in an electrochemical cell do not get oxidized or reduced with a significant current at the working electrode polarized close to their standard potential. Let us consider for example the case of glucose that can be oxidized into gluconolactone, releasing two electrons and two protons (oxidation of an hydroxyl to a carbonyl group) at approximately $E^{0\prime} = -0.56$ V *vs.* Ag/AgCl at pH 7. Addition of glucose in an aqueous buffered electrolyte results in virtually no oxidation current at most working electrode materials, even at high overpotentials. The lack of glucose oxidation current, or thereof its much lower than expected intensity, demonstrates the high activation energy barrier of this redox reaction involving not only electrons but also protons. Let us note that the nature of the electrode or of its surface may have a profound effect on the electron transfer kinetics. An illustrative example is given by the fast electron transfer of the ferri-/ferrocyanide couple on fractured glassy carbon whereas it is very slow on highly ordered pyrolytic graphite.[2] Two carbon electrodes with very different interfaces hence yield opposite response in term of electron transfer rate for the same redox couple. A cathodic example related to the case of glucose discussed above concerns the four electrons and four protons reduction of dioxygen to water (thermodynamic equilibrium potential of ca. $E^{0\prime} = +0.59$ V *vs.* Ag/AgCl at pH 7). It is well known that this reaction is particularly difficult even at the best electrode catalysts like platinum. Even at this metal, significant cathodic current density do not start after at least several hundreds of millivolt overpotential. The same holds true for the reverse reaction, namely

Ox/Red	**$E^{0\prime}$ (Volt *vs.* Ag/AgCl at pH 7)**
O_2/H_2O	+0.59 V
Type 1 Cu(II)/Cu(I)	+0.58 V in laccase from *T. Versicolor*
$ABTS^{\bullet}/ABTS$	+0.48 V
$FcOH/FcOH^+$	+0.18 V
$FAD/FADH_2$	−0.32 V in glucose oxidase
CO_2/Glucose	−0.40 V
NAD+/NADH	−0.52 V
Gluconolactone/Glucose	−0.56 V

Chart 1. Selection of relevant redox couples and their thermodynamic formal potential.

the oxidation of dioxygen to water, which also occurs at large anodic overpotential at a platinum electrode. Let us note here that the same platinum electrode gives very fast electron transfer kinetic for the H^+/H_2 redox couple but slow electron transfer rate for the O_2/H_2O interconversion. Hence the value of the thermodynamic oxidant and reductant of choice for a fuel cell like glucose (fuel or reductant with a low $E^{0\prime} = -0.56$ V) and oxygen (oxidant with a high $E^{0\prime} = +0.59$ V) with $\Delta E^{0\prime} = 1.15$ V is kinetically limited by the large overpotentials of the anodic and cathodic reaction at electrodes. To be useful in applications like fuel cells or sensors, the electrode reactions need to be catalyzed. As discussed next, enzymes are suitable as biological catalysts for these reactions since these molecules (glucose and O_2) are present in biological fluids and involved in biochemical reactions.

3.2.1 Redox reactions catalyzed by oxidoreductases at electrodes

Since the transformation of biological molecules is catalyzed by enzymes, redox chemical reactions may be in principle catalyzed by oxidoreductases at electrodes. Oxidoreductases catalyze redox reactions. Like all enzymes, they are proteins with an active site where catalysis occurs. Glucose oxidase (GOx) catalyzes the oxidation of glucose to gluconolactone using dioxygen as the natural electron acceptor. However, as for most enzymes and oxidoreductases, the active site of GOx containing the redox active prosthetic group flavine adenine dinucleotide (FAD) is too deeply buried (more than ca. 2 nm) and protected within the polypeptide backbone to permit significantly fast heterogeneous electron transfer with the electrode surface. To circumvent this limitation, the solution is to resort to redox mediators that shuttle electrons between the active site of the enzyme and the electrode. A redox mediator is a redox active molecule with a standard potential suitable for a thermodynamically favorable electron transfer with the active site of the enzyme ($\Delta G^{0\prime} < 0$). This means that for GOx where the active site has an apparent standard potential (pH 7) of $E^{0\prime} = -0.32$ V *vs.* Ag/AgCl, a redox mediator with $E^{0\prime}$ at least 50–100 mV greater (less negative) than -0.32 V is sufficient for a driving force warranting fast and favorable electron transfer with FAD. Other requirements for a suitable redox mediator are as follows. It must be stable in its two redox states and soluble in the medium of study (usually aqueous electrolytes). Its size and structure must be compatible with the size and physico-chemical properties of the protein channels in which it has to diffuse in and out. Evidently, any redox mediator must display an electrochemically reversible electron transfer with the electrode surface. Such requirements are met by ferrocene methanol with $E^{0\prime} = +0.18$ V *vs.* Ag/AgCl given here as an illustrative example. In an electrolyte containing the substrate (glucose), the enzyme (GOx) and the redox mediator (ferrocene methanol), at an electrode polarized above the $E^{0\prime}$ of the redox mediator, a catalytic anodic current is obtained due to the oxidation of glucose at the enzyme, mediated by the ferrocenyl derivative. This allows glucose determination[3] or the design of an anode for enzymatic biofuel cells.[4] For this latter application, the redox mediator should have a standard potential as close as that of the enzyme active site to secure the largest electromotive force (keeping in mind that about 50 mV driving force is necessary to yield a significant rate, *i.e.* current). For biofuel cells implantable *in vivo*, the enzyme-redox mediator system should be immobilized onto the electrode surface. Several approaches like crosslinking of the protein with redox polymers have been developed for that purpose, with the additional advantage of 3D catalysis. Since the enzyme is specific to a substrate, the biofuel cell then does not require a membrane.[4] Resorting to redox mediators may sometimes be avoided when the enzyme is small enough so that its active site is addressable electrochemically or when the active site is located at the outer surface of the protein. If this is not the case, protein engineering may be applied as in GOx de-glycosylation for example. GOx is indeed highly glycosylated and sugars contribute to the overall size and surface properties of the protein, shielding its active site from the electrode. The chemical removing of sugars from GOx results in a smaller and/or more flexible protein that has been shown to directly exchange electrons between an electrode surface and the FAD/$FADH_2$ active site but also provide a mediator-less catalytic anodic current in the presence of glucose.[5] Catalysis occurs, however, only at the electrode surface as opposed to the diffusion layer when enzymes are in solution with redox mediators or immobilized at the electrode surface with redox polymers. The latter discussion dealing with anodic enzymatic catalysis taking GOx as an example may be paralleled by a similar section on oxidoreductases at the cathode. Dioxygen reduction is interesting, especially in the context of enzymatic biofuel cells, as this molecule is free and widely available, it may be reduced into innocuous water and the standard potential of this redox couple is high ($E^{0\prime} = +0.59$ V *vs.* Ag/AgCl at pH 7). High overpotentials, however, limit its use (commonly over several hundreds of millivolts, even on materials like platinum). Laccase enzymes reduce oxygen to water at a trinuclear copper cluster (comprised of a mononuclear type 2 and binuclear type 3 sites) coupled to the oxidation and depolymerization of lignin from rotting wood at a mononuclear type 1 copper site.[6] At a laccase oxygen-reducing enzymatic cathode, the enzyme accepts electrons from the cathode via its type 1 copper site mediated by a redox shuttle with $E^{0\prime}$ at least 50 mV below that of the T1 site. 2,20-Azino-bis(3-ethylbenzothiazoline-6-sulfonic acid diammonium salt (ABTS) with $E^{0\prime} \approx 0.48$ V *vs.* Ag/AgCl is suitable for this purpose (Figure 1(b)). The type 1 copper site displays an apparent standard potential ranging from about 0.2 to 0.6 V *vs.* Ag/AgCl at pH 7 depending of the species.

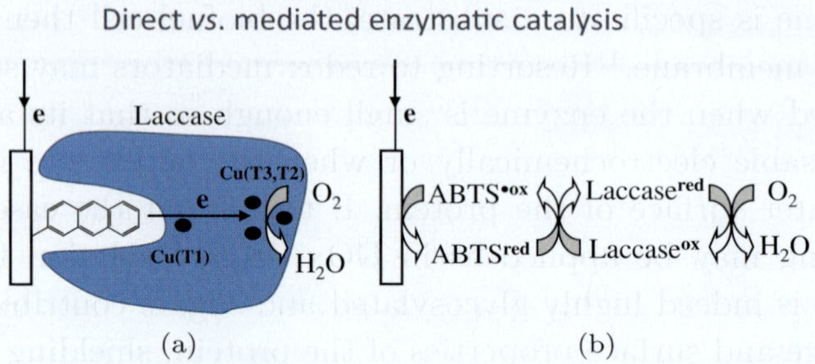

Figure 1. Comparison between direct[8] and mediated oxygen reduction catalysis[7] by laccase at a cathode.

Laccase from *Trametes versicolor* has a remarkably high type 1 copper site $E^{0\prime}$ ($E^{0\prime} = +0.58$ V *vs.* Ag/AgCl at pH 7) nearly equal to that of the O_2/H_2O redox couple. This translates to a driving force close to zero ($\Delta G^{0\prime} \approx 0$). For that reason, the enzymatic catalysis at pH 7 is inefficient and improves with the decrease of pH (*i.e.* with the increase of the driving force between the reduction of oxygen and the oxidation of the type one copper site).[7] As for the case of GOx discussed above, it is possible to design direct transfer between laccase and a cathode. This time, electrode engineering has been used to covalently modify carbon materials with a structural mime of lignin, namely an anthracene moiety, so that laccase is both well oriented and attached at the surface in a plug and socket fashion to ensure efficient direct electron transfer (DET) and mediator-less enzymatic catalysis (Figure 1(a)).[8] Again, compared with 3D catalytic hydrogel or mediated enzymatic catalysis in solution that use the depth of the diffusion layer, the catalytic reaction is here strictly limited to the surface of the electrode. Whether enzymatic electrodes are used for sensing or for generating electricity, one of the main restrictions for their long-term use is their short stability in time. Even if oxidoreductases are not all as sensitive toward denaturation, their fate is to inevitably evolve toward less activity. One line of research in this field is to understand and establish the cause of enzyme denaturation and find strategies to increase their stability, especially when immobilized. In any case the half-life of an enzymatic system should always be reported whether the oxidoreductase is in solution or immobilized at the electrode. The cost of producing (unstable) enzymes represents another drawback for these systems. To circumvent this, it is possible for some setups to maintain a bioreactor with microorganisms producing constant flow of fresh and active enzymes,[9] or engineer microbes so that the enzyme catalyst is expressed at their outer membrane.[10] For devices intended for *in vivo* implantation, the additional issue of biocompatibility must be addressed.[11] This often results in embedding the device into a biocompatible case or membrane. Despite these promising low-power applications, enzymatic fuel cells are limited by the relatively short life time of the catalyst. This is not the case of microbial electrodes, as discussed next.

3.2.2 Redox reactions catalyzed by living microorganisms at electrodes

The mere fact that living microbes can catalyze redox reactions at electrodes, self-connect, be active throughout the thickness of a biofilm, and, most of all, last forever if fed and not sterilized may come as a surprise to the researcher painfully trying to connect oxidoreductases to an electrode. The beauty of electroactive bacteria in the context of bioelectrodes is that one is only slightly diverting a natural metabolic pathway. The respiration of anodic electroactive bacteria does not rely on the reduction of dioxygen or other soluble oxidants but use extracellular solids as electron acceptors such as minerals containing iron (III) oxides, for example. Bacterial extracellular electron transfer may occur at potentials well up-stream of the electron transfer chain, which is important for application as microbial anodes in microbial fuel cells. In bacteria, as in eukaryotic cells and their mitochondria, a diverse range of organic substrates are oxidized stepwise and completely to carbon dioxide through the metabolic cycle. The electrons released by these oxidation reactions are stored in electron transporting molecules such as reduced nicotinamide adenine dinucleotide (NADH), forming the reducing power of the cell. NADH gets oxidized in the cytoplasm by NADH dehydrogenase, a membrane protein, partly anchored at the inner membrane. When oxygen is the acceptor, the electrons follow the classical path through

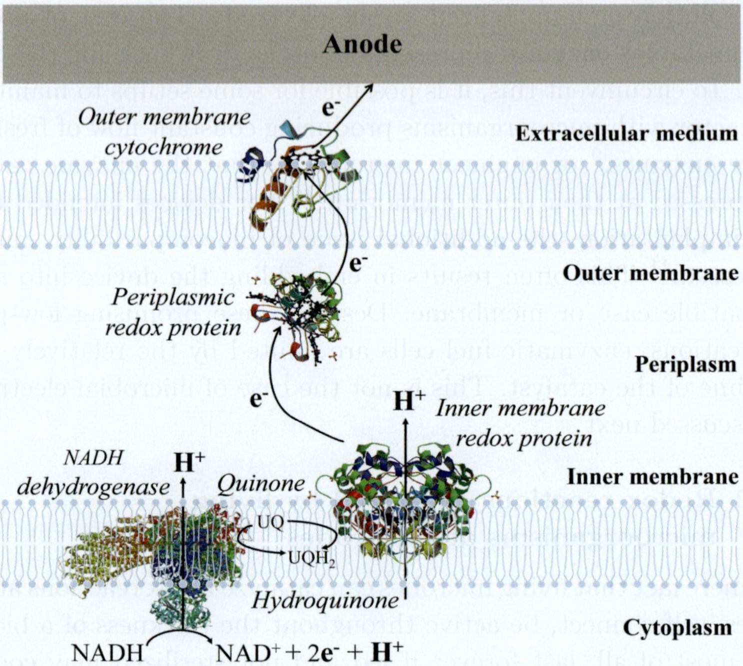

Figure 2. A simplified model of the electron transfer across the cell membranes of Gram-negative electroactive bacteria like *G. sulfurreducens*.

the different redox relays in the inner membrane (in eukaryotes: ubiquinone, succinate dehydrogenase, ubiquinol-cytochrome *c* reductase and cytochrome *c*) up to cytochrome *c* oxidase (also partly anchored at the inner membrane) where oxygen gets reduced in the cytoplasm. The path is different in electroactive bacteria as the electrons are accepted by solids (minerals or anodes) located outside the cell. A simplified model of the electron transfer across the cell membrane can be shown in Figure 2. Electrons flow through NADH-dehydrogenase to different membranes and soluble periplasmic redox proteins up to *c*-type outer membrane cytochrome (OMC).[12] OMC have been identified as the ultimate redox relays of the respiratory chain of some electroactive bacteria. They are mono or multi-heme proteins and are anchored to the outer membrane of the cell with part of their structure and prosthetic group exposed toward the extracellular medium. Figure 3 shows the structure of OmcF from

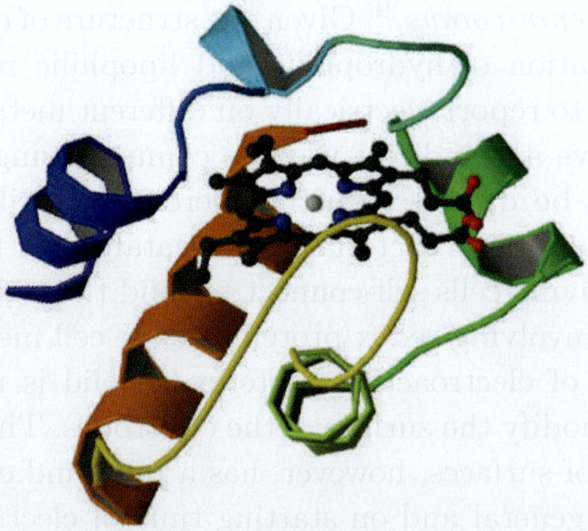

Figure 3. Close-up on the structure of OmcF from *G. sulfurreducens* (PDB code 3CU4). Note that this particular OMC is not the most critical for extracellular electron transfer but is the only one structurally characterized to date.[13]

Geobacter sulfurreducens.[13] It is assumed that fast and DET between the outer membrane cytochrome and the electron-accepting solid is possible. This is supported by gene deletion study that shows that species lacking specific cytochromes are not efficiently connected to anodes. The archetypical genus claimed to display such exocellular DET is *Geobacter*. This mechanism may also occur through conducting protein-based *pili* containing redox relays or conducting networks of undefined and debated nature.[14] These pili are important as they also play a key role in the conduction and catalytic activity of cells in the whole thickness of the biofilm. An alternative mechanism exemplified and found in the *Shewanella* genus is that the microorganism produce natural redox mediators (*i.e.* flavins) that link the anode and the outer membrane.[15] For *Geobacter*, it is possible that some *c*-type outer membrane cytochromes leach out in the exocellular medium and play the role of soluble redox mediators. Other connection mechanisms to an anode involve adding artificial redox mediators to the electrolyte or metabolic production and excretion by the microorganism of a product that can be oxidized at the electrode. For example, urea can be irreversibly oxidized at 0.6 V *vs.* Ag/AgCl and has been identified to be responsible for the anodic activity of the yeast *Arxula adeninivorans*.[16] Given the structure of eukaryotic yeast cells, a combination of hydrophilic and lipophilic redox mediators have been used to report electrically on different metabolic phases.[17] These alternative and indirect ways to connect living cells and electrodes will not be discussed further here. The striking differences between enzymatic and bacterial electrocatalysis is that stable and self-renewable living cells self-connect to solid through the metabolic redox network involving redox proteins inside cell membranes. Since the connection of electroactive bacteria to solid is natural, it may seem futile to modify the surface of the electrodes. The physicochemical properties of surfaces, however, has a profound effect on biofilm colonization in general and on starting time of electroactive biofilm in particular.[18,19] Its effects may be pro- or anti-"biofouling", the former being desired for electroactive biofilm development. In addition, tuning the biofilm-electrode properties by surface modification may be important for improving electron transfer kinetics between the biofilm and the electrode. When an electroactive biofilm develops onto an electrode from a mixed inoculum, say a soil sample, it becomes naturally enriched in electroactive species like *Geobacter* if fed with acetate even if no detectable amount of this strain is found in the inoculum. The selection of particular species directed by the surface state has not been demonstrated so far, but it is possible that carefully crafted interfaces with specific recognition elements will achieve this goal in the future. The connection of electroactive species is also documented for reduction of species like oxygen[20,21] or nitrate.[22] No definitive proof of DET at cathodes has been provided so far. However, inspired by the mechanism at anodes, a hypothesis can be put forward. Chemotroph microorganisms use the oxidation of extracellular electron donors, *e.g.* Fe(II), and the reduction of oxidants, *e.g.* O_2 or NO_3^-, as a source of energy. If the electron donor is a solid like a mineral, a similar direct transfer mechanism deduced from that at the anode can be hypothesized. A redox protein at the outer membrane of cathodic electroactive bacteria may be the entry point for electrons downstream in the electron transport chain close to the reduction potential of the oxidant. Within chemotroph, some species are also autotroph for carbon, meaning that the electrons from the cathode catalyze the reduction of the oxidant (*e.g.* O_2 to H_2O) but also permit the reduction of CO_2 by the cell that makes its own fixed carbon. The advantages of living bacteria over isolated enzymes are their stability in time, the wide substrate range and full oxidation of organic substrates to CO_2, their self-renewability, self-connection to electrodes and catalysis throughout the thickness of the biofilm. From a fundamental point of view, the direct anodic or cathodic connection of bacteria and biofilm is the most intriguing. Surface modification and tuning of the electrode interface properties is a common field of both enzymatic and microbial electrodes. For biofilms, however, the modification may have complementary or adverse effect on different target: development time, electrical performance, and speciation. As a concluding section, a short overview of applications of these biofuel cells is addressed next.

3.2.3 Applications of enzymatic and microbial electrodes

Catalytic bioelectrodes can be used as biosensors for the determination of specific substrate in the biomedical field (enzymatic glucose biosensors for patients with a diabetes condition) or as alert devices monitoring the state of a natural environment (the current delivered by microbial anodes being proportional to the metabolizable carbon present in streams, for example). The bioelectrodes may be part of a fuel cell delivering power to low-power demanding electronic devices. The Graal in enzymatic fuel cells is *in vivo* implanting devices for powering microelectronic biomedical implants as body fluids contain glucose and oxygen. Prototypes have already been developed and tested in animals, but issues related to time stability and biocompatibility remain to be solved for practical applications. As discussed above, microbial electrodes have the advantage of being stable in time but are unlikely to be considered for *in vivo* implantation with the exception of the gut and intestine (digestive system) already rich in bacteria. Microbial fuel cell have been proposed to treat wastewater from town or industries but the technological constraints leading to high internal resistance in the prototype have led to limited success so far. Passive and simple design low power microbial fuel cells

(a)

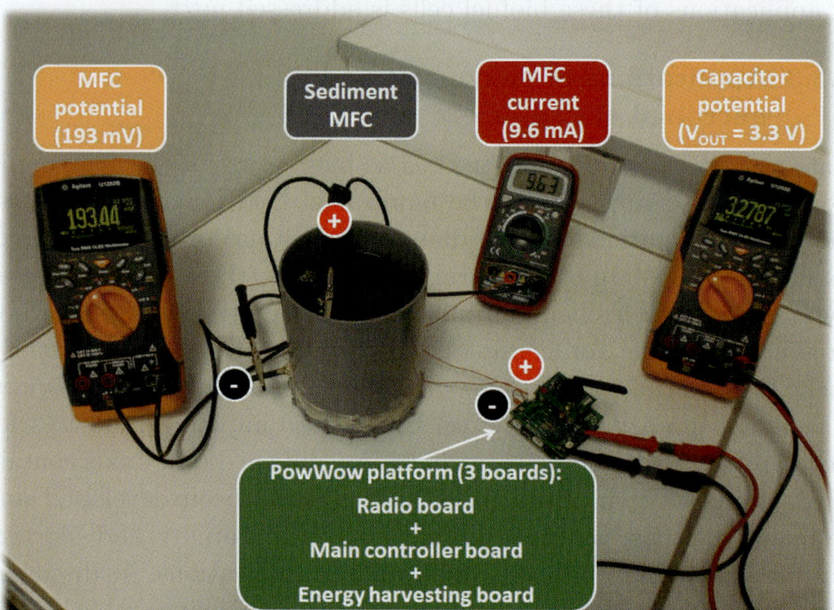

(b)

Figure 4. Example of a low-cost "artificial lake" membrane-less sediment fuel cell that can power a radio transmission system (from Ref. 23). The anode is a graphite rod and graphite granules at the bottom of the tube and the graphite felt cathode is floating at the water–air interface.

have been shown to be useful to charge capacitors or rechargeable batteries of electronic apparatus (see Figure 4 for an example of a low-cost "artificial lake" membrane-less sediment fuel cell powering a radio transmission system).[23] However, these devices are still prototypes for low-power niche applications.[23,24] Since the power recovered by microbial fuel cell is low, scale-up difficult and electricity cheap, the field of microbial electrodes has started to shift its focus towards bioelectrosynthesis where bacteria and cheap electricity convert small molecules into more profitable compounds.[25]

3.3 Conclusions

This short introductory overview has provided a perspective on the continuum of the catalysis of redox reactions at electrodes from materials to living cells. Catalysis may be brought about first by the nature of the electrode and second by the state of its interface. When oxidoreductases are used as catalyst a redox mediator is most often needed but DET can be engineered by specific alteration of the protein or of the electrode. Resorting to living microbial electrode catalysts such as bacteria or yeast solves in principle the problem of the instability of the enzymes as they are self-renewable. Among the mechanisms accounting for electrical connection of microorganisms to electrode, the DET linking the respiratory metabolism of some Gram-negative bacteria to the electrode surface is the most interesting. Similar DET is also most probably found in some microbial cathodes as well. In this field, surface modification and interface optimization have been shown to play a role in the development and performance of microbial electrodes. The applications of bioelectrodes are in sensoring, powering, and electrosynthesis. The glucose determination with single-use enzymatic electrodes represent the great success of oxidoreductases in the biomedical field. The use and value of enzymatic biofuel cells powering implanted (low power) biomedical devices remain to be demonstrated. Microbial electrodes may also be used for sensing, although most reliably as alerting devices. Electricity production coupled to wastewater treatment is an appealing potential application that has to date stumbled on scale up problems due in part on the low conductivity of waste effluents. Bioelectrosynthesis has only started to demonstrate its value with simple molecule production. The advent of some of these promising applications will depend on the fundamental progress made in all the complementary fields of bioelectrocatalysis and that are developed in the other chapters of this book.

List of Abbreviations

ABTS$^{\bullet}$/ABTS	oxidized and reduced state of 2,20-azino-bis(3-ethylbenzo-thiazoline-6-sulfonic acid
Ca.	circa
e.g.	*exempli gratia*
FAD/FADH$_2$	oxidized (quinone) and reduced state (with two electrons and two protons to the hydroquinone) of flavin adenine dinucleotide
FcOH	ferrocenemethanol (synonym: hydroxymethyl ferrocene)
i.e.	*id est*
NAD+/NADH	oxidized and reduced state (with two electrons and one proton) of nicotinamide adenine dinucleotide
OMC	outer membrane cytochrome
PDB	protein data bank

References

1. A. J. Bard and L. R. Faulkner (2001). *Electrochemical Methods: Fundamentals and Applications*, 2nd Edition, Wiley.
2. P. Kissinger and W. R. Heineman (1996). Laboratory techniques in electroanalytical chemistry, in R. L. McCreery and K. K. Cline (eds.), *Carbon Electrodes*, 2nd Edition, CRC Press, p. 298.
3. A. Heller and B. Feldman (2008). Electrochemical glucose sensors and their applications in diabetes management. *Chemical Reviews* **108**, 2482–2505.
4. A. Heller (2004). Miniature biofuel cells. *Physical Chemistry Chemical Physics* **6**, 209–216.
5. O. Courjean, F. Gao and N. Mano (2009). Deglycosylation *of* glucose oxidase for direct and efficient glucose electrooxidation on a glassy carbon electrode, *Angewandte Chemie International Editon* **48**, 5897–5899.

6. S. M. Jones and E. I. Solomon (2015). Electron transfer and reaction mechanism of laccases. *Cellular and Molecular Life Science* **72**, 869–883.
7. F. Barrière, Y. Ferry, D. Rochefort and D. Leech (2004). Targetting redox polymers as mediators for laccase oxygen reduction in a membrane-less biofuel cell. *Electrochemistry Communications* **6**, 237–241.
8. C. F. Blanford, R. S. Heath and F. A. Armstrong (2007). Efficient electrocatalytic oxygen reduction by the 'blue' copper oxidase, laccase, directly attached to chemically modified carbons. *Chemical Communications* **17**, 1710–1712.
9. S. Sané, K. Richter, S. Rubenwolf, N. J. Matschke, C. Jolivalt, C. Madzak, R. Zengerle, J. Gescher and S. Kerzenmacher (2014). Using planktonic microorganisms to supply the unpurified multi-copper oxidases laccase and copper efflux oxidases at a biofuel cell cathode. Bioresource Technology **158**, 231–238; (2015). Erratum to "Using planktonic microorganisms to supply the unpurified multi-copper oxidases laccase and copper efflux oxidases at a biofuel cell cathode", **180**, 386–387.
10. S. Fishilevich, L. Amir, Y. Fridman, A. Aharoni and L. Alfonta (2009). Surface display of redox enzymes in microbial fuel cells. *Journal of the American Chemical Society* **131**, 12052–12053.
11. A. Zebda, S. Cosnier, J.-P. Alcaraz, M. Holzinger, A. Le Goff, C. Gondran, F. Boucher, F. Giroud, K. Gorgy, H. Lamraoui and P. Cinquin (2013). Single glucose biofuel cells implanted in rats power electronic devices. *Scientific Reports* **3**, 1516.
12. A. Sydow, T. Krieg, F. Mayer, J. Schrader and D. Holtmann, Electroactive bacteria — molecular mechanisms and genetic tools. *Applied Microbiology and Biotechnology* **98**, 8481–8495.
13. P.R. Pokkuluri, Y.Y. Londer, S.J. Wood, N.E. Duke, L. Morgado, C.A. Salgueiro and M. Schiffer (2009). OmcF, Outer membrance cytochrome F from *Geobacter sulfurreducens*. *Proteins* **74**, 266–270.
14. M.V. Ordóñez, G.D. Schrott, D.A. Massazzaa and J.P. Busalmen (2016). The relay network of Geobacter biofilms. *Energy Environment Science* **9**, 2677–2681.
15. E. Marsili, D.B. Baron, In.D. Shikhare, D. Coursolle, J.A. Gralnick and D.R. Bond (2008). *Shewanella* secretes flavins that mediate extracellular electron transfer. *Proceedings of the National Academy of Sciences* **105**, 3968–3973.
16. N.D. Haslett, F.J. Rawson, F. Barrière, G. Kunze, N. Pasco, R. Gooneratne and K.H.R. Baronian (2011). Characterisation of yeast microbial fuel cell with the yeast *Arxula adeninivorans* as the biocatalyst. *Biosensors and Bioelectronics* **26**, 3742–3747.
17. K.H. Baronian, A.J. Downard, R.K. Lowen and N. Pasco (2002). Detection of two distinct substrate-dependent catabolic responses in yeast cells using a mediated electrochemical method. *Applied Microbiology and Biotechnology* **60**, 108–113.

18. M. Picot, L. Lapinsonnière, M. Rothballer and F. Barrière (2011). Graphite anode surface modification with controlled reduction of specific aryl diazonium salts for improved microbial fuel cells power output. *Biosensors and Bioelectronics* **28**, 181–188.
19. K. Guo, A. Prévoteau, S.A. Patil and K. Rabaey (2015). Engineering electrodes for microbial electrocatalysis. *Current Opinion in Biotechnology* **33**, 149–156.
20. A. Ter Heijne, D.P.B.T.B. Strik, H.V.M. Hamelers, C.J.N. Buisman (2010). Cathode potential and mass transfer determine performance of oxygen reducing biocathodes in microbial fuel cells. *Environmental Science & Technology* **44**, 7151–7156.
21. M. Rothballer, M. Picot, T. Sieper, J.B.A. Arends, M. Schmid, A. Hartmann, N. Boon, C.J.N. Buisman, F. Barrière and D.P.B.T.B. Strik (2015). Monophyletic group of unclassified γ-Proteobacteria dominates in mixed culture biofilm of high-performing oxygen reducing biocathode. *Bioelectrochemistry* **106**, 167–176.
22. N. Pous, C. Koch, J. Colprim, S. Puig and F. Harnisch (2014). Extracellular electron transfer of biocathodes: Revealing the potentials for nitrate and nitrite reduction of denitrifying microbiomes dominated by *Thiobacillus sp.* *Electrochemistry Communications* **49**, 93–97.
23. Y.R.J. Thomas, M. Picot, A. Carer, O. Berder, O. Sientieys and F. Barrière (2013). A single sediment-microbial fuel cell powering a wireless telecommunication system. *Journal of Power Sources* **241**, 703–708.
24. L.M. Tender, S.A. Gray, E. Groveman, D.A. Lowy, P. Kauffman, J. Melhado, R.C. Tyce, D. Flynn, R. Petrecca and J. Dobarro (2008). The first demonstration of a microbial fuel cell as a viable power supply: Powering a meteorological buoy. *Journal of Power Sources* **179**, 571–575.
25. S. Gildemyn, K. Verbeeck, R. Slabbinck, S. J. Andersen, A. Prévoteau and K. Rabaey (2015). Integrated production, extraction and concentration of acetic acid from CO2 through microbial electrosynthesis. *ES&T Letters* **2**, 325–328.

Chapter 4

Electron Transfer Between Bacteria and Electrodes

Lucie Semenec, Sanja Aracic, Elizabeth R. Mathews, and Ashley E. Franks*

Department of Physiology, Anatomy and Microbiology, School of Life Sciences, La Trobe University, Melbourne, Victoria 3086, Australia
**A.Franks@latrobe.edu.au*

4.1 Introduction: Electromicrobiology and Microbial Electrochemical Systems

4.1.1 Introducing electromicrobiology

Electromicrobiology encompasses the fields of microbiology, electrochemistry and engineering.[1–4] It focuses on the study of electrochemically active microorganisms (EAMs), microbes with the ability to interact with electrodes. EAMs can utilize an electrode as an electron donor or acceptor and are generally referred to as exoelectrogens.[5] EAMs can be further subdivided to electricigens, microbes that are capable of completely oxidizing organic compounds to carbon dioxide with electrons being transferred extracellularly to an electrode[6]; and electrotrophs, microbes that can accept electrons from electrodes for the reduction of terminal electron acceptors.[7]

EAMs are utilized to carry out a range of reactions in devices referred to as microbial electrochemical systems (MESs). MESs are also known as microbial electrochemical technologies and can often be used interchangeably with bioelectrochemical systems (BESs).[8,9] MESs can be further divided into microbial fuel cells (MFCs).[10,11] and microbial electrolysis cells (MECs). MFCs utilize microorganisms to convert organic compounds into electrical current and generate net power output.[3,12] MECs require net power input via poising the electrode potential for microbially driven synthesis of hydrogen and other commodity chemicals via microbial electrosynthesis.[2,13–15] In this chapter, MECs will refer solely to MESs that catalyze ME reactions at the electrode to synthesize products. Furthermore, an extension of MFC technology includes microbial desalination cells (MDCs) utilizing microbial electrolysis (ME) to desalinate water, and benthic or sediment microbial fuel cell (SMFCs), where MFCs are placed in underwater sediments and produce net power.

MESs are generally comprised of the same basic components: a working electrode and a counter-electrode. MESs may be single-chambered microbial fuel cells (SCMFCs) or two-chambered MFCs (2CMFCs) where a proton exchange membrane (PEM) separates the two electrodes. The anode of an MES is surrounded by a medium containing organic materials that act as electron sources. The EAMs oxidize the organic material and pass the electrons to the anode, which is why they are also referred to as anode-respiring bacteria. Electricigens can respire by utilizing the electrode as the final electron acceptor in pure culture. In the case of mixed communities, other microorganisms can participate in the reaction by fermenting complex organic substrates whose products can be used as a carbon and electron donor source by the anode-respiring bacteria. After utilizing the anode as the terminal electron acceptor electrons donated by the EAM travel to the cathode to combine with the protons that have physically relocated to form the resultant products.[3,15] EAMs utilize various mechanisms for extracellular electron transfer (EET) including endogenous redox mediators and shuttles for mediated transfer or pili and cytochromes for direct transfer[16,17] (discussed in Section 2.1). The efficiency of MESs at producing electricity and synthesizing products is dependent on the catalysis of the redox reaction at the electrode, the conductivity of the electrolyte and the dispersion of ions throughout the cell chambers.[15] Ion dispersion may become diffusion limited, leading to a loss of power generation or a greater power requirement to complete the reactions.[15] This chapter will explore the architecture and composition of different MESs and describe the currently characterized mechanisms of electron transfer that EAMs use to interact with electrodes. Also, strategies for improvement to various performance parameters of MESs will be discussed along with new applications of these technologies in the future.

4.1.2 Microbial fuel cells

MFCs have been extensively studied from an engineering, electrochemical and microbiological standpoint over the last couple of decades.[18] The main difference between MFCs and MECs/MDCs is that MFC devices were initially developed to harvest the electrical current generated from the microbial oxidation of organic substrates. This discovery has led to the use of MFCs in small-scale electricity production and more extensively in the successful remediation of land and water polluted with organic and inorganic contaminants.[2] MFCs come in a range of different architectures depending on their final application.

4.1.2.1 *Two-chamber microbial fuel cells*

The two-chamber MFC (2CMFC) resembles traditional electrochemical cells and has been a commonly used MFC configuration.[1,19] It consists of two chambers, anode and cathode, that are separated by a PEM, which allows protons to pass between the two chambers (Figure 1(a)). Within the anode chamber, the microbial community oxidizes organic substrates and pass their electrons to the anode via a number of EET mechanisms. The protons produced during substrate oxidation migrate via the PEM to the cathode chamber where they combine with the final electron acceptor and receive electrons transferred from the anode, thus completing the redox reaction.[12] These MFCs are used to study the ability of different EAM communities to degrade substrates in power generation and bioremediation applications. 2CMFCs can also be used to identify new EAM species, capable of forming biofilms on poised electrodes.[12,20,21] The two-chamber

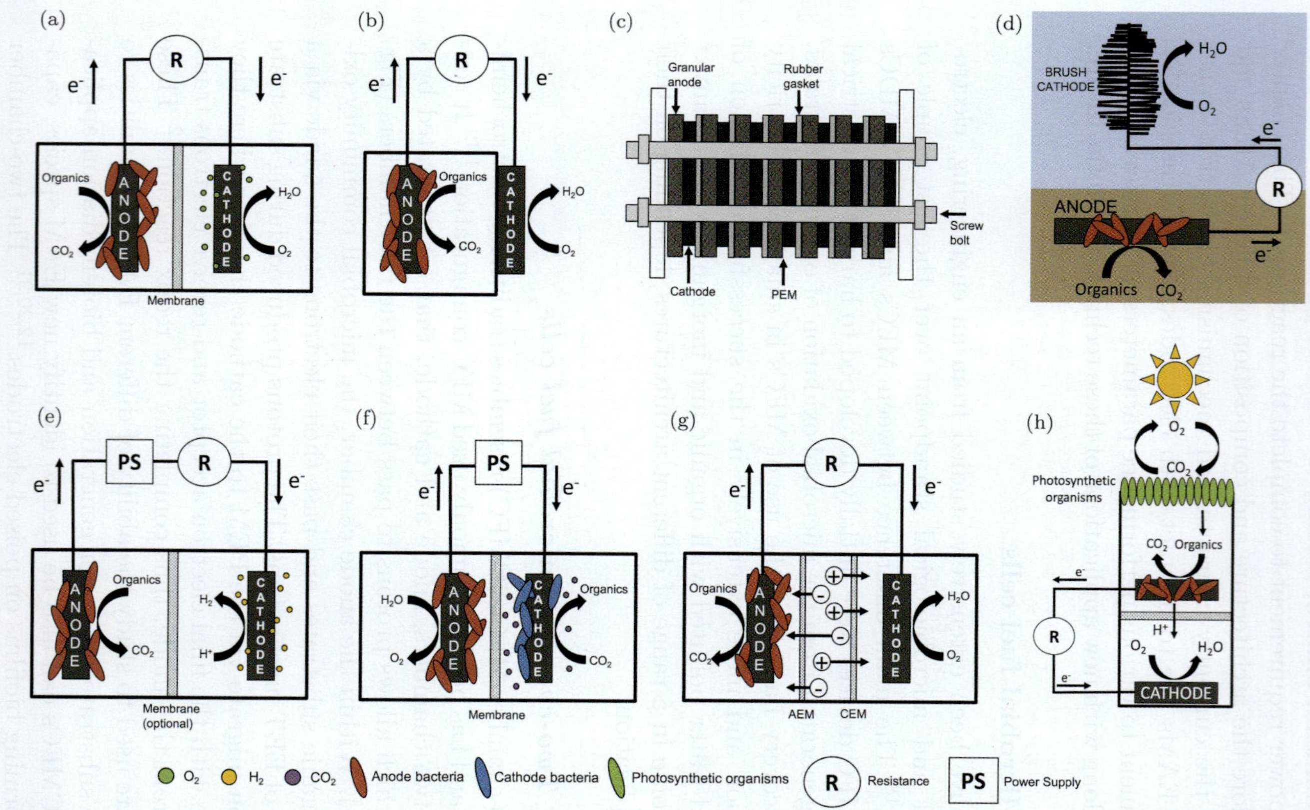

Figure 1. Schematic representations of different MES architecture. (a) 2-chamber microbial fuel cell, (b) single-chamber microbial fuel cell, (c) stacked microbial fuel cell, (d) benthic or sediment microbial fuel cell, (e) microbial electrolysis cell, (f) MEC synthesizing organic compounds, (g) desalination cell, (h) representative microbial solar cell. Diagrams adapted from Refs. 1–3.

design is advantageous due to its ease of setup and ability to be run in batch mode configuration, particularly in a laboratory setting. However, the disadvantages of this design are the lack of electrode surface area and consequently lower power densities generated, and the inability to efficiently scale up the system.[1,12] In most scaled-up studies the designs consist of either stacked electrodes or tubular single chambers that can be set up in series.[22] These tubular SCMFCs have the advantage of using an air-cathode.[12,22] Unlike a traditional aqueous cathode, which generates less power, air cathodes use oxygen as the final electron acceptor and do not require a separate chamber to function.[12] This type of cathode configuration is a design advantage of SCMFCs.

4.1.2.2 *Single-chamber microbial fuel cells*

As mentioned earlier, although similar in concept to 2CMFCs, the SCMFC design offers considerable advantages including reduction in cost, higher volumetric power density and ease of assembly. The simpler architecture means that the cathode does not require its own chamber and is directly exposed to air, thus eliminating the need for a catholyte solution and an aeration system (Figure 1(b)).[1,12,23] Furthermore, due to the absence of a cathode chamber, SCMFCs do not require a separating PEM, significantly reducing the internal resistance within the system, increasing the maximum power density achieved due to higher cathode potentials[3,24] and reducing production costs.[1] However, the exposed air cathode in PEM-free SCMFCs creates some challenges due to the increased diffusion of oxygen to the anode. One limiting factor was the reduction in coulombic efficiency (CE) as a result of the aerobic oxidation of substrates by facultative anaerobes in the anode chamber.[24] Additionally, short-circuiting may occur as a result of oxygen, thus limiting the activity of the anaerobes found in the anode chamber. These drawbacks can be greatly minimized with application of a thin cloth to the water exposed side of the air cathode, which reduces oxygen diffusion into the anodic chamber and leads to higher CEs.[23]

4.1.2.3 *Stacked microbial fuel cells*

Stacked MFCs evolved from SCMFCs in a bid to increase the power output of MFCs.[25] They consist of a number of SCMFCs sandwiched together and connected either in series or parallel.[26] In the basic setup, the stacked MFC consists of an anode separated from a cathode by a PEM. This configuration is repeated several times with each cell separated by a rubber gasket and then the whole arrangement is bolted together (Figure 1(c)).[1,26] An influent rich in organics is pumped into the anode chamber and a catholyte mixture is circulated throughout the cathode chamber.[26] The main advantage of the stacked setup is that it allows for increased current output and CE, particularly if the cells are connected in parallel.[1,25] However, if one of the cells in stacked MFCs, especially series-stacked MFCs, experiences fuel starvation, where bacterial activity is lost, the stack reverses in polarity.[26,27] This leads to an overall decrease in the final voltage produced. To mitigate the loss in voltage, continuous feeding of substrate to the stacked cells is required. Since there is an eventual limitation to the maximum power output possibly attained by SCMFCs, stacking several together in series or parallel is the most promising for larger scale power output and commercialization.[27,28] Extensive research is being conducted into using stacked MFCs for wastewater treatment and power generation of small devices.[25,29]

4.1.2.4 *Sediment and benthic microbial fuel cells*

The first sediment MFC (SMFC) that could harvest energy from organics in marine sediment was created by Reimers and Tender.[30,31] These systems are also known as benthic unattended generators (BUGs).[30,32] SMFCs have been used in marine environments to power remote sensing devices, where a low constant power output is generated by the microbial oxidation of substrates in the anaerobic sediment coupled with the reduction of oxygen in the overlying water column.[2,33] These fuel cells often employ a graphite anode and cathode, the most effective being a graphite "brush style cathode", which provides a greater surface area for microbial reactions to occur (Figure 1(d)).[2,12,33] SMFCs do not require additional electron mediators to generate electrical current since they are powered by the metabolism of *in situ* microbial communities inhabiting the sediment.[2] Benthic and SMFCs can also utilize the sulfides produced during anaerobic microbial respiration as electron donors.[34] However, due to the low levels of organics present in most marine sediments, substrate limitation does impede higher levels of power production.[35] This problem can be addressed by positioning SMFCs in areas of high organic flux such as beneath fish farms, near agricultural or wastewater runoff and other such nutrient-rich environments.[2] While increasing power production, sediments with high levels of organics can reduce the life span of the MFC due to the build-up of sulfide oxidation products on the anode surface.[32,35] Oxidation of this accumulated elemental sulfur to sulfate by bacteria such as *Desulfobulbus* spp. can help increase current density at the anode.[11] The ability of microorganisms to generate power from a variety of different organic sources and wastewater demonstrates the dual potential of future MFCs to generate power while utilizing organic wastes.

Attempts have been made to use SMFCs to power small electrical sensors. While this has worked for low-powered monitoring equipment, it has not been as successful for powering standard electronic equipment.[31,36] The problem lies with the inability to generate cell potentials greater than 800 mV.[36] These problems are partially overcome with the development of a power management system (PMS), which cannot only store electricity using a capacitor but increase the available potential using a charge pump DC–DC converter to boost the potential.[36] The resulting configuration was able to power a wireless temperature sensor for over a year. Electronically scaled-up SMFCs also utilize a PMS to increase their power output.[37] SMFCs connected in parallel generate more power than those connected in series, however, electronically scaled-up SMFCs produce more power than SMFCs connected in parallel.[1,25,37] These scaled-up SMFCs consisted of four smaller SMFCs connected to a four-channel PMS as opposed to one large SMFC connected to a single-channel PMS. However, despite the increase in power generation, SMFCs are yet to generate 1 W.[37]

4.1.3 Microbial electrolysis cells

Unlike standard MFCs, MECs require net energy input to electrosynthesize valuable products such as hydrogen, methane, and butanol, which can be harnessed as biofuels or commodity products (Figure 1(e)).[2,14,15,38–40] It is well understood that electricigens can use an electrode as a final electron acceptor while respiring anaerobically.[41,42] This ability to use electrodes as an electron acceptor has raised the question of whether electrodes can also be used as electron donors under certain conditions.[41] ME technology is utilized in MECs, where microbial redox reactions occurring at the electrode, particularly the cathode, are the reverse of those occurring at the anode in MFCs. Rather than donate electrons to the electrode, microorganisms, known as electrotrophs, receive electrons from the electrode, which they use to reduce CO_2 and form useful multi-carbon extracellular by-products such as biofuels (Figure 1(f)).[14,38,43] This technology not only provides a means of generating commodity chemicals but also can potentially reduce the use of valuable arable land in biofuel production.[38]

In a process known as electromethanogenesis, cathode biofilm microorganisms catalyze the reduction of CO_2 to methane using electrons provided by the cathode. This process may be enhanced by electricigens such as *Geobacter sulfurreducens* at the bioanode,[39] where they oxidize substrates such as acetate and donate resulting electrons to the anode. To optimize electromethanogenesis, the transfer of electrons from the cathode to the microorganism can be improved by using metal catalysts such as Au, Pd and Ni to reduce the activation energy.[44,45] In addition to methane, CO_2 can be reduced by acetogens to form other valuable commodity chemicals such as glycerol, acetate, formate and 2-oxobutyrate.[38,46] However, only a limited number of cathode utilizing electrotrophs have been described at present, namely *G. sulfurreducens*, *Geobacter metallireducens*, *Clostridium ljungdahlii* and *Sporomusa ovata*.[43] Another issue with MECs has been to prevent hydrogen formation by holding the cathode at a stable potential using a potentiostat. This is an energy-consuming process particularly in large-scale applications. Recent studies simplify MEC design by removing both the potentiostat and the membrane while still providing efficient electrosynthesis.[47] Membrane-free MEC designs (Figure 1(e)) are preferred as they are more feasible and less costly, particularly if hydrogen production can be improved by using a continuous rather than batch mode operation.[40,47]

4.1.4 Microbial desalination cells

Interest in MDCs has increased as global potable water supplies have decreased. Traditional desalination from seawater requires a considerable energy input (approximately 1.8–2.2 kWh/m^3).[48] MDCs use electricigens to produce an electrical potential in the anode chamber from the degradation of organic matter. This electrical potential is able to desalinate water by driving ion transport through ion-exchange membranes. Organic matter in wastewater can potentially provide approximately 1.8–2.1 kWh/m.[3] Therefore, using wastewater as influent provides the organics and energy needed to drive the desalination process as well as providing wastewater treatment.[48] The basic architecture of a MDC consists of the anode, desalting and cathode chambers (Figure 1(g)). The anode chamber contains an electrode and a microbial community comprising of electricigens that oxidize organics and transfer electrons to the anode. At the cathode, oxygen is often used as the final electron acceptor and water is the resultant product formed. Between the anode and desalting chamber is an anion exchange membrane (AEM) and between the desalting chamber and the cathode is a cation exchange membrane (CEM). When current is generated by the anode, protons are released and prevented from leaving the anodic compartment by the AEM.[49] This results in negatively charged ions flowing from the central chamber to the anode to balance the positive charge.[49] The reverse happens at the cathode, resulting in the desalination of water in the desalting chamber.[48,49] This ionic exchange can be magnified by stacking several of these membrane pairs and alternating the desalting cells, referred to as dilute cells with the concentrate cells.[48] The process does not require power input to drive desalination, achieving 90% desalination of water with salt concentrations as high as 35 g/L.[49] This technology promises to be an effective and feasible means of water desalination.

4.1.5 Microbial solar cells

Microbial solar cells (MSCs) is the collective term for MESs harnessing energy from a photosynthetic reaction.[3] These systems utilize photosynthetic organisms such as cyanobacteria and other phototrophic bacteria, algae, and plants in combination with EAMs to drive electrical current or produce synthetic compounds (Figure 1(h)).[50] These cells have numerous applications including the synthesis of biofuels without the need to commandeer large areas of arable land.[51] Depending on the photosynthetic organism being utilized, MSCs can be categorized into three groups: plant microbial solar cells (PMSCs), bacterial microbial solar cells (BMSCs), and algal microbial solar cells (AMSCs).[3] Each of these MSCs undergo the same four principles of operation including photosynthesis and generation of organic matter, transport of organic matter to the anodic compartment, oxidation of the organic matter at the anode by electricigens, and finally the cathodic reduction of oxygen or other final electron acceptors.[3,50]

4.1.5.1 *Plant microbial solar cells*

PMSCs are the most commonly studied MSCs, whereby a living plant fuels the microbial community with carbon-rich rhizodeposits, which increases the MFC power output. For instance, by incorporating rice plants into a regular sediment MFC, a seven-fold power increase was obtained.[50,52] However, efforts to replicate these results have been mixed, most likely due to various environmental factors including microbial community composition, fuel cell design and soil structure.[50] Further improvements to power output in PMSCs have been obtained by incorporating an air biocathode and algae to the catholyte solution. The biocathode contained a mixed population of microorganisms that catalyzed the reduction of oxygen on the graphite cathode, and the algae provided oxygen to the biocathode community via photosynthesis.[53] Furthermore, addition of compost to PMSCs constructed with rice plants doubled the power output.[54] Other plants that have been effectively utilized in PMSCs include *Canna indica*, where the presence of the plant influenced the diversity of the *in situ* microbial community and will be discussed further in Section 4.3.3.[55]

4.1.5.2 *Bacterial microbial solar cells*

BMSCs undergo photosynthesis via phototrophic bacterial biofilms, most of which are comprised of a mixed community of cyanobacteria and other phototrophs.[50] Although these bacteria require light in order to undergo photosynthesis and hence release organic substrates and nutrients into the cell, their power generation is often linked to a heterotrophic bacterial community, which increases current production when light is unavailable.[56] It is hypothesized that due to the overexposure to light, oxygen accumulation produced by phototrophs occurs. This can inhibit the strictly anaerobic electricigens and promote facultative anaerobes to switch to aerobic respiration. This hypothesis explains why current production is reduced in extended periods of illumination.[56] Therefore, BMSCs operate most efficiently when light exposure to the cell is removed before oxygen build-up occurs, allowing these heterotrophic electricigens to utilize the organics produced during photosynthesis and continue to anaerobically respire through the electrode.[56] It is hence evident that a balanced synergistic interaction between phototrophs and heterotrophs plays a pivotal role in the BMSC function. This synergistic ability of the microbial community gives BMSCs some advantages to MESs in that they allow the bacterial communities to utilize both day and night conditions to produce power. Furthermore, although BMSCs produce less overall power than other types of MESs; the fact that they are initially powered by photosynthesis means that availability of organic matter is not a significant limiting factor.[57]

4.1.5.3 *Algal microbial solar cells*

Algae have often been investigated for energy production due to their high growth rate and biomass, using only a seventh of the land area needed for terrestrial biofuel crops.[51,58] AMSCs utilize algae as the main photosynthetic organism to produce organics and store energy in the form of biofuels such as methane and hydrogen.[58] The utilization of algae for hydrogen production traces as far back to 1942.[59,60] However, the extreme sensitivity of the hydrogenase enzyme to oxygen still limits AMSCs as a productive biotechnology.[59] Further research is warranted to make AMSCs a competitive biotechnology for power generation.[51,58,61]

4.1.6 Summary

The different types of MFCs and MESs show great potential for future production of renewable energy. Another advantage of these systems is the ability to utilize wastewater and other organics while concurrently producing energy, thus providing a means of cleaning up waste and producing power. The microbial communities found in these devices are the progenitors of the energy being generated and the interactions that occur between the members of the microbial community as well as within the MES itself will be discussed in the following section and Section 4.3.2.

4.2 Mechanisms of Interaction

4.2.1 Introduction

Electron transfer mechanisms are crucial for cell growth and maintenance because all living cells need to dispose of electrons liberated during substrate degradation. This section reviews the mechanisms of EET utilized by electroactive microorganisms. EET interactions can be between homogeneous or heterogeneous populations of EAMs and can involve soluble electron acceptors, which are too large to enter the cell, or insoluble electron acceptors such as electrodes and metal oxides.[62,63] Soluble electron acceptors that can enter the cell are reduced by cytoplasmic or periplasmic reductases and do not require outer-membrane proteins. Conversely, dissimilatory metal-reducing bacteria (DMRB) need to transfer electrons to and from the respiratory electron transport chain, located in the plasma membrane, to an electron acceptor located outside the cell.[63,64] These interactions occur under anoxic conditions by phylogenetically diverse microorganisms.[63,65] Thus far, classes of Proteobacteria, Firmicutes, and Acidobacteria phyla as well as some microalgae, yeast, and fungi have been reported to exchange electrons with extracellular electron acceptors.[66] Although DMRB are phylogenetically diverse, the majority of research has focused on Gram-negative bacteria. Current understanding of EET mechanisms and the proteins or molecules involved is mostly limited to single-species electroactive bacterial biofilms of *G. sulfurreducens* and *Shewanella oneidensis*.[67] Species of *Geobacter* and *Shewanella* are ubiquitous in aquatic and sediment environments and have diverse respiration strategies.[68]

EET can be mediated directly by outer membrane *c*-type cytochromes (OMCs), pili, or a conductive biofilm matrix, or indirectly via redox-active molecules such as mediators.[69] Biofilm-forming microorganisms, such as *G. sulfurreducens*, mediate EET via direct contact of the cell membrane (referred to as direct extracellular electron transfer (DEET)) or via long outer membrane appendages (referred to as pili-associated extracellular electron transfer (PEET)) with the extracellular electron acceptor.[70,71] In addition to performing DEET via OMCs, *S. oneidensis* can also perform indirect EET via endogenously secreted flavins, referred to as shuttle-mediated extracellular electron transfer (SEET) (or mediator-associated EET, MET).[72] DEET and PEET mechanisms enable higher rates of electron transfer than SEET because they are not dependent on the diffusion of the redox mediators.[73] The genotype of microorganisms, as well as whether they are in pure culture or in a consortium, is a key biological parameter affecting EET.[73] With the exception of *G. sulfurreducens* cultures, mixed bacterial communities generally produce higher current densities than pure cultures.[74–76] Furthermore, biofilm-forming microorganisms produce higher current densities than planktonic-based systems and higher currents are generally generated by Gram-negative bacteria than Gram-positive bacteria.[73] The four known EET mechanisms — DEET, PEET, SEET, and direct interspecies electron transfer (DIET) — and the components involved are described in the following subsections.

4.2.2 Direct extracellular electron transfer

The biogeochemical cycling of metals and nutrients is largely dependent on microorganisms capable of performing DEET.[77] DEET is a form of microbial respiration, which has been reported in species of *Geobacter*, *Shewanella*, and relatively few species of Gram-positive bacteria namely those from *Clostridium*, *Thermincola*, and *Thermoanaerobacter* genera.[56,78–80] DEET occurs via physical contact of the cell membrane proteins with extracellular electron acceptors. Microorganisms capable of DEET possess membrane-bound electron transport proteins that transfer electrons from the quinone pool in the inner membrane to the cell surface, where the OMCs transfer electrons to an extracellular electron acceptor. OMCs are heme-containing proteins commonly found in bacteria.[81] Although numerous OMCs have been identified, relatively few have been implicated in EET. Thus it is plausible that many of the uncharacterized OMCs are involved in alternative electron transport pathways yet to be identified.[82] DEET to electrodes is generally limited by the cell density of the microbial electrogenic biofilm formed on the surface of the electrode, however, there are exceptions to this rule.[83]

4.2.2.1 *Direct extracellular electron transfer by Geobacter spp.*

G. sulfurreducens is commonly found in subsurface environments where it generates energy (ATP) by oxidization of organic compounds such as acetate to CO_2 using metal oxides as terminal electron acceptors.[84] Electrons released by oxidation of nutrients in the cytoplasm are transferred via the membrane-bound NADH dehydrogenase through a network of *c*-type cytochromes to the extracellular electron acceptor.[85,86] *G. sulfurreducens* contains over 100 putative *c*-type cytochromes.[84] It is proposed that the inner-membrane associated cytochrome MacA transfers electrons to the periplasmic cytochromes of the PpcA family (PpcA, PpcB, PpcD, and PpcE), which consequently transfer electrons to OMCs (Figure 2).[85] The reported roles of OMCs are mostly complementary, however, there are some exceptions where the roles reported are contradictory. Some

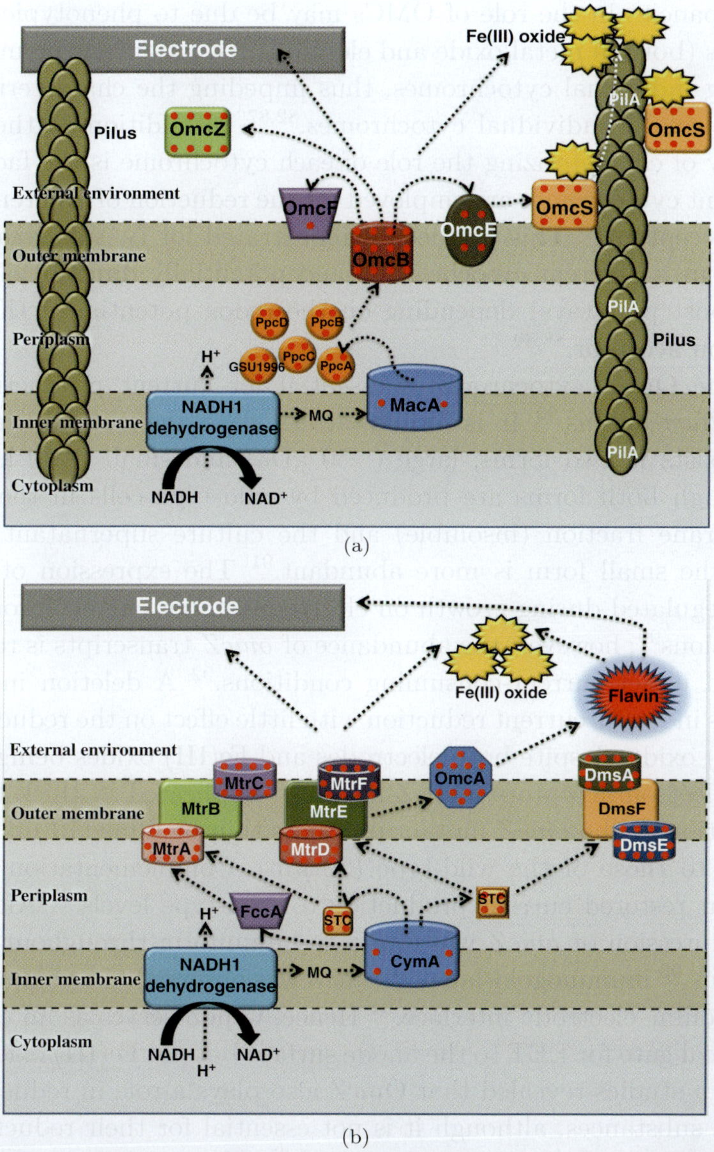

Figure 2. (a) Mechanisms of extracellular electron transfer by *Geobacter sulfurreducens*. The dotted arrows represent the proposed electron flow between proteins and cytochromes. The red dots represent the heme groups. MQ, menaquinone pool. (b) Mechanisms of extracellular electron transfer in *Shewanella ondeidensis*. The dotted arrows represent the proposed electron flow between proteins and cytochromes. The red dots represent the heme groups. MQ, menaquinone pool. Adapted from Ref. 85.

discrepancies in the role of OMCs may be due to phenotypic adaptations (both in metal oxide and electrode reduction) of the mutants lacking individual cytochromes, thus impeding the characterization of the role of individual cytochromes.[82,87] In addition to the complexity of characterizing the role of each cytochrome is the fact that different cytochromes are employed for the reduction of different electron acceptors.[43] This has been demonstrated for *G. sulfurreducens*, which uses different cytochromes (and potentially different electron transport pathways) depending on the redox potential of the final electron acceptor.[88,89]

The OmcZ cytochrome is essential for current production by *G. sulfurreducens*.[90] It is loosely attached to the outer membrane and exists in two forms: large (∼50 kDa) and small (∼30 kDa).[91] Although both forms are produced by wild-type cells in the outer membrane fraction (insoluble) and the culture supernatant (soluble), the small form is more abundant.[91] The expression of *omcZ* is upregulated during growth on electrodes under current-producing conditions[90]; however, the abundance of *omcZ* transcripts is reduced 10-fold under current-consuming conditions.[92] A deletion in *omcZ* results in >90% current reduction with little effect on the reduction of Fe(III) oxide, despite both electrodes and Fe(III) oxides being insoluble electron acceptors. OmcZ has been implicated in thick biofilm formation as the *omcZ* mutant produces thin biofilms (10 μm) compared to those of the wild-type (22 μm). Complementation of the mutant restored current production to wild-type levels.[90] Although the expression of *omcZ* was found to be uniform throughout anode biofilms,[93] immunogold-labeled OmcZ was observed to accumulate at the biofilm–electrode interface.[94] Hence, OmcZ serves as an electrochemical gate for EET to the anode surface but not Fe(III) oxide.[90,91] *In vitro* studies revealed that OmcZ also plays a role in reduction of humic substances, although it is not essential for their reduction as *omcZ* mutants still reduced humics.[91,95] It is proposed that OmcZ participates in homogenous EET between individual OmcZ proteins along the cell's outer membrane and by cell-to-cell contact.[91] The interaction of OmcZ with other membrane-bound mediators as well as pili is still unclear, but other proteins are likely to assist in the transfer of electrons across the electrode–biofilm interface.[71] OmcZ (or homologes of OmcZ) is only found in *Geobacter* spp., which possibly accounts for lower current densities observed with other members of the *Geobacteraceae* family, like *Pelobacter* spp.[96]

Microarray and genetic analysis, as well as mutational and complementation studies, have implicated OmcB in the reduction of both soluble Fe(III) citrate and insoluble Fe(III) oxides, but not current production[87,97,98] OmcB is embedded in the outer membrane and partially exposed to the outer surface. It is an important component in EET to Fe(III) oxides in *G. sulfurreducens*, however, whether it serves as an Fe(III) reductase or an intermediate electron carrier to the terminal Fe(III) reductase is unclear.[98] OmcB possibly accepts electrons from a variety of periplasmic electron carriers (PpcA or hydrogenases). Despite the lack of correlation with current production, cyclic voltammetry revealed that OmcB mediates heterogeneous EET across the biofilm–electrode interface.[71] Transcriptomic analysis of the anode biofilm showed increased expression of *omcB*, suggesting that OmcB might play an indirect role in EET to the anode, even though *omcB* deletants do not affect current production.[90] More recently, OmcB was demonstrated to increase in abundance towards the outer layers of the *G. sulfurreducens* biofilm, where availability of electron acceptor is more limiting.[99] OmcB forms two trans-outer membrane porin–cytochrome protein complexes with OmcC.[99] Both complexes transfer electrons across the outer membrane to extracellular electron acceptors. Other periplasmic *c*-type cytochromes (OmaB and OmaC) and porin-like outer membrane proteins (OmbB and OmbC) can also be associated with the OmcB–OmcC complexes. Orthologs of OmcB have been identified in other *Geobacteraceae* but not in phylogenetically distinct Fe(III)-reducing microorganisms.[98] The absence of OmcB in these organisms suggests an alternative mechanism of EET to that of *Geobacteraceae*.

One of the most abundant *c*-type cytochromes of *G. sulfurreducens* is OmcS. It is a hexaheme *c*-type cytochrome, which is localized along the pili and is easily sheared from the outer surface.[100] Although the role of OmcS has been extensively studied, there are numerous questions remaining. OmcS is a promiscuous redox protein that is able to perform EET to a wide range of electron acceptors.[101] It is essential for the reduction of metal oxides (Fe(III) and Mn(IV)) but is not required for reduction of soluble electron acceptors (Fe(III) citrate or fumarate) and current production.[102,103] A possible explanation for these observations is that an *omcS* mutant overproduces OmcZ.[71] OmcS has been reported to play a role in EET to electrodes[87] and also to humic substances.[95] OmcS possibly facilitates DEET from pili to Fe(III) oxides because the spacing between OmcS molecules is too great for electron hopping between OmcS cytochromes along the pili.[100,104] The abundance of *omcS* transcripts was reported to increase with power production,[87] which was later contradicted by a report that the *omcS* transcripts decreased with increased power production.[90] OmcS is more highly expressed in thin biofilms than in planktonic cells, but is down-regulated in thick biofilms.[87]

Another OMC, which can be easily sheared from the outer surface of *G. sulfurreducens*, is OmcE, which is localized in the extracellular matrix.[105] Transcriptional, genetic and microarray analyses show that *omcE* expression is upregulated in current producing anode biofilms, suggesting its involvement in EET to electrodes.[87,90] Although *omcE* mutants generate lower power outputs initially, they are not significantly affected in current production over longer periods of time.[87,90] This adaptation is proposed to be due to the increased production of OmcS by *omcE* mutants, thus compensating for their loss of OmcE. Therefore, OmcE is required for reduction of metal oxides but is not essential for reduction of electrodes.[71,97] OmcE possibly facilitates EET to pili, which then transfer electrons directly to the extracellular electron acceptor (metal oxide).[106]

Deletion of the *omcF* gene that encodes the low molecular weight cytochrome, OmcF, impairs the ability of *G. sulfurreducens* to generate current. Nonetheless, OmcF is not directly involved in EET to electrodes. Instead, it affects the surface localization of other cytochromes (OmcB and other uncharacterized cytochromes).[97]

OmcF is also required for transcriptional activation of the *omcB* gene. Other cytochromes such as OmcG and OmcH are possibly involved in translation or stabilization of the OmcB protein in the outer membrane. Deletions of *omcG* and *omcH* result in decreased levels of OmcB (post-transcriptionally) and consequently the rate of Fe(III) oxide reduction.[107] Introduction of *omcG* or *omcH in trans* restores both the ability to reduce Fe(III) oxide and wild-type levels of OmcB expression. In the absence of OmcG and OmcH, OmcB is accumulated and consequently degraded in the periplasmic space. Numerous other OMCs also do not play a role in EET to electrodes or metal oxides, including OmcT[87,105] and OmcC.[98] It is possible that the quantity and diversity of OMCs are more important in DEET than the presence of a specific OMC because OMCs are not only involved in EET but also in transcriptional and posttranscriptional regulation or processing.[107]

Geobacter spp. keep their OMCs in an oxidized state so that they can serve as electron acceptors for periplasmic cytochromes. The mechanism of DEET involves the transfer of electrons to extracellular electron acceptors with different redox potentials in a thermodynamically favorable manner. In order to remain feasible, the reduction of different extracellular electron acceptors may resort to utilization of redox proteins with a specific range of reduction potentials via EET pathway selection.[17,88] Although several members of the *Geobacteraceae* family have been reported to utilize DEET, not all contain *c*-type cytochromes. While species of *Geobacter* and *Desulfuromonas* contain numerous *c*-type cytochromes, species of *Pelobacter* do not, yet are still capable of reducing Fe(III) oxides.[108] The reason for these differences may be attributed to yet unidentified EET components being utilized in novel pathways.

4.2.2.2 *Direct extracellular electron transfer by Shewanella spp.*

Similar to *G. sulfurreducens*, *S. oneidensis* can also perform DEET to extracellular electron acceptors via its *c*-type cytochromes.[109]

S. oneidensis MR-1 strain encodes ∼42 putative *c*-type cytochromes with most (80%) located in the outer membrane.[110,111] The oxidation of organics by *S. oneidensis* releases electrons to the quinone pool via NADH dehydrogenase.[85] Several *c*-type cytochromes are responsible for the electron transfer from the quinone pool to OMCs and then to extracellular electron acceptors. Unlike *G. sulfurreducens*, *S. oneidensis* has a significantly smaller number of cytochromes capable of mediating DEET, mainly employing its inner membrane associated tetraheme cytochrome, CymA, and the MtrABC complex, which spans the outer membrane.

CymA acts as a hub, receiving electrons from the menaquinone pool in the cytoplasmic membrane and distributing them to various periplasmic redox proteins by forming transient protein complexes.[112] CymA is essential for EET to several electron acceptors including Fe(III) oxide, Mn(IV) oxide, and nitrate.[112] A deletion of *cymA* caused more than 80% reduction in current generation. The small tetraheme cytochrome *c* (STC) and the flavocytochrome c_3 FccA accept electrons from CymA and transfer them to OMCs.[112] In some instances, STC and FccA transfer electrons to cytochrome MtrA, which is the periplasmic component of the MtrABC complex associated with the outer membrane. MtrA is required for the stability of MtrB, a trans-membrane protein, and is essential for localization and interaction of MtrC and OmcA (another OMC, which can act as a reductase for external substrates). A deletion of *mtrA* and *mtrC* resulted in 15% and >90% reduction of current production, respectively, whereas the *omcA* deletant (*omcA* encodes a lipoprotein) exhibited comparable current production with the wild-type strain. The metal-reducing enzyme OmcA binds directly to the surface of Fe(III) oxide.[109]

A single mutation in *mtrABC* did not completely eliminate EET. Consequently several *mtrABC*-homologues have been identified. MtrF, the homologue of MtrC and OmcA, has a higher reducing capacity of flavins and soluble electron acceptors than either OmcA or MtrC *in vitro*. STC and FccA can also transfer electrons to MtrD and DmsE, which are the periplasmic components of the outer membrane complex MtrDEF and DmsEFA. Only anaerobically grown

S. oneidensis produces OMCs. Unlike most OMCs characterized in *G. sulfurreducens*, the OMCs of *S. oneidensis* are capable of accepting electrons.

4.2.2.3 *Direct extracellular electron transfer in Gram-positive bacteria*

Even though the majority of research on DEET has been done on Gram-negative bacteria, Gram-positive bacteria are the dominant DMRB in some environments.[113] Gram-positive bacteria were initially thought not to be able to perform DEET because of their thick peptidoglycan cell wall, however, DEET has since been reported in several Gram-positive bacteria including *Thermincola potens* strain JR,[78] *Thermoanaerobacter* sp. strain BSB-33,[79] *Thermincola ferriacetica*[56] and *Clostridium pasteurianum.*[114] Although Gram-positive bacteria interact with anodes via DEET, none have been found yet to produce higher currents than *G. sulfurreducens* and *S. oneidensis.* DEET in some Gram-positive bacteria is possibly facilitated through conductive cell walls or redox-active proteins linking the inner membrane to the cell surface.[115]

In the case of *T. potens*, no soluble redox-active components were detected by cyclic voltammetry in the medium of the anode, ruling out SEET.[115] Sequencing of the *T. potens* genome revealed 32 putative cytochrome genes, some of which were later shown to be localized to the cell wall.[116] Confocal microscopy revealed cells in contact with the electrode were disproportionately viable relative to the rest of the biofilm and lastly, no correlation between biofilm thickness and power production was observed.[78] Trypsin-shaving LC-MS and surface-enhanced Raman spectroscopy demonstrated that a number of OMCs are expressed and localized to the cell wall or cell surface during growth of *T. potens* on Fe(III) oxide.[115] Furthermore, cyanide inhibited the precipitation of silver metal on the cell surface, thus demonstrating the involvement of surface-localized redox-active heme proteins in metal reduction. Taken together, these biochemical and biophysical observations demonstrated that EET across the cell wall of *T. potens* occurs via cytochromes and that the cells in contact with the electrode are primarily responsible for current generation.

4.2.2.4 *Direct extracellular electron transfer in other microorganisms*

Detailed reports of microorganisms other than *Geobacter* and *Shewanella* performing DEET are relatively rare. Nonetheless, DEET has been demonstrated in relatively few Gram-positive bacteria (discussed in the subsection above) and phylogenetically diverse bacteria including *Geoalkalibacter subterraneus*,[117] *Rhodoferax ferrireducens*, *Desulfobulbus propionicus*,[118] and *Geothrix fermentans*. Reports of microorganisms being able to utilize electrodes as direct electron donors for anaerobic respiration have led to increased focus on organisms with this ability. *G. sulfurreducens* was demonstrated to reduce fumarate to succinate with an electrode as an electron donor over a decade ago.[41] Although the mechanism of DEET is still not well characterized yet, several other bacteria have been reported of oxidizing extracellular electron donors. For example, *Acidithiobacillus ferrooxidans* can oxidize Fe(II) whereby electrons travel across the cell wall from the surface of the outer membrane across the periplasm to the plasma membrane.[119] The biofilms at the cathode surface are mostly single layered, therefore, it is likely that these microorganisms perform DEET. A better understanding of the mechanisms of EET at the cathode would possibly reduce the cost of existing MFCs or could lead to new applications of MFCs.

4.2.3 Pili-associated extracellular electron transfer

G. sulfurreducens is capable of producing relatively thick (~50 μm) biofilms whereby the cells on the outer layers are capable of transferring electrons to the electrode surface.[75] The specific mechanism of multi-layer biofilms that *G. sulfurreducens* utilizes is being extensively studied and debated.[120–122] A strong correlation between biofilm growth and the abundance of *G. sulfurreducens* pili has been reported.[123] These pili are capable of extending the range of electron transfer from the cell surface to the surface of Fe(III) oxides and electrodes using extracellular cytochromes as electrochemical gates. Pili have been associated with OmcS cytochromes, which were proposed to facilitate EET from the pili to Fe(III) oxides. Cells at substantial distance from the electrode are metabolically active and able to oxidize organic substrates and donate electrons through a conductive biofilm matrix to the electrode.[124] OmcS associated with pili has been suggested to also have a capacitor role in PEET.[100] While the heme groups in cytochromes have been proposed to be responsible for capacitance instead of conductivity in long-range EET,[104] electron transfer in live electrode grown biofilms is a thermally activated process with incoherent conductivity indicative of EET involving hemes of *c*-type cytochromes.[122] Magnetite can compensate for the lack of a pilin-associated *c*-type cytochrome in EET,[125] while a *G. sulfurreducens* strain expressing *Pseudomonas* type IV pili localizes OmcS on pili but is deficient in Fe(III) oxide reduction and current production.[103]

G. sulfurreducens pili consist of PilA monomers, encoded by the *pilA* gene. Their proposed conductivity has been attributed to the overlapping π-orbitals of aromatic amino acids.[102,126] Deletion of *pilA* in *G. sulfurreducens* results in an inability to reduce insoluble electron acceptors such as Fe(III) oxides even though cells could attach to Fe(III) oxides. Complementation with a functional copy of the *pilA* gene supplied *in trans* restored Fe(III) oxide reduction.[90] The *Geobacter* pilin has a highly conserved N-terminal domain of type IV bacterial pilins, which function in inner membrane insertion, signal processing, and pilin polymerization, forming the central helical core of the pilus filament.[70] The pilus apparatus is anchored in the periplasm and outer membrane; thus they can accept electrons from periplasmic and/or outer membrane EET proteins if conductive. The presence of pili has been associated with the development of thicker electroactive biofilms and thus higher anode performance. A 10-fold increase in current production was observed upon nanowire formation of *G. sulfurreducens.*[127] In the case of *G. sulfurreducens* KN400, which produces thin biofilms, high current production has been associated with overexpression of pili.[83] Although not extensively investigated, conductive nanowires have also been implicated in PEET in oxygenic phototrophic cyanobacterium *Synechocystis* sp. and *Pelotomaculum thermopropionicum.*[128]

4.2.4 Shuttle-mediated extracellular electron transfer mechanism

Although DEET is the most efficient means of EET, microorganisms often grow under conditions where they are not in direct contact with the electron acceptors. In thick mixed species biofilms, the majority of cells are not in direct contact with an insoluble electron acceptor. In these niches, microorganisms may produce endogenous electron-shuttling compounds or use externally available (exogenous) shuttling compounds.[129] The production of endogenous redox mediators often occurs via secondary metabolic pathways. Endogenous mediators serve as reversible terminal electron acceptors, transferring electrons from the bacterial cell to a solid electron acceptor or into the aerobic layers of the biofilm where it becomes reoxidized and is again available for subsequent redox processes.[69] Organisms capable of performing SEET can transfer electrons to electron acceptors >50 μm from the cell surface. Small amounts of endogenous mediators are sufficient for efficient EET; however, the production of the mediators themselves is energetically expensive, thus resulting in biological losses. Only a handful of endogenous mediators have been identified, for example, pyocyanin and phenazine-1-carboxamide in *Pseudomonas aeruginosa*, and quinone-type redox shuttles in *S. oneidensis*.[130,131] In addition to secondary metabolites, reduced primary metabolites formed as a consequence of anaerobic metabolism can be utilized for EET.[132,133] The production of reduced primary metabolites is closely associated with substrate degradation either as a consequence of respiration (reduction of terminal electron acceptors) or via fermentation (oxidation of the reduced products).[69]

SEET can also involve small mobile molecules (*e.g.* humic substances, quinones, formate, hydrogen, acetate, and thiol-containing molecules like cysteine) capable of undergoing redox cycling.[129] They serve as terminal electron acceptors and once reduced can transfer electrons to extracellular electron acceptors, thus becoming reoxidized. Humics are ubiquitous organic compounds present in aquatic and soil environments; they are derived from degradation of microbial and plant matter. The quinone moieties within humic substances are the redox active centers, which function as electron shuttles. All iron-reducing bacteria and archaea, as well as some fermenting bacteria, can reduce humics. Once reduced, the humics can transfer electrons to Fe(III) oxides, which is likely to be the preferred EET mechanism in the environment. Extracellular humics have been shown to enhance the rate of bacterial Fe(III) oxide reduction. Several bacteria have been reported of utilizing exogenous electron shuttles, including species of *Shewanella* and *Geothrix.*[134,135] Currently, no *Geobacter* strains have been reported to secrete redox-active molecules that could serve as electron shuttles.

4.2.4.1 *Shuttle-mediated extracellular electron transfer via endogenous redox mediators*

Numerous microorganisms have been reported to secrete redox mediators, however, relatively few have been studied in terms of the ability of these redox mediators to serve as electron shuttles. Intact biofilms of *S. oneidensis* MR-1 were grown on poised electrodes and analyzed using electrochemical techniques. Differential pulse voltammetry and cyclic voltammetry detected redox-active molecules (riboflavin and riboflavin-5-phosphate) in supernatants from the biofilm reactors, demonstrating that *S. oneidensis* MR-1 release soluble molecules that shuttle electrons between the outer cell surface and extracellular electron acceptors.[72] Removal of riboflavin resulted in a 70% reduction of EET to electrodes. Marsili *et al.* (2008)[72] also demonstrated that riboflavin adsorbed to a range of insoluble electron acceptors, thus confirming its ability to act as a metal chelator. A mutant of *menC*, which lacks the *o*-succinylbenzoic acid synthase, an enzyme involved in the biosynthesis of menaquinone, was complemented with upon addition of the isolated extracellular mediator. Menaquinone is a lipophilic naphthoquinone that can freely diffuse in membranes and shuttles electrons between proteins of low redox potential in the respiratory electron transport chain. The Mtr respiratory pathway is essential for the reduction of flavins, MtrC requires flavin to reduce insoluble Fe(III) oxide.[136]

Riboflavin and flavin mononucleotide interact with OMCs, forming flavin–OMC complexes, which transfer electrons across the cell membrane.[137]

Another well-characterized microorganism that produces endogenous electron shuttles for reduction of insoluble electron acceptors is *P. aeruginosa.* The phenazine-based pyocyanin is a blue-pigmented aromatic molecule that is produced by *P. aeruginosa* and several other species of *Pseudomonas.* Phenazines have several important functions in quorum sensing and signaling and also enable survival in anaerobic conditions by functioning as an extracellular electron shuttle.[138] Cultures of pyocyanin-producing bacteria change from blue to colorless upon reaching anaerobic conditions; the blue pigment undergoes a reversible reduction into leukopyocyanin, which is readily oxidized by oxygen. Introduction of oxygen results in restoration of the blue color. Iron acquisition by *P. aeruginosa* and electron transport in methanogens also suggest that phenazines have electron-shuttling activity.

Other microorganisms capable of producing riboflavin include *Helicobacter pylori* and *Pichia guilliermondii.* Numerous other microorganisms have thus far been reported of secreting flavins, which mediate SEET. Environmentally relevant surfaces are often coated by electroactive flavins that may affect interactions with bacterial redox-active molecules. It is likely that flavins not only serve as electron shuttles but also as metal chelators.

4.2.4.2 *Shuttle-mediated extracellular electron transfer via exogenous redox mediators*

While planktonic cells of *S. oneidensis* and *Pseudomonas* spp. can synthesize their own redox-active mediators allowing them to perform SEET, some bacteria such as *Escherichia coli* require exogenous mediator molecules to respire insoluble electron acceptors. A large number of compounds, both organic and inorganic, have been reported to mediate EET from microbial cells to electrodes. For example, the Gram-positive, *Desulfitobacterium* sp. use external shuttling molecules such as humic acids or their analogs to perform EET.[139] Furthermore, although electrochemically active on their own, *G. sulfurreducens* and *Shewanella putrefaciens* can utilize solid-phase humic substances as electron shuttles to reduce Fe(III) oxide. The regular addition of exogenous redox mediators to MFCs that do not contain microorganisms capable of endogenous redox mediator production is required to produce low current densities. This becomes very costly in large-scale applications and since many synthetically produced mediators such as neutral red, thionin and Fe(III) EDTA are toxic, this option is also not environmentally friendly.[1] Mediator-less MFCs, which instead employ EAMs that produce their own electron shuttles, are thus a more attractive option. For example, the addition of the phenazine-based mediators such as phenazine-1-carboximde produced by *Pseudomonas* sp. enabled a mediator-less MFC containing the Gram-positive bacterium, *Brevibacillus* sp. to generate electrical current, which it was incapable of on its own.[140] Similarly, *Desulfitobacterium hafniense* strain DCB2 can also transfer electrons using exogenous shuttles.[139] Electron shuttles provide a synergic strategy for microbial members in current generation, however, phenazines can act as virulence factors or cell signals impacting microbial community compositions and biofilm formation.[141]

4.2.5 Direct interspecies electron transfer

In the environment, microbial communities are capable of directly transferring electrons to each other via DIET. DIET is dependent on two species exchanging electrons in order to metabolize a particular substrate. Defined cocultures can evolve over time to produce dual-species aggregates.[142] Unlike indirect EET, which is either mediated via the well-known process of interspecies hydrogen transfer or through formate mediators,[143] DIET involves direct transfer of electrons between two species without the need for external mediators. In interspecies hydrogen transfer, one microorganism reduces protons while the other oxidizes H_2. In contrast, DIET conserves energy for microorganisms by eliminating multiple steps necessary to produce and consume H_2. Methanogenic communities rely on

DIET or HIT to function and undergo anaerobic digestion processes. Electron-donating microorganisms reduce protons to H_2 and methanogens oxidize the H_2 with the reduction of CO_2 to methane (formate can substitute for H_2 as the electron carrier). DIET and HIT has been demonstrated between relatively few bacteria including *G. metallireducens* and *G. sulfurreducens*,[142] *G. metallireducens* and *Methanosarcina barkeri*,[143] *G. metallireducens* and *Methanosaeta* (*Methanotrix*) *harundinacea*,[144] *Pelotomaculum thermopropionicum* and *Methanothermobacter thermautotrophicus*,[145] *Methanobacillus omelianskii*, and *Methanobacterium ruminantium*.[146] Each instance of DIET is unique and requires the utilization of different EET components. For instance, the cocultures containing *G. metallireducens* and *M. barkeri* required pili in order to perform DIET, whereas interaction between *G. metallireducens* and *G. sulfurreducens* depended on OmcS expression.[142,147] It is proposed that the pili of *G. sulfurreducens* and *G. metallireducens* form a conductive matrix within aggregates and cytochromes (particularly OmcS) to facilitate EET to and from the pili and cells.[142]

M. barkeri is unable to metabolize ethanol, however, continued transfer of the co-culture with *G. metallireducens* resulted in metabolism of ethanol into methane in less than 7 days. *M. barkeri* is capable of directly receiving electrons from *G. metallireducens* but can also accept electrons in the form of H_2 from syntrophic partners like *Desulfovibrio* and *Pelobacter*. DIET requires cell-to-cell contact; therefore, it is not surprising that deletion of *pilA* or OMCs in *G. metallireducens* resulted in failure of the *M. barkeri*–*G. metallireducens* coculture to metabolize ethanol or produce methane. The electrons released from ethanol oxidation by *G. metallireducens* are subsequently transferred directly to *M. barkeri*, *G. sulfurreducens* and *M. harundinacea* via DIET. The fact that at least two major genera of methanogens have evolved the capacity for DIET suggests that there are conditions in some soils and sediments in which DIET confers a selective advantage. A better understanding of the interspecies microbial interactions is required to predict and control products of electrosynthesis.

4.2.6 Summary

The field of electromicrobiology has attracted increasing attention due to its potential applications in energy generation, bioremediation (biodegradation), and biosynthesis (biofuel production). The extent to which the EET mechanisms are exclusive or overlap is still unclear. It is noteworthy that the mechanism of EET to anodes in some instances is consistent among species of the same genera such as *Thermincola* spp. and *Geobacter*, but not others, such as *Pyrobaculum* and *Shewanella*. A better understanding of the mechanisms is required to explain why dissimilatory metal reduction is not always indicative of anode reduction and vice versa. For instance, *Ochrobactrum anthropi* produces electricity in MFCs but cannot reduce Fe(III) oxide.[148]

The yield and the efficiency of power production in MFCs are largely influenced by the interactions between electroactive microbial communities. However, non-EAMs can also be found within the electroactive biofilms, which may be involved in the breakdown of complex organics. Minimizing biofilm formation by the non-EAMs may improve performance of MFCs. A clearer role of non-EAMs may lead to improved efficiency of EET in mixed communities, which may lead to increased yield in biofuel production. Adjusting aeration and substrate levels can reduce the number of non-EAMs. Methanogens, reducers of sulfate, nitrate or iron, fermentative microorganisms, aerobes and hydrogen-utilizing autotrophs are known to reduce the performance of MFC systems. To optimize the electrochemical performance of EAM, alteration of operating physical, chemical, and biological parameters is necessary, and is discussed in Section 4.3.

4.3 Strategies for Microbial Electrochemical System Improvement

4.3.1 Introduction

Improvement of MESs is multi-faceted and depends on their particular application: power production, bioremediation, wastewater treatment, and biofuel/commodity substrate production. Numerous studies have investigated the following areas: enhancing power output, optimizing substrate removal efficiency as measured by chemical oxygen demand (COD), and/or biochemical oxygen demand, minimizing cost of materials, up-scaling for industrial scale applications, and increasing production and efficiency of biofuels/chemicals. Improvements in one area do not always equate to improvements in others. For example, increasing power output does not necessarily equal increased COD removal efficiency and the conditions, which are required for operating a system at the maximum power point are different to those required for attaining maximum substrate removal.[149] Therefore desired outcomes need to be considered during design and testing. For example, MESs incorporated in wastewater treatment facilities should be optimized for improved removal of organic substrates, since this is the main goal of wastewater treatment, instead of focusing on optimizing maximum power output.[149]

What is true for these systems, however, is that they face similar challenges, including electrode limitation, electrochemical losses, ohmic/internal resistance, PEM limitations, and microbially driven losses.[150–152] This section will summarize the biological, chemical, and physical strategies that can be used to address these limitations of the MES function. Biological improvements include selecting the appropriate microorganisms in defined and mixed communities in MESs as well as enhancing favorable traits of electricigens through genetic engineering. The physical strategies include varying electrode material and setup, minimizing PEM limitations, and controlling the temperatures in these systems. Lastly, chemical strategies involving adjustment of the anolyte and catholyte substrates, maintenance of pH and solution conductivity will be discussed.

4.3.2 Biological strategies

4.3.2.1 *Selection of microorganisms in microbial electrochemical systems*

Microorganisms play a pivotal role in the function and efficiency of MESs. They are responsible for the transfer of electrons from substrates to the anode, and in MESs with a biocathode microorganisms are ultimately responsible for the oxidation of cathodes. Use of microorganisms in MESs can significantly increase the rate of electrolysis by reducing activation energy, from requiring applied voltages higher than 1.6 V in non-microbial electrolysis to voltages theoretically as low as 0.14 V in the presence of microbes and obtaining nearly 100% hydrogen yield.[153,154] Previous methods for enhancing EET between microorganisms involved the addition of exogenous chemical redox mediators such as neutral red, thionin, and anthrahydroquinone-2,6-disulfonate to facilitate shuttle-mediated transport between microorganisms and the electrode (as discussed in Section 4.2.4.2). However, with the discovery of DIET, enrichment for the appropriate syntrophic partners in mediator-less systems has potential as a more sustainable and cost-effective strategy. Furthermore, *P. aeruginosa* and *S. oneidensis* are capable of producing their own electron shuttles that can mediate EET to the electrode in mediator-less systems (described in Section 4.2.4).

4.3.2.1.1 *Studying defined cultures to learn more about electricigens*

Although metagenomic approaches have elucidated which microorganisms predominate in certain *in situ* MES environments,[155,156] it is important to test these microorganisms in pure or defined cocultures to fully understand the roles they play in MES function. There is no definite rule as to whether increased diversity of microbial species enhances power output. However, certain microorganisms, such as electricigens, have an enhanced ability to generate electrical current. *G. sulfurreducens* strain KN400, with enhanced current production, was selected for and isolated after growth for 5 months on an electrode surface.[83] This strain lead to a greater than five-fold increase in current density as compared to the wild-type *G. sulfurreducens* strain. Additionally, KN400 had lower internal resistance and mass transfer limitations. These advantages may be attributed to its increased production of microbial nanowires and decreased production of OMCs. Thus, by studying defined cultures of electricigens, we are better able to understand what synthetic engineering approaches could be used to enhance MES function.

4.3.2.1.2 *Microbial communities that enhance the function of microbial electrochemical systems*

Mixed-culture MFCs can be categorized into those studied within a laboratory setting, where a mixed-species inoculum from a particular source is applied, such as 2CMFCs, and those constructed *in situ* and comprised of native microbial communities where they are subject to the natural environmental conditions present in that niche, such as benthic or sediment MFCs as described in Section 4.1.2.4. In both settings, the microbial community composition, enriched by the presence of electrodes, depends on the available substrates present in the environment, the setup of the MFC, and the external abiotic conditions such as pH, temperature, light exposure, and salinity. Additionally, the advantage and disadvantage of the presence of each microorganism in the system depends on the application of the particular MES. For instance, the community that enhances biofuel methane production from wastewater is different from the community that enhances pollutant degradation in contaminated soils. Microbial communities that are beneficial for distinct applications of MESs will be discussed, including those within the rhizosphere of PMSCs, contaminated soils, and anaerobic wastewater reactors.

Exposure of environmental microbial communities to an electrode enriches an increased population of electricigens that can mediate EET to the electrode.[76] In addition to the selective nature of the electrode, plants have also been demonstrated to influence microbial community composition in PMSCs. The microbial community from a PMSC, utilizing *Canna indica* wetland plants as the sole photosynthesizers, displayed increased diversity over time in the presence of the plant *vs.* no plant MSC controls.[55] Since the plant continuously adds nutrients to the soil via rhizodeposits, it is not surprising that an increase in available nutrients supports a greater metabolic range of microorganisms. Interestingly, the presence of *C. indica* also led to higher current densities. This enhanced current production likely resulted from syntrophic degradation of rhizodeposits by fermenters of the *Anaerolineaceae* family and electricigens such as *Geobacter* spp. as demonstrated via pyrosequencing and genomic community analysis. Such a predominance of electricigens does not occur in PMSCs grown in non-oligotrophic conditions containing complex polymer rhizodeposits along with oxygen and nitrate as available electron acceptors. Under such conditions, denitrifying bacteria and to a smaller extent, acetoclastic methanogens, compete with electricigens for electron donors as has been shown in *Glyceria maxima* PMSCs.[157] Thus, power output can be improved by limiting growth of electricigen competitors, which is simply done by choosing a plant that sustains growth in anaerobic conditions. Furthermore, addition of compost improves power output of rice PMSCs; however, it is yet to be determined how compost addition affects the microbial community composition.[54]

The structure of microbial communities in soil is also influenced by environmental contaminants. The presence of hydrocarbon contaminants in soil causes a decline in microbial community diversity.[158–160] However, the electrode has a selective positive effect on community structure and function. A carbon cloth anode and biochar anode were implanted into a constructed total petroleum hydrocarbon (TPH) contaminated soil environment.[158] Hydrocarbon degrading function improved significantly, by more than 200%, when the anodes were connected as compared to open-circuit controls where anodes were unconnected. Furthermore, TPH degradation rates were highest near the anode surface, indicating the anode was effectively selecting for hydrocarbon-degrading microorganisms with a higher capacity for pollutant degradation than the native population without an electrode present. This improved TPH removal function was corroborated with structural community analysis, which indicated a separation in grouping between the unconnected control reactors to the connected reactors. The community in TPH-contaminated soil without an electrode present predominantly comprised of common hydrocarbon oxidizing bacteria such as *Actinobacteria*. However, in contaminated soils with electrodes present, a shift in the community to electricigens capable of degrading hydrocarbons, which included species of *Ochrobactrum*, *Pseudomonas* and *Comamonas*, occurred.[158]

Electricigenic communities not only aid in remediation of environmental pollutants, but also are key players in the anaerobic digestion of organic wastes in wastewater treatment reactors. Anaerobic degradation of wastewater is preferred to aerobic treatment due to its greater cost efficiency, higher organic removal rates, and resultant biofuel production in the form of methane or hydrogen, especially since the introduction of upflow anaerobic sludge blanket reactors.[161–164] The utilization of electrodes in various anaerobic wastewater treatments such as electrodeposition, electrooxidation, electroflotation, and electrocoagulation further enhances basic anaerobic wastewater reactor functions while concurrently generating a low level of power output that helps offset operating costs.[165,166] Since it was shown that SCMFCs could also recover energy from organic wastewater degradation in the form of direct electrical current,[166] complementation of standard anaerobic digestion with MFC technology in treating wastewater is being increasingly considered.[167] Utilization of MFCs for wastewater treatment is particularly useful in conditions where standard anaerobic digestion fails such as in low substrate concentrations where temperatures fall below 20°C.[168] SCMFCs applied to wastewater treatment and power production were used to generate power in sucrose wastewater containing different levels of sulfate and electrolytes. CE was initially 25% but fell to 5% after 220 days due to increased methane production.[169] Increased sulfate concentration increased power density by 10%; however, this was also coupled with a loss of CE due to increased COD removal.[169] Thus, it is critical to control the population of methanogens and sulfidogens (anaerobes reducing sulfur to hydrogen sulfide) in these systems to maintain proper function.

Specific microbial communities also play a major role in achieving sufficient hydrogen production in MES bioreactors. The maximum hydrogen yield per glucose molecule via anaerobic bacterial fermentation is 4 mol hydrogen per mol of glucose, which is considerably lower than the stoichiometric potential yield of 12 mol hydrogen/mol glucose and is too low to be an economically feasible alternative to chemical and electrochemical hydrogen production.[170] Therefore, the inhibition of hydrogen consumers, such as homoacetogens and hydrogenotrophic methanogens in MES bioreactors, is essential to allow for hydrogen production. This can be done via heat killing microorganisms in the inocula albeit spore-forming fermenters, adjusting to higher dilution rates or maintenance of a low reactor pH.[171–174] The resulting microbial populations comprise of hydrogen-producing and mainly spore-forming fermenters such as *Clostridiaceae*, *Streptococcaceae*, *Thermoanaerobacteriacea*, *Lachnospiraceae*, and *Sporolactobacillaceae*. In addition to bacterial hydrogen production via dark fermentation, cyanobacteria (blue-green algae) as well as green algae are capable of biophotolysis of water into hydrogen and photoheterotrophs such as purple non-sulfur bacteria can also produce hydrogen from the photodecomposition of organic acids such as acetate.[175] Dark fermentation of glucose to produce hydrogen normally leads to organic acid by-products such as acetic, butyric, and propionic acid. Since photoheterotrophic bacteria can further undergo photo-fermentation of these organic acids also yielding hydrogen, combining anaerobic hydrogen forming microorganisms with photosynthetic bacteria can improve hydrogen yields.[176–178] Such hybrid MECs as bioelectrochemically assisted microbial reactors (BEAMRs) and photosynthetic microbial electrochemical cell (PMEC) reactors have been developed for this exact purpose, with PMECs still being in the concept stage of development.[175,177] Although BEAMRs require an input of voltage to function (>0.2 V), it is still more feasible than conventional water electrolysis reactions, which require applied voltages of >1.23 V.[179] BEAMRs utilize electricigens at the anode to oxidize organic substrates into hydrogen ions and CO_2. The protons are transported via the PEM to the cathode where they are combined with electrons from the cathode to make hydrogen with the help of the applied potential. BEAMRs only exploit dark fermentation processes using MEC technology to produce hydrogen, whereas PMECs would additionally utilize photoheterotrophs to ferment organic acid by-product of dark fermentation into hydrogen, thus improving hydrogen yields.

Anaerobic wastewater reactors harbor highly diverse microbial communities that fluctuate based on substrates present, operating conditions of the reactors and the MES setup within the reactor.[180] Interestingly, when switching from MFC to MEC function, the microbial community shifts.[181] Despite this complexity, several recurring patterns in wastewater microbial compositions have emerged. Methanogenic bacteria, in particular acetoclastic methanogens from the *Methanosaeta* genus and *Methanosarcina* genus, tend to dominate in wastewater environments including those from brewery waste,[182,183] distillery waste,[164] sludge and manure waste,[184,185] agricultural waste[186–188] and forestry waste.[189,190] However, they are not as predominant in sewage waste,[191] where sulfur-oxidizing bacteria and sulfate-reducing bacteria (SRB) tend to predominate. When sulfate-rich industrial wastewater is processed in anaerobic reactors, SRB proliferates and reduce sulfates to sulfur and hydrogen sulfide. These by-products of reduction are toxic to surrounding microorganisms and minimize growth of methanogens.[192] Additionally, SRBs directly compete with methanogens and homoacetogens for electron donors and substrates.[193] Being able to control the population of SRBs is of particular importance in sulfate-rich environments but currently not many strategies exist to do so. Electricigens, particularly *Geobacter* spp., are also commonly dominant members of wastewater communities. The direct electron exchange interaction between species of *Geobacter* and *Methanosaeta* has been implied to occur in brewery wastewater digesters.[183] DIET has also been demonstrated to occur between *G. metallireducens* and *Methanosarcina barkeri*[147] as well as *G. metallireducens* and *M. harundinacea*[144] as described in Section 4.2.5.

4.3.2.2 *Genetic engineering of electricigenic microorganisms*

Improving microbial function could theoretically increase the EET rate of electricigens by enhancing the ability of redox proteins to pass electrons to their terminal electron acceptor, the anode. This can be achieved via genetic engineering. Through targeted deletions of specific nucleotides in electricigen redox transfer proteins, and observing the effect these deletions have on electrical conductivity, it is possible to track which mutations enhance EET.

As there are various proteins along the electron transport chain of well-characterized electrogens,[85] many candidates for genetic manipulation exist (Figure 2). *G. sulfurreducens* contains over 110 *c*-type cytochrome genes, of which several are localized in the outer membrane; OmcS, OmcT, OmcB, OmcC, OmcE and OmcF[84,105,194,195] or the extracellular matrix; OmcZ[91] of which OmcS localizes on pili.[100] Each of these redox proteins has a unique reduction potential range, as determined via redox titration analysis that confers them the ability to oxidize or reduce a specific range of substrates (Tables 1 and 2). Thus, the *c*-type cytochromes in *G. sulfurreducens* are tailored to reduce electron acceptors within a specific range of reduction potentials. For instance, ImcH *c*-type cytochrome of *G. sulfurreducens* is an inner membrane redox protein that can only reduce electron acceptors with a redox potential above 0.1 V.[88] This phenomenon appears to work similarly for electron consumption in *G. sulfurreducens.* The periplasmic monoheme *c*-type cytochrome, PccH, has reduction potential values of -35 to $+1$ mV, which make it suitable for consumption of electrons from biocathodes to electron acceptors with more positive reduction potentials such as fumarate.[196]

It is clear from the above examples that microbial redox proteins involved in EET determine whether a particular electron donor or acceptor can be utilized. Fine-tuning these proteins through genetic modification is a sound strategy for improving electron transfer processes between microorganisms and electrodes. Single-site mutations of the PpcA *c*-type cytochromes within the periplasmic space (Figure 2) of *G. sulfurreducens* led to discovery of strains with enhanced electron transfer rates.[197] These strains maintained the structure of the PpcA proteins, however, were able to transfer electrons to the anode at lower reduction potentials. Another novel approach to enhancing electron transfer rates involves producing mutant strains that have an increased demand for ATP synthesis. Metabolic engineering of *G. sulfurreducens* was undertaken, where the hydrolytic F_1 domain of the outer membrane F_0F_1 (H^+)-ATP synthase protein was made to express in the presence of IPTG.[198] Thus in the presence of IPTG, the strain drained the available ATP

Table 1. Characteristics of *G. sulfurreducens* proteins involved in extracellular electron transfer.

Protein	Redox potential*	Molecular weight (kDa)	Number of hemes	Localization	Role in EET
OmcB	−190 mV[210]	~85[195,210]	12[98]	OM[99,195,210]	Involved in Fe(III) oxide reduction[98] and Fe(III) and ferrihydrite reduction[211]
OmcC	ND	80[211]	12[98]	OM[211]	Involved in Fe(III) and ferrihydrite reduction[211]
OmcE	ND	27.9[105]	4[105]	OM[105]	Involved in Fe(III) oxide and Mn(IV) oxide reduction[105]
OmcF	+180 mV[212]	9–11[85,194,212]	1[194,212]	OM[194,212]	Involved in transcriptional regulation of genes important for Fe(III), U(VI), and electrode reduction[97,194,213]
OmcS	−360 to −40 mV (midpoint −212 mV)[101]	51.7[105]	6[105]	OM[100,105]	Involved in Fe(III) oxide and Mn(IV) oxide reduction[105]
OmcZ	−420 to −60 mV (midpoint −220 mV)[91]	30 and 50[91]	8[91]	OM and ECM[91]	Important for anode reduction[90]
OmpB	N/A	140[214]	None	OM[195]	Involved in Fe(III) oxide reduction[214]
PccH	−24 mV (±5 mV)[215]	15[196]	1[196]	P[92,196]	Cathode (donor)[92]
PpcA/B/C/D/E	−175 to −116 mV[216]	10[217]	3[216–218]	P[216,217]	Important for acetate-dependent Fe(III) reduction[217]
MacA	−241 to −138 mV[219]	36[220]	2[219,220]	IM/P[220,221]	Involved in transcriptional regulation of OmcB and PpcA, which are important for Fe(III) reduction. Also required for U(VI) reduction[213]

Note: Table has been restricted to include only the most well-characterized proteins.
Abbreviations: IM = inner membrane; OM = outer membrane; P = periplasm; ND = not determined; N/A = not applicable.
*Redox potentials were measured at pH 7.0 against a standard hydrogen electrode.

Table 2. Characteristics of *S. oneidensis* proteins involved in extracellular electron transfer.

Protein	Redox potential*	Molecular weight (kDa)	Number of hemes	Localization	Role in EET
OmcA	−180 to −400 mV[222]	83[223,224]	10[224–226]	OM[223,227,228]	Involved in Mn(IV) oxide reduction[226]; important for reduction of anode;[229,230] and flavins[230]
MtrA	−100 to −400 mV[231]	35[232,233]	10[231,234]	P[235] and OM[232]	Involved in soluble Fe(III) reduction[231]; important for Fe(III) and Mn(IV) oxide reduction[235–238]; important for reduction of anode[230,236] and flavins[230]
MtrB	N/A	75[239]	None[111]	OM[239]	Involved in Fe(III) reduction;[239,240] important for reduction of anode[230,236,238] and flavins[230]
MtrC (OmcB)	+100 to −400 mV[241]	75[235]	10[235]	OM/ECM[227,228,235]	Involved in Mn(IV) oxide reduction[226]; important for reduction of anode;[229,230] and flavins[230]
MtrD	ND	40[225]	10[231]	P[231] & OM[230,242]	Minor role in Fe(III) oxide reduction[230,238]
MtrE	N/A	77[225]	None[111]	OM[239]	May be involved in DMSO reduction[243]
MtrF	+100 to −400 mV[244]	76[245]	10[244]	OM[244]	Important for Fe(III) oxide and flavin reduction[230,244]
CymA	∼−350 to 0 mV[246]	21[247]	4[222,247]	IM/P[112,247,248]	Involved in electrode, Mn(IV) and Fe(III) oxide reduction;[236,247] involved in DMSO reduction[248,249]
DmsA/B/E/F	E (−90 to −360 mV)[250]	E(38)[251]	A(10),[85] B(none), E(10),[250] F(none)	A/B(OM),[249] F(OM),[243,249] E(P/OM)[243,249]	Required for DMSO reduction[249]; DmsE has a minor role in Fe(III) oxide reduction[230,238]
STC (CctA)	∼−400 to 0 mV (midpoint −233 mV)[246,252]	12[111]	4[245]	P[245,253]	Minor role in Fe(III) oxide reduction[230,238]

Note: Table has been restricted to include only the most well-characterized proteins.
Abbreviations: IM = inner membrane; OM = outer membrane; P = periplasm; ND = not determined; N/A = not applicable.
*Redox potentials were measured at pH 7.0 against a standard hydrogen electrode.

more quickly, which led to an increased demand for cellular respiration, and hence electron transfer rates.

Genetic engineering of electricigens can be utilized to not only enhance electrical current production at the anode, it can also be used to enhance product synthesis at the cathode. A variety of *P. putida* strains have been engineered to generate commercially valuable compounds such as aromatic phenols,[199] rhamnolipids,[200] and styrenes.[201,202] *P. putida* is favored over the known electricigen, *P. aeruginosa,* for commercialization due to its non-pathogenic nature. For example, a strain of *P. putida* KT2440, has been engineered to produce rhamnolipids, which are naturally produced by *P. aeruginosa* and are commercially valuable biosurfactants.[200] However, since *P. putida* is an obligate aerobe, aeration of the bioreactors that facilitate bioproduction of rhamnolipids is required and leads to foaming, which interferes with the synthesis reaction. To bypass this limitation, a mutant strain of *P. putida* was constructed to contain the core genes required for anaerobic respiration by its close relative *P. aeruginosa,* which is a facultative anaerobe.[203] The introduction of seven phenazine genes, *phzA-G*, which encode redox mediator molecules, allowed *P. putida* to grow in oxygen-limited conditions.[203] The rate of rhamnolipid synthesis has been enhanced in *P. aeruginosa* via overexpression of *rhlA*. Interestingly, this enhanced strain also led to improved electrical output in MFCs.[204]

4.3.3 Physical strategies

4.3.3.1 *Varying electrode materials to enhance power*

Efficient exchange of electrons between microorganisms and the electrode surface is essential to enhancing the functionality of MESs. From improving organic electron donor consumption at the anode, which increases electrical current output, to providing the cathode with additional electrons, enhancing ME. Energy losses in the system occur due to internal resistance and diffusion limitations. The diffusion limitations exist due to the flat structure of electrodes and can be minimized by increasing their surface area. Thus, choosing a porous electrode material increases the maximum power output.

Characteristics of efficient electrode materials include biocompatibility, microbial and chemical stability, conductivity, and cost efficiency. The most commonly used electrode materials include graphite, carbon, and stainless steel. However, other materials such as copper and silver that are known as antimicrobial metals promote the growth of electricigens and have the benefit of being highly conductive, making them attractive for large-scale applications.[205] Other novel materials also include biochar salvaged from forestry and compressed milling residue, which also promote syntrophic electron exchange between microorganisms via DIET.[206,207]

In addition to choosing an appropriate electrode material for optimal biofilm formation, the properties of the material chosen can be enhanced by various methods. Acid treatment of graphite felt anodes increases hydrogen and peptide bonding between amide groups of outer membrane cytochromes and carboxyl groups formed on the electrode surface.[208] Flame-oxidation of stainless steel felt anodes increases their surface area by forming iron-oxide nanoparticles, leading to thicker biofilm formation and consequently maximum current densities of nearly five-fold greater than carbon felt anodes.[209]

Furthermore, choosing the optimal size of electrode can have significant impacts on power output.

4.3.3.2 *Varying electrode setup*

4.3.3.2.1 *Electrode spacing and numbers of electrodes*

Decreasing the distance between anode and cathode can significantly change power output in MFCs. For instance, when distance was decreased by 2 cm, the maximum power density increased by more than 1.5-fold.[254] This is attributed to a lowering of the internal resistance. Increasing the number of cathodes with respect to anodes in SCMFCs with air cathodes also helps with increasing power output. For instance, increasing the number of cathodes from 1 to 4 in an air cathode SCMFC increased power output by five-fold.[255] Furthermore, stacked cells are capable of enhancing power output of MFCs.[256] As described in Section 4.1.2.3, stacked MFCs can be connected in series or parallel. Arranging the anodes in series can lead to inconsistent voltages over time at increasing current densities due to microbial limitations.[26]

4.3.3.2.2 *Microbial fuel cell electrode size matters*

Varying the size ratio of the anode to cathode in MFCs can affect power output. Using a larger cathode than anode surface area can decrease the internal resistance within the system and thus improve power output. This is due to the greater contribution to internal resistance by the cathode than the anode.[257] For instance, increasing the cathode surface area by three-fold led to a 24% increase in the power generated.[258] Miniaturizing MFCs is also a way to improve surface area to volume ratios of the electrodes.[259,260] Microscale MFCs hold potential for their application in robotics, medical devices, and powering small devices. They are also able to solve certain limitations of milliliter-sized MFCs such as ohmic losses, concentration overpotentials, and polarization.[261]

4.3.3.2.3 *Cathode catalysts vs. biocathodes*

Conventional non-microbial fuel cells traditionally employ a platinum catalyst to enhance hydrogen production using dissolved oxygen as a final electron acceptor. Although platinum has also been used in MFCs, it is too expensive for the low current densities they generate. Alternately, carbon cathodes in the presence of ferricyanide oxidizers have been used instead to increase power output, however, as with platinum catalyst-coated cathodes, the use of ferricyanide is expensive as they require continual replenishment.[12] To solve this problem, enzymatic[262,263] and microbial[264–266] biocathodes have been developed that employ cheaper electrode materials like graphite. The main issue with enzymatic biocathodes is that they are not self-regenerating systems and can become unstable over time.[267] Conversely, microbial biocathodes are self-sustaining systems. The first microbial biocathode comprised of a culture of *Desulfovibrio vulgaris* that utilized methyl viologen as a redox mediator.[264,266] However, use of mediators is costly and often the mediators are toxic, thus strategies for mediator-less biocathodes are preferred for large-scale applications.[268] The first mediator-less microbial biocathode for hydrogen production used a three-step process to enrich for EAMs capable of hydrogen production. The first step selects for EAMs that can oxidize acetate and hydrogen. Second, a more refined selection for only hydrogen oxidizers is applied and lastly, the electrode polarity is reversed to switch the hydrogenase activity in the direction of hydrogen production instead of oxidation.[265] Since then, several microbial biocathodes have been constructed using a similar process and fall into three major categories: two-chamber full biological MEC biocathode, two-chamber half biological MEC biocathode and single-chamber full biological MEC biocathode.[269] Most of these studies have consisted of a mixed community of microorganisms forming the biocathode, with dominant phyla comprising of *Proteobacteria*, *Firmicutes*, and *Bacteroidetes*.[270–272] Only several bacteria including *G. sulfurreducens*,[273] *Desulfovibrio paquesii*,[274] and *Desulfovibrio* G11[271] have been confirmed to produce hydrogen in MECs as pure cultures thus far. As additional hydrogen producers are discovered, the establishment of defined co-cultures to enhance yields can be tested.

4.3.3.3 *Proton exchange membrane limitations*

The PEM prevents transfer of bacteria between anode and cathode chambers, allowing the osmotic flow of protons between the chambers in 2CMFCs and minimizing the leakage of substrates between the two compartments. For every electron that is donated to the anode, a proton must be transferred from the anode to the cathode chamber via the PEM to maintain current flow.[150] As with electrodes, the PEM is subject to several challenges including fouling, reduced oxygen transfer rates, substrate losses, and cation transport and accumulation.[150–152] In addition to these challenges, the presence of a membrane can lead to increased ohmic resistance and potential losses due to the large pH gradient generated by the membrane.[275] Furthermore, the presence of a membrane leads to declines in energy efficiency due to selection of unwanted microorganisms in the system.[276] Therefore, due to the high cost of PEMs, their use in large-scale applications such as wastewater treatment is not currently practical.[277] Other separator materials that function as PEMs include salt bridges, ion exchange membranes, glass fibers, composite membranes, and porous materials such as ceramics, earthen pot, nylon filter, J-cloth, glass fiber mat, and non-woven cloth.[278] The challenges of currently used PEMs in MESs and strategies for addressing these issues will be discussed in this subsection.

4.3.3.3.1 *Fouling of proton exchange membranes*

Fouling is a major limiting factor to MFC performance as it prevents the efficient transport of cations from the anode to the cathode, thereby reducing cathode potential.[152] This phenomenon is the result of microbial colonization on the PEM surface by microorganisms within an extracellular polymeric substance matrix. Application of an electric field can disrupt the fouling.[279] However, varying the PEM material is a more effective method of preventing fouling. Currently used PEMs within MESs come in three varieties: CEM, AEM, and ultrafiltration membrane. CEMs made from Nafion material have been the most commonly used separators in MFCs,[150] however, engineering efforts toward the construction of new materials in the past decade has been fruitful. A Fe_3O_4/polyethersulfone (PES) nanocomposite material, particularly at 15% Fe_3O_4 nanoparticle concentrations, resulted in nearly 30% higher power output and less biofouling than the standard Nafion 117 CEMs.[280] The Fe_3O_4 nanoparticles increased conductivity of the membrane leading to its higher power densities achieved and the nanocomposite material also had a smoother surface as visualized by scanning electron microscopy and atomic force microscopy (AFM) than Nafion 117, leading to lower rates of biofilm formation on the PEM surface. Lastly, pretreatment of the membrane by boiling, washing in hydrogen peroxide and then acid treating can also decrease biofouling.[281]

4.3.3.3.2 *Oxygen transfer rates*

PEMs need to prevent the leakage of oxygen from the cathode to the anode chamber. If oxygen leaks through to the anode, the activity of strictly anaerobic electricigens becomes inhibited and decreases the CE of the MFC.[151] Oxygen within the anode also allows for aerobic microbial respiration processes to consume substrate that would normally be exclusively available to anaerobic electricigens. Thus, selection of an appropriate membrane material that minimizes oxygen transfer rates is critical to maintaining high CE. Comparison of Nafion with size selective ultrafiltration membranes and Ultrex CMI-7000 found that Nafion had the highest oxygen mass transfer coefficient of $K_O = 1.3 \times 10^{-4}$ cm/s.[282] Furthermore, J-cloth has a higher oxygen transfer rate ($K_O = 290 \times 10^{-5}$ cm/s) than glass fiber separators due to their highly porous and hydrophilic nature and are also not a preferred choice of PEM.[283]

4.3.3.3.3 *Substrate losses*

As mentioned above, substrate losses can occur as a result of oxygen leakage from the cathode into the anode chamber through the PEM. When oxygen enters the anode, facultative anaerobic bacteria are able to overtake the existing community and metabolize the substrates otherwise available to anode-respiring bacteria via aerobic respiration. Furthermore, oxygen leakage can inhibit growth of some of the strictly anaerobic microorganisms within the anode chamber.[24] Substrate losses lead to decreased CE and power output, making PEMs less attractive for large-scale treatment of wastewater and air-cathodes a more viable option.

4.3.3.3.4 *Cation transport and accumulation*

The PEM is important for the selective transport of protons from the anode to the cathode chamber. However, Nafion membranes permit the flow of other cationic species such as Mg^{2+}, Ca^{2+}, NH_4^+ Na^+, and K^+ ions, which are present in 10^5 times higher concentrations in wastewater than H^+ protons and thus are preferentially transported over H^+ protons.[284] This causes over-acidification of the cathode, reduced proton transport to the cathode, and increased conductivity in the cathode, ultimately leading to the reduction of power output in the MFC.[285] To mitigate the competition of cations other than protons at the Nafion PEM, other materials have been developed. Nanofiber-reinforced composite PEM, based on sulfonated polyether ether ketone, greatly increases selectivity for H^+ protons over competing cations and thus increases hydrogen yields when applied in MECs.[286]

4.3.3.3.5 *Ohmic resistance and potential losses*

To minimize the internal ohmic resistance of MESs and thus increase their power output, the surface area of the PEM must not be smaller than that of the anode and cathode.[287] When the PEM surface area is increased with respect to that of electrodes, the power output increases incrementally. Using CEMs and AEMs in SCMFCs, it was determined that pH gradients along the membranes due to transport of ions other than protons and hydroxyl ions also lead to potential losses.[275] Although the AEM configuration led to less potential losses due to pH gradients, these types of membranes caused higher cathode overpotentials than CEMs. Thus, modifying separator membrane size and composition can have significant impacts on minimizing resistance and potential losses in BESs.

4.3.3.3.6 *Controlling growth of performance limiting microorganisms*

Membrane-separated MFCs are most commonly required in MECs where it is important to prevent leakage of hydrogen produced in the cathode from combining with the CO_2 formed at the anode.[153] Membrane-less MECs used in wastewater treatment are more common when methane is the main target of production and tend to decrease the occurrence of acetogenesis.[276] While homoacetogens at the anode can contribute to power output, they are less favored near the cathode where they consume the desired hydrogen product and lead to increases in COD.[276] A membrane made of activated carbon nanofiber/Nafion increases power outputs in MFCs as compared with general Nafion materials. These novel materials reduce the migration of microorganisms between the anode and cathode and oxygen transfer from the cathode to the anode chamber.[288] Hence, depending on the application of the MES, the choice of whether or not to use a separating membrane and the type of membrane can greatly influence the desired outcomes.

4.3.3.4 *Effects of temperature on microorganisms and reaction rates*

The metabolic activity of bacteria is influenced by temperature and can be modeled as $\theta^{(T-20)}$, with T being the temperature (°C) and $\theta = 1.20$ representing microbial growth in anoxic conditions and $\theta = 1.094$, representing microbial growth of heterotrophs.[289] Since every microorganism has its own distinct range of temperatures at which it can grow, varying BES temperature is potentially a way of shifting the microbial community within the system. The performance of MFCs is highly dependent on their operating temperature.[254,290,291] Increasing temperatures in the range of 20–35°C has a positive effect on the COD removal efficiency of the system and on maximum power outputs obtained.[290] On the contrary, decreasing temperature of SCMFCs from 32°C to 20°C reduces power output as a result of a decreased cathode potential above a certain threshold of current density (>0.24 mA/cm^2).[254] Decreasing temperatures from 32°C to 20°C did not have any effect on the anode potential within the tested current density range of 0.11–0.36 mA/cm.^{2} Decreasing temperature in MFCs also contributes to longer lag phases of growth in the anode chamber, thus delaying start-up times for the MFCs.[291] Lastly, the CE decreases from 43% to 8% when lowering the temperature from 30°C to 20°C.[291]

Mesophilic and thermophilic are two basic types of anaerobic digestion and each function at a specific range of temperatures of 30–38°C and 49–57°C, respectively. For the majority of anaerobic digestion processes, activity is gradually hindered at temperatures falling below 30°C and above 39°C. However, some MFCs can also be operated under thermophilic conditions of 50°C, which was first demonstrated in a SCMFC treating artificial wastewater while concurrently generating power.[292]

4.3.4 Chemical strategies

4.3.4.1 *The right medium for enhanced performance*

Electron donor substrates need to be considered in the anode chamber and appropriate electron acceptors for the cathode chamber in order to optimize MFC performance. Inorganic electron acceptors include oxygen, nitrates, metallic ions and sulfates.[293] Feeding of substrate to the MFC microbial community can either be done through fed-batch mode; where substrate is replenished at certain repeated intervals during culture growth, or continuous mode; in which substrate is continually flowing through the system. Fed-batch mode is mostly employed in the laboratory setting[166] whereas continuous mode is the preferred choice for *in situ* MFCs.[294]

4.3.4.1.1 *Choosing the right anode substrates*

Many different substrates have been used in MFCs and play a major role in determining power output.[295] This is due to the selective nature these substrates play in enriching particular populations of microorganisms as each microorganism has a specific set of organic and inorganic compounds it can metabolize. Therefore, the type of substrate fed to the microbial community in the MFC impacts, which microorganisms dominate the population.[296] Anodes fed with formate tend to enrich for *Acetobacterium* sp., *Geobacter* sp., and *Arcobacter* sp.[297] However, when fed with glucose, *Enterobacter* sp. are enriched.[298] Since substrates play a selective role in determining microbial community composition, they also affect the maximum power output attained by the MFC. For example, brewery wastewater SCMFCs fed with glucose, butyrate, propionate, acetate, or a mixture of the four, each resulted in different maximum power outputs and each had a distinct microbial composition.[299] The substrate provided affects the power output since it provides a source of carbon and energy to the microorganisms that consume it. The more nutrient-rich and accessible the substrate, the more energy the microorganism acquires from it and hence the amount of energy that is transferred to the anode is greater. Therefore, the chemical composition and concentration of substrate are important when optimizing microbial metabolic processes in MFCs.[300]

The choice of substrate also depends on the desired function of the MES, be it power production, bioremediation, chemical synthesis, or biofuel production. If the desired function is production of power, then acetate can be used,[301] which is a preferred electron donor and carbon source for electricigens such as *Desulfuromonas* and *Geobacter* spp.[302–304] However, methanogens and sulfidogens are also adept at acetate oxidation in anaerobic conditions.[305] When the desired role for the MFC is bioremediation of contaminated environments, electrodes are placed into the contaminated environment and the microorganisms that predominate are adept at consuming these contaminants. Phenol and indole, which are toxic by-products of the coking industry, can be utilized as substrates for electrical current generation in an MFC.[298,306]

4.3.4.1.2 *Catholyte solution selection*

Various electron acceptors can be added to the cathode chamber in non-air cathode-type MFCs. Depending on the reduction potential of an electron acceptor used, the amount of power generated can vary. Oxygen is a strong oxidizing agent and results in high power outputs when used as a final electron acceptor. Although hexacynoferrate can also be used as an electron acceptor, oxygen is preferred due to its sustainability. However, when permanganate is used as the cathodic electron acceptor, it can produce up to 11.3-fold higher maximum power density than oxygen.[307]

Although the cathode serves as an electron donor to the microbial biocathode community, cathode substrates must also provide a carbon source to these microorganisms. The choice of carbon source for the cathode is important as it can select for microorganisms that predominate in the cathode biofilm.[270] Since the goal of MECs is to produce biofuels such as hydrogen or methane, it is important to promote the growth of methanogens or hydrogen-generating microorganisms and inhibit the growth of hydrogen or methane-consuming microorganisms. By enriching the cathode communities from pond sediment and wastewater reactors with bicarbonate, the presence of acetogens and hydrogenotrophic methanogens declines.[308]

4.3.4.2 *Maintenance of optimal pH*

Various factors must be considered when trying to understand the effects of pH on MES performance. Protons are generated through oxidation of organic compounds via the electrogenic biofilm community on the anode. These protons must then traverse from the anode surface, through the anode biofilm, toward the PEM and finally into the cathode chamber. Current densities are dependent on substrate diffusion and utilization in the anode biofilm, the flow of electrons to the anode and the diffusion and migration of protons out of the anode biofilm.[127,130,309] The accumulation of protons in the biofilm is known to cause pH gradients within the biofilm. The limiting effect of pH on microbial activity has also been shown in *G. sulfurreducens* anode biofilms. The pH within electrogenic biofilms varies throughout, with the lowest pH levels found within the center.[310] This lowered pH of 6 as compared to 7 near the outer surfaces of the biofilm is also correlated to lower metabolic levels of *G. sulfurreducens.* The strong pH gradient that exists within electrogenic biofilms on anode surfaces has a detrimental effect on the bioelectrochemical performance of the MFC.[309] To enhance proton efflux out of the anode biofilm and thus improve current production, a buffering medium can be implemented.[309]

Most well-studied electricigens have been isolated from anaerobic environments, like ocean sediments, where they are able to undergo dissimilatory Fe(III) or Mn(IV) reduction.[311] Their ability to transfer electrons extracellularly to these insoluble electron acceptors makes them good candidates for EET to electrodes as electron acceptors.[30] As such, they have mainly been isolated from neutrophilic environments and most known electricigens are thus restricted to a pH range that is neutral.[312] Indeed when treating pH-neutral wastewater with an electrogenic microbial biofilm, the microbial activity of the biofilm was only functional at a narrow pH range of 6–9.[313] Using flow-cytometry to examine the microbial community formed at the anode of this wastewater inoculum, it was found to fluctuate with the changing pH of the water. The changing microbial composition correlated with changes in electrogenic biofilm activity and thus current densities and CEs were also dependent on pH. This is because the redox potentials of electron transfer proteins depended on the solution pH. The highest current densities and CEs occurred at pH 7, which resembled the original wastewater inoculum pH of 6.7. Therefore, it is important to maintain pH values within the MES that are closest to the environmental pH from which the inocula originated. The activity of the anode biofilm showed an adaptive reversibility of function in response to pH. For instance, when the pH was 7, the current density was highest, which plummeted down to nearly 0 μA cm^2 when the pH was switched to 5 and quickly (less than 24 h) reverted to a high value when the pH was set back to 7.[313] Such dependent response of current density to pH was also observed in other studies.[290,314]

It would be beneficial to have acid-tolerant microorganisms in the center of the electrogenic biofilms to mitigate any losses from EET to the anode. The first acidophile to be utilized as a biocatalyst in MFCs was *Acidiphilium cryptum*.[312] This microorganism could function anaerobically at a pH as low as 3.5. Another study found an alkaliphilic electricigen, *Corynebacterium* sp. strain MFC03.[315] It is valuable to classify electricigens with greater pH tolerance ranges so that they may be employed in MFCs with wide pH ranges.

4.3.4.3 *Solution conductivity*

Conductivity in the electrode solution of MFCs impacts the ion transport rates within the anode or cathode chambers. Increasing conductivity in the anode chamber solution enhances the transfer of electron donors to electricigens and thus improves electrical current production.[261] Addition of sodium chloride to anolyte reduces internal resistance and thus overall enhances maximum power output.[316] Similarly, addition of sodium chloride to both anolyte and catholyte solutions also showed a decline in the internal resistance of the MFC and hence an improvement in power density was attained.[254] Furthermore, increasing the ionic strength from 100 to 400 mM increased power from 720 to 1330 mW/m^2 as a result of reduced internal resistance.[254] Electrolyte concentration in the solution can also be a limiting factor in power output. For instance, when the phosphate buffer concentration was decreased from 200 to 50 mM, the internal resistance in an MFC increased from 47.3% to 78.2%.[257]

4.3.5 Summary

Through continued multi-disciplinary efforts to understand MESs, their performance in a multitude of applications can be optimized. In the discipline of biology, an understanding of how mixed communities of microorganisms can be controlled effectively to catalyze redox reactions and enhance EET to electrodes is still needed. Although research has elucidated many of the EET components, there are still additional components to be discovered and in a greater number of potential electricigens. The alternative EET pathways utilized to reduce specific electron acceptors are still not well characterized. Additionally, the syntrophic and competitive interactions that exist between microorganisms in these systems warrant further study. This would enable construction of an interactome map that would help predict how different microorganisms are contributing or limiting MES performance. On the physical front, many engineering efforts have helped establish which MES architectures as well as electrode and PEM materials perform optimally. However, further insight into how to make these systems viable for both nano and large-scale applications is still required. Lastly, a deeper understanding of the chemical barriers existing in these systems and how substrates, pH and conductivity can be controlled is necessary for the future commercialization of MESs.

List of Abbreviations

2CMFC	two-chamber microbial fuel cell
AEM	anion exchange membrane
AFM	atomic force microscopy
AMSC	algal microbial solar cell
BEAMR	bioelectrochemically assisted microbial reactor

BES	bioelectrochemical system
BMSC	bacterial microbial solar cell
CE	coulombic efficiency
CEM	cation exchange membrane
COD	chemical oxygen demand
DEET	direct extracellular electron transfer
DIET	direct interspecies electron transfer
DMRB	dissimilatory metal-reducing bacteria
EAM	electrochemically active microorganism
EET	extracellular electron transfer
MDC	microbial desalination cell
ME	microbial electrosynthesis
MEC	microbial electrolysis cell
MET	mediator-associated extracellular electron transfer
MES	microbial electrochemical system
MFC	microbial fuel cell
MSC	microbial solar cell
OMC	outer membrane *c*-type cytochrome
PEET	pili-associated extracellular electron transfer
PEM	proton exchange membrane
PMS	power management system
PMSC	plant microbial solar cell
PMEC	photosynthetic microbial electrochemical cell
SCMFC	single-chamber microbial fuel cell
SEET	shuttle-mediated extracellular electron transfer
SMFC	sediment microbial fuel cell
SRB	sulfate-reducing bacteria
STC	small tetraheme cytochrome *c*
TPH	total petroleum hydrocarbon

References

1. Z. Du, H. Li and T. Gu (2007). A state of the art review on microbial fuel cells: A promising technology for wastewater treatment and bioenergy. *Biotechnology Advances* **25**(5), 464–482.
2. A.E. Franks and K.P. Nevin (2010). Microbial fuel cells, a current review. *Energies* **3**(5), 899–919.

3. H. Wang and Z.J. Ren (2013). A comprehensive review of microbial electrochemical systems as a platform technology. *Biotechnology Advances* **31**(8), 1796–1807.
4. D.R. Lovley (2012). Electromicrobiology. *Annual Review of Microbiology* **66**, 391–409.
5. B.E. Logan (2009). Exoelectrogenic bacteria that power microbial fuel cells. *Nature Reviews Microbiology* **7**(5), 375–381.
6. D.R. Lovley (2006). Bug juice: Harvesting electricity with microorganisms. *Nature Reviews Microbiology* **4**, 497–508.
7. D.R. Lovley (2011). Powering microbes with electricity: direct electron transfer from electrodes to microbes. *Environmental Microbiology Reports* **3**(1), 27–35.
8. R.A. Rozendal, H.V. Hamelers, K. Rabaey, J. Keller and C.J. Buisman (2008). Towards practical implementation of bioelectrochemical wastewater treatment. *Trends in Biotechnology* **26**(8), 450–459.
9. U. Schröder, F. Harnisch and L.T. Angenent (2015). Microbial electrochemistry and technology: terminology and classification. *Energy and Environmental Science* **8**(2), 513–519.
10. G. Tayhas, G.T.R. Palmore and G.M. Whitesides (1994). Microbial and enzymatic biofuel cells. in M.E. Himmel, J.O. Baker and R.P. Overend (Eds.), *Enzymatic Conversion of Biomass for Fuels Production: Developed from a Symposium at the 205th National Meeting of the American Chemical Society*, Denver, Colorado, March 28–April 2, 1993, 566, Washington: American Chemical Society, pp. 271–290.
11. L.M. Tender, C.E. Reimers, H.A. Stecher, D.E. Holmes, D.R. Bond, D.A. Lowy, K Pilobello, S.J. Fertig and D.R. Lovley (2002). Harnessing microbially generated power on the seafloor. *Nature Biotechnology* **20**, 821–825.
12. B.E. Logan, B. Hamelers, R.A. Rozendal, U. Schröder, J. Keller, S. Freguia, P. Aelterman, W. Verstraete and K. Rabaey (2006). Microbial fuel cells: Methodology and technology. *Environmental Science and Technology* **40**(17), 5181–5192.
13. W.O. Khunjar, A. Sahin, A.C. West, K. Chandran and S. Banta (2012). Biomass production from electricity using ammonia as an electron carrier in a reverse microbial fuel cell. *Plos One* **7**(9), e44846.
14. K.P. Nevin, T.L. Woodard, A.E. Franks, Z.M. Summers and D.R. Lovley (2010). Microbial electrosynthesis: Feeding microbes electricity to convert carbon dioxide and water to multicarbon extracellular organic compounds. *mBio* **1**(2), e00103–10.
15. K. Rabaey and R.A. Rozendal (2010). Microbial electrosynthesis-revisiting the electrical route for microbial production. *Nature Reviews Microbiology* **8**(10), 706–716.
16. R. Kumar, L. Singh and A.W. Zularisam (2016). Exoelectrogens: Recent advances in molecular drivers involved in extracellular electron transfer and strategies used to improve it for microbial fuel cell applications. *Renewable and Sustainable Energy Reviews* **56**, 1322–1336.

17. L. Semenec and A.E. Franks (2015). Delving through electrogenic biofilms: From anodes to cathodes to microbes. *AIMS Bioengineering* **2**(3), 222–248.
18. Z. He (2012). Microbial fuel cells: Now let us talk about energy. *Environmental Science and Technology* **47**(1), 332–333.
19. L. Xu, Y. Zhao, L. Doherty, Y. Hu and X. Hao (2016). The integrated processes for wastewater treatment based on the principle of microbial fuel cells: A review. *Critical Reviews in Environmental Science and Technology* **46**(1), 60–91.
20. D.R. Bond (2010). Electrodes as electron acceptors, and the bacteria who love them. In A. Loy, M. Mandl and L. Barton (eds.), *Geomicrobiology: Molecular and Environmental Perspective*, Netherlands: Springer, pp. 385–399.
21. C.M. Cordas, L.T. Guerra, C. Xavier and J.J. Moura (2008). Electroactive biofilms of sulphate reducing bacteria. *Electrochimica Acta* **54**(1), 29–34.
22. A. Janicek, Y. Fan and H. Liu (2014). Design of microbial fuel cells for practical application: A review and analysis of scale-up studies. *Biofuels* **5**(1), 79–92.
23. Y. Fan, H. Hu and H. Liu (2007). Enhanced Coulombic efficiency and power density of air-cathode microbial fuel cells with an improved cell configuration. *Journal of Power Sources* **171**(2), 348–354.
24. H. Liu and B.E. Logan (2004). Electricity generation using an air-cathode single chamber microbial fuel cell in the presence and absence of a proton exchange membrane. *Environmental Science and Technology* **38**(14), 4040–4046.
25. M. Rahimnejad, A. Ghoreyshi, G. Najafpour, H. Younesi and M. Shakeri (2012). A novel microbial fuel cell stack for continuous production of clean energy. *International Journal of Hydrogen Energy* **37**(7), 5992–6000.
26. P. Aelterman, K. Rabaey, H.T. Pham, N. Boon and W. Verstraete (2006). Continuous electricity generation at high voltages and currents using stacked microbial fuel cells. *Environmental Science and Technology* **40**(10), 3388–3394.
27. S.-E. Oh and B.E. Logan (2007). Voltage reversal during microbial fuel cell stack operation. *Journal of Power Sources* **167**(1), 11–17.
28. S. Choi and J. Chae (2012). An array of microliter-sized microbial fuel cells generating 100 μW of power. *Sensors and Actuators A Physical* **177**, 10–15.
29. G. Papaharalabos, J. Greenman, C. Melhuish and I. Ieropoulos (2015). A novel small scale microbial fuel cell design for increased electricity generation and waste water treatment. *International Journal of Hydrogen Energy* **40**(11), 4263–4268.
30. D.R. Lovley (2006). Microbial fuel cells: novel microbial physiologies and engineering approaches. *Current Opinion in Biotechnology* **17**(3), 327–332.
31. C.E. Reimers, L.M. Tender, S. Fertig and W. Wang (2001). Harvesting energy from the marine sediment-water interface. *Environmental Science and Technology* **35**(1), 192–195.

32. C. Reimers, P. Girguis, H. Stecher, L. Tender, N. Ryckelynck and P. Whaling (2006). Microbial fuel cell energy from an ocean cold seep. *Geobiology* **4**(2), 123–136.
33. Y. Gong, S.E. Radachowsky, M. Wolf, M.E. Nielsen, P.R. Girguis and C.E. Reimers (2011). Benthic microbial fuel cell as direct power source for an acoustic modem and seawater oxygen/temperature sensor system. *Environmental Science and Technology* **45**(11), 5047–5053.
34. N. Ryckelynck, H.A. Stecher III and C.E. Reimers (2005). Understanding the anodic mechanism of a seafloor fuel cell: Interactions between geochemistry and microbial activity. *Biogeochemistry* **76**(1), 113–139.
35. F. Rezaei, T.L. Richard, R.A. Brennan and B.E. Logan (2007). Substrate-enhanced microbial fuel cells for improved remote power generation from sediment-based systems. *Environmental Science and Technology* **41**(11), 4053–4058.
36. C. Donovan, A. Dewan, D. Heo and H. Beyenal (2008). Batteryless, wireless sensor powered by a sediment microbial fuel cell. *Environmental Science and Technology* **42**(22), 8591–8596.
37. T. Ewing, P.T. Ha, J.T. Babauta, N.T. Tang, D. Heo and H. Beyenal (2014). Scale-up of sediment microbial fuel cells. *Journal of Power Sources* **272**, 311–319.
38. K.P. Nevin, S.A. Hensley, A.E. Franks, Z.M. Summers, J. Ou, T.L. Woodard, O.L. Snoeyenbos-West and D.R. Lovley (2011). Electrosynthesis of organic compounds from carbon dioxide is catalyzed by a diversity of acetogenic microorganisms. *Applied Environmental Microbiology* **77**(9), 2882–2886.
39. M. Villano, G. Monaco, F. Aulenta and M. Majone (2011). Electrochemically assisted methane production in a biofilm reactor. *Journal of Power Sources* **196**(22), 9467–9472.
40. M. Zhou, H. Wang, D.J. Hassett and T. Gu (2013). Recent advances in microbial fuel cells (MFCs) and microbial electrolysis cells (MECs) for wastewater treatment, bioenergy and bioproducts. *Journal of Chemical Technology and Biotechnology* **88**(4), 508–518.
41. K.B. Gregory, D.R. Bond and D.R. Lovley (2004). Graphite electrodes as electron donors for anaerobic respiration. *Environmental Microbiology* **6**(6), 596–604.
42. M.A. Rosenbaum and A.W. Henrich (2014). Engineering microbial electrocatalysis for chemical and fuel production. *Current Opinion in Biotechnology* **29**, 93–98.
43. L. Semenec and A.E. Franks (2014). The microbiology of microbial electrolysis cells. *Microbiology Australia* **35**(4), 201–206.
44. T. Zhang, H. Nie, T.S. Bain, H. Lu, M. Cui, O.L. Snoeyenbos-West, A.E. Franks, K.P. Nevin, T.P. Russell and D.R. Lovley (2013). Improved cathode materials for microbial electrosynthesis. *Energy and Environmental Science* **6**(1), 217–224.
45. J. Wei, P. Liang and X. Huang (2011). Recent progress in electrodes for microbial fuel cells. *Bioresource Technology* **102**(20), 9335–9344.

46. L. Soussan, J. Riess, B. Erable, M.-L. Délia and A. Bergel (2013). Electrochemical reduction of CO_2 catalysed by *Geobacter sulfurreducens* grown on polarized stainless steel cathodes. *Electrochemistry Communications* **28**, 27–30.
47. C.G. Giddings, K.P. Nevin, T. Woodward, D.R. Lovley and C.S. Butler (2015). Simplifying microbial electrosynthesis reactor design. *Frontiers in Microbiology* **6**(468), 1–6.
48. Y. Kim and B.E. Logan (2013). Microbial desalination cells for energy production and desalination. *Desalination* **308**, 122–130.
49. X. Cao, X. Huang, P. Liang, K. Xiao, Y. Zhou, X. Zhang and B.E. Logan (2009). A new method for water desalination using microbial desalination cells. *Environmental Science and Technology* **43**(18), 7148–7152.
50. D.P. Strik, R.A. Timmers, M. Helder, K.J. Steinbusch, H.V. Hamelers and C.J. Buisman (2011). Microbial solar cells: Applying photosynthetic and electrochemically active organisms. *Trends in Biotechnology* **29**(1), 41–49.
51. G.C. Dismukes, D. Carrieri, N. Bennette, G.M. Ananyev and M.C. Posewitz (2008). Aquatic phototrophs: Efficient alternatives to land-based crops for biofuels. *Current Opinion in Biotechnology* **19**(3), 235–240.
52. L.D. Schamphelaire, L.V.D. Bossche, H.S. Dang, M. Höfte, N. Boon, K. Rabaey and W. Verstraete (2008). Microbial fuel cells generating electricity from rhizodeposits of rice plants. *Environmental Science and Technology* **42**(8), 3053–3058.
53. K. Wetser, E. Sudirjo, C.J. Buisman and D.P. Strik (2015). Electricity generation by a plant microbial fuel cell with an integrated oxygen reducing biocathode. *Applied Energy* **137**, 151–157.
54. M. Moqsud, J. Yoshitake, Q. Bushra, M. Hyodo, K. Omine and D. Strik (2015). Compost in plant microbial fuel cell for bioelectricity generation. *Waste Management* **36**, 63–69.
55. L. Lu, D. Xing and Z.J. Ren (2015). Microbial community structure accompanied with electricity production in a constructed wetland plant microbial fuel cell. *Bioresource Technology* **195**, 115–121.
56. C.W. Marshall and H.D. May (2009). Electrochemical evidence of direct electrode reduction by a thermophilic Gram-positive bacterium, *Thermincola ferriacetica*. *Energy and Environmental Science* **2**(6), 699–705.
57. Y. Zou, J. Pisciotta, R.B. Billmyre and I.V. Baskakov (2009). Photosynthetic microbial fuel cells with positive light response. *Biotechnology and Bioengineering* **104**(5), 939–946.
58. S.B. Velasquez-Orta, T.P. Curtis and B.E. Logan (2009). Energy from algae using microbial fuel cells. *Biotechnology and Bioengineering* **103**(6), 1068–1076.
59. M. Rosenbaum, U. Schröder and F. Scholz (2005). Utilizing the green alga *Chlamydomonas reinhardtii* for microbial electricity generation: A living solar cell. *Applied Microbiology and Biotechnology* **68**(6), 753–756.
60. H. Gaffron and J. Rubin (1942). Fermentative and photochemical production of hydrogen in algae. *Journal of General Physiology* **26**(2), 219–240.

61. D.P. Strik, H. Terlouw, H.V. Hamelers and C.J. Buisman (2008). Renewable sustainable biocatalyzed electricity production in a photosynthetic algal microbial fuel cell (PAMFC). *Applied Microbiology and Biotechnology* **81**(4), 659–668.
62. G.E. Brown, V.E. Henrich, W.H. Casey, D.L. Clark, C. Eggleston, A. Felmy, D.W. Goodman, M. Grätzel, G. Maciel and M.I. McCarthy (1999). Metal oxide surfaces and their interactions with aqueous solutions and microbial organisms. *Chemical Reviews* **99**(1), 77–174.
63. A.J.M. Stams, F.A.M. De Bok, C.M. Plugge, V. Eekert, H.A. Miriam, J. Dolfing, G. Schraa (2006). Exocellular electron transfer in anaerobic microbial communities. *Applied and Environmental Microbiology* **8**(3), 371–382.
64. B.A. Haddock and C.W. Jones (1977). Bacterial respiration. *Bacteriological Reviews* **41**(1), 47–99.
65. D.R. Lovley (2013). Dissimilatory Fe(III)- and Mn(IV)-reducing prokaryotes. In E. Rosenberg, E.F. DeLong, S. Lory, E. Stackebrandt and F. Thompson (eds.), *The Prokaryotes: Prokaryotic Physiology and Biochemistry*, Berlin Heidelberg: Springer, pp. 287–308.
66. R. Kumar, L. Singh, Z.A. Wahid and M.F.M. Din (2015). Exoelectrogens in microbial fuel cells toward bioelectricity generation: a review. *International Journal of Energy Research* **39**(8), 1048–1067.
67. A. Sydow, T. Krieg, F. Mayer, J. Schrader and D. Holtmann (2014). Electroactive bacteria-molecular mechanisms and genetic tools. *Applied Microbiology and Biotechnology* **98**(20), 8481–8495.
68. Y. Yang, M. Xu, J. Guo and G. Sun (2012). Bacterial extracellular electron transfer in bioelectrochemical systems. *Process Biochemistry* **47**(12), 1707–1714.
69. U. Schröder (2007). Anodic electron transfer mechanisms in microbial fuel cells and their energy efficiency. *Physical Chemistry Chemical Physics* **9**(21), 2619–2629.
70. G. Reguera, K.D. McCarthy, T. Mehta, J.S. Nicoll, M.T. Tuominen and D.R. Lovley (2005). Extracellular electron transfer via microbial nanowires. *Nature* **435**, 1098–1101.
71. H. Richter, K.P. Nevin, H.F. Jia, D.A. Lowy, D.R. Lovley and L.M. Tender (2009). Cyclic voltammetry of biofilms of wild type and mutant *Geobacter sulfurreducens* on fuel cell anodes indicates possible roles of OmcB, OmcZ, type IV pili, and protons in extracellular electron transfer. *Energy and Environmental Science* **2**(5), 506–516.
72. E. Marsili, D.B. Baron, I.D. Shikhare, D. Coursolle, J.A. Gralnick and D.R. Bond (2008). *Shewanella* secretes flavins that mediate extracellular electron transfer. *Proceedings of the National Academy of Sciences of the United States of America* **105**(10), 3968–3973.
73. A.P. Borole, G. Reguera, B. Ringeisen, Z.-W. Wang, Y. Feng and B.H. Kim (2011). Electroactive biofilms: Current status and future research needs. *Energy and Environmental Science* **4**(12), 4813–4834.

74. S. Ishii, K. Watanabe, S. Yabuki, B.E. Logan and Y Sekiguchi (2008). Comparison of electrode reduction activities of *Geobacter sulfurreducens* and an enriched consortium in an air-cathode microbial fuel cell. *Applied and Environmental Microbiology* **74**, 7348–7355.
75. K.P. Nevin, H. Richter, S.F. Covalla, J.P. Johnson, T.L. Woodard, A.L. Orloff, H. Jia, M. Zhang and D.R. Lovley (2008). Power output and columbic efficiencies from biofilms of *Geobacter sulfurreducens* comparable to mixed community microbial fuel cells. *Environmental Microbiology* **10**, 2505–2514.
76. K. Rabaey, N. Boon, S.D. Siciliano, M. Verhaege and W. Verstraete (2004). Biofuel cells select for microbial consortia that self-mediate electron transfer. *Applied and Environmental Microbiology* **70**, 5373–5382.
77. L. Shi, D.J. Richardson, Z. Wang, S.N. Kerisit, K.M. Rosso, J.M. Zachara and J.K. Fredrickson (2009). The roles of outer membrane cytochromes of *Shewanella* and *Geobacter* in extracellular electron transfer. *Environmental Microbiology Reports* **1**(4), 220–227.
78. K.C. Wrighton, J.C. Thrash, R.A. Melnyk, J.P. Bigi, K.G. Byrne-Bailey, J.P. Remis, D. Schichnes, M. Auer, C.J. Chang and J.D. Coates (2011). Evidence for direct electron transfer by a Gram-positive bacterium isolated from a microbial fuel cell. *Applied and Environmental Microbiology* **77**(21), 7633–7639.
79. D.C. Bhowmick, B. Bal, N.S. Chatterjee, A.N. Ghosh and S. Pal (2009). A low-GC Gram-positive *Thermoanaerobacter*-like bacterium isolated from an Indian hot spring contains Cr(VI) reduction activity both in the membrane and cytoplasm. *Journal of Applied Microbiology* **106**(6), 2006–2016.
80. D.R. Lovley (2011). Live wires: Direct extracellular electron exchange for bioenergy and the bioremediation of energy-related contamination. *Energy and Environmental Science* **4**(12), 4896–4906.
81. M. Schütz, M. Brugna, E. Lebrun, F. Baymann, R. Huber, K.-O. Stetter, G. Hauska, R. Toci, D. Lemesle-Meunier and P. Tron (2000). Early evolution of cytochrome bc complexes. *Journal of Molecular Biology* **300**(4), 663–675.
82. C. Leang, L.A. Adams, K.-J. Chin, K.P. Nevin, B.A. Methe, J. Webster M.L. Sharma and D.R. Lovley (2005). Adaptation to disruption of the electron transfer pathway for Fe(III) reduction in *Geobacter sulfurreducens. Journal of Bacteriology* **187**(17), 5918–5926.
83. H. Yi, K.P. Nevin, B.-C. Kim, A.E. Franks, A. Klimes, L.M. Tender and D.R. Lovley (2009). Selection of a variant of *Geobacter sulfurreducens* with enhanced capacity for current production in microbial fuel cells. *Biosensors and Bioelectronics* **24**(12), 3498–3503.
84. B.A. Methe, K.E. Nelson, J.A. Eisen, I.T. Paulsen, W. Nelson, J.F. Heidelberg, D. Wu, M. Wu, N. Ward and M.J. Beanan (2003). Genome of *Geobacter sulfurreducens*: Metal reduction in subsurface environments. *Science* **302**(5652), 1967–1969.
85. T.C. Santos, M.A. Silva, L. Morgado, J.M. Dantas and C.A. Salgueiro (2015). Diving into the redox properties of *Geobacter sulfurreducens*

cytochromes: A model for extracellular electron transfer. *Dalton Transactions* **44**(20), 9335–9344.

86. T.S. Magnuson, A.L. Hodges-Myerson and D.R. Lovley (2000). Characterization of a membrane-bound NADH-dependent Fe^{3+} reductase from the dissimilatory Fe^{3+}-reducing bacterium *Geobacter sulfurreducens. FEMS Microbiology Letters* **185**(2), 205–211.
87. D.E. Holmes, S.K. Chaudhuri, K.P. Nevin, T. Mehta, BA Methé, A. Liu, JE Ward, T.L. Woodard, J. Webster and D.R. Lovley (2006). Microarray and genetic analysis of electron transfer to electrodes in *Geobacter sulfurreducens. Environmental Microbiology* **8**(10), 1805–1815.
88. C.E. Levar, C.H. Chan, M.G. Mehta-Kolte and D.R. Bond (2014). An inner membrane cytochrome required only for reduction of high redox potential extracellular electron acceptors. *mBio* **5**(6), e02034–14.
89. L. Zacharoff, C.H. Chan and D.R. Bond (2016). Reduction of low potential electron acceptors requires the CbcL inner membrane cytochrome of *Geobacter sulfurreducens. Bioelectrochemistry* **107**, 7–13.
90. K.P. Nevin, B.-C. Kim, R.H. Glaven, J.P. Johnson, T.L. Woodard B.A. Methé, R.J. DiDonato Jr., S.F. Covalla, A.E. Franks and A. Liu (2009). Anode biofilm transcriptomics reveals outer surface components essential for high density current production in *Geobacter sulfurreducens* fuel cells. *Plos One*, **4**(5), e5628.
91. K. Inoue, X. Qian, L. Morgado, B.-C. Kim, T. Mester, M. Izallalen, C.A. Salgueiro and D.R. Lovley (2010). Purification and characterization of OmcZ, an outer-surface, octaheme *c*-type cytochrome essential for optimal current production by *Geobacter sulfurreducens. Applied and Environmental Microbiology* **76**(12), 3999–4007.
92. S.M. Strycharz, R.H. Glaven, M.V. Coppi, S.M. Gannon, L.A. Perpetua, A. Liu, K.P. Nevin and D.R. Lovley (2011). Gene expression and deletion analysis of mechanisms for electron transfer from electrodes to *Geobacter sulfurreducens.* Bioelectrochemistry **80**(2), 142–150.
93. A.E. Franks, R.H. Glaven and D.R. Lovley (2012). Real-time spatial gene expression analysis within current-producing biofilms. *ChemSusChem* **5**(6), 1092–1098.
94. K. Inoue, C. Leang, A.E. Franks, T.L. Woodard, K.P. Nevin and D.R. Lovley (2011). Specific localization of the *c*-type cytochrome OmcZ at the anode surface in current-producing biofilms of *Geobacter sulfurreducens. Environmental Microbiology Reports* **3**(2), 211–217.
95. J.W. Voordeckers, B.-C. Kim, M. Izallalen and D.R. Lovley (2010). Role of *Geobacter sulfurreducens* outer surface *c*-type cytochromes in reduction of soil humic acid and anthraquinone-2, 6-disulfonate. *Applied and Environmental Microbiology* **76**(7), 2371–2375.
96. H. Richter, M. Lanthier, K.P. Nevin and D.R. Lovley (2007). Lack of electricity production by *Pelobacter carbinolicus* indicates that the capacity for Fe (III) oxide reduction does not necessarily confer electron transfer ability to fuel cell anodes. *Applied and Environmental Microbiology* **73**(16), 5347–5353.

97. B.C. Kim, B.L. Postier, R.J. Didonato, S.K. Chaudhuri, K.R. Nevin and D.R. Lovley (2008). Insights into genes involved in electricity generation in *Geobacter sulfurreducens* via whole genome microarray analysis of the OmcF-deficient mutant. *Bioelectrochemistry* **73**, 70–75.
98. C. Leang, M.V. Coppi and D.R. Lovley (2003). OmcB, a *c*-type polyheme cytochrome, involved in Fe(III) reduction in *Geobacter sulfurreducens. Journal of Bacteriology* **185**(7), 2096–2103.
99. C.S. Stephen, E.V. LaBelle, S.L. Brantley and D.R. Bond (2014). Abundance of the multiheme *c*-type cytochrome OmcB increases in outer biofilm layers of electrode-grown *Geobacter sulfurreducens. Plos One* **9**(8), e104336.
100. C. Leang, X. Qian, T. Mester and D.R. Lovley (2010). Alignment of the *c*-type cytochrome OmcS along pili of *Geobacter sulfurreducens. Applied and Environmental Microbiology* **76**(12), 4080–4084.
101. X. Qian, T. Mester, L. Morgado, T. Arakawa, M.L. Sharma, K. Inoue, C. Joseph, C.A. Salgueiro, M.J. Maroney and D.R. Lovley (2011). Biochemical characterization of purified OmcS, a *c*-type cytochrome required for insoluble Fe(III) reduction in *Geobacter sulfurreducens. Biochimica et Biophysica Acta* **1807**(4), 404–412.
102. M. Vargas, N.S. Malvankar, P.-L. Tremblay, C. Leang, J.A. Smith, P. Patel, O. Synoeyenbos-West, K.P. Nevin and D.R. Lovley (2013). Aromatic amino acids required for pili conductivity and long-range extracellular electron transport in *Geobacter sulfurreducens. mBio* **4**(2), e00105–13.
103. X. Liu, P.-L. Tremblay, N.S. Malvankar, K.P. Nevin, D.R. Lovley and M. Vargas (2014). A *Geobacter sulfurreducens* strain expressing *Pseudomonas aeruginosa* type IV pili localizes OmcS on pili but is deficient in Fe(III) oxide reduction and current production. *Applied and Environmental Microbiology* **80**(3), 1219–1224.
104. N.S. Malvankar, M.T. Tuominen and D.R. Lovley (2012). Lack of cytochrome involvement in long-range electron transport through conductive biofilms and nanowires of *Geobacter sulfurreducens. Energy and Environmental Science* **5**(9), 8651–8659.
105. T. Mehta, M.V. Coppi, S.E. Childers and D.R. Lovley (2005). Outer membrane *c*-type cytochromes required for Fe(III) and Mn(IV) oxide reduction in *Geobacter sulfurreducens. Applied and Environmental Microbiology* **71**(12), 8634–8641.
106. L. Shi, T.C. Squier, J.M. Zachara and J.K. Fredrickson (2007). Respiration of metal (hydr)oxides by *Shewanella* and *Geobacter*: A key role for multihaem *c*-type cytochromes. *Molecular Microbiology* **65**(1), 12–20.
107. B.-C. Kim, X. Qian, C. Leang, M.V. Coppi and D.R. Lovley (2006). Two putative *c*-type multiheme cytochromes required for the expression of OmcB, an outer membrane protein essential for optimal Fe(III) reduction in *Geobacter sulfurreducens. Journal of Bacteriology* **188**(8), 3138–3142.
108. D.R. Lovley, E.J. Phillips, D.J. Lonergan and P.K. Widman (1995). Fe(III) and S^0 reduction by *Pelobacter carbinolicus. Applied and Environmental Microbiology* **61**(6), 2132–2138.

109. Y. Xiong, L. Shi, B. Chen, M.U. Mayer, B.H. Lower, Y. Londer, S. Bose, M.F. Hochella, J.K. Fredrickson and T.C. Squier (2006). High-affinity binding and direct electron transfer to solid metals by the *Shewanella oneidensis* MR-1 outer membrane *c*-type cytochrome OmcA. *Journal of the American Chemical Society* **128**(43), 13978–13979.
110. J.F. Heidelberg, I.T. Paulsen, K.E. Nelson, E.J. Gaidos, W.C. Nelson, T.D. Read, J.A. Eisen, R. Seshadri, N. Ward and B. Methe (2002). Genome sequence of the dissimilatory metal ion–reducing bacterium *Shewanella oneidensis. Nature Biotechnology* **20**(11), 1118–1123.
111. T.E. Meyer, A.I. Tsapin, I. Vandenberghe, L, De Smet, D. Frishman, K.H. Nealson, M.A. Cusanovich and J.J. Van Beeumen (2004). Identification of 42 possible cytochrome *c* genes in the *Shewanella oneidensis* genome and characterization of six soluble cytochromes. *Omics: A Journal of Integrative Biology* **8**(1), 57–77.
112. J.M. Myers and C.R. Myers (2000). Role of the tetraheme cytochrome CymA in anaerobic electron transport in cells of *Shewanella putrefaciens* MR-1 with normal levels of menaquinone. *Journal of Bacteriology* **182**(1), 67–75.
113. K.C. Wrighton, P. Agbo, F. Warnecke, K.A. Weber, E.L. Brodie, T.Z. DeSantis, P. Hugenholtz, G.L. Andersen and J.D. Coates (2008). A novel ecological role of the Firmicutes identified in thermophilic microbial fuel cells. *The ISME Journal* **2**(11), 1146–1156.
114. O. Choi, T. Kim, H.M. Woo and Y. Um (2014). Electricity-driven metabolic shift through direct electron uptake by electroactive heterotroph *Clostridium pasteurianum. Scientific Reports* **4**, 1–10.
115. H.K. Carlson, A.T. Iavarone, A. Gorur, B.S. Yeo, R. Tran, R.A. Melnyk, R.A. Mathies, M. Auer and J.D. Coates (2012). Surface multiheme *c*-type cytochromes from *Thermincola potens* and implications for respiratory metal reduction by Gram-positive bacteria. *Proceedings of the National Academy of Sciences of the United States of America* **109**(5), 1702–1707.
116. K.G. Byrne-Bailey, K.C. Wrighton, R.A. Melnyk, P. Agbo, T.C. Hazen and J.D. Coates (2010). Complete genome sequence of the electricity-producing "*Thermincola potens*" strain JR. *Journal of Bacteriology* **192**(15), 4078–4079.
117. A.A. Carmona-Martínez, M. Pierra, E. Trably and N. Bernet (2013). High current density via direct electron transfer by the halophilic anode respiring bacterium *Geoalkalibacter subterraneus. Physical Chemistry Chemical Physics* **15**(45), 19699–19707.
118. D.E. Holmes, D.R. Bond and D.R. Lovley (2004). Electron transfer by *Desulfobulbus propionicus* to Fe (III) and graphite electrodes. *Applied and Environmental Microbiology* **70**(2), 1234–1237.
119. T. Yamanaka and Y. Fukumori (1995). Molecular aspects of the electron transfer system which participates in the oxidation of ferrous ion by *Thiobacillus ferrooxidans. FEMS Microbiology Letters* **17**(4), 401–413.
120. N.S. Malvankar, M.T. Tuominen and D.R. Lovley (2012). Comment on "On electrical conductivity of microbial nanowires and biofilms" by SM

Strycharz-Glaven, R.M. Snider, A. Guiseppi-Elie and LM Tender, Energy Environ. Sci., 2011, 4, 4366. *Energy and Environmental Science* **5**(3), 6247–6249.

121. S.M. Strycharz-Glaven and L.M. Tender (2012). Reply to the 'Comment on "On electrical conductivity of microbial nanowires and biofilms"' by NS Malvankar, MT Tuominen and DR Lovley, *Energy and Environmental Science* **5**(3), 6250–6255.
122. M.D. Yates, J.P. Golden, J. Roy, S.M. Strycharz-Glaven, S. Tsoi, J.S. Erickson, M.Y. El-Naggar, S.C. Barton and L.M. Tender (2015). Thermally activated long range electron transport in living biofilms. *Physical Chemistry Chemical Physics* **17**(48), 32564–32570.
123. G. Reguera, R.B. Pollina, J.S. Nicoll and D.R. Lovley (2007). Possible nonconductive role of *Geobacter sulfurreducens* pilus nanowires in biofilm formation. *Journal of Bacteriology* **189**(5), 2125–2127.
124. A.E. Franks, K.P. Nevin, R.H. Glaven and D.R. Lovley (2010). Microtoming coupled to microarray analysis to evaluate the spatial metabolic status of *Geobacter sulfurreducens* biofilms. *The ISME Journal* **4**(4), 509–519.
125. F. Liu, A.E. Rotaru, P.M. Shrestha, N.S. Malvankar, K.P. Nevin and D.R. Lovley (2015). Magnetite compensates for the lack of a pilin-associated *c*-type cytochrome in extracellular electron exchange. *Environmental Microbiology* **17**(3), 648–655.
126. N.S. Malvankar, M. Vargas, K. Nevin, P.-L. Tremblay, K. Evans-Lutterodt, D. Nykypanchuk, E. Martz, M.T. Tuominen and D.R. Lovley (2015). Structural basis for metallic-like conductivity in microbial nanowires. *mBio* **6**(2), e00084–15.
127. G. Reguera, K.P. Nevin, J.S. Nicoll, S.F. Covalla, T.L. Woodard and D.R. Lovley (2006). Biofilm and nanowire production leads to increased current in *Geobacter sulfurreducens* fuel cells. *Applied and Environmental Microbiology* **72**(11), 7345–7348.
128. Y.A. Gorby, S. Yanina, J.S. McLean, K.M. Rosso, D. Moyles, A. Dohnalkova, T.J. Beveridge, I.S. Chang, B.H. Kim and K.S. Kim (2006). Electrically conductive bacterial nanowires produced by *Shewanella oneidensis* strain MR-1 and other microorganisms. *Proceedings of the National Academy of Sciences of the United States of America* **103**(30), 11358–11363.
129. M.E. Hernandez and D.K. Newman (2001). Extracellular electron transfer. *Cellular and Molecular Life Sciences* **58**(11), 1562–1571.
130. K. Rabaey, N. Boon, M. Hofte and W. Verstraete (2005). Microbial phenazine production enhances electron transfer in biofuel cells. *Environmental Science and Technology* **39**(9), 3401–3408.
131. H. Von Canstein, J. Ogawa, S. Shimizu and J.R. Lloyd (2008). Secretion of flavins by *Shewanella* species and their role in extracellular electron transfer. *Applied and Environmental Microbiology* **74**(3), 615–623.
132. K.L. Straub and B. Schink (2004). Ferrihydrite-dependent growth of *Sulfurospirillum deleyianum* through electron transfer via sulfur cycling. *Applied and Environmental Microbiology* **70**(10), 5744–5749.

133. M. Rosenbaum, F. Zhao, U. Schröder and F. Scholz (2006). Interfacing electrocatalysis and biocatalysis with tungsten carbide: A high-performance, noble-metal-free microbial fuel cell. *Angewandte Chemie International Edition* **45**(40), 6658–6661.
134. D.P. Lies, M.E. Hernandez, A. Kappler, R.E. Mielke, J.A. Gralnick and D.K. Newman (2005). *Shewanella oneidensis* MR-1 uses overlapping pathways for iron reduction at a distance and by direct contact under conditions relevant for biofilms. *Applied and Environmental Microbiology* **71**(8), 4414–4426.
135. K.P. Nevin and D.R. Lovley (2002). Mechanisms for accessing insoluble Fe (III) oxide during dissimilatory Fe (III) reduction by *Geothrix fermentans. Applied and Environmental Microbiology* **68**(5), 2294–2299.
136. C.M. Ding, M.L. Lv, Y. Zhu, L. Jiang and H. Liu (2015). Wettability-regulated extracellular electron transfer from the living organism of *Shewanella loihica* PV-4. *Angewandte Chemie International Edition* **127**(5), 1466–1471.
137. A. Okamoto, S. Kalathil, X. Deng, K. Hashimoto, R. Nakamura and K.H. Nealson (2014). Cell-secreted flavins bound to membrane cytochromes dictate electron transfer reactions to surfaces with diverse charge and pH. *Scientific Reports* **4**, 1–8.
138. Y. Wang, S.E. Kern and D.K. Newman (2010). Endogenous phenazine antibiotics promote anaerobic survival of *Pseudomonas aeruginosa* via extracellular electron transfer. *Journal of Bacteriology* **192**(1), 365–369.
139. C. Milliken and H. May (2007). Sustained generation of electricity by the spore-forming, Gram-positive, *Desulfitobacterium hafniense* strain DCB2. *Applied Microbiology and Biotechnology* **73**(5), 1180–1189.
140. N. Boon, P. Aelterman, P. Clauwaert, L. De Schamphelaire, L. Vanhaecke, K. De Maeyer, M. Höfte, W. Verstraete and K. Rabaey (2008). Metabolites produced by *Pseudomonas* sp. enable a Gram-positive bacterium to achieve extracellular electron transfer. *Applied Microbiology and Biotechnology* **77**(5), 1119–1129.
141. H.T. Pham, N. Boon, P. Aelterman, P. Clauwaert, L. De Schamphelaire P. Van Oostveldt, K. Verbeken, K. Rabaey and W. Verstraete (2008). High shear enrichment improves the performance of the anodophilic microbial consortium in a microbial fuel cell. *Microbial Biotechnology* **1**(6), 487–496.
142. Z.M. Summers, H.E. Fogarty, C. Leang, A.E. Franks, N.S. Malvankar and D.R. Lovley (2010). Direct exchange of electrons within aggregates of an evolved syntrophic coculture of anaerobic bacteria. *Science* **330**(6009), 1413–1415.
143. A.-E. Rotaru, P.M. Shrestha, F. Liu, T. Ueki, K. Nevin, Z.M. Summers and D.R. Lovley (2012). Interspecies electron transfer via hydrogen and formate rather than direct electrical connections in cocultures of *Pelobacter carbinolicus* and *Geobacter sulfurreducens. Applied and Environmental Microbiology* **78**(21), 7645–7651.
144. A.-E. Rotaru, P.M. Shrestha, F. Liu, M. Shrestha, D. Shrestha, M. Embree, K. Zengler, C. Wardman, K.P. Nevin and D.R. Lovley (2014). A new model

for electron flow during anaerobic digestion: Direct interspecies electron transfer to *Methanosaeta* for the reduction of carbon dioxide to methane. *Energy and Environmental Science* **7**(1), 408–415.
145. S. Ishii, T. Kosaka, K. Hori, Y. Hotta and K. Watanabe (2005). Coaggregation facilitates interspecies hydrogen transfer between *Pelotomaculum thermopropionicum* and *Methanothermobacter thermautotrophicus*. *Applied and Environmental Microbiology* **71**(12), 7838–7845.
146. M.P. Bryant, E.A. Wolin, M.J. Wolin and R.S. Wolfe (1967). *Methanobacillus omelianskii*, a symbiotic association of two species of bacteria. *Archives of Microbiology* **59**, 20–31.
147. A.-E. Rotaru, P.M. Shrestha, F. Liu, B. Markovaite, S. Chen, K.P. Nevin and D.R. Lovley (2014). Direct interspecies electron transfer between *Geobacter metallireducens* and *Methanosarcina barkeri*. *Applied and Environmental Microbiology* **80**(15), 4599–4605.
148. Y. Zuo, D. Xing, J.M. Regan and B.E. Logan (2008). Isolation of the exoelectrogenic bacterium *Ochrobactrum anthropi* YZ-1 by using a U-tube microbial fuel cell. *Applied and Environmental Microbiology* **74**(10), 3130–3137.
149. H.M. Wang, J.D. Park and Z.J. Ren (2015). Practical energy harvesting for microbial fuel cells: A review. *Environmental Science and Technology* **49**(6), 3267–3277.
150. K.J. Chae, M. Choi, F.F. Ajayi, W. Park, I.S. Chang and I.S. Kim (2008). Mass transport through a proton exchange membrane (Nafion) in microbial fuel cells. *Energy and Fuels* **22**(1), 169–176.
151. M. Rahimnejad, G. Bakeri, M. Ghasemi and A. Zirepour (2014). A review on the role of proton exchange membrane on the performance of microbial fuel cell. *Polymers for Advanced Technologies* **25**(12), 1426–1432.
152. J. Xu, G.P. Sheng, H.W. Luo, W.W. Li, L.F. Wang and H.Q. Yu (2012). Fouling of proton exchange membrane (PEM) deteriorates the performance of microbial fuel cell. *Water Research* **46**(6), 1817–1824.
153. B.E. Logan, D. Call, S. Cheng, H.V.M. Hamelers, T.H.J.A. Sleutels, A.W. Jeremiasse and R.A. Rozendal (2008). Microbial electrolysis cells for high yield hydrogen gas production from organic matter. *Environmental Science and Technology* **42**(23), 8630–8640.
154. R.A. Rozendal, H.V.M. Hamelers, G.J.W. Euverink, S.J. Metz and C.J.N. Buisman (2006). Principle and perspectives of hydrogen production through biocatalyzed electrolysis. *International Journal of Hydrogen Energy* **31**(12), 1632–1640.
155. L. Kiseleva, S.K. Garushyants, H. Ma, D. Simpson, V. Fedorovich, M.F. Cohen and I. Goryanin (2015). Taxonomic and functional metagenomic analysis of anodic communities in two pilot-scale microbial fuel cells treating different industrial wastewaters. *Journal of Integrative Bioinformatics* **12**(1): 273, 1–15.
156. A. Kouzuma, T. Kasai, G. Nakagawa, A. Yamamuro, T. Abe and K. Watanabe (2013). Comparative metagenomics of anode-associated microbiomes developed in rice paddy-field microbial fuel cells. *Plos One* **8**(11), e77443.

157. R.A. Timmers, M. Rothballer, D.P. Strik, M. Engel, S. Schulz, M. Schloter, A. Hartmann, B. Hamelers and C. Buisman (2012). Microbial community structure elucidates performance of *Glyceria maxima* plant microbial fuel cell. *Applied Microbiology and Biotechnology* **94**(2), 537–548.
158. L. Lu, T. Huggins, S. Jin, Y. Zuo and Z.J. Ren (2014). Microbial metabolism and community structure in response to bioelectrochemically enhanced remediation of petroleum hydrocarbon-contaminated soil. *Environmental Science and Technology* **48**(7), 4021–4029.
159. N. Popp, M. Schlomann and M. Mau (2006). Bacterial diversity in the active stage of a bioremediation system for mineral oil hydrocarbon-contaminated soils. *Microbiology Society* **152**, 3291–3304.
160. M.C. Rosano-Hernandez, H. Ramirez-Saad and L. Fernandez-Linares (2012). Petroleum-influenced beach sediments of the Campeche Bank, Mexico: Diversity and bacterial community structure assessment. *Journal of Environmental Management* **95**, S325–S331.
161. L.T. Angenent, K. Karim, M.H. Al-Dahhan and R. Domiguez-Espinosa (2004). Production of bioenergy and biochemicals from industrial and agricultural wastewater. *Trends in Biotechnology* **22**(9), 477–485.
162. G. Lettinga, A.F.M. Vanvelsen, S.W. Hobma, W. Dezeeuw and A. Klapwijk (1980). Use of the upflow sludge blanket (USB) reactor concept for biological wastewater-treatment, especially for anaerobic treatment. *Biotechnology and Bioengineering* **22**(4), 699–734.
163. T. Narihiro and Y. Sekiguchi (2007). Microbial communities in anaerobic digestion processes for waste and wastewater treatment: a microbiological update. *Current Opinion in Biotechnology* **18**(3), 273–278.
164. A. Van Haandel, J. De Vrieze, W. Verstraete and V.S. Dos Santos (2014). *Methanosaeta* dominate acetoclastic methanogenesis during high-rate methane production in anaerobic reactors treating distillery wastewaters. *Journal of Chemical Technology and Biotechnology* **89**(11), 1751–1759.
165. G.H. Chen (2004). Electrochemical technologies in wastewater treatment. *Separation and Purification Technology* **38**(1), 11–41.
166. H. Liu, R. Ramnarayanan and B.E. Logan (2004). Production of electricity during wastewater treatment using a single chamber microbial fuel cell. *Environmental Science and Technology* **38**(7), 2281–2285.
167. T.H. Pham, K. Rabaey, P. Aelterman, P. Clauwaert, L. De Schamphelaire, N. Boon and W. Verstraete (2006). Microbial fuel cells in relation to conventional anaerobic digestion technology. *Engineering in Life Sciences* **6**(3), 285–292.
168. W. Verstraete, F. Morgan-Sagastume, S. Aiyuk, M. Waweru, K. Rabaey and G. Lissens (2005). Anaerobic digestion as a core technology in sustainable management of organic matter. *Water Science and Technology* **52**(1–2), 59-66.
169. K. Kubota, W. Yoochatchaval, T. Yamaguchi and K. Syutsubo (2010). Application of a single-chamber microbial fuel cell (MFC) for organic

wastewater treatment: Influence of changes in wastewater composition on the process performance. *Sustainable Environment Research* **20**(6), 347–351.
170. H. Liu, S. Grot and B.E. Logan (2005). Electrochemically assisted microbial production of hydrogen from acetate. *Environmental Science and Technology* **39**(11), 4317–4320.
171. C.C. Chen, C.Y. Lin and J.S. Chang (2001). Kinetics of hydrogen production with continuous anaerobic cultures utilizing sucrose as the limiting substrate. *Applied Microbiology and Biotechnology* **57**(1–2), 56–64.
172. H.H.P. Fang, H. Liu and T. Zhang (2002). Characterization of a hydrogen-producing granular sludge. *Biotechnology and Bioengineering* **78**(1), 44–52.
173. S.E. Oh, S. Van Ginkel and B.E. Logan (2003). The relative effectiveness of pH control and heat treatment for enhancing biohydrogen gas production. *Environmental Science and Technology* **37**(22), 5186–5190.
174. Y. Ueno, S. Haruta, M. Ishii and Y. Igarashi (2001). Microbial community in anaerobic hydrogen-producing microflora enriched from sludge compost. *Applied Microbiology and Biotechnology* **57**(4), 555–562.
175. D. Das and T.N. Veziroglu (2008). Advances in biological hydrogen production processes. *International Journal of Hydrogen Energy* **33**(21), 6046–6057.
176. Y. Asada and J. Miyake (1999). Photobiological hydrogen production. *Journal of Bioscience and Bioengineering* **88**(1), 1–6.
177. S. Bensaid, B. Ruggeri and G. Saracco (2015). Development of a photosynthetic microbial electrochemical cell (PMEC) reactor coupled with dark fermentation of organic wastes: Medium term perspectives. *Energies* **8**(1), 399–429.
178. N. Kataoka, A. Miya and K. Kiriyama (1997). Studies on hydrogen production by continuous culture system of hydrogen-producing anaerobic bacteria. *Water Science and Technology* **36**(6–7), 41–47.
179. K. Kinoshita, Electrochemical Society (1992). *Electrochemical Oxygen Technology*. New York, Wiley.
180. B. St-Pierre and A.D.G. Wright (2014). Comparative metagenomic analysis of bacterial populations in three full-scale mesophilic anaerobic manure digesters. *Applied and Environmental Microbiology* **98**(6), 2709–2717.
181. P.D. Kiely, R. Cusick, D.F. Call, P.A. Selembo, J.M. Regan and B.E. Logan (2011). Anode microbial communities produced by changing from microbial fuel cell to microbial electrolysis cell operation using two different wastewaters. *Bioresource Technology* **102**(1), 388–394.
182. E.E. Diaz, A.J.A. Stams, R. Amils and J.L. Sanz (2006). Phenotypic properties and microbial diversity of methanogenic granules from a full-scale upflow anaerobic sludge bed reactor treating brewery wastewater. *Applied and Environmental Microbiology* **72**(7), 4942–4949.
183. M. Morita, N.S. Malvankar, A.E. Franks, Z.M. Summers, L. Giloteaux, A.E. Rotaru, C. Rotaru and D.R. Lovley (2011). Potential for direct interspecies electron transfer in methanogenic wastewater digester aggregates. *mBio* **2**(4). e00159–11.

184. D. Karakashev, D.J. Batstone and I. Angelidaki (2005). Influence of environmental conditions on methanogenic compositions in anaerobic biogas reactors. *Applied and Environmental Microbiology* **71**(1), 331–338.
185. Y. Yang, K. Yu, Y. Xia, F.T.K. Lau, D.T.W. Tang, W.C. Fung, H.H.P. Fang and T. Zhang (2014). Metagenomic analysis of sludge from full-scale anaerobic digesters operated in municipal wastewater treatment plants. *Applied Microbiology and Biotechnology* **98**(12), 5709–5718.
186. S.L. Garcia, K. Jangid, W.B. Whitman and K.C. Das (2011). Transition of microbial communities during the adaption to anaerobic digestion of carrot waste. *Bioresource Technology* **102**(15), 7249–7256.
187. M. Tabatabaei, M.R. Zakaria, R.A. Rahim and A.D.G. Wright, Y. Shirai, N. Abdullah, K. Sakai, S. Ikeno, M. Mori, N. Kazunori, A. Sulaiman and M.A. Hassan (2009). PCR-based DGGE and FISH analysis of methanogens in an anaerobic closed digester tank for treating palm oil mill effluent. *Electronic Journal of Biotechnology* **12**(3), 1–12.
188. A.M. Ziganshin, J. Liebetrau, J. Proter and S. Kleinsteuber (2013). Microbial community structure and dynamics during anaerobic digestion of various agricultural waste materials. *Applied Microbiology and Biotechnology* **97**(11), 5161–5174.
189. A.P. Buzzini, I.K. Sakamoto, M.B. Varesche and E.C. Pires (2006). Evaluation of the microbial diversity in an UASB reactor treating wastewater from an unbleached pulp plant. *Process Biochemistry* **41**(1), 168–176.
190. K. Roest, H.G. Heilig, H. Smidt, W.M. De Vos, A.J. Stams and A.D. Akkermans (2005). Community analysis of a full-scale anaerobic bioreactor treating paper mill wastewater. *Systematic and Applied Microbiology* **28**(2), 175–185.
191. V. Gomez-Alvarez, R.P. Revetta and J.W.S. Domingo (2012). Metagenome analyses of corroded concrete wastewater pipe biofilms reveal a complex microbial system. *BMC Microbiology* **12**(122), 1–14.
192. G.F. Parkin, M.A. Sneve and H. Loos (1991). Anaerobic filter treatment of sulfate-containing wastewaters. *Water Science and Technology* **23**(7–9), 1283–1291.
193. S.V. Mohan, N.C. Rao, K.K. Prasad and P.N. Sarma (2005). Bioaugmentation of an anaerobic sequencing batch biofilm reactor (AnSBBR) with immobilized sulphate reducing bacteria (SRB) for the treatment of sulphate bearing chemical wastewater. *Process Biochemistry* **40**(8), 2849–2857.
194. B.C. Kim, C. Leang, Y.H.R. Ding, R.H. Glaven, M.V. Coppi and D.R. Lovley (2005). OmcF, a putative *c*-type monoheme outer membrane cytochrome required for the expression of other outer membrane cytochromes in *Geobacter sulfurreducens*. *Journal of Bacteriology* **187**(13), 4505–4513.
195. X.L. Qian, G. Reguera, T. Mester and D.R. Lovley (2007). Evidence that OmcB and OmpB of *Geobacter sulfurreducens* are outer membrane surface proteins. *FEMS Microbiology Letters* **277**(1), 21–27.
196. J.M. Dantas, D.M. Tomaz, L. Morgado and C.A. Salgueiro (2013). Functional characterization of PccH, a key cytochrome for electron transfer from

electrodes to the bacterium *Geobacter sulfurreducens*. *FEBS Letters* **587**(16), 2662–2668.

197. J.M. Dantas, L. Morgado, M. Aklujkar, M. Bruix, Y.Y. Londer, M. Schiffer, P.R. Pokkuluri and C.A. Salgueiro (2015). Rational engineering of *Geobacter sulfurreducens* electron transfer components: a foundation for building improved *Geobacter*-based bioelectrochemical technologies. *Frontiers in Microbiology* **6**(752), 1–15.
198. M. Izallalen, R. Mahadevan, A. Burgard, B. Postier, R. Didonato, J. Sun, C.H. Schilling and D.R. Lovley (2008). *Geobacter sulfurreducens* strain engineered for increased rates of respiration. *Metabolic Engineering* **10**(5), 267–275.
199. N.J.P. Wierckx, H. Ballerstedt, J.A.M. De Bont and J. Wery (2005). Engineering of solvent-tolerant *Pseudomonas putida* S12 for bioproduction of phenol from glucose. *Applied and Environmental Microbiology* **71**(12), 8221–8227.
200. A. Wittgens, T. Tiso, T.T. Arndt, P. Wenk, J. Hemmerich, C. Muller, R Wichmann, B. Kupper, M. Zwick, S. Wilhelm, R. Hausmann, C. Syldatk, F. Rosenau and L.M. Blank (2011). Growth independent rhamnolipid production from glucose using the non-pathogenic *Pseudomonas putida* KT2440. *Microbial Cell Factories* **10**(80), 1–17.
201. S. Verhoef, N. Wierckx, R.G.M. Westerhof and J.H. De Winde and H.J. Ruijssenaars (2009). Bioproduction of p-hydroxystyrene from glucose by the solvent-tolerant bacterium *Pseudomonas putida* S12 in a two-phase water-decanol fermentation. *Applied and Environmental Microbiology* **75**(4), 931–936.
202. J. Volmer, C. Neumann, B. Buhler and A. Schmid (2014). Engineering of *Pseudomonas taiwanensis* VLB120 for constitutive solvent tolerance and increased specific styrene epoxidation activity. *Applied and Environmental Microbiology* **80**(20), 6539–6548.
203. S. Schmitz, S. Nies, N. Wierckx, L.M. Blank and M.A. Rosenbaum (2015). Engineering mediator-based electroactivity in the obligate aerobic bacterium *Pseudomonas putida* KT2440. *Frontiers in Microbiology* **6**(284), 1–13.
204. T. Zheng, Y.S. Xu, X.Y. Yong, B. Li, D. Yin, Q.W. Cheng, H.R. Yuan and Y.C. Yong (2015). Endogenously enhanced biosurfactant production promotes electricity generation from microbial fuel cells. *Bioresource Technology* **197**, 416–421.
205. A. Baudler, I. Schmidt, M. Langner, A. Greiner and U. Schröder (2015). Does it have to be carbon? Metal anodes in microbial fuel cells and related bioelectrochemical systems. *Energy and Environmental Science* **8**(7), 2048–2055.
206. S.S. Chen, A.-E. Rotaru, P.M. Shrestha, N.S. Malvankar, F.H. Liu, W. Fan, K.P. Nevin and D.R. Lovley (2014). Promoting interspecies electron transfer with biochar. *Scientific Reports* **4**(5019), 1–7.

207. T. Huggins, H.M. Wang, J. Kearns, P. Jenkins and Z.J. Ren (2014). Biochar as a sustainable electrode material for electricity production in microbial fuel cells. *Bioresource Technology* **157**, 114–119.
208. C.S. Kang, N. Eaktasang, D.Y. Kwon and H.S. Kim (2014). Enhanced current production by *Desulfovibrio desulfuricans* biofilm in a mediator-less microbial fuel cell. *Bioresource Technology* **165**, 27–30.
209. K. Guo, B.C. Donose, A.H. Soeriyadi, A. Prevoteau, S.A. Patil, S. Freguia, J.J. Gooding and K. Rabaey (2014). Flame oxidation of stainless steel felt enhances anodic biofilm formation and current output in bioelectrochemical systems. *Environmental Science and Technology* **48**(12), 7151–7156.
210. T.S. Magnuson, N. Isoyama, A.L. Hodges-Myerson, G. Davidson, M.J. Maroney, G.G. Geesey and D.R. Lovley (2001). Isolation, characterization and gene sequence analysis of a membrane-associated 89 kDa Fe(III) reducing cytochrome *c* from *Geobacter sulfurreducens*. *Biochemical Journal* **359**, 147–152.
211. Y.M. Liu, Z.M. Wang, J. Liu, C. Levar, M.J. Edwards, J.T. Babauta, D.W. Kennedy, Z. Shi, H. Beyenal, D.R. Bond, T.A. Clarke, J.N. Butt, D.J. Richardson, K.M. Rosso, J.M. Zachara, J.K. Fredrickson and L. Shi (2014). A trans-outer membrane porin-cytochrome protein complex for extracellular electron transfer by *Geobacter sulfurreducens* PCA. *Environmental Microbiology Reports* **6**(6), 776–785.
212. R.R. Pokkuluri, Y.Y. Londer, S.J. Wood, N.E.C. Duke, L. Morgado, C.A. Salgueiro and M. Schiffer (2009). Outer membrane cytochrome *c*, OmcF, from *Geobacter sulfurreducens*: High structural similarity to an algal cytochrome c_6. *Proteins: Structure, Function and Bioinformatics* **74**(1), 266–270.
213. E.S. Shelobolina, M.V. Coppi, A.A. Korenevsky, L.N. DiDonato, S.A. Sullivan, H. Konishi, H.F. Xu, C. Leang, J.E. Butler, B.C. Kim and D.R. Lovley (2007). Importance of *c*-type cytochromes for U(VI) reduction by *Geobacter sulfurreducens*. *BMC Microbiology* **7**(16): 1–15.
214. T. Mehta, S.E. Childers, R. Glaven, D.R. Lovley and T. Mester (2006). A putative multicopper protein secreted by an atypical type II secretion system involved in the reduction of insoluble electron acceptors in *Geobacter sulfurreducens*. *Microbiology Society* **152**, 2257–2264.
215. J.M. Dantas, L.M. Campelo, N.E. Duke, C.A. Salgueiro and P.R. Pokkuluri (2015). The structure of PccH from *Geobacter sulfurreducens* — a novel low reduction potential monoheme cytochrome essential for accepting electrons from an electrode. *FEBS Journal* **282**(11), 2215–2231.
216. L. Morgado, M. Bruix, M. Pessanha, Y.Y. Londer and C.A. Salgueiro (2010). Thermodynamic characterization of a triheme cytochrome family from *Geobacter sulfurreducens* reveals mechanistic and functional diversity. *Biophysical Journal* **99**(1), 293–301.
217. J.R. Lloyd, C. Leang, A.L. Hodges Myerson, M.V. Coppi, S. Cuifo, B. Methe, S.J. Sandler and D.R. Lovley (2003). Biochemical and genetic characterization of PpcA, a periplasmic *c*-type cytochrome in *Geobacter sulfurreducens*. *Biochemical Journal* **369**(Pt 1), 153–161.

218. P.R. Pokkuluri, Y.Y. Londer, X. Yang, N.E.C. Duke, J. Erickson, V. Orshonsky, G. Johnson and M. Schiffer (2010). Structural characterization of a family of cytochromes c_7 involved in Fe(III) respiration by *Geobacter sulfurreducens*. *Biochimica Et Biophysica Acta-Bioenergetics* **1797**(2), 222–232.
219. J. Seidel, M. Hoffmann, K.E. Ellis, A. Seidel, T. Spatzal, S. Gerhardt, S.J. Elliott and O. Einsle (2012). MacA is a second cytochrome *c* peroxidase of *Geobacter sulfurreducens*. *Biochemistry* **51**(13), 2747–2756.
220. J.E. Butler, F. Kaufmann, M.V. Coppi, C. Nunez and D.R. Lovley (2004). MacA a diheme *c*-type cytochrorne involved in Fe(III) reduction by *Geobacter sulfurreducens*. *Journal of Bacteriology* **186**(12), 4042–4045.
221. B.C. Kim and D.R. Lovley (2008). Investigation of direct *vs.* indirect involvement of the *c*-type cytochrome MacA in Fe(III) reduction by *Geobacter sulfurreducens*. *FEMS Microbiology Letters* **286**(1), 39–44.
222. S.J. Field, P.S. Dobbin, M.R. Cheesman, N.J. Watmough, A.J. Thomson and D.J. Richardson (2000). Purification and magneto-optical spectroscopic characterization of cytoplasmic membrane and outer membrane multiheme *c*-type cytochromes from *Shewanella frigidimarina* NCIMB400. *Journal of Biological Chemistry* **275**(12), 8515–8522.
223. C.R. Myers and J.M. Myers (1997). Outer membrane cytochromes of *Shewanella putrefaciens* MR-1: Spectral analysis, and purification of the 83-kDa *c*-type cytochrome. *Biochimica Et Biophysica Acta-Biomembranes* **1326**(2), 307–318.
224. J.M. Myers and C.R. Myers (1998). Isolation and sequence of *omcA*, a gene encoding a decaheme outer membrane cytochrome *c* of *Shewanella putrefaciens* MR-1, and detection of *omcA* homologs in other strains of *S. putrefaciens*. *Biochimica et Biophysica Acta* **1373**(1), 237–251.
225. J. Borloo, L. De Smet, V. Beeumen and B. Devreese (2011). Bacterial two-hybrid analysis of the *Shewanella oneidensis* MR-1 multi-component electron transfer pathway. *Journal of Integrated OMICS* **1**(2), 260–267.
226. J.M. Myers and C.R. Myers (2001). Role for outer membrane cytochromes OmcA and OmcB of *Shewanella putrefaciens* MR-1 in reduction of manganese dioxide. *Applied and Environmental Microbiology* **67**(1), 260–269.
227. B.H. Lower, R. Yongsunthon, L. Shi, L. Wildling, H.J. Gruber, N.S. Wigginton, C.L. Reardon, G.E. Pinchuk, T.C. Droubay, J.F. Boily and S.K. Lower (2009). Antibody recognition force microscopy shows that outer membrane cytochromes OmcA and MtrC are expressed on the exterior surface of *Shewanella oneidensis* MR-1. *Applied and Environmental Microbiology* **75**(9), 2931–2935.
228. C.R. Myers and J.M. Myers (2003). Cell surface exposure of the outer membrane cytochromes of *Shewanella oneidensis* MR-1. *Letters in Applied Microbiology* **37**(3), 254–258.
229. D. Baron, E. LaBelle, D. Coursolle, J.A. Gralnick and D.R. Bond (2009). Electrochemical measurement of electron transfer kinetics by *Shewanella oneidensis* MR-1. *Journal of Biological Chemistry* **284**, 28865–28873.

230. D. Coursolle and J.A. Gralnick (2010). Modularity of the Mtr respiratory pathway of *Shewanella oneidensis* strain MR-1. *Molecular Microbiology* **77**(4), 995–1008.
231. K.E. Pitts, P.S. Dobbin, F. Reyes-Ramirez, A.J. Thomson, D.J. Richardson and H.E. Seward (2003). Characterization of the *Shewanella oneidensis* MR-1 decaheme cytochrome MtrA. *Journal of Biological Chemistry* **278**(30), 27758–27765.
232. M.A. Firer-Sherwood, N. Ando, C.L. Drennan and S.J. Elliott (2011). Solution-based structural analysis of the decaheme cytochrome, MtrA, by small-angle X-ray scattering and analytical ultracentrifugation. *Journal of Physical Chemistry B* **115**(38), 11208–11214.
233. C. Reyes, F. Qian, A. Zhang, S. Bondarev, A. Welch, M.P. Thelen and C.W. Saltikov (2012). characterization of axial and proximal histidine mutations of the decaheme cytochrome MtrA from *Shewanella* sp. strain ANA-3 and implications for the electron transport system. *Journal of Bacteriology* **194**(21), 5840–5847.
234. L. Shi, J.T. Lin, L.M. Markillie, T.C. Squier and B.S. Hooker (2005). Overexpression of multi-heme *c*-type cytochromes. *Biotechniques* **38**(2), 297–299.
235. A.S. Beliaev, D.A. Saffarini, J.L. McLaughlin and D. Hunnicutt (2001). MtrC, an outer membrane decahaem *c* cytochrome required for metal reduction in *Shewanella putrefaciens* MR-1. *Molecular Microbiology* **39**(3), 722–730.
236. O. Bretschger, A. Obraztsova, C.A. Sturm, I.S. Chang, Y.A. Gorby, S.B. Reed and D.E. Culley, C.L. Reardon, S. Barua, M.F. Romine, J. Zhou, A.S. Beliaev, R. Bouhenni, D. Saffarini, F. Mansfeld, B.H. Kim, J.K. Fredrickson and K.H. Nealson (2007). Current production and metal oxide reduction by *Shewanella oneidensis* MR-1 wild type and mutants. *Applied and Environmental Microbiology* **73**(21), 7003–7012.
237. R.S. Hartshorne, C.L. Reardon, D. Ross, J. Nuester, T.A. Clarke, A.J. Gates, P.C. Mills, J.K. Fredrickson, J.M. Zachara, L. Shi, A.S. Beliaev, M.J. Marshall, M. Tien, S. Brantley, J.N. Butt and D.J. Richardson (2009). Characterization of an electron conduit between bacteria and the extracellular environment. *Proceedings of the National Academy of Sciences of the United States of America* **106**(52), 22169–22174.
238. D.E. Ross, J.M. Flynn, D.B. Baron, J.A. Gralnick and D.R. Bond (2011). Towards electrosynthesis in *Shewanella*: energetics of reversing the Mtr pathway for reductive metabolism. *Plos One* **6**(2), e16649, 1–9.
239. A.S. Beliaev and D.A. Saffarini (1998). *Shewanella putrefaciens* mtrB encodes an outer membrane protein required for Fe(III) and Mn(IV) reduction. *Journal of Bacteriology* **180**(23), 6292–6297.
240. D.E. Ross, S.S. Ruebush, S.L. Brantley, R.S. Hartshorne, T.A. Clarke, D.J. Richardson and M. Tien (2007). Characterization of protein-protein interactions involved in iron reduction by *Shewanella oneidensis* MR-1. *Applied and Environmental Microbiology* **73**(18), 5797–5808.

241. R.S. Hartshorne, B.N. Jepson, T.A. Clarke, S.J. Field, J. Fredrickson, J. Zachara, L. Shi, J.N. Butt and D.J. Richardson (2007). Characterization of *Shewanella oneidensis* MtrC: A cell-surface decaheme cytochrome involved in respiratory electron transport to extracellular electron acceptors. *Journal of Biological Inorganic Chemistry* **12**(7), 1083–1094.
242. C. Bucking, F. Popp, S. Kerzenmacher and J Gescher (2010). Involvement and specificity of *Shewanella oneidensis* outer membrane cytochromes in the reduction of soluble and solid-phase terminal electron acceptors. *FEMS Microbiology Letters* **306**(2), 144–151.
243. D.W. Ding, J. Xu, L. Li, J.M. Xie and X. Sun (2014). Identifying the potential extracellular electron transfer pathways from a *c*-type cytochrome network. *Molecular Biosystems* **10**(12), 3138–3146.
244. T.A. Clarke, M.J. Edwards, A.J. Gates, A. Hall, G.F. White, J. Bradley, C.L. Reardon, L. Shi, A.S. Beliaev, M.J. Marshall, Z.M. Wang, N.J. Watmough, J.K. Fredrickson, J.M. Zachara, J.N. Butt and D.J. Richardson (2011). Structure of a bacterial cell surface decaheme electron conduit. *Proceedings of the National Academy of Sciences of the United States of America* **108**(23), 9384–9389.
245. A.I. Tsapin, I. Vandenberghe, K.H. Nealson, J.H. Scott, T.E. Meyer, M.A. Cusanovich, E. Harada, T. Kaizu, H. Akutsu, D. Leys and J.J. Van Beeumen (2001). Identification of a small tetraheme cytochrome *c* and a flavocytochrome *c* as two of the principal soluble cytochromes *c* in *Shewanella oneidensis* strain MR1. *Applied and Environmental Microbiology* **67**(7), 3236–3244.
246. M. Firer-Sherwood, G.S. Pulcu and S.J. Elliott (2008). Electrochemical interrogations of the Mtr cytochromes from *Shewanella*: Opening a potential window. *Journal of Biological Inorganic Chemistry* **13**(6), 849–854.
247. C.R. Myers and J.M. Myers (1997). Cloning and sequence of *cymA* a gene encoding a tetraheme cytochrome *c* required for reduction of iron(III), fumarate, and nitrate by *Shewanella putrefaciens* MR-1. *Journal of Bacteriology* **179**(4), 1143–1152.
248. C. Schwalb, S.K. Chapman and G.A. Reid (2003). The tetraheme cytochrome CymA is required for anaerobic respiration with dimethyl sulfoxide and nitrite in *Shewanella oneidensis. Biochemistry* **42**(31), 9491–9497.
249. J.A. Gralnick, H. Vali, D.P. Lies and D.K. Newman (2006). Extracellular respiration of dimethyl sulfoxide by *Shewanella oneidensis* strain MR-1. *Proceedings of the National Academy of Sciences of the United States of America* **103**(12), 4669–4674.
250. K.D. Bewley, M.A. Firer-Sherwood, J.Y. Mock, N. Ando, C.L. Drennan and S.J. Elliott (2012). Mind the gap: Diversity and reactivity relationships among multihaem cytochromes of the MtrA/DmsE family. *Biochemical Society* **40**, 1268.
251. M. Breuer, K.M. Rosso, J. Blumberger and J.N. Butt (2015). Multi-haem cytochromes in *Shewanella oneidensis* MR-1: structures, functions and

opportunities. *Journal of the Royal Society Interface* **12**(102), 20141117, 1–27.

252. A.I. Tsapin, K.H. Nealson, T. Meyers, M.A. Cusanovich, J. VanBeuumen and L.D. Crosby, B.A. Feinberg and C. Zhang (1996). Purification and properties of a low-redox-potential tetraheme cytochrome c_3 from *Shewanella putrefaciens*. *Journal of Bacteriology* **178**(21), 6386–6388.
253. M.N. Alves, S.E. Neto, A.S. Alves, B.M. Fonseca, A. Carrelo, I. Pacheco, C.M. Paquete, C.M. Soares and R.O. Louro (2015). Characterization of the periplasmic redox network that sustains the versatile anaerobic metabolism of *Shewanella oneidensis* MR-1. *Frontiers in Microbiology* **6**(665), 1–10.
254. H. Liu, S. Cheng and B.E. Logan (2005). Power generation in fed-batch microbial fuel cells as a function of ionic strength, temperature, and reactor configuration. *Environmental Science and Technology* **39**(14), 5488–5493.
255. S.A. Cheng and B.E. Logan (2011). Increasing power generation for scaling up single-chamber air cathode microbial fuel cells. *Bioresource Technology* **102**(6), 4468–4473.
256. E.A. Franzosa, X.C. Morgan, N. Segata, L. Waldron, J. Reyes, A.M. Earl, G Giannoukos, M.R. Boylan, D. Ciulla, D. Gevers, J. Izard, W.S. Garrett, A.T. Chan and C. Huttenhower (2014). Relating the metatranscriptome and metagenome of the human gut. *Proceedings of the National Academy of Sciences of the United States of America* **111**(22), e2329–e2338.
257. Y.Z. Fan, E. Sharbrough and H. Liu (2008). Quantification of the internal resistance distribution of microbial fuel cells. *Environmental Science and Technology* **42**(21), 8101–8107.
258. S. Oh, B. Min and B.E. Logan (2004). Cathode performance as a factor in electricity generation in microbial fuel cells. *Environmental Science and Technology* **38**(18), 4900–4904.
259. B.R. Ringeisen, E. Henderson, P.K. Wu, J. Pietron, R. Ray, B. Little, J.C. Biffinger and J.M. Jones-Meehan (2006). High power density from a miniature microbial fuel cell using *Shewanella oneidensis* DSP10. *Environmental Science and Technology* **40**, 2629–2634.
260. A. Shantaram, H. Beyenal, R. Raajan, A. Veluchamy and Z. Lewandowski (2005). Wireless sensors powered by microbial fuel cells. *Environmental Science and Technology* **39**(13), 5037–5042.
261. A. El Mekawy, H.M. Hegab, X. Dominguez-Benetton and D. Pant (2013). Internal resistance of microfluidic microbial fuel cell: Challenges and potential opportunities. *Bioresource Technology* **142**, 672–682.
262. S.V. Morozov, P.M. Vignais, V.L. Cournac, N.A. Zorin, E.E. Karyakina, A.A. Karyakin and S. Cosnier (2002). Bioelectrocatalytic hydrogen production by hydrogenase electrodes. *International Journal of Hydrogen Energy* **27**(11–12), 1501–1505.
263. H.R. Pershad, J.L.C. Duff, H.A. Heering, E.C. Duin, S.P.J. Albracht and F.A. Armstrong (1999). Catalytic electron transport in *Chromatium vinosum* [NiFe]-hydrogenase: Application of voltammetry in detecting redox-active centers and establishing that hydrogen oxidation is very fast even

at potentials close to the reversible H^+/H^{-2} value. *Biochemistry* **38**(28), 8992–8999.
264. E. Lojou, M.C. Durand, A. Dolla and P. Bianco (2002). Hydrogenase activity control at *Desulfovibrio vulgaris* cell-coated carbon electrodes: Biochemical and chemical factors influencing the mediated bioelectrocatalysis. *Electroanalysis* **14**(13), 913–922.
265. R.A. Rozendal, A.W. Jeremiasse, H.V.M. Hamelers and C.J.N. Buisman (2008). Hydrogen production with a microbial biocathode. *Environmental Science and Technology* **42**(2), 629–634.
266. H. Tatsumi, K. Takagi, M. Fujita, K. Kano and T. Ikeda (1999). Electrochemical study of reversible hydrogenase reaction of *Desulfovibrio vulgaris* cells with methyl viologen as an electron carrier. *Analytical Chemistry* **71**(9), 1753–1759.
267. E. Lojou and P. Bianco (2004). Electrocatalytic reactions at hydrogenase-modified electrodes and their applications to biosensors: From the isolated enzymes to the whole cells. *Electroanalysis* **16**(13–14), 1093–1100.
268. J.K. Jang, T.H. Pham, I.S. Chang, K.H. Kang, H. Moon, K.S. Cho and B.H. Kim (2004). Construction and operation of a novel mediator- and membrane-less microbial fuel cell. *Process Biochemistry* **39**(8), 1007–1012.
269. T. Jafary, W.R.W. Daud, M. Ghasemi, B.H. Kim, J.M. Jahim, M. Ismail and S.S. Lim (2015). Biocathode in microbial electrolysis cell; present status and future prospects. *Renewable and Sustainable Energy Reviews* **47**, 23–33.
270. E. Croese, A.W. Jeremiasse, I.P.G. Marshall, A.M. Spormann, G.J.W. Euveritik, J.S. Geelhoed, A.J.M. Stams and C.M. Plugge (2014). Influence of setup and carbon source on the bacterial community of biocathodes in microbial electrolysis cells. *Enzyme and Microbial Technology* **61**, 67–75.
271. E. Croese, M.A. Pereira, G.J.W. Euverink, A.J.M. Stams and J.S. Geelhoed (2011). Analysis of the microbial community of the biocathode of a hydrogen-producing microbial electrolysis cell. *Applied Microbiology and Biotechnology* **92**(5), 1083–1093.
272. C.W. Marshall, D.E. Ross, E.B. Fichot, R.S. Norman and H.D. May (2012). Electrosynthesis of commodity chemicals by an autotrophic microbial community. *Applied and Environmental Microbiology* **78**(23), 8412–8420.
273. J.S. Geelhoed and A.J.M. Stams (2011). Electricity-assisted biological hydrogen production from acetate by *Geobacter sulfurreducens*. *Environmental Science and Technology* **45**(2), 815–820.
274. F. Aulenta, L. Catapano, L. Snip, M. Villano and M. Majone (2012). Linking bacterial metabolism to graphite cathodes: Electrochemical insights into the H_2-producing capability of *Desulfovibrio* sp. *ChemSusChem* **5**(6), 1080–1085.
275. R.A. Rozendal, H.V.M. Hamelers, R.J. Molenkmp and J.N. Buisman (2007). Performance of single chamber biocatalyzed electrolysis with different types of ion exchange membranes. *Water Research* **41**(9), 1984–1994.

276. N. Xafenias and V. Mapelli (2014). Performance and bacterial enrichment of bioelectrochemical systems during methane and acetate production. *International Journal of Hydrogen Energy* **39**(36), 21864–21875.
277. B.E. Logan (2008). *Microbial Fuel Cells.* New Jersey, John Wiley & Sons.
278. S.M. Daud, B.H. Kim, M. Ghasemi and W.R.W. Daud (2015). Separators used in microbial electrochemical technologies: Current status and future prospects. *Bioresource Technology* **195**, 170–179.
279. Y. Tian, H. Li, L.P. Li, X.Y. Su, Y.B. Lu, W. Zuo and J. Zhang (2015). In-situ integration of microbial fuel cell with hollow-fiber membrane bioreactor for wastewater treatment and membrane fouling mitigation. *Biosensors and Bioelectronics* **64**, 189–195.
280. M. Rahimnejad, M. Ghasemi, G.D. Najafpour, M. Ismail, A.W. Mohammad, A.A. Ghoreyshi and S.H.A. Hassan (2012). Synthesis, characterization and application studies of self-made Fe_3O_4/PES nanocomposite membranes in microbial fuel cell. *Electrochimica Acta* **85**, 700–706.
281. M. Ghasemi, W.R.W. Daud, M. Ismail, M. Rahimnejad, A.F. Ismail, J.X. Leong, M. Miskan and K. Ben Liew (2013). Effect of pre-treatment and biofouling of proton exchange membrane on microbial fuel cell performance. *International Journal of Hydrogen Energy* **38**(13), 5480–5484.
282. J.R. Kim, S. Cheng, S.E. Oh and B.E. Logan (2007). Power generation using different cation, anion, and ultrafiltration membranes in microbial fuel cells. *Environmental Science and Technology* **41**(3), 1004–1009.
283. X.Y. Zhang, S.A. Cheng, X. Wang, X. Huang and B.E. Logan (2009). Separator characteristics for increasing performance of microbial fuel cells. *Environmental Science and Technology* **43**(21), 8456–8461.
284. R.A. Rozendal, H.V.M. Hamelers and C.J.N. Buisman (2006). Effects of membrane cation transport on pH and microbial fuel cell performance. *Environmental Science and Technology* **40**(17), 5206–5211.
285. S. Ayyaru and S. Dharmalingam (2011). Development of MFC using sulphonated polyether ether ketone (SPEEK) membrane for electricity generation from waste water. *Bioresource Technology* **102**(24), 11167–11171.
286. K.J. Chae, K.Y. Kim, M.J. Choi, E.T. Yang, I.S. Kim, X.H. Ren and M. Lee (2014). Sulfonated polyether ether ketone (SPEEK)-based composite proton exchange membrane reinforced with nanofibers for microbial electrolysis cells. *Chemical Engineering Journal* **254**, 393–398.
287. S.E. Oh and B.E. Logan (2006). Proton exchange membrane and electrode surface areas as factors that affect power generation in microbial fuel cells. *Applied Microbiology and Biotechnology* **70**(2), 162–169.
288. M. Ghasemi, S. Shahgaldi, M. Ismail, Z. Yaakob and W.R.W. Daud (2012). New generation of carbon nanocomposite proton exchange membranes in microbial fuel cell systems. *Chemical Engineering Journal* **184**, 82–89.
289. C.P.L. Grady, G.T. Daigger and H.C. Lim (1999). *Biological Wastewater Treatment*, 2nd Editon. New York: Marcel Dekker.
290. G.S. Jadhav and M.M. Ghangrekar (2009). Performance of microbial fuel cell subjected to variation in pH, temperature, external load and substrate concentration. *Bioresource Technology* **100**(2), 717–723.

291. B. Min, O.B. Roman and I. Angelidaki (2008). Importance of temperature and anodic medium composition on microbial fuel cell (MFC) performance. *Biotechnology Letters* **30**(7), 1213–1218.
292. B.C. Jong, B.H. Kim, I.S. Chang, P.W. Liew, Y.F. Choo and G.S. Kang (2006). Enrichment, performance, and microbial diversity of a thermophilic mediatorless microbial fuel cell. *Environmental Science and Technology* **40**(20), 6449–6454.
293. C.S. He, Z.X. Mu, H.Y. Yang, Y.Z. Wang, Y. Mu and H.Q. Yu (2015). Electron acceptors for energy generation in microbial fuel cells fed with wastewaters: A mini-review. *Chemosphere* **140**, 12–17.
294. K. Rabaey, W. Ossieur, M. Verhaege and W. Verstraete (2005). Continuous microbial fuel cells convert carbohydrates to electricity. *Water Science and Technology*, **52**(1–2), 515–523.
295. D. Pant, G. Van Bogaert, L. Diels and K. Vanbroekhoven (2010). A review of the substrates used in microbial fuel cells (MFCs) for sustainable energy production. *Bioresource Technology* **101**(6), 1533–1543.
296. K.J. Chae, M.J. Choi, J.W. Lee, K.Y. Kim and I.S. Kim (2009). Effect of different substrates on the performance, bacterial diversity, and bacterial viability in microbial fuel cells. *Bioresource Technology* **100**(14), 3518–3525.
297. P.T. Ha, B. Tae and I.S. Chang (2008). Performance and bacterial consortium of microbial fuel cell fed with formate. *Energy and Fuels* **22**(1), 164–168.
298. Y. Luo, R. Zhang, G. Liu, J. Li, M. Li and C. Zhang (2010). Electricity generation from indole and microbial community analysis in the microbial fuel cell. *Journal of Hazardous Materials* **176**(1–3), 759–764.
299. J. Yu, Y. Park, B. Kim and T. Lee (2015). Power densities and microbial communities of brewery wastewater-fed microbial fuel cells according to the initial substrates. *Bioprocess and Biosystems Engineering* **38**(1), 85–92.
300. L.T. Angenent and B.A. Wrenn (2008). Optimizing mixed-culture bioprocessing to convert wastes into bioenergy. In J.D. Wall, C.S. Harwood and A. Demain (eds.), *Bioenergy*. Herndon: ASM Press, pp. 179–194.
301. J.Y. Lee, N.T. Phung, I.S. Chang, B.H. Kim and H.C. Sung (2003). Use of acetate for enrichment of electrochemically active microorganisms and their 16S rDNA analyses. *FEMS Microbiology Letters* **223**(2), 185–191.
302. D.R. Bond, D.E. Holmes, L.M. Tender and D.R. Lovley (2002). Electrode-reducing microorganisms that harvest energy from marine sediments. *Science* **295**, 483–485.
303. A.M. Speers and G. Reguera (2012). Electron donors supporting growth and electroactivity of *Geobacter sulfurreducens* anode biofilms. *Applied and Environmental Microbiology* **78**(2), 437–444.
304. T.H. Yang, M.V. Coppi, D.R. Lovley and J. Sun (2010). Metabolic response of *Geobacter sulfurreducens* towards electron donor/acceptor variation. *Microbial Cell Factories* **9**(90), 1–15.

305. R.K. Thauer, D. Mollerzinkhan and A.M. Spormann (1989). Biochemistry of acetate catabolism in anaerobic chemotropic bacteria. *Annual Review of Microbiology* **43**, 43–67.
306. H.P. Luo, G.L. Liu, R.D. Zhang and S. Jin (2009). Phenol degradation in microbial fuel cells. *Chemical Engineering Journal* **147**(2–3), 259–264.
307. S.J. You, Q.L. Zhao, J.N. Zhang, J.Q. Jiang and S.Q. Zhao (2006). A microbial fuel cell using permanganate as the cathodic electron acceptor. *Journal of Power Sources* **162**(2), 1409–1415.
308. L. Jourdin, S. Freguia, B.C. Donose and J. Keller (2015). Autotrophic hydrogen-producing biofilm growth sustained by a cathode as the sole electron and energy source. *Bioelectrochemistry* **102**, 56–63.
309. C.I. Torres, A.K. Marcus and B.E. Rittmann (2008). Proton transport inside the biofilm limits electrical current generation by anode-respiring bacteria. *Biotechnology and Bioengineering* **100**(5), 872–881.
310. A.E. Franks, K.P. Nevin, H.F. Jia, M. Izallalen, T.L. Woodard and D.R. Lovley (2009). Novel strategy for three-dimensional real-time imaging of microbial fuel cell communities: Monitoring the inhibitory effects of proton accumulation within the anode biofilm. *Energy and Environmental Science* **2**(1), 113–119.
311. D.R. Lovley, D.E. Holmes and K.P. Nevin (2004). Dissimilatory Fe(III) and Mn(IV) reduction. *Advances in Microbial Physiology* **49**, 219–286.
312. A.P. Borole, H. O'Neill, C. Tsouris and S. Cesar (2008). A microbial fuel cell operating at low pH using the acidophile *Acidiphilium cryptum*. *Biotechnology Letters* **30**(8), 1367–1372.
313. S.A. Patil, F. Harnisch, C. Koch, T. Hubschmann, I. Fetzer, A.A. Carmona-Martinez, S. Muller and U. Schröder (2011). Electroactive mixed culture derived biofilms in microbial bioelectrochemical systems: The role of pH on biofilm formation, performance and composition. *Bioresource Technology* **102**(20), 9683–9690.
314. S. Ishii, Y. Hotta and K. Watanabe (2008). Methanogenesis versus electrogenesis: Morphological and phylogenetic comparisons of microbial communities. *Bioscience, Biotechnology, and Biochemistry* **72**(2), 286–294.
315. M. Liu, Y. Yuan, L.X. Zhang, L. Zhuang, S.G. Zhou and J.R. Ni (2010). Bioelectricity generation by a Gram-positive *Corynebacterium* sp. strain MFC03 under alkaline condition in microbial fuel cells. *Bioresource Technology* **101**(6), 1807–1811.
316. Z.J. Wang, T. Lee, B. Lim, C. Choi and J. Park (2014). Microbial community structures differentiated in a single-chamber air-cathode microbial fuel cell fueled with rice straw hydrolysate. *Biotechnology for Biofuels* **7**(9), 1–10.

Section II

Electrodes for Enzymatic Electrochemical Systems

Chapter 5

Architectures of Enzyme Electrodes Using Redox Mediators

Victoria Flexer[*,†,¶], Antonin Prévoteau[‡], and Nicolas Brun[§]

[*]*Centro de Investigación y Desarrollo en Materiales Avanzados y Almacenamiento de Energía de Jujuy-CIDMEJu (CONICET-Universidad Nacional de Jujuy), Argentina*

[†]*Centro de Investigación y Transferencia Jujuy, CIT-JUJUY, Argentina*

[‡]*Centre for Microbial Ecology and Technology, Ghent University, Coupure Links 653, B-9000 Ghent, Belgium*

[§]*Institut Charles Gerhard de Montpellier, UMR5253 CNRS, ENSCM, Université de Montpellier, France*

[¶]*vflexer@unju.edu.ar*

5.1 Introduction

An incredibly large amount of scientific publications and patents have been published proposing different electrode configurations for mediated enzyme electrodes. A whole book would not be enough to list all the different possibilities reported to date. Some important examples will be presented in this chapter, including a longer discussion on two of the most iconic systems: redox hydrogels and layer-by-layer electrostatic self-assemblies. Interestingly, while the detailed architectures are almost endless, they can all be classified in one of seven general strategies, according to which components are either immobilized on the electrode surface or free to diffuse in solution.

Here, we will concentrate on enzyme–mediator interactions. In order to focus our attention on the enzyme–mediator chemistry, the solid-state electrode will not be a particularly complex or purposely designed material. In the examples that we will discuss next, it will usually be a 2D flat electrode, often commercially available, that will act as a solid and electrically conductive support for enzyme and mediator. Some of the examples that will be presented here have later been incorporated into more complex electrode materials/architectures.[1–3] The basic principles that will be discussed here will mostly still hold for those electrodes as well.

Figure 1 summarizes the seven possible electrode architectures in which most mediated systems can be classified.

While attempting to build an enzyme electrode, the absolute must of any immobilization technique is that the enzyme does not lose its activity upon immobilization, as was already pointed out in Chapter 1. In the present chapter, we will discuss some immobilization strategies, and we will pay particular attention to techniques that make use of the redox mediator to attach the enzyme to the electrode surface. Not all immobilization strategies will yield results with the same degree of reproducibility. Moreover, not all techniques will produce the same degree of homogeneity within the enzyme–mediator film. Eventually, we have to call the reader's attention to the fact that extremely few examples have been reported of immobilization strategies that manage to control the enzyme orientation, be it toward the solid-state electrode, or toward the redox mediator, an added advantage that could potentially increase the rate of electron transfer.

5.2 Mediator in Solution

The early work on artificial redox mediators was carried out with soluble mediators, which are free to diffuse in solution.[4–8] This is schematically exemplified for an oxidase in Figures 1(a)–1(c). The easiest possible configuration is to leave both enzyme and mediator free to diffuse in solution, as depicted in Figure 1(a). Experiments with both mediator and enzyme dissolved in solution are particularly easy to perform, and do not require neither elaborate nor specialized equipment, nor other chemicals beyond enzymes, mediators and substrates. They are particularly well-fitted for studies of homogeneous kinetics (easy to control, simple theoretical treatment and good reproducibility). This simple configuration is also useful to quickly assess the compatibility and efficiency of an enzyme with new mediator molecules.[4,5,7–12]

A step forward is to immobilize the enzyme in a film at the electrode surface, but leave the mediator free to diffuse at the film–solution interface, which is schematized in Figure 1(b).[4,13–15] For example, Savéant *et al.*[4,13–15] reported extensive work on the construction of enzyme multi-layers via bioaffinity assays. The enzyme molecules were connected to the electrode surface via different soluble ferrocene derivatives.

Figure 1(c) schematically shows the example where only one enzyme monolayer is immobilized on the electrode surface, while the mediator is free to diffuse in solution.[16,17] Interesting results have been obtained for such a configuration when the catalytic activity of enzyme is retained after immobilization. In particular, the normalized rate (*i.e.* current) per enzyme is particularly high since they are located at the closest vicinity of the electrode surface regenerating the oxidized form of the mediator. The total current density is, however, limited because of the monolayer configuration.

The great advantage of using a dissolved mediator is that it maximizes the mass transfer ability of the mediator toward the enzyme, increasing the frequencies of favorable collisions for electron exchange. Also, the small size and high mobility of the dissolved mediator, as compared to immobilized mediators, likely favor an *ad hoc* approach to the prosthetic group of the enzyme, which is usually deeply buried inside the protein structure. A second advantage is that many commercially available molecules have been proved efficient as redox mediators, *i.e.* they are readily available for quick experiments, or for researchers willing to venture into the field for

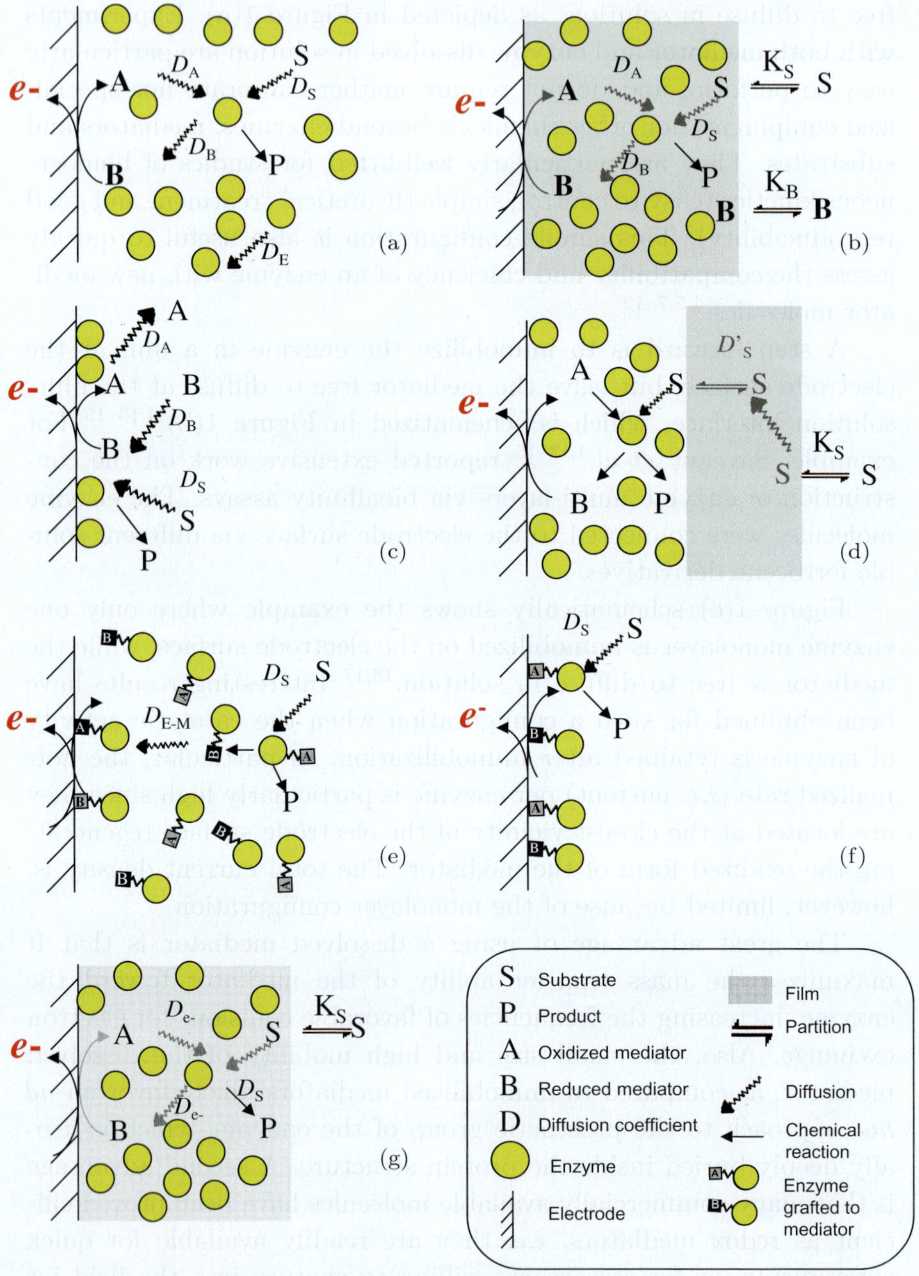
e-
A
B
D_A
D_B
S
D_S
P
D_E
(a)
K_S
K_B
(b)
(c)
D'_S
(d)
$D_{E\text{-}M}$
(e)
(f)
$D_{e\text{-}}$
(g)
S Substrate
P Product
A Oxidized mediator
B Reduced mediator
D Diffusion coefficient
Enzyme
Electrode
Film
Partition
Diffusion
Chemical reaction
Enzyme grafted to mediator

the first time. These include ferricyanide, different quinones, soluble derivatives of ferrocene, viologen and its derivatives, methylene blue, among others.[10,12,18–20]

A third advantage is the ease to assess the impact of the mediator concentration (either absolute concentrations or relative to the enzyme concentration). The limit is only given by the solubility of the mediator and the associated cost of a potentially large mediator concentration. Therefore, if solubility permits, it is easier to work at mediator concentrations that are large enough to secure that the kinetics are substrate-enzyme limited, and therefore, the electrochemical signal will be proportional to the substrate concentration in a wider concentration range (the mediator concentration is not limiting the kinetics). It is usually much more difficult to achieve large mediator concentrations with immobilized mediators, and hence, the amperometric signal becomes saturated in substrate at lower concentrations (lower $K_{MS,app}$, see Section 1.4 in Chapter 1), because the mediator concentration is also limiting the kinetics.

One of the main disadvantages of such a configuration is that much larger quantities of mediator are usually needed when working in typical electrochemical cells (10–100 mL). If the enzyme is also free in solution to diffuse (Figure 1(a)), large amounts of enzyme will be needed as well. This limitation has been partially overcome in disposable (single use) strips for glucose monitoring resorting to reaction compartments with volumes in the order of the microliter. In many commercially available strips (at time of writing this chapter), the working electrode is modified with a dry mixture of enzyme,

←

Figure 1. Possible electrode configurations for mediated enzyme electrodes. The electron transport direction is presented for an anodic process and does not modify the principles illustrated: (a) enzyme and mediator in solution, (b) immobilized enzyme (more than one monolayer) and mediator in solution, (c) immobilized enzyme monolayer and mediator in solution, (d) enzyme and mediator trapped behind a partially permeable membrane, (e) mediator grafted to enzyme in solution (chemical bond or modified by techniques of molecular biology), (f) mediator grafted to enzyme attached to the electrode surface (chemical bond or modified by techniques of molecular biology), and (g) enzyme and mediator coimmobilized within a film at the electrode surface.

mediator and other additives that solubilize when in contact with liquid blood, and the mediated enzymatic reaction takes place in solution. Evidently, this type of configuration only allows for one single experiment/measurement. Moreover, the mixing of sample and mediator/enzyme forbids its use *in vivo* (contamination).

5.3 Enzyme Trapped Behind a Membrane

The first attempts at immobilizing enzymes were based on the entrapment of a concentrated enzyme solution behind a membrane.[21–24] This is schematically shown in Figure 1(d). This can be accomplished by wetting a nylon porous net and fixing it against the electrode surface. The membrane must allow the enzymatic substrate to diffuse across. Generally, the small size of the mediators ($<10^3$ g mol^{-1}) will also let them cross the membrane. Moreover, depending on the membrane, the enzyme could also get partially lost into solution. It is therefore complex to efficiently coimmobilize enzyme and mediator behind a membrane without further immobilization techniques. However, the addition of a well-chosen membrane over any enzyme/mediator architecture[4] can be advantageous to selectively hinder the mass transfer of either interfering dissolved species, or enzymatic substrate for increasing its linear range of detection in the case of biosensor.

5.4 Enzymes with Grafted Redox Mediators

Another strategy has been the attachment of enzyme to the redox mediator to ensure their proximity.[25–31] This bonding can either be a non-specific chemical bonding, or the enzyme can be modified by techniques of molecular biology. This configuration is schematically shown in Figures 1(e) and 1(f). We understand by non-specific chemical bonding the use of classical organic synthesis techniques that unspecifically target certain chemical groups that are abundant in the outer protein structure, such as primary amine groups or sugar residues.

Another strategy to favor a strong interaction is to covalently attach the mediator to the solid electrode surface and to subsequently bind the enzyme to the mediator.[32,33] This is schematically shown in Figure 1(f). Willner *et al.* have been pioneers in pursuing this route, where they graft a molecule with a certain affinity to the electrode (for example a thiol, if working with a gold electrode). The mediator is grafted to the opposite end of this molecule, and the redox active center of the enzyme is grafted in turn to the mediator. The addition of the apoenzyme (the enzyme deprived from its active center) allows the reconstitution of the holoenzyme (functional enzyme) on the surface by affinity (exemplified in Figure 2). While elegant on the principle, this configuration is complex and only allows to work with a monolayer at the best and is therefore not able to provide substantial current densities.

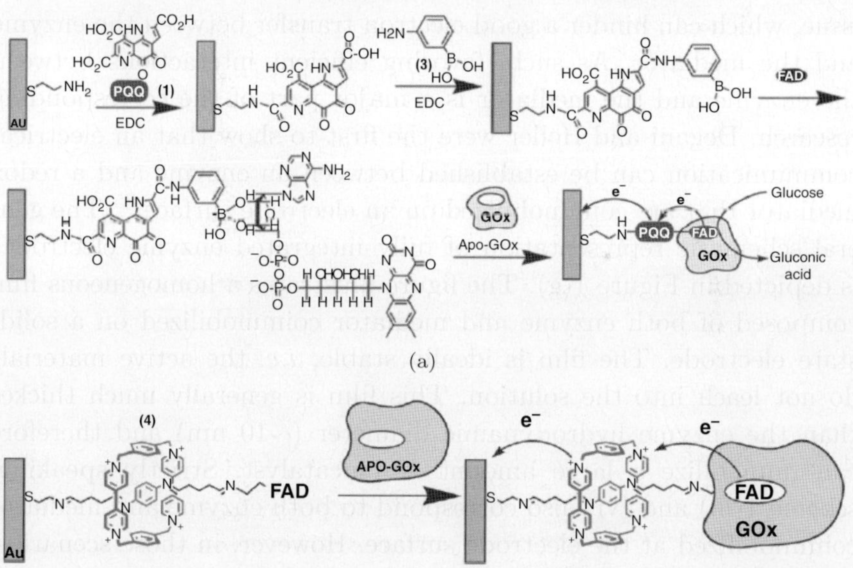

Figure 2. (a) Two different strategies for reconstituting an holoenzyme on an electrode surface. Assembly of a PQQ/FAD monolayer on a thiol-modified Au electrode via a boronic acid bridge, and the reconstitution of holo-glucose oxidase (GOx) on the FAD cofactor sites. (b) The reconstitution of GOx on an FAD cofactor that "stoppers" the cyclophane on the molecular wire. The redox enzyme is contacted with the electrode by means of the electrochemically redox unit along the wire. Reprinted with permission from Ref. 33. Copyright (2009) John Wiley and Sons.

5.5 Mediator and Enzyme Coimmobilized at the Electrode Surface

If we are looking into minimizing the quantities of enzyme and mediator to be used, in addition to avoiding sample contamination with enzyme/mediator, and if we would like to pursue long-term operation of enzyme electrodes, the answer is to coimmobilize the enzyme with the mediator. In the enzyme electrode literature, these configurations are typically referred to as *integrated systems*, *integrated enzyme electrodes*, or *fully integrated enzyme electrodes*.[34]

A successful enzyme immobilization technique is only the first step to design an efficient integrated enzyme electrode. The second step is to coimmobilize the redox mediator. The limited mobility due to the immobilization within the electrode matrix is a crucial issue, which can hinder a good electron transfer between the enzyme and the mediator. As such, favoring efficient interactions between the enzyme and the mediator is a major part of the corresponding research. Degani and Heller were the first to show that an electrical communication can be established between an enzyme and a redox mediator that are coimmobilized on an electrode surface.[35] The general schematic representation of fully integrated enzyme electrodes is depicted in Figure 1(g). The figure illustrates a homogeneous film composed of both enzyme and mediator coimmobilized on a solid-state electrode. The film is ideally stable, *i.e.* the active materials do not leach into the solution. This film is generally much thicker than the enzyme hydrodynamic diameter (∼10 nm) and therefore can immobilize a large amount of biocatalyst. Strictly speaking, scheme 1(iii) and (vi) also correspond to both enzyme and mediator coimmobilized at the electrode surface. However, in those scenarios, only an enzyme monolayer is immobilized. In the general case summarized by scheme 1(vii), we would like to discuss the particular case where a large amount of enzyme, *i.e.* more than a monolayer is immobilized at the electrode surface.

The enzymatic substrate undergoes a partition between the bulk solution and the active film, which is represented by partition coefficient K_S. The substrate (if sufficiently concentrated) can diffuse along the whole film to reach all enzyme molecules, regardless of their proximity to the bulk solution. Enzyme and mediator have only limited mobility within the film, the degree of limited mobility being defined by the particular type of immobilization technique that we will describe shortly. For an efficient enzyme layer, the partition coefficient for the substrate should ideally be close to unity. The enzyme/mediator film should be permeable enough to allow for fast diffusion of the enzymatic substrate deep enough within the film. Moreover, the enzyme product(s) should be able to diffuse out of the film.

In most literature reports (see Sections 5.5.1 and 5.5.2 for specific examples), the redox mediator is covalently grafted to an otherwise non-redox polymeric chain. As such, the mediator cannot diffuse *stricto sensu* across the enzyme film, although a certain level of hydration should allow for local mobility of the flexible polymeric chain. As a result, reduced and oxidized redox mediators periodically encounter each other and can exchange charge by incoherent *electron hopping* (electron self-exchange). Sequential self-exchange steps allow the long distance electron transport from an enzyme to an electrode surface (illustrated in Figure 3). This electron transfer mechanism can be described as diffusional such as initially described by the Dahms–Ruff model for redox compounds in solution.[36–39] The model has been extended to the particular example of supramolecular assemblies of electroactive species with limited mobility by Blauch and Savéant.[40,41] Indeed in these type of systems, the mobility of the redox mediators is restrained around an equilibrium position, usually not allowing the redox mediator to interact beyond its nearest neighboring species. This "diffusional" mobility is driven by the concentration gradient of reduced mediator (*i.e.* electrons). This gradient is generated by the enzymatic catalysis (electron source) and the heterogeneous electron transfer at the solid-state electrode (electron sink). This electron diffusion ability can be quantitatively assessed by an apparent electron diffusion coefficient, which has been noted in Figure 1(g) as D_{e-}.

An extremely large field of research has focused on developing different types of enzyme and mediator coimmobilization strategies.

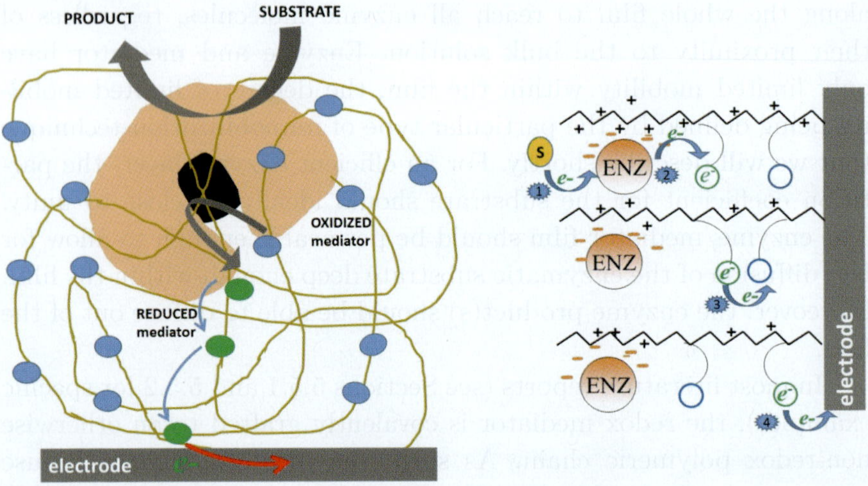

Figure 3. Two schematic representations of the mediated electron transfer within an enzyme–polymer coimmobilized system. The substrate (S) oxidation is catalyzed by the enzyme (step 1) which transfers electrons to a redox mediator in its oxidized form (step 2). Electron self-exchange between adjacent redox centers is driven by a redox gradient (step 3) until the heterogeneous electron transfer to the electrode takes place (step 4).

The early work on coimmobilization proposed simple techniques, such as coadsorption of enzyme and mediator[42]; or its coadsorption with the aid of carbon paste.[43–45] While appreciated for their simplicity, the lack of strong bonding between the electroactive species and the electrode surface generally implies a slow release of enzyme and/or mediator into the surrounding solution. This will lead to a loss of electrochemical activity and sample contamination.

More advanced enzyme/mediator coimmobilization strategies that have successively been reported by different groups include

(i) layer-by-layer assembly by bioaffinity reactions, including antigen-antibody[46] and avidin–biotin[15,20,47];
(ii) Langmuir–Blodgett films[48];
(iii) layer-by-layer electrostatic self-assembly[49–61];
(iv) immobilization within the matrix of an electropolymerized conducting polymer[62]; and
(v) redox hydrogels.[1–3,27,63–78]

The first three techniques are recognized for a very precise control over the electrode architecture and the enzyme/polymer organization within the film. They allow for a precise control in the design and spatial distribution of mediator and enzyme. They even allow for a spatial confinement of different enzymes and/or mediators within a restricted region of the immobilized film, *i.e.* at a specific distance of the solid-state electrode, when a multi-enzyme/multi-mediator systems is constructed. Conversely, the last two strategies are less controlled and sometimes present lower level of reproducibility in terms of performances. However, several specific characteristics make redox hydrogels an *ad hoc* architecture to achieve high current densities (*vide infra*). Finally, it is important to stress the point that none of the listed techniques allow for spatial orientation of the enzyme, neither with respect to the solid-state electrode, nor with respect to the redox mediator.

In Sections 5.5.1 and 5.5.2, we will concentrate on a more detailed discussion on hydrogels, and layer-by-layer electrostatic self-assembly. Conducting polymers and the analogy between 3D structures formed by classical and modern hydrogels will be discussed in Chapter 8. While hundreds of scientific reports have been written independently on either redox hydrogels or layer-by-layer electrostatic self-assemblies, only a very few pages have been written comparing the two more common mediator/enzyme coimmobilization structures, or highlighting their common points. Not surprisingly, the two techniques share the same electron transfer mechanism. This is why we choose to introduce it as a generic mechanism not pertaining to any specific immobilization technique. Following the same argumentation, in addition to incorporating the same enzymes into the coimmobilized film, the two techniques also make use of extremely similar polymeric structures as fundamental building blocks. Some redox polymers, such as poly(allylamine) (PAH)-[Os(4,4′-dimethyl-2,2′-bipyridine)$_2$Cl]$^{+/2+}$,[50,79] and PAH-ferrocene[53,80] have even been alternatively reported as being able to form both hydrogels and layer-by-layer self-assembled structures. Finally, we should recall that in both techniques the polymer has a double role, it is both responsible for electron transfer from the enzyme to the solid-state-electrode and it is a key building block acting as a *cement* or *glue* immobilizing the enzyme on the electrode surface.

Following the historic reports, we will first introduce redox hydrogels, and we will continue with layer-by-layer self-assemblies. While there is no doubt that the former has been the technique producing the highest reported current densities in the field, the latter offers a large degree of flexibility and variability in the possibility of complex and ordered structures that can be built-up. Both techniques share the great advantage of extremely reduced costs, without the requirement of expensive or complex equipment for their implementation. Finally, they are amenable for automation and are potentially industrially scalable.

5.5.1 Redox hydrogels

After the pioneering work on redox mediation in immobilized systems,[35] it was again the group directed by Adam Heller at the University of Austin, Texas, that came out with the idea of immobilizing enzymes and redox mediators in redox hydrogels.[81–84] A redox hydrogel can be described as a 3D network of hydrophilic polymeric chains where the redox mediator(s) has been covalently grafted. The enzyme is entrapped within the 3D network. Redox hydrogels allow for the immobilization of enzymes within a film mostly hydrated, favoring excellent mass transfer of dissolved substrates and products. Furthermore, redox hydrogels are still able to efficiently transfer electrical charges via the self-exchange mechanisms described above. Redox hydrogels can be described as the perfect blend of immobilized and mediated systems. Many of the highest catalytic current densities on fully integrated systems have been reported on systems based on redox hydrogels.[1,64,70,73,85,86] If redox hydrogels are still very popular as an immobilization and mediation strategy almost 30 years after they were first introduced, it is because of the wide range of advantages they present:

(i) A large proportion of the enzymes immobilized throughout the volume of the redox hydrogel are electrically wired to the electrode surface. Therefore, large amounts of enzyme can be deposited on the electrode surface, allowing for high biocatalytic current densities.

(ii) It is a rather *soft* immobilization technique. Because the hydrogel does not hold the enzyme in a tight configuration or emplacement, and many of its properties closely resemble those of aqueous media, the enzyme is not forced into conformational changes or deformations that could eventually negatively affect its catalytic activity.

(iii) Enzyme and redox moieties conserve a certain mobility around their equilibrium position, likely favoring collisions frequencies in adequate orientations and therefore electron transfer reaction rates (enzyme-mediator, and mediator-mediator).

(iv) Because the hydrogel has a high water content, substrate and product diffusion is excellent within the film. In addition, the substrate partition coefficient between the bulk solution and the film is very close to one. However, note that this is also true for interfering species potentially present in solution.

(v) Despite the resemblance with an aqueous media and the (relative) associated mobility of species, all components are chemically bonded to the polymeric structure. Therefore, enzyme and mediator leaching from the film is reduced to a minimum, avoiding both contamination of the sample and loss of activity of the enzymatic electrode.

5.5.1.1 *Redox hydrogel components*

The key components of the redox hydrogel are the redox polymer, the enzyme, and a cross-linking agent. In order to prepare redox-hydrogels, the three components are vigorously mixed and deposited on the surface of a solid-state electrode where they are let to slowly *dry*. The cross-linking reaction is produced upon water evaporation, which roughly takes 12–24 h. The enzyme will be chosen according to the desired final application (sensing a given analyte, electricity production in a biofuel cell operating under certain conditions, etc.). The nature and proportion of the polymer(s) and reticulating agent allow for further tuning of the redox hydrogel.

We have already discussed in Chapter 1 that the selection of enzymes most often used to prepare enzyme electrodes is relatively limited. Conversely, the large selection of redox mediators is almost unlimited. While some researchers, including companies manufacturing commercial glucose sensors, choose to work with their own enzymes produced by techniques of molecular biology,[19,87–90] for those who do not willing, or are not able to do so, there is quite a large selection of commercial manufacturers and retailers of purified enzymes. However, when we look for redox mediators that are chemically grafted to a polymer or macrostructure (*i.e.* the mediators needed for coimmobilized electrodes), there are much less commercially available options. This means that each research group, or industrial company chemically synthesizes its own redox mediators according to their needs. None of the redox mediators that will be described in Sections 5.5.1.1.1 are commercially available.

5.5.1.1.1 *Polymer*

The polymeric backbone is the fundamental scaffold of the hydrogel. By definition, polymers used to form redox hydrogels must be hydrophilic and water soluble. Polymers that have repeatedly been reported to yield good results include pure or copolymers of polyvinylimidazole (PVI), polyvinylpyridine (PVP), polyacrylamide (PAA), and poly(allylamine) (PAH).[64] The redox mediator is in turn *grafted* to the polymeric backbone. This chemical bonding is produced in a synthetic step upstream to the electrode modification with all the other components (see below). Polymers with grafted redox centers such as the ones we are describing are often called in the literature *redox polymers.* Several families of redox compounds have been reported to have been successfully grafted to different polymeric backbones. Organic molecules, such as quinones, have been reported. However, the corresponding redox hydrogels showed low apparent diffusion coefficients (D_{e-}, see above), while the chemical stability of quinones is limited.[64] Conversely, organometallic compounds have repeatedly demonstrated good performance as grafted mediators. Among these, osmium-based complexes have stood out

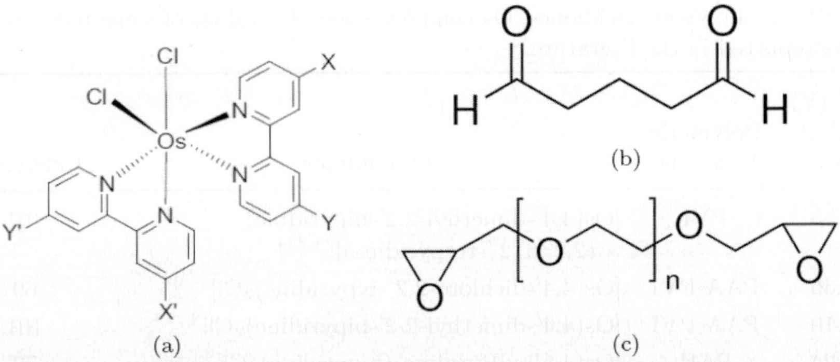

Figure 4. (a) Chemical structure of glutaraldehyde. (b) Chemical structure of poly(ethylene glycol) diglycidyl ether (PEGDGE). (c) Example of an osmium complex commonly used as a redox mediator when grafted to a polymeric backbone. This example shows two chloride and two bipyridine ligands. It is possible to graft complementary groups in positions X and/or Y. The nature of these ligands (especially their electrophilicity) affects the $E^{0\prime}$ of the complex which can therefore be tailored for specific purpose.

and have been repeatedly used by several research groups both for biosensor and enzymatic biofuel cell applications.[64,72,77–79,91–97]

Osmium complexes commonly display a coordination number of 6, and ligands often include derivatives of pyridine or imidazole. This is schematically shown in Figure 4(a). Chemically modifying these ligands with different functionalities (chloro, amino, methoxy, etc.) shift the formal potential (E^0) value of the complex by modifying the electrophilicity surrounding the central osmium ion. This offers the opportunity to synthetize complexes with an extremely wide range of E^0, with reports ranging from -0.43 to $+0.8$ V.[77,98] With such a wide potential range, these osmium complexes have been used as mediators for a large variety of enzymes, both oxidases and reductases.

Considering that the nature of the ligands shifts the E^0, grafting the redox complex to a polymeric chain via one of the ligands will further affect the E^0 value. Moreover, E^0 will evolve differently depending on whether the complex is grafted by either one or two ligand positions. On the contrary, when the redox complex is attached via a long molecular spacer or *tether* (see below), E^0 is not substantially

Table 1. Polymer backbones, Os-complexes and E^0 values of some redox polymers reported in the literature.

$E^{0\prime}$ (V) *vs* Ag/AgCl	Polymeric backbone	Os complex	References
+0.55	PVI	[Os(4,4′-dimethyl-2,2′-bipyridine)(2,2′,6′,2″-terpyridine)]$^{2+/3+}$	101
+0.36	PAA-PVI	[Os(4,4′-dichloro-2,2′-bipyridine)$_2$Cl]$^{+/2+}$	69
+0.10	PAA-PVI	[Os(4,4′-dimethyl-2,2′-bipyridine)$_2$Cl]$^{+/2+}$	101
+0.24	PAH	[Os(4-alkylPyridine)(bipyridine)$_2$Cl]$^{+/+2}$	79
−0.07	PVP	[Os(4,4′-dimethoxy-2,2′-bipyridine)$_2$Cl]$^{+/2+}$	68
−0.16	PVI	[Os(4,4′-diamino-2,2′-bipyridine)$_2$Cl]$^{+/2+}$	67
−0.19	PVP	[Os(N,N′-dialkylated-2,2′-biimidazole)$_3$]$^{2+/3+}$	73

modified. The choice of a complex with a specific E^0 will depend on several considerations depending on the *wired* enzyme and on the targeted application. For example, low E^0 values are generally preferred for a modified anode in order to (i) increase the operating voltage of a biofuel cell or (ii) avoid the direct oxidation of interfering species in biosensors. However, the choice of a complex with too low E^0 would decrease its kinetics for collecting the electrons of an enzyme. Furthermore, decreasing E^0 has been shown to increase exponentially the O_2 reduction on the complex in aerobic conditions.[99] This last phenomenon could also lead to substantial issues for both aforementioned applications.[100] Table 1 presents a few examples of osmium-based redox polymers that have been reported for redox hydrogel applications. The reader should appreciate the large variation in E^0, upon producing minimal changes on the ligands. In addition, some complexes present charge +1/+2, while others present charge +2/+3 (depending on the oxidation state of the osmium center being +2 or +3, respectively). Once grafted, the complex adds its positive charge to the already cationic backbone (at circumneutral pH). Cationic polymers have been purposely selected to favor electrostatic interactions with many enzymes that are negatively charged at physiological pH, such as GOx.[63,64]

If osmium complexes have become so extremely popular, this is because, in addition to the wide potential window to choose from,[77,98] they present all the required characteristics of an efficient redox mediator: they are equally stable in the reduced and oxidized state; they efficiently transfer electrons with a large variety of enzymes; they show high self-exchange rate constant which favors redox conduction within the hydrogel; they perform (quasi) reversible electron transfer with different electrode surfaces (low activation overpotentials) and their E^0 values are mostly insensitive to pH. On the other hand, we should mention their relative high price, as well as their potential toxicity, if they would leach from the electrode surface, for an *in vivo* application. Even though thorough studies would have to be performed before approval by the relevant regulatory agencies, we should remark that the amounts of osmium required are minimal, and osmium tetraoxide has been routinely used in some medical treatments.[102,103]

It is possible to optimize the polymeric chain in order to improve the hydrogel performance. For example, amines in the polymeric backbone can be quaternized, which will increase the linear charge density on the polymer, and thus will increase the polymer hydrophilicity. This will improve the hydration of the whole hydrogel 3D structure, which will induce better flexibility and will therefore favor contact between enzyme and redox centers.[96] There is, however, a very delicate equilibrium and a compromise to find in the fraction of amine sites to be quaternized. Amine functions are also responsible for the reticulation reaction that will covalently bound polymer and enzyme together. Therefore, quaternization reactions should not take place on all available amine sites. Otherwise, the lack of available sites for reticulation will result in poor bonding of the different active components and partial polymer/enzyme leaching into the solution.

Since the first report on redox hydrogels, probably the most important breakthrough in the field has been the introduction of a spacer between the Os redox center and the main polymeric backbone. By addition of a 13-atom spacer, Mao *et al.* observed an unprecedented increase in the apparent electron diffusion coefficient, D_{e-}.[73] Figure 5(a) shows the molecular structure of the first reported

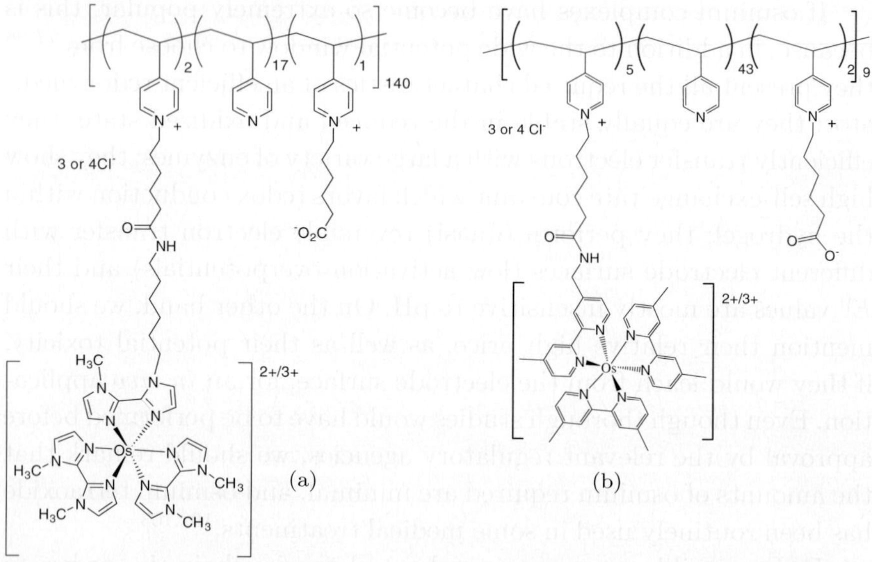

Figure 5. Molecular structure of two different redox polymers[73,85] showing long spacer arms separating the redox center from the polymer backbone. (a) PVP-[Os(*N*,*N*-dialkylated-2,2-biimidazole)$_3$]$^{2+/3+}$. (b) PVP-(4,4-dimethyl-2,2-bipyridine)2(4-aminomethyl-4-methyl-2,2-bipyridine)]$^{2+/3+}$.

polymer introducing such a spacer, which also includes a certain proportion of quaternized amines that will additionally increase the hydrophilicity of the polymer and the resulting hydrogel. An apparent electron diffusion coefficient of 5.8×10^{-6} cm^2 s^{-1},[73] a value in the same order of magnitude of diffusion coefficients for small soluble species in aqueous solutions was reported for hydrogels prepared with GOx and this polymer. The long and flexible *tethers* allow the redox centers to sweep electrons from large volume elements within the hydrogel 3D structure and to efficiently collect electrons from enzymes. Figure 5(b) shows the molecular structure of an analogous polymer designed following the same concept of a spacer arm or long tether. In this case, an eight-atom spacer was introduced and the ligands around the Os center were chosen so as to obtain a much higher E^0 value, so that the polymer would be suitable for wiring of laccase enzymes. An apparent electron diffusion coefficient of

7.6×10^{-7} cm^2 s^{-1} has been reported for hydrogels built-up using this polymer.[85]

5.5.1.1.2 *Cross-linker*

It is necessary to covalently bind the enzyme to the redox polymer in order to avoid leakage of the biocatalyst and therefore guarantee a certain performance stability. The most frequently used cross-linkers are glutaraldehyde and PEGDGE. The chemical structures of these molecules are depicted in Figures 4(b) and 4(c) respectively. PEGDGE is an oligomer, and it is commercially available with different average chain lengths, most commonly 200, 400 and 600 g mol^{-1}. The cross-linking reaction occurs between the carboxylic (glutaraldehyde) or epoxy (PEGDGE) group, and both amino groups (enzymes) and heterocycles (polymeric backbone or redox complexes). Depending on the chemical identity, the chain length, and the concentration of the reticulating agent within the enzymatic/polymeric/cross-linker mixture, the 3D hydrogel structure will be more or less flexible and hydrated, impacting its swelling ratio, substrate and product mass transport, as well as electron transport properties.

5.5.2 Layer-by-layer electrostatic self-assembly

The technique of electrostatic layer-by-layer self-assembly was first proposed by Decher *et al.*[104,105] It was originally proposed for polyelectrolytes, namely for charged linear polymers. Although it was not long before it was realized that enzymes can also be considered ionic macromolecules, and hence they could also be used as building blocks for self-assembled layer-by-layer electrodes, Hodak *et al.*[53] were the first ones to prove that enzymes immobilized on layer-by-layer structures could be electrically wired and efficiently transduce a biochemical process into an amperometric response. A schematic representation of the processes involved to build up a layer-by-layer electrostatically self-assembled film is shown in Figure 6(a). The process starts by chemically modifying an electrode so as to confer it a certain degree a surface charge. This initial charge can be produced by chemical oxidation, self-assembling of charged thiols or silanes, hydrazine

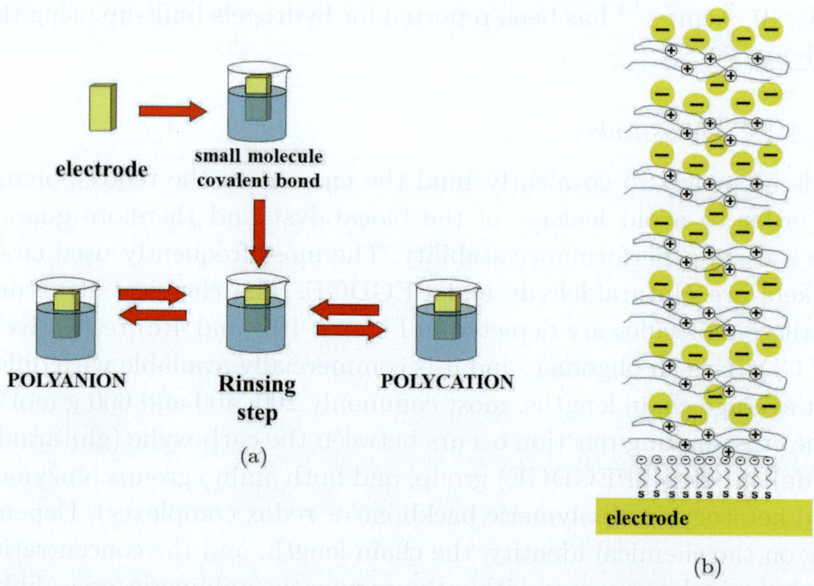

Figure 6. (a) Schematic of the layer-by-layer film deposition process using a flat electrode and beakers. The process starts with charging the electrode surface by chemical modification. Each electrode modification step is followed by a washing step. The layer-by-layer process is accomplished by successive immersions of the electrode in a solution containing a polycation and a polyanion, with washing steps in between. The four steps are the basic build-up sequence for the simplest film architecture $(A/B)_n$. The build-up of more complex structures requires only additional beakers and a different adsorption sequence. (b) Simplified molecular scheme of layer-by-layer (seven bilayers) architecture started by the modification of the electrode with a negatively charged thiol, followed by the alternate deposition of a cationic polymer and a negatively charged enzyme.

reaction, etc. The electrode is washed after this first chemical reaction so that only charges that are chemically bonded to the electrode surface remain (*i.e.* no chemicals will leach from the electrode upon immersion of the electrode in the next solution).

Let us assume that the initial surface charge is negative, as depicted in Figure 6(b). Next, the electrode is dipped in a solution containing a polyion with the opposite charge. The polycation will electrostatically adsorb on the electrode. Consecutively, the electrode is washed (in either pure water or a diluted buffer solution) and the non-specifically adsorbed molecules are washed away. The self-assembly process produces charge overcompensation. Therefore, after adsorption of the first layer of polycation, the initial charge is reversed. The electrode bears now an overall positive charge. If we now dip this electrode in a solution containing a polyion of opposite charge, a new electrostatic adsorption step will be produced, once more, with concomitant charge overcompensation,[104–106] *i.e.* in the present example, the electrode bears now overall negative charge. The process is thus repeated, by sequentially alternating dipping steps in solutions of polyelectrolytes with opposite charges with washing steps. Figure 6(b) is a schematic representation of the resulting layer-by-layer architecture for the specific example of a negatively charged enzyme, a positively charged polymer, and an initial charged self-assembled layer of a thiol with negative charge on the opposite end. Following the general nomenclature in the layer-by-layer jargon, the structure shown in Figure 6(b) would be denoted: electrode-material/short molecule/(polymer/enzyme)$_7$, for example Au/MPS/(PAH-Os/GOx)$_7$, if the solid state electrode would be gold, the short molecule imparting the first charge would be mercaptopropancsulfonate (MPS), and if polymer and enzyme would, respectively, be PAH-Os and GOx, see Section 5.5.2.1.

By alternating immersion steps in solutions of polyelectrolytes of opposite charge, and washing steps, researchers can build up structures of the desired thickness, *i.e.* structures of the desired number of layers. Moreover, more complex architectures can be built up by alternating more than two components (one polyanion and one polycation). In order to achieve multi-component ordered, complex structures, each component must be adsorbed at a different deposition step, and always respecting the rule of opposite charge on successive dipping steps.

Finally, the schematic representation of segregated polymer and enzyme depicted in Figure 6(b) is indeed an over simplistic representation of the real structure. There is ample experimental evidence pointing toward an advance interpenetration of the individual layers.[106,107] This is indeed most important, in view of the properties and functionality of the layered structure. An efficient enzyme–mediator electron transfer would probably not be possible, or it would be extremely inefficient, if the layers would not show an important degree of interpenetration.

5.5.2.1 *An iconic example: The PAH-Os/GOx system*

A model system that has been largely studied is PAH-Os/GOx. Most of the studies on the PAH-Os/GOx system have been carried out on gold electrodes. Calvo *et al.* initially modified the gold substrate by the covalent bonding of a short thiol molecule. It has been known for many years that thiol groups have a very strong affinity for Au, Ag and Cu, among other metals. When gold is immersed in a dilute aqueous or ethanolic solution of thiols, these will spontaneously self-assemble on ordered thin layers.[108] By choosing a hydrocarbon chain with the right functional group in the opposite end to the thiol, it is easy to impart charge (or any desired chemical functionality) to the metallic surface. Calvo *et al.*[50] proposed to modify gold with a short, three-carbon thiol modified with a negative group, MPS ($HSCH_2CH_2CH_2SO_3H$). A short thiol is ideally suited for an amperometric electrode. Long thiols form very well organized and ordered structures, which might become something close to an electrical insulator and considerably slow down, if not completely avoid the electron transfer from mediator to the solid state electrode.[109] Conversely, short thiols form more disordered layers[110] with many defects, that while conferring charge to the surface, they still allow for efficient electron transfer towards the gold underlying surface.

We have already introduced the enzyme GOx in Chapters 1 and 3. GOx has an isoelectric point of 4.2. This means that in solutions of pH above that value, it will be negatively charged, the higher the pH, the higher the electrical charge on the enzyme surface.

PAH-Os denotes a weak and commercially available polyelectrolyte, poly(allylamine), to which an Os redox mediator, [Os(4-alkylpyridine)(bipyridine)$_2$Cl]$^{+/+2}$ has been covalently grafted. Almost every single word that we have written in Section 5.5.1.1.1 applies as well to the description of redox mediators/polymers used to build up layer-by-layer enzyme electrodes, and therefore we will not repeat those concepts. Let us simply recall that most molecular wires for mediator/enzyme coimmobilized systems are not commercially available, and therefore they must be chemically synthesized in the laboratory where they will be further employed. This is also the case for PAH-Os, which again consists of a commercially available non-redox polyelectrolyte to which a six-ligand laboratory-synthesized Os complex is chemically grafted.[111]

Figure 7 clearly demonstrates how the increasing number of adsorption cycles produces an almost linear increment in the amount of adsorbed active materials, as it is to be expected from a technique that successively adds up more building blocks on each dipping cycle. Figure 7(a) shows how the redox charge in the absence of enzymatic substrate increases with the amount of assembled layers. The shape of the cyclic voltammograms is in good agreement with that of a thin layer of electroactive species adsorbed on an electrode surface and performing reversible electron transfer, with almost no potential difference for the anodic and cathodic peaks, and the 90 mV peak width at medium height ($n = 1$). Therefore, the amount of redox

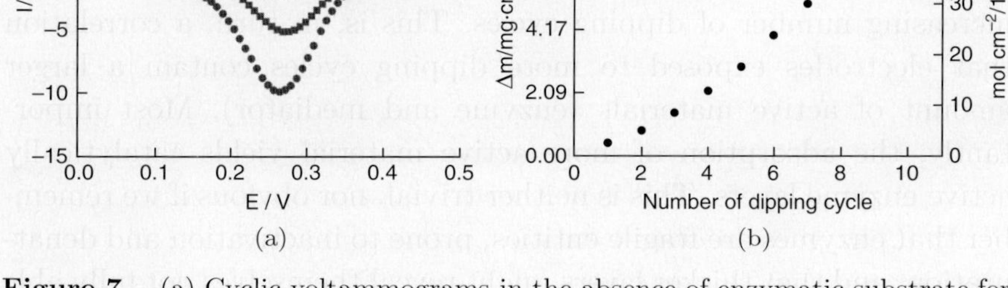

Figure 7. (a) Cyclic voltammograms in the absence of enzymatic substrate for $(\text{PAH-Os/GOx})_n$ multi-layers. Triangles, $n = 2$; squares, $n = 4$; circles, $n = 6$. Reprinted with permission from Ref. 50. Copyright (2005) American Chemical Society. (b) Total amount of adsorbed GOx (from a 1 μM solution) as determined by QCM for $(\text{PAH-Os/GOx})_n$ multi-layers, *vs* number of dipping cycles. Reprinted with permission from Ref. 49. Copyright (2005) The Royal Society of Chemistry.

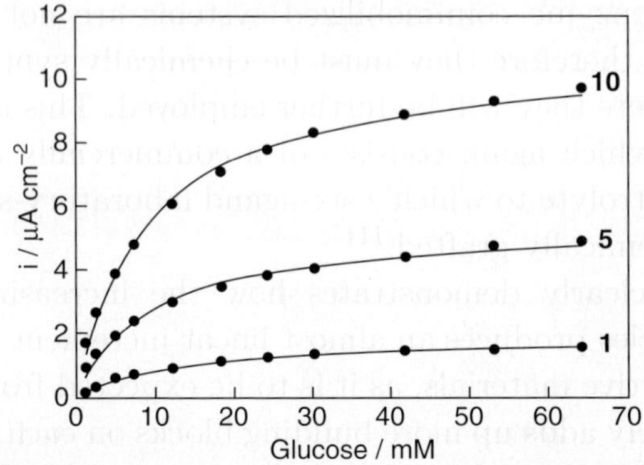

Figure 8. Calibration curves as a function of bulk glucose concentration for layer-by-layer multi-layered structures of (PAH-Os/GOx)$_n$. n is variable and shown to the right of each of the calibration curves. Reprinted with permission from Ref. 49. Copyright (2005) The Royal Society of Chemistry.

charge is linearly related to the total amount of PAH-Os polymer that has been deposited on the electrode. Figure 7(b) shows that total amount of adsorbed enzyme (GOx) increases in every dipping cycle, as determined by quartz crystal microbalance (QCM).

Figure 8 depicts calibration curves (current density as a function of enzymatic substrate concentration) for (PAH-Os/GOx)$_n$ multilayers prepared by electrode exposure to a different number of dipping cycles (namely 1, 5, and 10). The figure shows a clear increase in catalytic current for the whole glucose concentration range with increasing number of dipping cycles. This is, in turn, a correlation that electrodes exposed to more dipping cycles contain a larger amount of active materials (enzyme and mediator). Most importantly, the adsorption of more active material yields catalytically active enzyme layers. This is neither trivial, nor obvious if we remember that enzymes are fragile entities, prone to inactivation and denaturation, and that thicker layers might reveal themselves not fully able to deal with mass transfer phenomena within the film.

A very interesting phenomenon that takes place when working with weak polyelectrolytes is that the surface charge on both polymers and enzymes will be a function of the pH of the dipping solutions.[112,113] While extreme pH variations are not encouraged

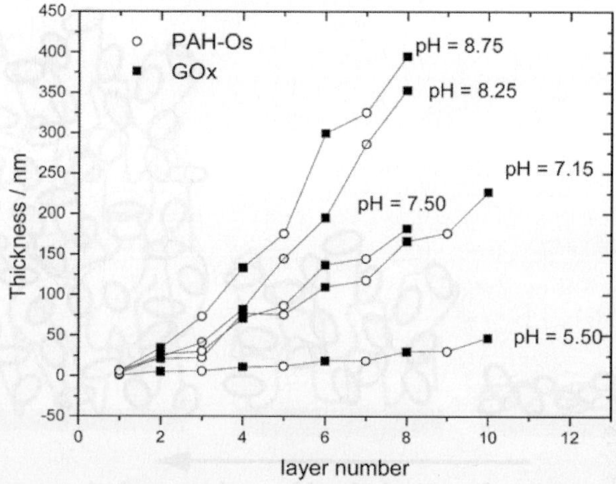

Figure 9. Ellipsometric thickness as a function of the number of self-assembled layers for $(\text{PAH-Os})_n(\text{GOx})_m$ grown from PAH-Os solutions of different pH, as depicted to the right of the curves. Reprinted with permission from Ref. 61. Copyright (2006) American Chemical Society.

because of the risk of enzyme denaturation, some degree of freedom is still possible, and the impact of changing thc pH by only one unit can be astounding. Figure 9 is an example of such behavior. The curves show the ellipsometric thickness increase of PAH-Os/GOx multi-layers that have been formed by dipping an electrode in solutions of constant pH for GOx, and varying pH for PAH-Os. PAH-Os is a week polyelectrolyte whose charge density is a function on the degree of protonation of the amine groups. An increasing pH leads to a decrease of the linear charge density of PAH-Os. At slightly acidic pH, the polymeric chains will be heavily charged, while at pH above 8, charge on polymeric chains will be limited. As reported by Rubner *et al.*,[112,113] it is to be expected that there is a thickness transition from very thin adsorbed layers to abnormally thick films as the pH of PAH-Os dipping solution is increased. At a critical high surface charge density, the entropic penalty for spreading a chain into a flat conformation on the surface is overcome by the enthalpic gain to the free energy of adsorption. However, at lower charge density in the polyelectrolyte, this electrostatic contribution to the free energy of adsorption is not sufficient to yield a full spreading of the polymeric

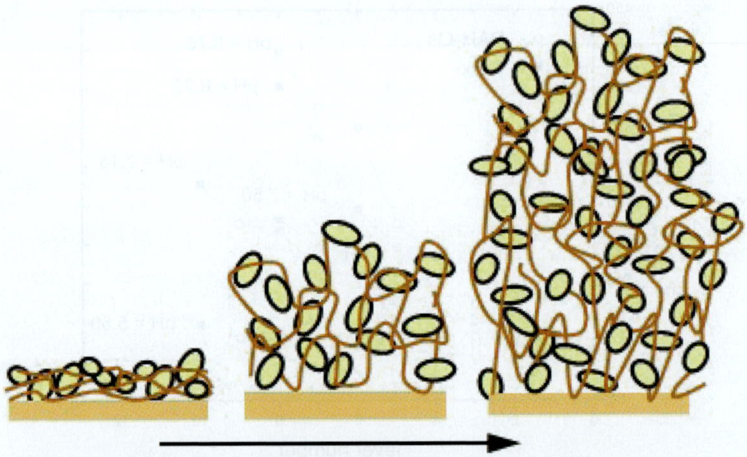

Figure 10. Schematic representation of multi-layers of $(PAH\text{-}Os/GOx)_3$ grown from PAH-Os dipping solutions of increasing pH and therefore carrying polyelectrolyte of different linear charge density.

chains on the electrode surface. At higher pH, poly(allylamine) polymeric chains carry less positive charge per unit length and a thicker layer is formed with a high segmental population of *loops*. The polymer is already partially coiled in solution, and in addition, longer polymer segments need to be adsorbed in each deposition step to compensate, revert and overcompensate the surface negative charge of GOx adsorbed on the previous dipping step. This leads to the formation of *loops* and *tails* with many polymer segments that will be exposed to the GOx solution in the following adsorption step.

The picture that emerges from the above discussion and the full experimental evidence presented in the original research article[61] for the structure and thickness of the $(GOx/PAH\text{-}Os)_3$ multi-layers is schematically shown in Figure 10. For weak polyelectrolytes such as PAH-Os, control over the multi-layer and molecular organization of the self-assembled redox polymer wire and adsorbed polyanionic GOx can be achieved by simple adjustment of the pH of the PAH-Os dipping solutions, which controls the linear charge density of the adsorbing PAH-Os and the charge density of the previously adsorbed GOx layers.

Two types of multi-layer growth profiles have been observed by ellipsometry and AFM depending on the adsorbing PAH-Os solution pH: At pH <7.5, the enzyme contributes the most to the film thickness, while at pH >7.5, both polymer and enzyme contribute substantially to the film thickness. Therefore, the spatial distribution of PAH-Os and GOx within the self-assembled structure can be significantly changed by varying either the pH and/or the ionic strength of the PAH-Os solution used during the self-assembling process. The use of high pH (or high ionic strength) allows the incorporation of more Os-modified polymer and enzyme as well as enhancement of the [Os]/[GOx], *i.e.* the ratio of the respective concentrations.

From the above discussion, it is clear that two different and complimentary approaches can be used to grow thicker enzyme/polymer layers. We can alternatively increase the number of layers to be adsorbed on a solid-state electrode, and/or we modify the pH (and/or increase the ionic strength) of the polymer dipping solution toward values where it will show weak linear charge density.

5.5.2.2 *How far can we push?*

With the objective to move across the borders from the limiting cases derived in the theoretical model proposed by Bartlett and Pratt,[114] we built electrodes of increasing number of layers going beyond 1 μm thickness.[51] Because we were looking into accelerating the deposition process, we chose to work with polymer solutions of high pH, so as to produce thick layers at every deposition step, as we have discussed just above.

Figure 11(a) depicts calibration curves for (PAH-Os/GOx)$_n$ layers of increasing thickness (increasing number of dipping cycles). Surprisingly, while the catalytic current consistently shows larger values with increasing thickness up to a thickness value of about 500 nm, it almost flattens out in a plateau beyond that value. Figure 11(b) is a plot of the catalytic current values at glucose saturation as a function of the total Os charge as measured by cyclic voltammetry at very low scan rate in the absence of enzymatic substrate. The results suggest that beyond a certain thickness value, important concentration

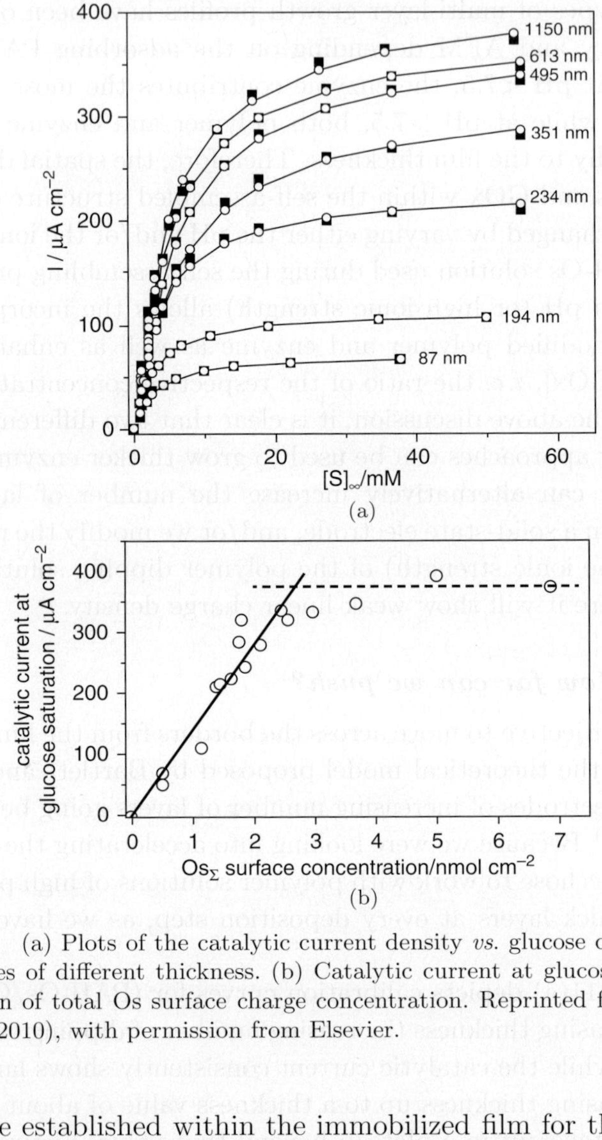

Figure 11. (a) Plots of the catalytic current density *vs.* glucose concentration for electrodes of different thickness. (b) Catalytic current at glucose saturation as a function of total Os surface charge concentration. Reprinted from Ref. 51. Copyright (2010), with permission from Elsevier.

profiles are established within the immobilized film for the oxidized and reduced mediator. Consequently, a concentration profile is, in turn, established for the oxidized enzyme. The oxidized mediator, which is constantly regenerated at the electrode surface, is completely consumed before arriving at the outer edge of the immobilized layer. Therefore, in the absence of oxidized redox mediator, the enzyme molecules immobilized beyond this given distance cannot transfer electrons to the redox matrix and do not generate current. The fact that the increased amount of redox centers is indeed observable by cyclic voltammograms in the absence of enzymatic substrates suggests that there is no problem with the proposed electron transport mechanism within the film. The unbalance between the enzymatic catalytic oxidation of the substrate and the redox mediation arises because of the fast enzyme/substrate/mediator kinetics, which adds to the fact that the substrate has a diffusion coefficient, which is three orders of magnitude higher than the apparent diffusion coefficient of the redox centers (10^{-6} *vs.* 10^{-9} $cm^2\ s^{-1}$).[51]

5.5.2.3 *More complex structures*

As mentioned above, one of the elegant possibilities that becomes possible through layer-by-layer electrostatic self-assembly is the possibility of orderly arrange more than two components on the electrode surface. For example, Wolosiuk *et al.* alternatively combined layers of holo and apo enzyme (with and without the prosthetic group) at different distances from the electrode surface so as to analyze the redox charge transport from the redox enzyme to the electrode along nanometric distances.[115] Calvo *et al.* proposed bi-enzymatic structures composed alternatively of GOx and soy bean peroxidase, and of GOx and glucose dehydrogenase.[116] We have also studied the possibility of alternatively capping the whole immobilized structure by enzyme, redox polymer, or non-reactive polymer, to study the effect of the upper-most layer on the overall catalytic current.[117]

A certain limitation of the layer-by-layer electrostatic self-assembly technique is that, to the best of our knowledge, its efficiency has not yet been proved to build enzymatic architectures within porous electrodes. It is yet to be proven whether mass transport within a highly porous electrode would be fast enough to allow the build-up of these type of structures. Conversely, layer-by-layer electrostatic self-assembly has been proved efficient to build layers on variable supports such as nanoparticles[118] and very rough electrodes.[119]

5.6 The Concept of Wired Enzyme

There is a common denominator to enzyme electrodes where both enzyme and mediator are coimmobilized at the electrode surface. It is extremely difficult to get most of the enzymes contained in the immobilized layer to electrically communicate with the electrode surface. In the enzyme electrode jargon, the terminology *wired enzyme*[53] or *active enzyme*[13] is used to describe the enzyme molecules that are effectively connected to the electrode surface, *i.e.* that effectively interact with the redox mediator, and are brought back to their original redox state, so they are able to further react with another substrate molecule. The amount of *wired enzyme* is most of the time different from the amount of *total enzyme*, or the amount of *immobilized enzyme.*

Following the concept of *wired enzyme*, redox mediators are often called *molecular wires.*

Let us further develop the concept of wired enzyme. This is a concept that is usually applied to enzyme electrodes where the mediator is coimmobilized with the enzyme, but it actually applies as well to electrodes of immobilized enzyme that directly transfer electrons to the electrode surface. When enzyme electrodes are prepared, the total amount of enzyme deposited at the electrode surface can be determined by different physical techniques. Most commonly, the total amount of enzyme is determined either by QCM[50,61] or by ^{125}I radioactive labeling.[17] By physical characterization techniques, we refer to the fact that these techniques do not determine whether the enzyme is biochemically active, or whether it is effectively wired to the electrode surface.

Although the amperometric output of an enzyme electrode cannot be predicted for all enzyme films, it is possible to predict that response for certain limiting cases (see Section 1.4 in Chapter 1).[114] Under those circumstances, it is often found that the measured current is well below the predicted current output if all the enzyme molecules conserved their catalytic activity and were effectively wired to the electrode surface. Divergences from the predicted output are important, with experimental catalytic currents sometimes below 10% of the predicted current. Even though a proportion of the deposited enzymes might lose their activity through the immobilization process, this reason itself seems insufficient to justify the considerable loss of catalytic current. The persistence of the enzymatic activity of most immobilized enzymes has indeed been proved by the addition of a soluble redox mediator that is able to diffuse within the immobilized enzyme layer, react with enzymes less efficiently *wired*, and further react at the solid-state electrode, producing a considerable increase of the original current.[53] This proves that the electron transfer step from the enzyme to the immobilized redox mediator is, at least partially, limiting the catalytic current. The kinetics might be limited by a rather low concentration of immobilized redox mediator, as well as by enzyme/mediator limited mobility within the film. The fact that the mediator is physically anchored to a polymeric backbone highly restrains its mobility. Thus, if the electrode build-up process did not leave the redox centers close enough to the prosthetic group of the enzyme, they cannot efficiently collect electron(s) from the enzyme.

It is not always straightforward to identify which proportion of the total enzyme is effectively wired. Indeed, inefficient wiring of the enzyme is a waste of resources, and the unwired enzymes partially fill up the film with non-conductive, non-active moieties making the whole system less efficient. Unfortunately, not all authors openly recognize these limitations of their results. Therefore, most reports do not make a difference between wired and total enzyme, and do not open a discussion on the matter. Since it is not always possible to analytically predict the amperometric output of an enzyme electrode, an increment in catalytic current upon addition of a soluble mediator does not necessarily indicate that more enzymes have been efficiently wired. Indeed, a higher catalytic current upon addition of a soluble mediator could simply be a consequence of a more efficient electron transfer process, in the sense of the soluble mediator might display a higher second order catalytic constant k (see Section 1.4 in Chapter 1).

List of Abbreviations

AFM	atomic-force microscopy
Au	gold electrode
ENZ	enzyme
FAD	flavin adenine dinucleotide
GOx	enzyme glucose oxidase
MPS	mercaptopropanesulfonate
PAA	polyacrylamide
PAH	polyallylamine
PEGDGE	poly(ethyleneglycol) diglycidyl ether
PVI	polyvinylimidazole
PVP	polyvinylpyridine
QCM	quartz crystal microbalance
PQQ	pyrroloquinoline quinone
S	substrate

References

1. V. Flexer, N. Brun, M. Destribats, R. Backov and N. Mano (2013). A novel three-dimensional macrocellular carbonaceous biofuel cell. *Physical Chemistry Chemical Physics* **15**(17), 6437–6445.
2. V. Flexer, N. Brun, R. Backov and N. Mano (2010). Designing highly efficient enzyme-based carbonaceous foams electrodes for biofuel cells. *Energy and Environmental Science* **3**(9), 1302–1306.
3. V. Flexer, N. Brun, O. Courjean, R. Backov and N. Mano (2011). Porous mediator-free enzyme carbonaceous electrodes obtained through Integrative Chemistry for biofuel cells. *Energy & Environmental Science* **4**(6), 2097–2106.
4. A.E.G. Cass, G. Davis, G.D. Francis, H. Allen, O. Hill, W.J. Aston, I. John Higgins, E.V. Plotkin, L.D.L. Scott and A.P.F. Turner (1984). Ferrocene-mediated enzyme electrode for amperometric determination of glucose. *Analytical Chemistry* **56**(4), 667–671.
5. P.N. Bartlett and K.F.E. Pratt (1995). A study of the kinetics of the reaction between ferrocene monocarboxylic acid and glucose oxidase using the rotating-disc electrode. *Journal of Electroanalytical Chemistry* **397**(1–2), 53–60.
6. C. Bourdillon, C. Demaille, J. Moiroux and J.M. Savéant (1993). New insights into the enzymic catalysis of the oxidation of glucose by native and recombinant glucose oxidase mediated by electrochemically generated one-electron redox cosubstrates. *Journal of American Chemical Society* **115**(1), 2–10.

7. E. Liaudet, F. Battaglini and E.J. Calvo (1990). Electrochemical study of sulphonated ferrocenes as redox mediators in enzyme electrodes. *Journal of Electroanalytical Chemistry* **293**(1–2), 55–68.
8. K. Yokoyama and Y. Kayanuma (1998). Cyclic voltammetric simulation for electrochemically mediated enzyme reaction and determination of enzyme kinetic constants. *Analytical Chemistry* **70**(16), 3368–3376.
9. V. Flexer, M.V. Ielmini, E.J. Calvo and P.N. Bartlett (2008). Extracting kinetic parameters for homogeneous [Os(bpy)2ClPyCOOH]+ mediated enzyme reactions from cyclic voltammetry and simulations. *Bioelectrochemistry* **74**(1), 201–209.
10. C. Bourdillon, C. Demaille, J. Moiroux and J.M. Savéant (1993). New insights into the enzymatic catalysis of the oxidation of glucose by native and recombinant glucose oxidase mediated by electrochemically generated one-electron redox cosubstrates. *Journal of the American Chemical Society* **115**(1), 2–10.
11. N. Anicet, C. Bourdillon, C. Demaille, J. Moiroux and J.-M. Savéant (1996). Catalysis of the electrochemical oxidation of glucose by glucose oxidase and a single electron cosubstrate: Kinetics in viscous solutions. *Journal of Electroanalytical Chemistry* **410**(2), 199–202.
12. P. Alzari, N. Anicet, C. Bourdillon, J. Moiroux and J.M. Savéant (1996). Molecular recognition of artificial single-electron acceptor cosubstrates by glucose oxidase? *Journal of the American Chemical Society* **118**(28), 6788–6789.
13. C. Bourdillon, C. Demaille, J. Moiroux and J.M. Savéant (1994). Step-by-step immunological construction of a fully active multilayer enzyme electrode. *Journal of the American Chemical Society* **116**(22), 10328–10329.
14. C. Bourdillon, C. Demaille, J. Moiroux and J.M. Savéant (1995). Catalysis and mass transport in spatially ordered enzyme assemblies on electrodes. *Journal of the American Chemical Society* **117**(46), 11499–11506.
15. N. Anicet, C. Bourdillon, J. Moiroux and J.M. Savéant (1998). Electron transfer in organized assemblies of biomolecules. Step-by-step avidin/biotin construction and dynamic characteristics of a spatially ordered multilayer enzyme electrode. *Journal of Physical Chemistry B* **102**(49), 9844–9849.
16. M.E.G. Lyons (2003). Mediated electron transfer at redox active monolayers. Part 4: Kinetics of redox enzymes coupled with electron mediators. *Sensors* **3**, 19–42.
17. C. Bourdillon, C. Demaille, J. Gueris, J. Moiroux and J.M. Savéant (1993). A fully active monolayer enzyme electrode derivatized by antigen-antibody attachment. *Journal of American Chemical Society* **115**(26), 12264–12269.
18. K. Kano (2002) Redox potentials of proteins and other compounds of bioelectrochemical interest in aqueous solutions. *Review of Polarography* **48**(1), 29–46.
19. F. Durand, C. Stines-Chaumeil, V. Flexer, I. André and N. Mano (2010). Designing a highly active soluble PQQ-glucose dehydrogenase for efficient

glucose biosensors and biofuel cells. *Biochemical and Biophysical Research Communications* **402**(4), 750–754.
20. N. Anicet, A. Anne, C. Bourdillon, C. Demaille, J. Moiroux and J.M. Savéant (2000). Electrochemical approach to the dynamics of molecular recognition of redox enzyme sites by artificial cosubstrates in solution and in integrated systems. *Faraday Discussions* **116**, 269–279.
21. G.G. Guilbault and G.J. Lubrano (1973). An enzyme electrode for the amperometric determination of glucose. *Analytica Chimica Acta* **64**(3), 439–455.
22. J.R. Mor and R. Guarnaccia (1977). Assay of glucose using an electrochemical enzymatic sensor. *Analytical Biochemistry* **79**(1–2), 319–328.
23. D.L. Williams, A.R. Doig and A. Korosi (1970). Electrochemical-enzymatic analysis of blood glucose and lactate. *Analtical Chemistry* **42**(1), 118–121.
24. É. Lojou and P. Bianco (2004). Membrane electrodes for protein and enzyme electrochemistry. *Electroanalysis* **16**(13–14), 1113–1121.
25. Y. Degani and A. Heller (1987). Direct electrical communication between chemically modified enzymes and metal electrodes. 1. Electron transfer from glucose oxidase to metal electrodes via electron relays, bound covalently to the enzyme. *Journal of Physical Chemistry* **91**(6), 1285–1289.
26. Y. Degani and A. Heller (1988). Direct electrical communication between chemically modified enzymes and metal electrodes. 2. Methods for bonding electron-transfer relays to glucose oxidase and D-amino-acid oxidase. *Journal of the American Chemical Society* **110**(8), 2615–2620.
27. W. Schuhmann, T.J. Ohara, H.L. Schmidt and A. Heller (1991). Electron-transfer between glucose-oxidase and electrodes via redox mediators bound with flexible chains to the enzyme surface. *Journal of American Chemical Society* **113**(4), 1394–1397.
28. P.N. Bartlett, R.G. Whitaker, M.J. Green and J. Frew (1987). Covalent binding of electron relays to glucose-oxidase. *Chemical Communications* **20**, 1603–1604.
29. E.S. Ryabova, V.N. Goral, E. Csoregi, B. Mattiasson and A.D. Ryabov (1999). Coordinative approach to mediated electron transfer: Ruthenium complexed to native glucose oxidase. *Angewandte Chemie International Edition* **38**(6), 804–807.
30. A. Badia, R. Carlini, A. Fernandez, F. Battaglini, S.R. Mikkelsen and A.M. English (1993). Intramolecular electron-transfer rates in ferrocene-derivatized glucose-oxidase. *Journal of American Chemical Society* **115**(16), 7053–7060.
31. F. Battaglini, P.N. Bartlett and J.H. Wang (2000). Covalent attachment of osmium complexes to glucose oxidase and the application of the resulting modified enzyme in an enzyme switch responsive to glucose. *Analytical Chemistry* **72**(3), 502–509.
32. I. Willner, E. Katz and B. Willner (1997). Electrical contact of redox enzyme layers associated with electrodes: Routes to amperometric biosensors. *Electroanalysis* **9**(13), 965–977.

33. I. Willner, Y.M. Yan, B. Willner and R. Tel-Vered (2009). Integrated enzyme-based biofuel cells. A review. *Fuel Cells* **9**(1), 7–24.
34. A.J. Bard (1994). *Integrated Chemical Systems: A Chemical Approach to Nanotechnology.* Wiley.
35. Y. Degani and A. Heller (1989). Electrical communication between redox centers of glucose oxidase and electrodes via electrostatically and covalently bound redox polymers. *Journal of the American Chemical Society* **111**(6), 2357–2358.
36. H. Dahms (1968). Electronic conduction in aqueous solution. *Journal of Physical Chemistry* **72**(1), 362–364.
37. I. Ruff and V.J. Friedrich (1971). Transfer diffusion. I. Theoretical. *Journal of Physical Chemistry* **75**(21), 3297–3302.
38. I. Ruff, V.J. Friedrich, K. Demeter and K. Csillag (1971). Transfer diffusion. II. Kinetics of electron exchange reaction between ferrocene and ferricinium ion in alcohols. *Journal of Physical Chemistry* **75**(21), 3303–3309.
39. H.S. White, J. Leddy and A.J. Bard (1982). Polymer films on electrodes. 8. Investigation of charge-transport mechanisms in nafion polymer modified electrodes. *Journal of the American Chemical Society* **104**(18), 4811–4817.
40. D.N. Blauch and J.M. Savéant (1992). Dynamics of electron hopping in assemblies of redox centers. Percolation and diffusion. *Journal of the American Chemical Society* **114**(9), 3323–3332.
41. D.N. Blauch and J.M. Savéant (1993). Effects of long-range electron transfer on charge transport in static assemblies of redox centers. *Journal of Physical Chemistry* **97**(24), 6444–6448.
42. S.M. Zakeeruddin, D.M. Fraser, M.K. Nazeeruddin and M. Grätzel (1992). Towards mediator design: Characterization of tris-(4,4′-substituted-2,2′-bipyridine) complexes of iron(II), ruthenium(II) and osmium(II) as mediators for glucose oxidase of Aspergillus niger and other redox proteins. *Journal of Electroanalytical Chemistry* **337**(1–2), 253–283.
43. J. Wang, J.W. Mo, S. Li and J. Porter (2001). Comparison of oxygen-rich and mediator-based glucose-oxidase carbon-paste electrodes. *Analytica Chimica Acta* **441**(2), 183–189.
44. T. Ikeda, T. Shibata, S. Todoriki, M. Senda and H. Kinoshita (1990). Amperometric response to reducing carbohydrates of an enzyme electrode based on oligosaccharide dehydrogenase. Detection of lactose and α-amylase. *Analytica Chimica Acta* **230**(C), 75–82.
45. E.V. Ivanova, V.S. Sergeeva, J. Oni, C. Kurzawa, A.D. Ryabov and W. Schuhmann (2003). Evaluation of redox mediators for amperometric biosensors: Ru-complex modified carbon-paste/enzyme electrodes. *Bioelectrochemistry* **60**(1–2), 65–71.
46. E.J. Calvo, C. Danilowicz, C.M. Lagier, J. Manrique and M. Otero (2004). Characterization of self-assembled redox polymer and antibody molecules on thiolated gold electrodes. *Biosensors and Bioelectronics* **19**(10), 1219–1228.
47. N. Anicet, C. Bourdillon, J. Moiroux and J.M. Savéant (1999). Step-by-step avidin-biotin construction of bienzyme electrodes. Kinetic analysis of

the coupling between the catalytic activities of immobilized monomolecular layers of glucose oxidase and hexokinase. *Langmuir* **15**(19), 6527–6533.
48. H. Tsuji and K. Mitsubayashi (1997). An amperometric glucose sensor with modified langmuir-blodgett films. *Electroanalysis* **9**(2), 161–164.
49. E.J. Calvo, C.B. Danilowicz and A. Wolosiuk (2005). Supramolecular multilayer structures of wired redox enzyme electrodes. *Physical Chemistry Chemical Physics* **7**(8), 1800–1806.
50. E.J. Calvo, R. Etchenique, L. Pietrasanta, A. Wolosiuk and C. Danilowicz (2001). Layer-by-layer self-assembly of glucose oxidase and Os(Bpy)2Cl PyCH2NH-poly(Allylamine) bioelectrode. *Analytical Chemistry* **73**(6), 1161–1168.
51. V. Flexer, E.J. Calvo and P.N. Bartlett (2010). The application of the relaxation and simplex method to the analysis of data for glucose electrodes based on glucose oxidase immobilised in an osmium redox polymer. *Journal of Electroanalytical Chemistry* **646**(1–2), 24–32.
52. V. Flexer, K.F.E. Pratt, F. Garay, P.N. Bartlett and E.J. Calvo (2008). Relaxation and Simplex mathematical algorithms applied to the study of steady-state electrochemical responses of immobilized enzyme biosensors: Comparison with experiments. *Journal of Electroanalytical Chemistry* **616**(1–2), 87–98.
53. J. Hodak, R. Etchenique, E.J. Calvo, K. Singhal and P.N. Bartlett (1997). Layer-by-layer self-assembly of glucose oxidase with a poly(allylamine) ferrocene redox mediator. *Langmuir* **13**(10), 2708–2716.
54. S.F. Hou, H.Q. Fang and H.Y. Chen (1997). An amperometric enzyme electrode for glucose using immobilized glucose oxidase in a ferrocene attached poly(4-vinylpyridine) multilayer film. *Analytical Letters* **30**(9), 1631–1641.
55. S.A. Miscoria, J. Desbrieres, G.D. Barrera, P. Labbé and G.A. Rivas (2006). Glucose biosensor based on the layer-by-layer self-assembling of glucose oxidase and chitosan derivatives on a thiolated gold surface. *Analytica Chimica Acta* **578**(2), 137–144.
56. N. Ferreyra, L. Coche-Guérente and P. Labbé (2004). Construction of layer-by-layer self-assemblies of glucose oxidase and cationic polyelectrolyte onto glassy carbon electrodes and electrochemical study of the redox-mediated enzymatic activity. *Electrochimica Acta* **49**(3), 477–484.
57. M.C. Rodríguez and G.A. Rivas (2004). Assembly of glucose oxidase and different polyelectrolytes by means of electrostatic layer-by-layer adsorption on thiolated gold surface. *Electroanalysis* **16**(20), 1717–1722.
58. X. Yan, H.F. Ji and Y. Lvov (2004). Modification of microcantilevers using layer-by-layer nanoassembly film for glucose measurement. *Chemical Physics Letters* **396**(1–3), 34–37.
59. M.V. Bracamonte, C. Yañez, S. Bollo, G.A. Rivas and N.F. Ferreyra (2014). Physico-chemical characterization of ferrocenyl-modified hyperbranched poly(ethylenimine) self-assembled multilayers. *Journal of Electroanalytical Chemistry* **712**, 124–131.

60. E.N. Primo, F.A. Gutierrez, M.D. Rubianes and G.A. Rivas (2015). Bamboo-like multiwalled carbon nanotubes dispersed in double stranded calf-thymus DNA as a new analytical platform for building layer-by-layer based biosensors. *Electrochimica Acta* **182**, 391–397.
61. V. Flexer, E.S. Forzani, E.J. Calvo, S.J. Ludueña and L.I. Pietrasanta (2006). Structure and thickness dependence of "molecular wiring" in nanostructured enzyme multilayers. *Analytical Chemistry* **78**(2), 399–407.
62. A. Vilkanauskyte, T. Erichsen, L. Marcinkeviciene, V. Laurinavicius and W. Schuhmann (2002). Reagentless biosensors based on co-entrapment of a soluble redox polymer and an enzyme within an electrochemically deposited polymer film. *Biosensors and Bioelectronics* **17**(11–12), 1025–1031.
63. A. Heller (2004). Miniature biofuel cells. *Physical Chemistry Chemical Physics* **6**(2), 209–216.
64. A. Heller (2006). Electron-conducting redox hydrogels: Design, characteristics and synthesis. *Current Opinion in Chemical Biology* **10**(6), 664–672.
65. A. Heller and B. Feldman (2008). Electrochemical glucose sensors and their applications in diabetes management. *Chemical Reviews* **108**(7), 2482–2505.
66. A. Heller and B. Feldman (2010). Electrochemistry in diabetes management. *Accounts of Chemical Research* **43**(7), 963–973.
67. H.H. Kim, N. Mano, Y. Zhang and A. Heller (2003). A miniature membraneless biofuel cell operating under physiological conditions at 0.5 V. *Journal of the Electrochemical Society* **150**(2), A209–A213.
68. N. Mano and A. Heller (2003). A miniature membraneless biofuel cell operating at 0.36 V under physiological conditions. *Journal of the Electrochemical Society* **150**(8), A1136–A1138.
69. N. Mano, H.-H. Kim, Y. Zhang and A. Heller (2002). An oxygen cathode operating in a physiological solution. *Journal of the American Chemical Society* **124**(22), 6480–6486.
70. N. Mano, F. Mao and A. Heller (2002). A miniature biofuel cell operating in a physiological buffer. *Journal of the American Chemical Society* **124**(44), 12962–12963.
71. N. Mano, F. Mao, W. Shin, T. Chen and A. Heller (2003). A miniature biofuel cell operating at 0.78 V. *Chemical Communications* **9**(4), 518–519.
72. F. Mao and A. Heller (2013). *Redox Polymers*. Google Patents.
73. F. Mao, N. Mano and A. Heller (2003). Long tethers binding redox centers to polymer backbones enhance electron transport in enzyme "wiring" hydrogels. *Journal of the American Chemical Society* **125**(16), 4951–4957.
74. P. Pinyou, A. Ruff, S. Pöller, S. Ma, R. Ludwig and W. Schuhmann (2016). Design of an Os complex-modified hydrogel with optimized redox potential for biosensors and biofuel cells. *Chemistry — A European Journal* **22**(15), 5319–5326.
75. P. Pinyou, A. Ruff, S. Pöller, S. Alsaoub, S. Leimkühler, U. Wollenberger and W. Schuhmann (2016). Wiring of the aldehyde oxidoreductase PaoABC to electrode surfaces via entrapment in low potential phenothiazine-modified redox polymers. *Bioelectrochemistry* **109**, 24–30.

76. P. Pinyou, A. Ruff, S. Pöller, S. Barwe, M. Nebel, N.G. Alburquerque, E. Wischerhoff, A. Laschewsky, S. Schmaderer, J. Szeponik, N. Plumeré and W. Schuhmann (2016). Thermoresponsive amperometric glucose biosensor. *Biointerphases* **11**(1), 1–7.
77. J.W. Gallaway and S.A.C. Barton (2008). Kinetics of redox polymer-mediated enzyme electrodes. *Journal of the American Chemical Society* **130**(26), 8527–8536.
78. J.W. Gallaway and S.A. Calabrese Barton (2009). Effect of redox polymer synthesis on the performance of a mediated laccase oxygen cathode. *Journal of Electroanalytical Chemistry* **626**(1–2), 149–155.
79. C. Danilowicz, E. Cortón, F. Battaglini and E.J. Calvo (1998). An Os(byp)2ClPyCH2NHPoly(allylamine) hydrogel mediator for enzyme wiring at electrodes. *Electrochimica Acta* **43**(23), 3525–3531.
80. E.J. Calvo, R. Etchenique, C. Danilowicz and L. Diaz (1996). Electrical communication between electrodes and enzymes mediated by redox hydrogels. *Analytical Chemistry* **68**(23), 4186–4193.
81. B.A. Gregg and A. Heller (1990). Cross-linked redox gels containing glucose oxidase for amperometric biosensor applications. *Analytical Chemistry* **62**(3), 258–263.
82. B.A. Gregg and A. Heller (1991). Redox polymer films containing enzymes. 1. A redox-conducting epoxy cement: Synthesis, characterization and electrocatalytic oxidation of hydroquinone. *Journal of Physical Chemistry* **95**(15), 5970–5975.
83. B.A. Gregg and A. Heller (1991). Redox polymer films containing enzymes. 2. Glucose oxidase containing enzyme electrodes. *Journal of Physical Chemistry* **95**(15), 5976–5980.
84. A. Heller (1990). Electrical wiring of redox enzymes. *Accounts of Chemical Research* **23**(5), 128–134.
85. V. Soukharev, N. Mano and A. Heller (2004). A four-electron O2-electroreduction biocatalyst superior to platinum and a biofuel cell operating at 0.88 V. *Journal of the American Chemical Society* **126**(27), 8368–8369.
86. V. Flexer and N. Mano (2014). Wired pyrroloquinoline quinone soluble glucose dehydrogenase enzyme electrodes operating at unprecedented low redox potential. *Analytical Chemistry* **86**(5), 2465–2473.
87. J. Gallaway, I. Wheeldon, R. Rincon, P. Atanassov, S. Banta and S.C. Barton (2008). Oxygen-reducing enzyme cathodes produced from SLAC, a small laccase from Streptomyces coelicolor. *Biosensors and Bioelectronics* **23**(8), 1229–1235.
88. F. Durand, S. Gounel, C.H. Kjaergaard, E.I. Solomon and N. Mano (2012). Bilirubin oxidase from Magnaporthe oryzae: An attractive new enzyme for biotechnological applications. *Applied Microbiology and Biotechnology* **96**(6), 1489–1498.
89. F. Durand, C.H. Kjaergaard, E. Suraniti, S. Gounel, R.G. Hadt, E.I. Solomon and N. Mano (2012). Bilirubin oxidase from *Bacillus pumilus*:

A promising enzyme for the elaboration of efficient cathodes in biofuel cells. *Biosensors and Bioelectronics* **35**(1), 140–146.

90. R. Ludwig, R. Ortiz, C. Schulz, W. Harreither, C. Sygmund and L. Gorton (2013). Cellobiose dehydrogenase modified electrodes: Advances by materials science and biochemical engineering. *Analytical and Bioanalytical Chemistry* **405**(11), 3637–3658.
91. P. Kavanagh and D. Leech (2013). Mediated electron transfer in glucose oxidising enzyme electrodes for application to biofuel cells: Recent progress and perspectives. *Physical Chemistry Chemical Physics* **15**(14), 4859–4869.
92. F. Barrière, P. Kavanagh and D. Leech (2006). A laccase-glucose oxidase biofuel cell prototype operating in a physiological buffer. *Electrochimica Acta* **51**(24), 5187–5192.
93. K. Habermüller, A. Ramanavicius, V. Laurinavicius and W. Schuhmann (2000). An oxygen-insensitive reagentless glucose biosensor based on Osmium-complex modified polypyrrole. *Electroanalysis* **12**(17), 1383–1389.
94. F. Gao, L. Viry, M. Maugey, P. Poulin and N. Mano (2010). Engineering hybrid nanotube wires for high-power biofuel cells. *Nature communications* **1**, 2.
95. Z. Liu, B.J. Feldman, F. Mao and A. Heller (2013). *Redox Polymers for Use in Analyte Monitoring*. Google Patents.
96. A. Aoki, R. Rajagopalan and A. Heller (1995). Effect of quaternization on electron diffusion coefficients for redox hydrogels based on poly(4-vinylpyridine). *Journal of physical chemistry* **99**(14), 5102–5110.
97. P.P. Joshi, S.A. Merchant, Y. Wang and D.W. Schmidtke (2005). Amperometric biosensors based on redox polymer-carbon nanotube-enzyme composites. *Analytical Chemistry* **77**(10), 3183–3188.
98. D.A. Guschin, J. Castillo, N. Dimcheva and W. Schuhmann (2010). Redox electrodeposition polymers: Adaptation of the redox potential of polymer-bound Os complexes for bioanalytical applications. *Analytical and Bioanalytical Chemistry* **398**(4), 1661–1673.
99. A. Prévoteau and N. Mano (2012). Oxygen reduction on redox mediators may affect glucose biosensors based on "wired" enzymes. *Electrochimica Acta* **68**, 128–133.
100. A. Prévoteau and N. Mano (2013). How the reduction of O_2 on enzymes and/or redox mediators affects the calibration curve of "wired" glucose oxidase and glucose dehydrogenase biosensors. *Electrochimica Acta* **112**, 318–326.
101. T. Chen, S. Calabrese Barton, G. Binyamin, Z. Gao, Y. Zhang, H.H. Kim and A. Heller (2001). A miniature biofuel cell. *Journal of the American Chemical Society* **123**(35), 8630–8631.
102. S. Calabrese Barton, J. Gallaway and P. Atanassov (2004). Enzymatic biofuel cells for implantable and microscale devices. *Chemical Reviews* **104**(10), 4867–4886.
103. X. Ayral (2001). Chemical synoviorthesis with osmic acid in haemophilia. *Haemophilia* **7**(2), 20–25.

104. G. Decher and J.D. Hong (1991). Buildup of ultrathin multilayer films by a self-assembly process, 1 consecutive adsorption of anionic and cationic bipolar amphiphiles on charged surfaces. Makromolekulare Chemie. *Macromolecular Symposia* **46**(1), 321–327.
105. G. Decher, J.D. Hong and J. Schmitt (1992). Buildup of ultrathin multilayer films by a self-assembly process: III. Consecutively alternating adsorption of anionic and cationic polyelectrolytes on charged surfaces. *Thin Solid Films* **210–211**(PART 2), 831–835.
106. G. Decher (1997). Fuzzy nanoassemblies: Toward layered polymeric multicomposites. *Science* **277**(5330), 1232–1237.
107. M. Lösche, J. Schmitt, G. Decher, W.G. Bouwman and K. Kjaer (1998). Detailed structure of molecularly thin polyelectrolyte multilayer films on solid substrates as revealed by neutron reflectometry. *Macromolecules* **31**(25), 8893–8906.
108. R.G. Nuzzo and D.L. Allara (1983). Adsorption of bifunctional organic disulfides on gold surfaces. *Journal of the American Chemical Society* **105**(13), 4481–4483.
109. F.T. Arce, M.E. Vela, R.C. Salvarezza and A.J. Arvia (1998). Dynamic characteristics of adsorbed monolayers of 1-dodecanethiol on gold (111) terraces from in-situ scanning tunneling microscopy imaging. *Electrochimica Acta* **44**(6–7), 1053–1067.
110. O. Azzaroni, M.E. Vela, H. Martin, A. Hernández Creus, G. Andreasen and R.C. Salvarezza (2001). Electrodesorption kinetics and molecular interactions at negatively charged self-assembled thiol monolayers in electrolyte solutions. *Langmuir* **17**(21), 6647–6654.
111. C. Danilowicz, E. Cortón and F. Battaglini (1998). Osmium complexes bearing functional groups: Building blocks for integrated chemical systems. *Journal of Electroanalytical Chemistry* **445**(1–2), 89–94.
112. S.S. Shiratori and M.F. Rubner (2000). pH-dependent thickness behavior of sequentially adsorbed layers of weak polyelectrolytes. *Macromolecules* **33**(11), 4213–4219.
113. D. Yoo, S.S. Shiratori and M.F. Rubner (1998). Controlling bilayer composition and surface wettability of sequentially adsorbed multilayers of weak polyelectrolytes. *Macromolecules* **31**(13), 4309–4318.
114. P.N. Bartlett and K.F.E. Pratt (1995). Theoretical treatment of diffusion and kinetics in amperometric immobilized enzyme electrodes Part I: Redox mediator entrapped within the film. *Journal of Electroanalytical Chemistry* **397**(1–2), 61–78.
115. E.J. Calvo and A. Wolosiuk (2005). Wiring enzymes in nanostructures built with electrostatically self-assembled thin films. *ChemPhysChem* **6**(1), 43–47.
116. E.J. Calvo, F. Battaglini, C. Danilowicz, A. Wolosiuk and M. Otero (2000). Layer-by-layer electrostatic deposition of biomolecules on surfaces for molecular recognition, redox mediation and signal generation. *Faraday Discussions* **116**, 47–65.

117. E.J. Calvo, V. Flexer, M. Tagliazucchi and P. Scodeller (2010). Effects of the nature and charge of the topmost layer in layer by layer self assembled amperometric enzyme electrodes. *Physical Chemistry Chemical Physics* **12**(34), 10033–10039.
118. P. Scodeller, V. Flexer, R. Szamocki, E.J. Calvo, N. Tognalli, H. Troiani and A. Fainstein (2008). Wired-enzyme core-shell Au nanoparticle biosensor. *Journal of the American Chemical Society* **130**(38), 12690–12697.
119. N.G. Tognalli, P. Scodeller, V. Flexer, R. Szamocki, A. Ricci, M. Tagliazucchi, E.J. Calvo and A. Fainstein (2009). Redox molecule based SERS sensors. *Physical Chemistry Chemical Physics* **11**(34), 7412–7423.

Chapter 6

Functional Electrodes for Enzymatic Electrosynthesis

Lin Zhang, Mathieu Etienne*, Neus Vilà, and Alain Walcarius

Laboratoire de Chimie Physique et Microbiologie pour l'Environnement (LCPME), UMR7564 CNRS — Université de Lorraine, 405 rue de Vandoeuvre, F-54600 Villers-lès-Nancy, France

**mathieu.etienne@univ-lorraine.fr*

6.1 Introduction

Enzymatic synthesis is an alternative and a very promising field of research for addressing some challenges in organic synthesis since it offers facile routes to selectively prepare some valuable products without suffering from side-reactions as in conventional organic synthesis methods.[1] Different classes of enzymes such as acylases, amidases, hydrolases, or cellulases are already used in the industrial production of antibiotics, herbicides, fuel alcohols, and pharmaceutical intermediates. In industrial processes, some enzymes like lipases or nitrilases have been produced on a large scale or have been used for the production of valuable products like enantiopure alcohols or *R*-mandelic acid.[2,3]

Proteins can be broadly categorized as either soluble or membrane-associated. Soluble proteins, which include enzymes, antibodies, regulatory proteins, and many others, reside in an aqueous environment, and thus are generally amenable to aqueous processing methods for immobilization. The solubility of these proteins arises owing to the presence of polar or charged amino acid residues on the exterior surface. Therefore, immobilization techniques for soluble proteins must provide a hydrated environment at a pH that does not alter the structure of proteins and does not have significant polarity differences relative to water. On the other hand, membrane-associated proteins contain either fully (intrinsic membrane proteins) or partially (extrinsic membrane proteins) embedded proteins within cellular lipid membranes.[4] Therefore, successful immobilization of membrane proteins must address two important issues. First, the method must allow retention of the tertiary folded structure of the protein as is the case for soluble proteins. Second, it must accommodate the phospholipid membrane structure, which is held together mainly through hydrophobic interactions[4] or replace these lipids by surfactant molecules. These issues are necessary to accommodate both the hydrophilic and hydrophobic parts of the protein.

Among the different enzymatic catalytic systems, redox enzymes are an important class of proteins that catalyze redox reactions. Enzymatic cofactors, which are involved in redox reactions, are used for transferring both electrons and protons. The structures of some of these cofactors are illustrated in Figure 1, including their different redox states. Flavin adenine dinucleotide (FAD), flavin mononucleotide (FMN), heme, or transition metals are associated with the enzymes. Nicotinamide adenine dinucleotide $(NAD(P)^+/NAD(P)H$ are dissociated from the enzymes.[5,6]

Efficient regeneration and reuse of the cofactors are required for large-scale synthetic applications.[7–9] Any method for regenerating cofactors has to be practical and should enable an enzymatic process to be inexpensive and convenient. The reaction used for regeneration has to proceed in high yield and the total turnover numbers (TTN) for the cofactor must be high. Cofactor regeneration will facilitate the thermodynamically and kinetically favorable product formation. Any reagents or by-products of the regenerative system must not interfere with isolation of the desired product. Methods including chemical, electrochemical, photochemical, microbial, and enzymatic reactions have all been developed for cofactor regeneration.[7–9] Chemical regeneration of cofactor can be realized by introducing reducing

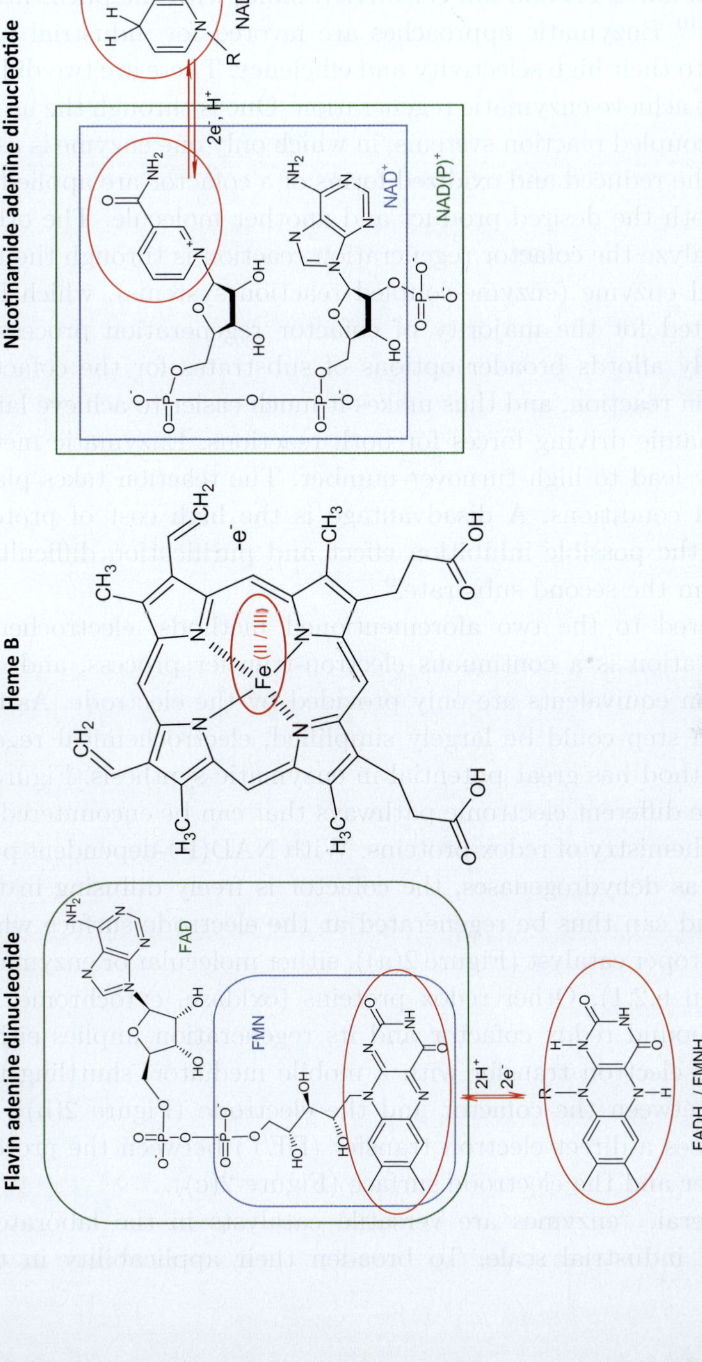

Figure 1. Structures of some cofactors with their different redox states: (a) nicotinamide adenine dinucleotide (phosphate); (b) flavin adenine dinucleotide (mononucleotide); (c) heme B.

agents (*e.g.* sodium dithionite, sodium borohydride, H_2, etc.) or chemical oxidants (*e.g.* O_2) into the system. However, this method suffers from low TTN and low selectivity, along with the purification difficulties.[10] Enzymatic approaches are favored for industrial processes due to their high selectivity and efficiency. There are two different ways to achieve enzymatic regeneration. One is through the use of substrate-coupled reaction systems, in which only one enzyme is used and both the reduced and oxidized forms of a cofactor are applied to catalyze both the desired product and another molecule. The other way to catalyze the cofactor regeneration reaction is through the use of a second enzyme (enzyme-coupled reaction systems), which has been adapted for the majority of cofactor regeneration processes. This usually affords broader options of substrates for the cofactor regeneration reaction, and thus makes it much easier to achieve large thermodynamic driving forces for both reactions. Enzymatic methods usually lead to high turnover number. The reaction takes place under mild conditions. A disadvantage is the high cost of protein as well as the possible inhibition effect and purification difficulties coming from the second substrate.[9]

Compared to the two aforementioned methods, electrochemical regeneration is a continuous electron-transfer process, and the regeneration equivalents are only provided by the electrode. As the purification step could be largely simplified, electrochemical regeneration method has great potential in enzymatic synthesis. Figure 2 shows three different electronic pathways that can be encountered in bioelectrochemistry of redox proteins. With NAD(P)-dependent proteins such as dehydrogenases, the cofactor is freely diffusing in the solution and can thus be regenerated at the electrode surface when using the proper catalyst (Figure 2(a)), either molecular or enzymatic (see Section 6.2.1). Other redox proteins (oxidase, cytochrome...) possess a bound redox cofactor and its regeneration implies either a mediated electron transfer with a mobile mediator, shuttling the electrons between the cofactor and the electrode (Figure 2(b)), or in some cases a direct electron transfer (DET) between the protein redox center and the electrode surface (Figure 2(c)).

In general, "enzymes are versatile catalysts in the laboratory and on an industrial scale. To broaden their applicability in the

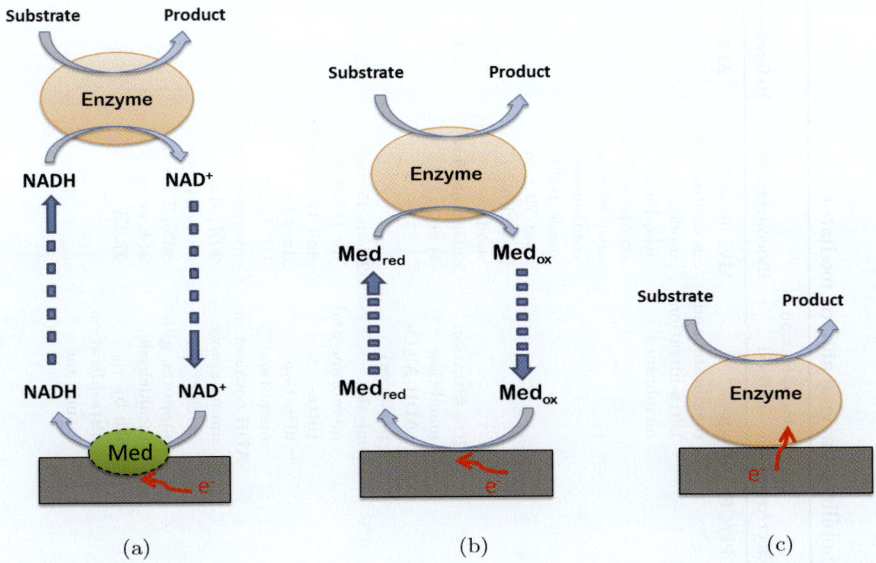

Figure 2. Schematic representation of three bioelectrochemical routes for redox enzymes: (a) electrochemical regeneration of diffusing cofactors for NAD(P)-dependent proteins using proper molecular catalyst or a second redox protein; (b) electrochemical regeneration of enzyme-bounded redox cofactors through a mobile mediator; (c) Direct electron transfer with the protein redox center at the electrode surface.

laboratory and to ensure their (re)use in manufacturing the stability of enzymes can often require improvement. Immobilization can address the issue of enzymatic instability. Immobilization can also help to enable the employment of enzymes in different solvents, at extreme pH and temperature values and exceptionally high substrate concentrations".[11] Furthermore, in electroenzymatic synthesis, the local concentration of enzymes on the electrode can be significantly increased, which can facilitate electron transfer reactions.[12] Therefore, immobilization of the enzymes on electrodes offers a possibility to improve the efficiency of the bioelectrochemical system. In addition, the immobilization of electron mediator can be advantageous in order to limit the quantity of molecule needed (increasing the total turnover), to promote their reusability and limit the need for further purification steps. Table 1 is a summary of some investigations carried out in the field of electroenzymatic synthesis

Table 1. Electroenzymatic synthesis involving immobilized enzyme and/or mediator.

Enzyme	Cofactor	Substrate	Product	Mediator	Co-enzymes	Electrode	Immobilization method	Bioconversion	References
ADH	NAD	Cyclohexanone	Cyclohexanol	MV/rhodium complex	LipDH	Pt/CF	Dialysis membrane and Ultra-filtration membranes	MV: up to 72% in continuous mode; Rhodium complex: 75% over 36 h in continuous mode 100% over 70 h in recycling mode	22,55
ADH	NAD	Alcohol	Aldehyde	—			Ultra-filtration membrane ($ADH-Al_2O_3$ particles)	Continuous-flow type	59
ADH	NAD	Cyclohexanone	Cyclohexanol	$([Rh (III) (C_2Me_5)(bpy)Cl]^+$		CF	Dialysis membrane and Ultra-filtration membrane	92% in 45 days for D-MER and 100% in 3 days for UF-MER	21
ADH	NAD	*p*-Chloroacetophenone	*p*-Chloro-(*R*)-phenylethanol	Polymer-bound Rh(bpy)			ADH covalent on amino-epoxy carrier supports, glutaraldehyde, Rh by ultra-filtration membrane	Conversion reach 90%, Rh recovery rate 86%, TTN = 214, ee = 97.3%	56

ADH	NAD	Butyraldehyde	Butanol	—	—	CF	Encapsulation in alginate–silicate gel	Faradaic efficiency ~40%	128
ADH	NADP	Acetophenone	(*R*)-phenylethanol	Cp*Rh(bpy)		CF	Covalent by glutardialdehyde	Complete conversion with TTN of 75,000 in one phase, twice in two-phase system, ee 99.9%. Stability of 36 h in MTBE organic solvent	23
ADH	NAD	2- and 3-Methylcyclohexanone	(1*S*, 2*S*)-(+)-2- and (1*S*, 3*S*)-(–)-3-Methylcyclohexanol	Methyl viologen	Diaphorase	GF	Polyacrylic acid cross-linking	Yield (49.8 and 50.2%), current efficiency (98.6 and 96.5%) and *ee*% (100% and 93.1%) in 8 h	27
AOx	FAD	Pyruvic acid	D-alanine	ADPy		GC	Covalent by glutaraldehyde	*ee* ~ 100%, current efficiency 97% in 10 h, 8.9 mmol dm^{-3}, TTN more than 36,000	85
CPO	Heme	Cinnamyl alcohol	Cinnamic aldehyde	O_2	—	Glassy carbon	DDAB surfactant film + chitosan	Yield is 51.8% TTN is 80,500, over 4 cycles	94

(*Continued*)

Table 1. (*Continued*)

Enzyme	Cofactor	Substrate	Product	Mediator	Co-enzymes	Electrode	Immobilization method	Bioconversion	References
CYP119	Heme	Nitrite	Ammonia	—	—	Plane pyrolytic graphite	DDAB surfactant film	Important role for protein stability in the selectivity of biocatalysts	88
CYP1A2 or CYP2E1	Heme	4-(Methylnitrosamino)-1-(3-pyridyl)-1-butanone	4-Hydroxy-1-(3-pyridyl)-1-butanone (HPB)		Cyt P450-reductase and cyt b5	Pyrolytic graphite disk electrodes	Layer-by-layer	Turnover rate: 57 ± 9 h^{-1} (mol HPB)$^{-1}$ (mol CYP2E1)$^{-1}$	93
Cys-DSDH	NAD	D-fructose	D-sorbitol	$[Cp^*Rh(bpy)Cl]^+$	—	CF	Covalent by 'thiol-ene' click chemistry	Conversion rate is 72%, faradaic efficiency is 77 %	58
Cys-GatDH	NAD	1,2-Hexanediol	Hydroxyacetone	4-carboxy-2,5,7-trinitrofluorenylidenmalonnitrile	—	Gold	Self-assembled monolayers	Proof of concept with cyclic voltammetry	69
CYP2E1, MUT261, MUT268	Heme	*p*-Nitrophenol	*p*-Nitrocatechol	—	—	Gold	SAMs + covalent by (DTME)	1.77 ± 0.64 (P450 2E1); 4.53 ± 0.44 (MUT261); 3.66 ± 0.35 (MUT268) μM	89

CYP102A1	Heme+ diflavin reductase	*p*-Nitrophenoxycarboxylic acid	*p*-Nitrophenolate		—	Pt and GC	Encapsulation in polypyrrole	Production rate of: 0.66 μM h^{-1} cm^{-2} (Pt) 0.13 μM h^{-1} cm^{-2} (GC) 25 h bioconversion	91
CYP102A1	Heme+ diflavin reductase	3-Phenoxytoluene	Toluene and phenol	—	—	Glassy carbon	Encapsulation in PEDOT/PSS polymer films	Reaction rate is 21.89 μM h^{-1} cm^{-2}	90
CYP1A2 + CYP3A4	Heme	Clopidogrel	Clopidogrel carboxylic acid	—	—	Au/chitosan/RGO nanosheets	Covalent, glutaraldehyde+carbonyldiimide	7.2% after 1 h	92
DSDH	NAD	D-sorbitol	D-fructose	Poly (methylene green)	—	MWCNTs-modified CF	Encapsulation in silica gel	TTN close to 60, conversion rate 60%, enzyme stability up to 2 weeks (in solution $t_{1/2} = 48$ h)	67
Enoate reductase	Iron-sulfur-FAD	(*E*)-2-methyl-3-phenyl-2-propenoate/(*E*)-2-methyl-2-butenoate	3-Phenyl-2-methylpropionic acid/2-methylbutanoic acid	MV	—	CF	Encapsulated in calcium alginate gels	Conversion is 95% and 99%, half-life increase from 8 h to 350 h when immobilized	125
EPMH									

(Continued)

Table 1. (*Continued*)

Enzyme	Cofactor	Substrate	Product	Mediator	Co-enzymes	Electrode	Immobilization method	Bioconversion	References
FAD	4-Ethyl-phenol	1-(4-Hydro-xyphenyl)-ethanol	Polyethy-leneglycol-modified ferrocene		CF	Ultrafiltration mem-brane	Conversion is 100% after 3,000 minutes, *ee* =88% TTN = 400,000	83	
FNR	FAD	NADPH	$NADP^+$	Ferrocene	—	Gold	SAMs + Histidine chelated with Cu^{2+}	Orientation of the protein to favor DET reactions	103
G6PDH	NAD	Glucose 6-phosphate	6-Phosphog-luconate	PMS	—	Rotating disc graphite	Covalent by glutaraldehyde or carbonyldi-imide	0.016 mol/day, current efficiency is 100%	68
GDH	NAD	Glucose	Gluconic acid	3,4-DHB	—	GF	Covalent by glutaralde-hyde+ carbonyldi-imide	TTN_{max} is 100 after 4 h	70
GDH	NAD	Glucose	Gluconic acid	—	—	GF	Nafion membrane	Half-life deactivation time of enzyme $t_{1/2} = 24.5$ h; Substrate consumption rates $rs = 59$ μmol min^{-1}	60

GOx (GOx+ HPR on cathode)	FAD	Glucose	Gluconic acid	TTF		3-D cross linked carbon NPs	Gelatin+ glutaraldehyde	47% after 7 h	127
GOx	FAD	Glucose	Gluconic acid	O_2	—	Pt/Carbon	Encapsulation in polypyrrole film	45% conversion after 8 h	80
GOx	FAD	Glucose	Gluconic acid	O_2	—	Pt	Encapsulation in polypyrrole film	Different thickness give yied 23–45% after 8h	81
GOx	FAD	Glucose	Gluconic acid	Benzo-quinone	—	CF	Covalent by carbodiimide	1.5 g gluconate/h; Conversion <40%, Stability increase 50 times	82
GOx	FAD	Glucose	Gluconic acid	O_2		Pt grid	Dialysis membrane	Transformation ratio up to 30% after 3h with membrane cut off 12–14 kDa	79
GOx	FAD	Glucose	Gluconic acid	O_2	—	Pt / CF	Dialysis membrane	Conversion 30% after 3 h	22
LDH	NAD	Pyruvate or α-Ketobutyrate	(L)-lactic acid (2a) or (L)-2-hydroxybutyric acid	MV	LipDH	Pt	Dialysis membrane	Reaction >95% complete after 8 days, *ee* >98%	44
LDH	NAD	Pyruvate	Lactate	FAD	FDH	Gold	Proton exchange membrane	Complete transformation in batch recycling mode, TTN =1	50

(Continued)

Table 1. *(Continued)*

Enzyme	Cofactor	Substrate	Product	Mediator	Co-enzymes	Electrode	Immobilization method	Bioconversion	References
LDH	NAD	Pyruvate	Lactate	MV	LipDH	Porous graphite	Cation exchange membrane	70% in a semi-batch for 24 h, 50% in a simple batch mode for 200 h	46,51
LDH	NAD	Pyruvate	Lactate	MV	LipDH	RVC	Nafion coating; LDH cross-linked by glutaraldehyde in solution	In batch: cross-linked, LDH yield 70%, 10% lost in stability over 25 days. Soluble LDH 50%, $t_{1/2} = 2.5$ day	45
LDH	NAD	Pyruvate	Lactate	FAD/*p*FAD	LipDH	GC	Nafion coating	TTN is 123/43, Stoichiometric yield is 58/30%,	47
LDH	NAD	Pyruvate	Lactate	MV	LipDH	RVC	Nafion coating	Complete after 8 days, current efficiency is close to 100%	48
LDH	NAD	Pyruvate	Lactate	MV	LipDH	Porous graphite	Nafion coating	70% in a semi-batch mode for 24 h, 50% for 200 h in simple batch mode	49
LDH	NAD	Pyruvate	Lactate	MV	LipDH	Carbon	Nafion coating (MV-LipDH)	Yield is 60%, current efficiency is 660 and 107 nmol cm $^{-2}$h^{-1} for lactate and malate, active for 43 days	52
LDH	NAD	Pyruvate	L-lactate	MV derivative	VAPOR	CF	Covalent by carbodiimide	148 nmol cm^{-2} in 20 h	54

LDH	NAD	Pyruvate	Lactate	PV	LipDH	GC	Covalent by glutaraldehyde	35% in 62 h, TTN is 38	53
MDH		Oxaloacetate	Malate						
(*S*)-MDH	FMN	(*S*)-mandelic acid	Phenylglyoxylic acid	Ferrocenedimethanol	—	GC/CF	Encapsulation in silica film	94% of conversion after 9 h, faradaic efficiency of 91%	84
RDH	NAD	D/L-glucitol	D-fructose/ D-sorbose	Ferrocene-dimethanol	Diaphorase	GC	Encapsulation in silica gel	Higher oxidation peak with D-glucitol	78
Tyr	Cu	L-tyrosine	L-3,4-dihydroxyphenylalanine	—	—	Tyr/ carbon NPs/ polypyrrole	Carbodiimide + polymerization with pyrrole	Conversion is 95.9% in 4 h and productivity 47.27 mgl^{-1} h^{-1}	99
						CF	Nafion coating	Conversion is 44.4% in 4 h and 21.93 mgl^{-1} h^{-1}	
Tyr (well-dispersed)	Cu	L-tyrosine	L-3,4-dihydroxyphenylalanine	—	—	Tyr/ carbon nanopowder/ polypyrrole	Covalent by carbodiimide	Conversion rate up to 77.7% in 5 h, productivity is 15,300 mgl^{-1} h^{-1};	100

Note:

Abbreviations: ADH, alcohol dehydrogenase; AO*x*, Amino acid oxidase; ADPy, L-aminopropyl-1-methyl-4,4′-dipyridine; CF, carbon felt; CPO, chloroperoxidase; CYP, cytochrome P450; Cys-DSDH, cysteine-tagged dehydrogenase; Cys-GatDH, cysteine-tagged galactitol dehydrogenase; 3,4-DHB, 3,4-dihydroxybenzaldehyde; DSDH, D-sorbitol dehydrogenase; DTME, dithio-bismaleimidoethane; EPMH, 4-ethylphenol methylenehydroxylase; FDH, formate dehydrogenase; FNR, ferredoxin: $NADP^+$ reductase; G6PDH, glucose-6-phosphate dehydrogenase; GC, glassy carbon; GDH, glucose Dehydrogenase; GF, graphite felt; GOx, glucose oxidase; LDH, L-lactate dehydrogenase; LipDH, lipoamide dehydrogenase; MV, methyl viologen; NPs, nanoparticles; PCMH, *p*-cresol methylenehydroxylase; pFAD, polymerized FAD film; PMS, phenazine methosulfate; PV, polyviologen; Pt, Platinium; RDH, ribitol dehydrogenase; RVC, reticulated vitreous carbon; S-MDH, (*S*)-mandelate dehydrogenase; VAPOR, viologen accepting pyridine nucleotide oxidoreductase; TTF, tetrathiafulvalene; Tyr, tyrosinase.

with immobilized redox enzymes or mediators on electrodes, providing the enzymes involved and their cofactor, the target reactions, the mediators used, the electrode materials, the strategies for protein or cofactor immobilization as well as some information available about the bioconversion process. In the first section of this chapter, we will discuss the most typical target reactions that have been used in electroenzymatic synthesis for different classes of redox enzymes according to their redox cofactor: NAD(P)-dependent-redox enzymes, flavoenzymes, and metal complex-based cofactors from cytochromes or tyrosinase (heme, copper, ...). The purpose is not to provide an exhaustive view of the bioelectrochemistry of these redox proteins that have been covered by specific reviews in others fields such as enzymatic biosensor[13] and enzymatic biofuel cell,[14] but to provide a detailed description of their application in electroenzymatic synthesis, focusing on the strategies used to immobilize the enzyme and/or the mediator.

6.2 Enzymatic Systems, Cofactor Recycling and Target Reactions

6.2.1 NAD(P)-dependent-redox enzymes

Both nicotinamide adenine dinucleotide phosphate ($NADP^+$/NADPH) and nicotinamide adenine dinucleotide (NAD^+/NADH) have been used in electroenzymatic synthesis. These different forms of the nicotinamide adenine dinucleotide cofactor display similar electrochemical properties. For this reason, the discussion will be focused on the regeneration of NAD^+ or NADH and all considerations and conclusions could be, in principle, applied to $NADP^+$ and NADPH.

The electrochemistry of NAD^+/NADH is highly irreversible, and the oxidation of NADH or the reduction of NAD^+ at a bare electrode only occurs at high overpotential. Several efficient methods have been developed for cofactor regeneration. Synthesis where the cofactor NAD^+/NADH has to be regenerated to its oxidized/reduced state can be carried out with direct, indirect, and enzyme-coupled electrochemical cofactor regeneration.[6] Direct regeneration means that the species to be regenerated itself reacts at the electrode; however, it can result in inactivation of the cofactor.[15] Indirect regeneration means that a mediator catalyst acts as an electron shuttle between the electrode and the cofactor and decreases the overpotential of the electrochemical reaction. Different molecules are involved in catalyzing either the oxidation of NADH or the reduction NAD^+. If a second enzyme is used, *e.g.* diaphorase, the process is called enzyme-coupled electrochemical cofactor regeneration. With few exceptions, it will be discussed in the section related to electrode materials (Section 6.3); direct regeneration is not used in practice, so the discussion will be mainly focused on indirect regeneration and enzyme-coupled electrochemical cofactor regeneration. Both NADH and NAD^+ regeneration will be discussed in the following sections for electrosynthesis applications.

6.2.1.1 *NADH regeneration by NAD^+ electrochemical reduction*

In NADH regeneration process, NAD^+ reduction occurs through two discrete $1e^-$-steps with a radical protonation process concerted with the second electron transfer.[15] The first electron transfer to NAD^+ is a reversible process. However, the dimerization of the NAD radical is more favorable than the second electron transfer in a direct electrochemical reduction[16] and the enzymatically inactive NAD^+-dimers can be formed, resulting in the degradation of the electroregeneration system.[6,17,18] In consequence, it is critical to avoid the formation of electrochemically inactive NAD-dimers, and indirect NADH regeneration methods have been developed.

The non-enzymatic indirect regeneration of NADH consists of using a mediator catalyst which can transfer two electrons in one step.[18] As the formal potential of the NAD^+/NADH redox couple is -0.56 V (*vs.* SCE), which means catalytic reduction of NAD^+ into NADH is thermodynamically favorable when the potential is more negative than -0.56 V,[15] to avoid direct electrochemical regeneration, the operating window of the mediator should be in the potential window between -0.56 and -0.9 V. Many efforts have been made to find the appropriate mediator.

$Rh(bpy)_3$ was the first kind of rhodium-based mediator used for indirect electrochemical regeneration of NADH. The synthesis of cyclohexanol from cyclohexanone catalyzed by alcohol dehydrogenase in the presence of this mediator was carried out[19] and its electropolymerized form on the electrode was also tested.[20] Later on, the mediator selectivity has been highly improved up to 99% when the pentamethyl cyclopentadienyl (Cp^*) ligand was incorporated. After the first electroenzymatic synthesis from pyruvate to lactate in the presence of $[Cp^*Rh(bpy)Cl]^+$ catalyzed by D-lactate dehydrogenase, it was later combined with alcohol dehydrogenase to synthesize different products (*e.g.* cyclohexanol,[21,22] (*S*)-4-phenyl-2-butanol,[18] (*R*)-phenylethanol,[23] L-malate[24]). Besides, $[Cp^*Rh(phen)Cl]^+$ with a 1,4-phenanthroline ligand instead of 2′,2-bipyridine ligand was also proved to be efficient in electroenzymatic synthesis of L-glutamate.[25] Moreover, polymerized neutral red (NR) film (at a potential of −0.6 V *vs.* SCE) was also successfully applied to NADH regeneration.[26]

The enzymatic electrochemical regeneration of NADH consists on the combination of a NAD-dependent dehydrogenase with a second enzyme (usually a flavoenzyme) in the presence of a mediator to prevent the dimer formation. If one refers to Figure 2, the electron transfer pathway of the dehydrogenase is described by Figure 2(a), while the electron transfer pathway of the second enzyme is described by Figure 2(b). The most commonly used regeneration enzymes include diaphorase,[27–29] lipoamide dehydrogenase,[22,29,30] ferredoxin $NAD(P)^+$ reductase,[30,31] and AMAPORS (artificial mediator accepting oxidoreductases).[32] Then the electron mediators like viologen derivatives[22,27–32] and flavin[33,34] or quinone[35] have been used to mediate the electron transfer reaction from the electrode to the flavin adenine cofactor of the enzymatic catalyst. Electrons were then transferred to the NAD^+ cofactor in the enzyme catalytic site. Some redox enzymes can also directly communicate with electrode supports (as schematically drawn in Figure 2(c)) and thus stimulate the regeneration of the NADH cofactor. Hydrogenases (from *Rhodococcus opacus* and *Atcaligenes eutrophus* H16) have been successfully applied for the bioelectrocatalytic regeneration of NADH without the use of a redox-mediator[36–39] and have been applied to the bioconversion of alpha-ketoglutarate into L-glutamate catalyzed by an L-glutamate dehydrogenase.[40] However, the absence of protein confinement at the electrode surface was pointed out at that time as a limit of the experiment. More recently, isolated diaphorase fragment has been described as a catalyst for the regeneration of NADH in the presence of H_2 and hydrogenase on pyrolytic graphite[41,42] and DET reaction starts to be explored for NAD^+ – NADH interconversion by *Escherichia coli* flavohemoglobin.[43]

D-lactate is by far the most studied target. It is an important product in industrial manufacturing (*e.g.* cosmetic, food) that can be synthesized from pyruvate in the presence of D-lactate dehydrogenase.[44–54] The conversion of cyclohexanone to cyclohexanol,[21,22,55] acetophenone to (*R*)-phenylethanol,[23,56] and 2-methylcyclohexanone to (1*S*,2*S*)-(+)-2-methylcyclohexanol[27] have been reported. The synthesis of (L)/(*S*)-glutamate, which is a compound used as flavor enhancer and applied in pharmacology, have also been carried out with the redox enzyme glutamate dehydrogenase.[25,30,32,40,57] Another valuable product is D-sorbitol, which is widely used in medical, food, as well as cosmetic applications was electrochemically converted from D-fructose catalyzed by D-sorbitol dehydrogenase.[58]

6.2.1.2 *NAD^+ Regeneration by NADH electrochemical oxidation*

The potential of direct NADH oxidation is close to 0.9 V (*vs.* SCE).[6] The electroenzymatic synthesis using direct NAD^+ oxidation has been applied in the conversion of alcohol to aldehyde in the very early stages of electroenzymatic synthesis.[59] The production of gluconic acid from glucose catalyzed by glucose dehydrogenase using this direct electrochemical oxidation of NADH was also reported.[60] But, the overpotential for this reaction can be dramatically decreased in the presence of a mediator electrocatalyst, as commonly used in a biosensors.[13] It prevents the risk of side reaction that could occur when the electrode surface is poised at a high potential. If a mediator is introduced in NAD^+ regeneration, the mechanism of the mediated electrochemical regeneration of NAD^+ can be described in the following steps: first, the mediator is oxidized by applying a suitable potential; then the reduced NADH is oxidized by the oxidized mediator,[61] and the mediator itself will go back to its reduced form; finally, the reduced mediator will be oxidized again on the electrode, in the meantime, the oxidized NAD^+ will participate in the enzymatic reaction. An additional possibility to regenerate cofactors is through photoelectrochemical methods,[62,63] but this approach has not been applied in the field of electroenzymatic synthesis yet.

Different mediators have been used for indirect electrochemical regeneration of NAD^+ in electroenzymatic synthesis.[10,15,64] Two-electron mediators are the most commonly found, like bipyridine and phenanthroline metal complexes,[65,66] azine dyes like methylene green[67] or phenazine methosulfate,[68] 4-carboxy-2,5,7-trinitrofluorenyliden-malonnitrile,[69] and 3,4-dihydroxybenzaldehyde.[70] Another kind of mediators for NAD^+ regeneration are based on single-electron transfer processes like 2,2′-azinobis(3-ethylbenzothiazoline-6-sulfonate) (ABTS). Compared to two-electron based mediators, one-electron mediator was reported to be more stable under basic conditions, which could be suitable for some large-scale synthesis.[71] It is also possible to combine indirect regeneration of NAD^+ with an enzymatic regeneration step. For example, diaphorase has been applied to oxidize NADH using a variety of quinone compounds,[35] ferrocene derivatives,[72,73] or osmium redox polymers.[74]

By using different mediators as electron carriers, a variety of substrates catalyzed by different NAD-dependent enzymes are obtained. The production of gluconic acid catalyzed by glucose dehydrogenase[70] or glucose-6-phosphate dehydrogenase[68] was reported. Alcohol dehydrogenase has been applied on different substrates to regio-selectively oxidize alcohols to corresponding carbonyl group products (*e.g.* cyclohexanone,[65,72,75] 2-butanon[76]). Glycerol dehydrogenase was applied to synthesize (*S*)-phenylethane-1,2-diol.[77] D-sorbotol dehydrogenase was used for D-fructose production from D-sorbitol.[67] Moreover, there are examples suggesting the electrosynthesis of D-sorbose catalyzed by ribitol dehydrogenase[78] and hydroxyacetone with galactitol dehydrogenase.[69]

6.2.2 Flavoenzymes

Flavoenzymes contain a flavin nucleotide (FMN or FAD) as a prosthetic group. Glucose oxidase (GOx) is a typical flavoenzyme, which has been used for gluconic acid production in the presence of oxygen; however, hydrogen peroxide, which is consequently formed, could deactivate the redox enzymes. In some system, the conversion rate of glucose was significantly improved (from 4% to 30%[79]) by eliminating the H_2O_2 through electrochemical method, by which the H_2O_2 was oxidized back to O_2.[22,80,81] However, for the majority of flavoenzymes applied in electrosynthesis, the bioprocess was accomplished by electrochemical regeneration of cofactor FAD or FMN using an electron mediator (see Figure 2(b)). For example, instead of using oxygen, when benzoquinone was introduced into the system, the formed hydroquinone was electrochemically regenerated back to benzoquinone. The operational stability for gluconic acid production was increased in this case by a factor 50.[82] *p*-Cresol methylenehydroxylase and 4-ethylphenol methylenehydroxylase have been applied for synthesis of 4-hydroxybenzaldehyde or 1-(4-hydroxy phenyl)ethanol in the presence of ferrocene as electron mediator, a high TTN up to 400,000 and a high conversion rate up to 100% were achieved in the electrochemical bioreactor.[83] Ferrocenedimethanol was used for wiring membrane-bounded (*S*)-mandelate dehydrogenase (S-MDH), which is a FMN-associated enzyme and applied to the catalytic conversion of (*S*)-mandelic acid to phenylglyoxylic acid in order to separate (*S*)-mandelic and (*R*)-mandelic.[84] In electroenzymatic reduction, l-aminopropyl-1 methyl-4,4′-dipyridine (ADPy) amino acid oxidase (AO_x) was used in the production of D-alanine from pyruvic acid. The conversion was almost total and a TTN more than 36,000 was achieved for immobilized ADPy.[85]

6.2.3 Metalloproteins

Most of the metalloproteins used in electroenzymatic synthesis contain a heme prosthetic group with a central iron atom (see an illustration in Figure 1(c)). Cytochrome P450 (CYP) is a representative class of heme-containing monooxygenase enzymes responsible for oxidative metabolism of most drugs and xenobiotics.[86] For most

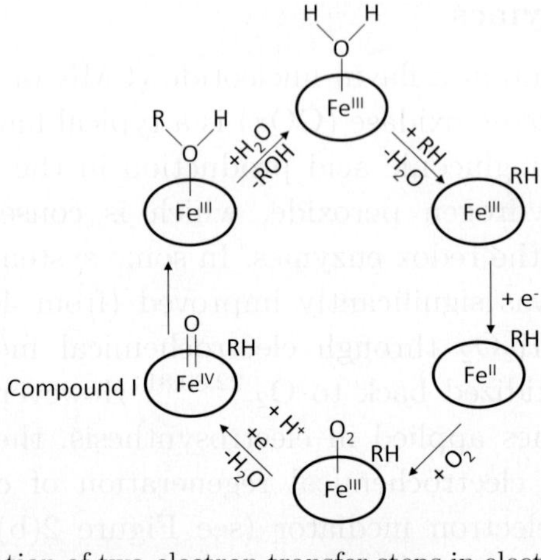

Figure 3. Illustration of two-electron transfer steps in electrochemical catalytic cycle of P450 enzymes. Adapted from Refs. 86 and 87.

applications, CYP needs to be immobilized on the electrode to benefit from DET in electrochemical regeneration as shown in Figure 2(c). The immobilization is not only necessary for stabilizing or recycling the enzyme (as for dehydrogenases and flavoenzymes). It is also a requirement for electrochemical recycling of the electroactive redox cofactor. CYP catalysis route is a two-electron transfer process, a catalytic cycle using electrochemistry as a driving force is illustrated in Figure 3. The first electron is used for heme reduction, and then the reduced iron group is associated with the co-substrate O_2. The second electron is further used for reducing the oxygenated heme, forming the target product in the presence of protons.[86,87]

CYP 119 was used for electrocatalytic reduction of nitrite to ammonia. This is a two-step electrocatalytic process with two redox peaks corresponding to the reduction of nitric oxide and production of ammonia.[88] Human CYP 2E1 enzyme was immobilized on gold electrode surface for synthesizing *p*-nitrocatechol by applying a potential of −500 mV (*vs.* Ag/AgCl) for 30 min.[89] Hydroxylation of 3-phenoxytoluene was catalyzed by immobilized CYP BM-3 by applying a potential of −0.7 V (*vs.* SCE), showing the possibility of using electrochemical method as electron donor.[90] The same protein was also used for synthesis of *p*-nitrophenolate by hydroxylation of *p*-nitrophenoxycarboxylic acid.[91] Not only single enzymes have been considered in electrosynthesis but also bi-enzymatic systems have been studied. Recently, a work combining two species of cytochrome P450, CYP1A2, and CYP3A4 on an electrode was described. Two redox peaks of the bi-enzyme complexes were observed at −0.531 V and −0.474 V, respectively. Clopidogrel carboxylic acid was then obtained as a metabolite.[92] Instead of using the direct electrochemical reduction of the P450 redox center, Rusling *et al.* reported the combination of a human cytochrome P450/cytochrome P450 reductase (CYP/CPR) in a bioelectronic film on pyrolytic graphite electrode. They developed the idea that electrons could be delivered from cytochrome P450 through CPR.[93]

Chloroperoxidase is another family of enzymes containing heme cofactor that catalyzes the chlorination of organic chemicals. In a CPO functionalized electrode, H_2O_2, which was used for catalytic oxidation of cinnamyl alcohol to produce aldehydes was constantly generated by the electrochemical reduction of O_2, giving a high yield of 52% and TTN of 80,500.[94] H_2O_2 can also be provided for this enzyme by a gas diffusion electrode (GDE).[95,96] The role of electrochemistry was not to provide electrons to the redox center of the enzyme, but to regenerate a co-substrate of the enzymatic reaction, regeneration that could be carefully tuned by electrochemistry. GDE with CPO catalyzed system has also been employed for monochlorodimedone chlorination,[97] and halogenation of the phenolic monoterpenes thymol and carvacrol.[98]

A final example of redox enzyme is tyrosinase, which contains a coupled binuclear copper active site that has been used as catalyst for hydroxylation of L-tyrosine to L-3,4-dihydroxyphenylalanine (L-DOPA). Tyrosinase catalyzed the ortho-hydroxylation of L-tyrosine to L-DOPA by its cresolase activity, and subsequently also catalyzed the oxidoreduction of L-DOPA to DOPA quinone by its catecholase activity. The by-product DOPA quinone was then electrochemically reduced to L-DOPA.[99,100] According to the authors, this electroenzymatic system showed the highest conversion rate and a highly enhanced productivity if compared to other approaches for L-DOPA synthesis reported previously.

6.2.4 Electrode materials

Several factors could limit the efficiency of a biosystem to be applied for a bioreactor,[101] and among them the electrode materials is an important factor. Metal, metal oxides, and carboceous materials were used as a flat or as a porous electrode.

6.2.4.1 *Metals*

The use of platinum electrodes is mostly encountered in membrane electroenzymatic reactors in which the enzymes stay in solution.[22,44,55,79] It has been reported that the adsorption of some hydrogenases on a platinum electrode would yield efficient catalysis for NAD^+ reduction.[37,102] The usage of platinum is expected to decrease the regeneration potential in the presence of *Alcaligenes eutrophus* H16 hydrogenase. However, the working potential window of platinum is quite limited since hydrogen evolution often occurs in the range of NAD^+ reduction. Another kind of applicable metal electrode is gold, even though it suffers from the similar hydrogen evolution problem in reduction reactions, it offers a possibility to modify the electrode surface by means of the formation of a self-assembled monolayer with thiol-containing molecules.[69,89,103] The use of macroporous gold electrodes was also suggested in order to take advantage of the high porosity, which would facilitate the mass transport through the electrode macrostructure to favor electrochemical bioconversion processes.[104–106] Finally, platinum and nickel nanoparticles deposited on a glassy carbon electrode were recently evaluated for NAD^+ reduction. The role of Pt and Ni was to provide "active" adsorbed hydrogen (Pt–H ads or Ni–H ads) for the fast protonation of electrochemically formed NAD-radical. The authors could demonstrate that up to a 100% recovery of 1,4-NADH from NAD^+ was possible on the produced electrodes at relatively low potentials, making them excellent candidates for regeneration of 1,4-NADH in industrial bioreactors.[107] The same research group also tested other electrode materials (Ti, Ni, Co, and Cd).[108]

6.2.4.2 *Metal oxides*

Tin (IV) oxide electrode was prepared using an anodization and annealing method and was used for mediator-free electrochemical cofactor regeneration. Both oxidation of NADH and reduction of NAD^+ were studied. The electrochemical oxidation of NADH and NAD(P)H were performed at −0.5 V *vs.* a Ag/AgCl reference electrode. The NAD^+ reduction was performed at −0.95 V. Only the NAD^+ regeneration was, however, applied to electroenzymatic reaction. The authors suggested that the method can be applied to various cofactor-dependent enzyme reactions for chemical production, biosensor, or enzyme fuel cell construction.[109] A thermally prepared iridium/ruthenium-oxide coating (Ir0.8Ru0.2-oxide) formed on a titanium substrate was also proposed as a possible electrode for direct electrochemical regeneration of enzymatically active 1,4-NADH from its oxidized form, NAD^+. The coating surface was characterized by "cracked mud" morphology, yielding a high surface roughness. The percentage of enzymatically active 1,4-NADH present in the product mixture was strongly dependent on the electrode potential, reaching a maximum (88%) at −1.70 V *vs.* a mercury/mercurous sulfate reference electrode.[110] The use of conductive vanadia–silica gels with encapsulated GDH was also reported for NADH regeneration, allowing complete conversion of alpha-ketoglutarate to L-glutamate.[57]

6.2.4.3 *Carbon materials*

Carbon materials are widely used due to their low cost, large potential window, relatively inert surface, and active electrochemical properties for redox reactions.[111] Compared to metal electrodes, different carbon materials exhibit different microstructure as well as surface functional groups, leading to different electrochemical behaviors. There are already reviews exhaustively explaining the interfacial as well as the electroanalytical properties of carbon materials.[111,112] Here, we only focus the discussion on the application of carbon materials as electrodes in the electroenzymatic synthesis with immobilized proteins.

Carbon paste electrodes (CPE) have also been used after its introduction in 1958. Basically, it is made from a mixture of carbon powder with a liquid non-electroactive binder. The carbon paste could be "modified" by adding one or more components (*e.g.* enzymes or mediators).[113] The CPEs are popular because they are easily obtained at minimal costs and certain electrode properties can be pre-determined by adding additional admixtures.[114] It can be applied for encapsulation of enzymes on electrode.[115] However, its properties differ from one preparation to another and this limits their applications.

The most conventional carbon electrode is glassy carbon (GC), its use is convenient since its surface is renewable by polishing. As it is a flat electrode with uniform functional groups on the surface, it is a good candidate to be covalently modified for bounding enzymes or molecules,[53,85] or deposited with uniform membranes,[47] nanoparticles[107] or films.[78,90,91] If the GC is produced as a foam, it is then called reticulated vitreous carbon (RVC), which also been used in electrochemical bioreactors.[45,48]

Porous carbon felt (CF) electrodes have been largely used electrode in the field of electroenzymatic synthesis, taking benefit from their low cost and large surface area, which lead to high current density and high conversion efficiency. CF electrodes have been used as working electrodes in electroenzymatic synthesis reactors with dialysis[21,22,55] or ultrafiltration membranes.[83] From another point of view, the functional groups on the surface make it an appropriate substrate for surface modification, which could be further used for different covalent reactions. CF was used for example as electrode support for immobilization of enzymes with carbodiimide or glutaraldehyde as linking agent.[23,54,82]

Furthermore, to increase the surface area as well as the biocatalytic property of the electrodes, carbon nanoparticles,[99,100] carbon nanotubes (CNTs),[67] graphene nanosheets,[116] and graphene oxide nanocomposite[92] allowed to improve significantly the efficiency of the electroenzymatic synthesis system by increasing the enzyme loading and/or by tuning some electrochemical properties.

6.3 Immobilization of Enzymes on Electrodes

In the past years, different strategies have been developed to build enzyme-immobilized electrodes.[12,117] There are several conditions that need to be fulfilled for enzyme immobilization. First, the activity of enzymes needs to be kept upon immobilization. Second, the stability of enzymes toward microenvironments on the electrode needs to be ensured. Usually, adsorption is the easiest method for enzyme immobilization, but it is inappropriate for bioconversion, because the electrosynthesis process requires long-term stability, especially when the solution is under continuous stirring, the adsorbed enzymes would be rapidly leached out into the solution. In the field of electro-enzymatic synthesis, the immobilization methods that have been typically used are membrane confinement in a small volume at a close distance from the electrode (Figure 4(a)), surface immobilization with chemical bonding (Figures 4(b) and 4(c)) and encapsulation (Figures 4(d)–4(f)). The following sections will provide a discussion of these different strategies.

6.3.1 Membrane confinement with membranes

Membrane confinement is a simple way to immobilize enzymes in an electrosynthesis bioreactor. The enzymes are confined in a solution near the electrode without any physical or chemical modification, but only soluble redox enzymes can be immobilized by this method. Generally, two kinds of membrane configuration are used for immobilization: dialysis or ultra-filtration membrane. The formation of a Nafion coating on an electrode surface can be considered as a surface confinement comparable to encapsulation and will be discussed in another section.

Commercial dialysis or ultra-filtration membranes have been used inside a bioreactor.[21,22,44,46,49–51,55,59,60,79,83] These membranes are made of different kinds of materials (*e.g.* cellulose,[21,22,44,55,79] hollow fiber,[59] and Nafion 117[45,46,50,51,60]). Generally speaking, the ability to immobilize the enzyme is dependent on the pore size of the membrane. Enzymes larger than the membrane cutoff value are confined in the membrane near electrode and small molecules can freely pass through the membrane.

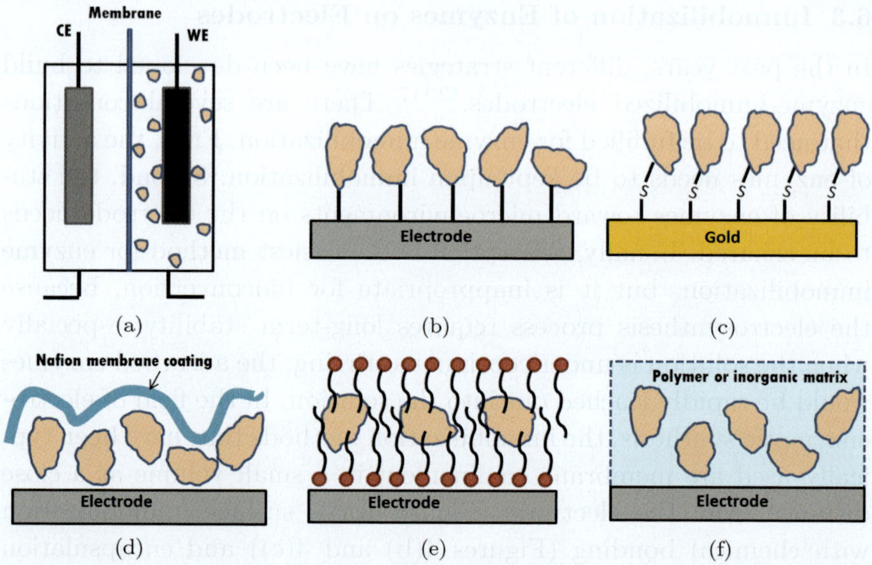

Figure 4. Typical enzyme immobilization methods on electrodes (a)–(f): (a) Membrane confinement; (b) covalent bonding; (c) "thiol–gold" self-assembly monolayer; (d) nafion coating; (e) surfactant film embedding; (f) encapsulation in solid matrix.

In a dialysis membrane electrochemical reactor (D-MER), the transport of compounds through the membrane was mainly due to the gradient of concentration, while in ultra-filtration membrane electrochemical reactor (UF-MER), solution passing through the membrane in a direction perpendicular to the electrode was driven by a pressure gradient. The influence of flowing mode on the transformation rate in a bioreactor has been studied. An ultrafiltration-membrane electrochemical reactor (UF-MER) (Figure 5(a)) was compared with D-MER (Figure 5(b)) in synthesis of cyclohexanol by alcohol dehydrogenase with [Cp(Me)$_5$Rh(bpy)Cl] as the mediator. The transformation rate in these two configurations is compared in Figure 5(c), leading to 92% in 45 days for D-MER and 100% in 3 days for UF-MER. UF-MER was the more efficient reactor configuration, and a limiting step in D-MER was observed: the inhibition of the enzyme by increasing concentration of the products at the end of the electrolysis.[21]

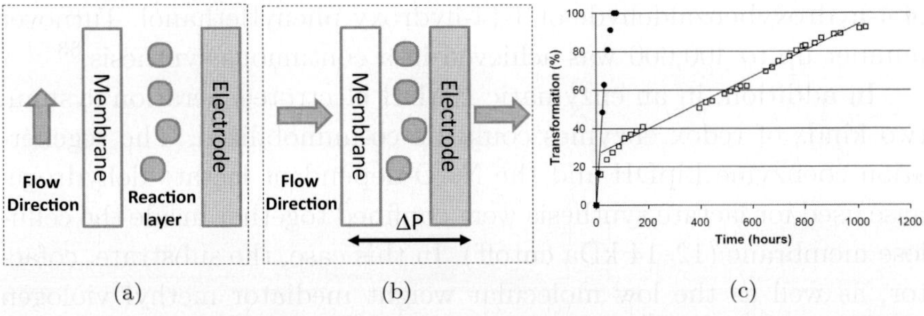

Figure 5. ((a),(b)) Schematic description of (a) dialysis-membrane electrochemical reactor (D-MER) and (b) ultra-filtration-membrane electrochemical reactor (UF-MER). (c) Comparison of the transformation of cyclohexanol obtained with the D-MER (square) and the UF-MER (circle) in a recycling mode with a rhodium complex as electrochemical mediator. Reprinted from Ref. 21.

In an early stage, membrane confinement of redox enzymes has been used in the bioreactor with direct electrochemical regeneration of NAD^+. Coughlin *et al.* use a continuous-flow reactor with ultra-filtration membrane to confine NAD-dependent alcohol dehydrogenase for direct electrochemical regeneration of NAD^+ in electrosynthesis of aldehyde from alcohol. Compared with dialysis, the ultra-filtration membrane was able to keep the cofactors and enzymes persistently in the continuous-flow process, while the products left the system through the membrane.[59] Later on, the confinement of glucose dehydrogenase in a dialysis Nafion membrane reactor for synthesis of gluconic acid with direct regeneration of NAD^+ was also achieved.[60] To avoid the large overpotential, which may cause inactivation of the cofactor, diffusing mediators were added in the system afterward to facilitate the electron transfer between electrodes and cofactors.[60]

When mediators were added into the system, the membrane could be used to confine only the redox enzymes,[21,22,50,55] or enzymes together with mediators.[44,46,51,83] The confinements of the mediators were achieved by coupling it with large size polymers in order to achieve a size larger than the cut-off of the membrane. In an electro-enzymatic membrane flow cell, ferrocene was, for example, modified with polymers and immobilized together with *p*-cresol methylenehydroxylase or 4-ethylphenol methylenehydroxylase for electrosynthesis of 4-hydroxybenzaldehyde or 1-(4-hydroxy phenyl)ethanol. Turnover number up to 400,000 was achieved in a continuous synthesis.[83]

In addition, in an enzymatic NADH electroregeneration system, two kinds of redox enzymes could be co-immobilized. The regeneration coenzyme LipDH and the NAD-dependent lactate dehydrogenase used for lactate synthesis were confined together inside the cellulose membrane (12–14 kDa cutoff). In this case, the substrate, cofactor, as well as the low molecular weight mediator methyl viologen could be freely transferred in the system. The substrate alpha-keto acid (2.5 mmol) was converted at 95% after 8 days with constant addition of new lipoamide dehydrogenase every 2 days, and the product was obtained with 98% enantiomeric excess, which was comparable to that obtained in a non-membrane system.[44]

6.3.2 Chemical bonding

Compared to adsorption, the leaching of enzymes from electrode was minimized with covalent bonding and long-term stability was ensured. Generally, the chemical reaction occurred between the functional groups on the electrode surface and active amino acid residues of the enzyme. But an obvious disadvantage of covalent binding is that enzyme itself is chemically modified when immobilized, if essential amino acid residues close to the enzyme's active site are involved in the chemical reaction, enzyme activity might be reduced. So when covalently immobilizing the enzymes on electrodes, physical and chemical nature of the electrode, the nature of the linkage or binding chemistry, and even enzyme orientation need to be considered.[118]

Different kinds of carbon materials (carbon felt,[23,54,82] graphene felt,[70] graphene oxide nanocomposite (RGO),[92] carbon nanopodwer,[99,100] rotating disc graphite,[68] and glassy carbon[53,85]) can be used for covalent immobilization of enzymes in the field of electroenzymatic synthesis. These carbon materials have advantages such as chemical and physical stability, large surface area, and rich surface functional groups. It is already reported that several chemical methods could be applied for covalent immobilization of enzymes,[119,120] but the chemical reactions that have been used in the electroenzymatic synthesis are quite limited. The generally used methods include carbodiimide activation,[54,82,99,100] glutardialdehyde cross-linking,[23,53,85] and thiol-ene click chemistry.[58] Sometimes the carbodiimide and glutardialdehyde are used together in two successive steps.[68,70,92] On gold electrode, self-assembled monolayers could be also involved.[69,89]

6.3.2.1 *Carbodiimide activation*

There are three steps to immobilize the enzymes on electrode surface by means of carbodiimide activation. As shown in Figure 6(a), first, the carbon surface is oxidized to create carboxylic groups, and then the carboxylic groups are activated by carbodiimide; meanwhile the active amino group from the enzyme is reacting.

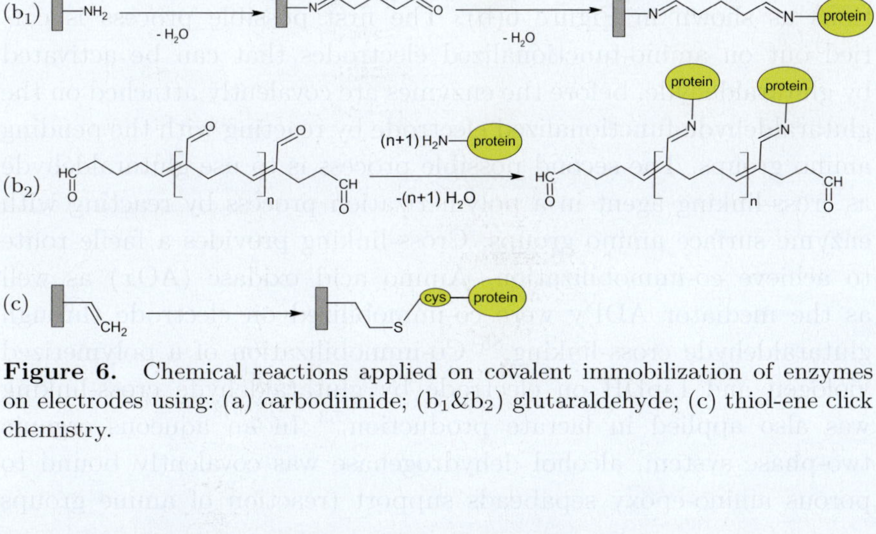

Figure 6. Chemical reactions applied on covalent immobilization of enzymes on electrodes using: (a) carbodiimide; (b_1&b_2) glutaraldehyde; (c) thiol-ene click chemistry.

For the production of gluconic acid from glucose GOx in the presence of benzoquinone in the solution as the mediator, the enzyme was covalently immobilized on the carbon felt electrode by carbodiimide.[82] Tyrosinase was immobilized to functionalized carbon nanoparticles via (1-ethyl-3-(3-dimethylaminopropyl) carbodiimide, before being integrated in a 3D bioelectrode.[99,100] In some systems, both mediator and enzymes could be covalently co-immobilized on the electrode by using carbodiimide. For the electrosynthesis of lactate, the coenzyme viologen-accepting pyridine nucleotide oxidoreductases (VAPORS) combined with viologen directives (DAPV) was co-immobilized on an electrode for NADH regeneration. Different immobilization order of both components were tested, either first covalent binding DAPV to VAPORS then followed by attachment of the DAPV-VAPORS to a functionalized carbon electrode, or covalent binding of the enzyme to viologen functionalized carbon electrode. These different covalent bonding strategies allowed optimization of the enzyme orientation. A relationship was drawn between electron transfer reactions and bioconversion efficiencies.[54]

6.3.2.2 *Glutaraldehyde cross-linking*

Another popular linking agent for covalent immobilization of enzymes is glutaraldehyde. There are two types of possible reactions as shown in Figure 6(b). The first possible process is carried out on amino-functionalized electrodes that can be activated by glutaraldehyde, before the enzymes are covalently attached on the glutaraldehyde-functionalized electrode by reacting with the pending amino groups. The second possible process is to use glutaraldehyde as cross-linking agent in a polymerization process by reacting with enzyme surface amino groups. Cross-linking provides a facile route to achieve co-immobilization. Amino acid oxidase (AOx) as well as the mediator ADPy were co-immobilized on electrode through glutaraldehyde cross-linking.[85] Co-immobilization of a polymerized viologen and LipDH on electrode by glutaraldehyde cross-linking was also applied in lactate production.[53] In an aqueous–organic two-phase system, alcohol dehydrogenase was covalently bound to porous amino-epoxy sepabeads support (reaction of amine groups from the protein with epoxy groups) and a subsequent cross-linking step with glutaraldehyde was used. The enzyme-support preparation was applied to the production of (R)-phenylethanol from acetophenone in the presence of Cp*Rh(bpy) as mediator for NADH regeneration.[23] The stability of the enzyme in a bioreactor was strongly increased after glutaraldehyde treatment, resulting in a half-life time of over 1200 h when stored at 30°C. This was a 60-fold increase in stability compared to the soluble enzyme.[121]

6.3.2.3 *Thiol–ene click reaction with cysteine-tagged proteins*

Thiol–ene reaction is an organic reaction between a thiol and an alkene to form an alkyl sulfide. This reaction has been used for covalent immobilization of cysteine-tagged D-sorbitol dehydrogenase on carbon felt electrode previously functionalized with the vinyl group by electrografting of a suitable diazonium species (Figure 6(c)). Taking advantage of the thiol group from the cysteine tag (cysteine moieties at the N terminus of the polypeptide chain), the enzymes were proved to maintain activity upon grafting. This electrode was applied to the bioconversion of D-fructose to D-sorbitol using $[Cp^*Rh(bpy)Cl]^+$ as a catalyst for NADH regeneration,[58] and a high faradaic efficiency of 77% was finally obtained. This reaction provides certainly an original pathway for covalent immobilization of other cysteine-tagged proteins in an active form for bioconversions.

6.3.2.4 *Self-assembled monolayer (SAM)*

In some applications, enzymes can be bound to a gold electrode surface with a self-assembled monolayer (SAM) due to the interaction between thiol groups and gold. Mak *et al.* reported the immobilization of CYP2E1 and its mutants on gold electrodes that were first modified with dithio-bismaleimidoethane (DTME) forming a SAM. Then the functionalized DTME electrode was allowed to react with the exposed cysteine group on the enzyme surface. The orientation of the protein was well controlled and better catalytic activity was kept in the presence of the molecular spacer provided by the SAM.[89]

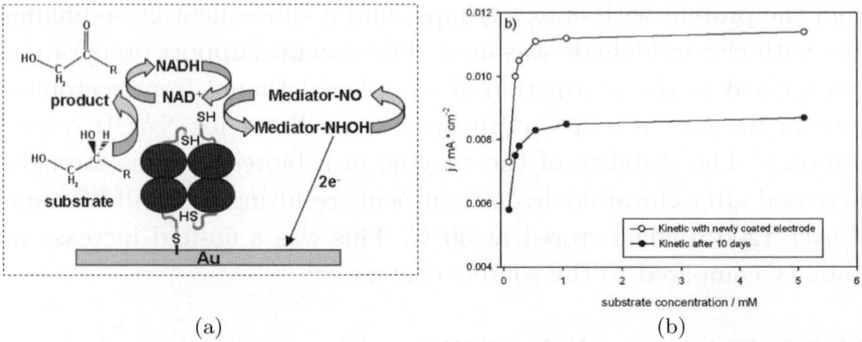

Figure 7. (a) Schematic principle of the immobilization of the cysteine-modified GatDH with electrochemical NAD^+ regeneration for enzymatic synthesis of hydroxyacetone; (b) Michaelis–Menten plots of the current at +500 mV *vs.* the substrate concentrations for GatDH-immobilized electrode before and after 10 days. Reprinted from Ref. 69.

A successful immobilization by direct reaction between the thiol group of a cysteine tag on the protein and a gold electrode was also reported. The schematic representation of this electrochemical system is shown in Figure 7(a). Galactitol dehydrogenase modified with a cysteine tag at the N-terminal of protein surface is immobilized on the surface of an gold electrode by SAM without losing the active configuration of the enzymes, the electroenzymatic oxidation from 1,2-hexanediol was achieved. This immobilized electrode had a good stability, and similar Michaelis–Menten plots were observed after 10 days immobilization at the gold surface (Figure 7(b)).[69]

Another kind of self-assembled immobilization is based on histidine group of protein. Histidine tag not only has been widely used in protein purification due to its prominent affinity, but also employed in enzyme immobilization by the interaction between histidine tag and transition metal ion (*e.g.* Cu^{2+}, Ni^{2+}, Co^{2+} and Zn^{2+}).[122] This self-assembled immobilization is a reversible process with controlled enzyme orientation. In a bioelectrochemical system for $NADP^+$ production, ferredoxin: $NADP^+$ reductase were immobilized on a modified gold electrode using a combination of nitrilotriacetic–Cu^{2+} reaction and SAM. The gold electrode was first covered with a self-assembled monolayer of thiols appended with nitrilotriacetic acid groups complexed with copper complex, and then the enzymes containing histidine were immobilized through its histidine residues.[103]

6.3.3 Encapsulation

The immobilization of proteins by encapsulation can be done either with conductive electrode materials (conducting polymers) or more often with insulating materials such as Nafion, alginate, silica gel, or hybrid materials.

6.3.3.1 *Nafion coating*

Nafion layer from a Nafion solution can be deposited on the top of enzyme-modified electrodes[45,47–49,52] as shown in Figure 4(c). This method was generally applied for co-immobilization of enzymes and mediators in electrosynthesis of lactate. The Nafion-coated electrode was prepared by first dipping an electrode into the solution containing the enzymes and mediators. After drying, the modified electrode was dipped into a polymer Nafion solution to get a Nafion membrane-coated electrode.

In an electroenzymatic synthesis of lactate, MV and LipDH have been confined together on the cathode by a Nafion coating membrane, allowing the extension of LipDH lifetime.[48] Compared to dialysis or ultrafiltration membrane reactors, this kind of coating membrane was easier to prepare, but it had the disadvantage of slow leakage of MV into the solution. As MV species are very toxic, this may bring in an undesirable contamination, a viologen derivative was first chemically attached to LipDH before immobilization on the electrode with a Nafion coating.[45,52]

Alternatively, the toxic viologen species could be replaced by a natural redox cofactor such as FAD. Both electropolymerized FAD or monomeric FAD were immobilized together with LipDH on the electrode by a Nafion coating, a relatively slower conversion rate of lactate was observed when using FAD in place of viologen, but the long-term stability of the enzyme was improved with electropolymerized FAD-immobilized electrode, and LipDH kept 80% of the initial activity after 30 days with constant performance.[47]

A comparison of immobilization with a Nafion coating and chemical bonding was recently reported by Min *et al.*[99] for the bioelectrochemical synthesis of L-DOPA (Figure 8). The recyclability of the electrode prepared with a Nafion coating (triangles) was much lower, decreasing by 20% after 10 cycles, than the one observed after glutaraldehyde grafting of tyrosinase on carbon nanoparticles and construction of the 3D bioelectrode with polypyrrole electropolymerization (circles).

6.3.3.2 *Surfactant and lipidic layers*

The embedding of membrane proteins into a surfactant film is another attempt for immobilization of enzymes for electroenzymatic synthesis (Figure 4(e)). The surfactant films usually consist of water-insoluble alkyl long-chain molecule like didodecyldimethylammonium bromide (DDAB). One approach is to mix the enzyme solution with surfactant vesicles dispersion, then the mixture solution is cast

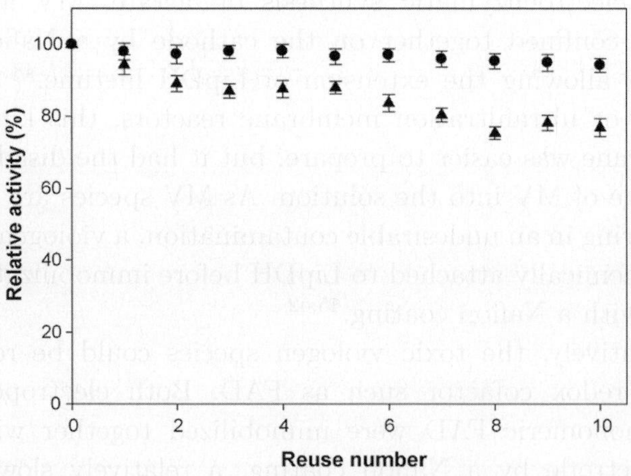

Figure 8. Stability of tyrosinase-immobilized electrode as function of reuse number (•, tyrosinase–carbon nanoparticle–polypyrrole composite electrode; ▲, tyrosinase-modified carbon felt electrode). Relative activity was measured by synthesized l-DOPA concentration for 4 h. Reprinted from Ref. 99.

on the electrode, and a multi-layer film is formed after drying.[123] A CYP119-DDBA film was prepared on an electrode by this approach, and its electrocatalytic response toward nitrite reduction to ammonia was studied.[88] Chloroperoxidase (CPO)-DDAB film functionalized electrode was also reported for electrooxidation of cinnamyl alcohol to produce cinnamic aldehyde. The lifetime of CPO was extended and a high production yield of 52% and a total turnover number of 80,500 for CPO were obtained.[94] Of related interest is the immobilization of lipid vesicles with embedded membrane proteins. A recent study was reported by Mazurenko *et al.* for the application of mandelate dehydrogenase to electrosynthesis. The lipid vesicles could be simply deposited on glassy carbon electrode, but the stability of the modified bioelectrode was greatly improved by combining this approach with encapsulation in a sol–gel silica matrix.[84] CYP/DDAB liquid crystal system could also be encapsulated in a hydrophobic sol–gel layer before further cross-linking with glutaraldehyde.[124]

6.3.3.3 *Alginate-based gel and sol–gel silica*

An early attempt on enzyme encapsulation is through calcium alginate gels.[125] A carbon-felt electrode was soaked in the solution containing enoate reductase and alginate, and then it was gelated by calcium ions. This ionotropic gel encapsulation approach not only increased the concentration of enzymes close to the electrode surface, but also significantly increased the half-life of the enzyme up to 350 h in electrochemistry cell under continuous operation.

Silica "sol–gel" process has been used for the immobilization of enzymes. The sol–gel process involves hydrolysis and condensation of silica precursors during which the enzymes are incorporated into the gel matrix.[120] This is a mild process for enzymes to retain their structures and activities, and the mechanical properties of silica matrix also facilitate its application in a flow-through reactor.[126] Another advantage of silica sol–gel encapsulation is the successful application of membrane-bound enzymes in electroenzymatic synthesis as discussed in the previous section.[84] When membrane-bounded (*S*)-mandelate dehydrogenase (S-MDH) was immobilized on electrode by silica gel, a conversion rate of 96% after 9 h and a high current efficiency of 91% was obtained in the presence of ferrocenedimethanol as a mediator. Vanadia–silica gels were also involved in the immobilization of glucose dehydrogenase.[57]

Recently, a proof-of-concept of an enzymatic electrosynthesis reactor containing an immobilized bioanode with D-sorbitol dehydrogenase encapsulated in silica matrix, and an oxygen gas-flow cathode was reported (Figure 9(a)).[67] The enzymes immobilized in the silica film show very good operational stability for two weeks, which was much better than that in solution (half-life period of 48 h at 28°). In simultaneous energy production mode, as shown in Figure 9(b), 60% D-sorbitol was converted to D-fructose. Gluconic acid production from glucose was also achieved in a membraneless, flow-through electroenzymatic reactor operating in the mode of co-generating chemicals and electrical energy.[127]

Hybrid gel can be used for the immobilization of redox proteins. Recently, a bioelectrode was prepared with alcohol dehydrogenases embedded in an alginate–silicate hybrid gel. This study is very intriguing as the authors argue for a DET reaction into the biocatalysts for the reduction of butyraldehyde to butanol[128]; however, the

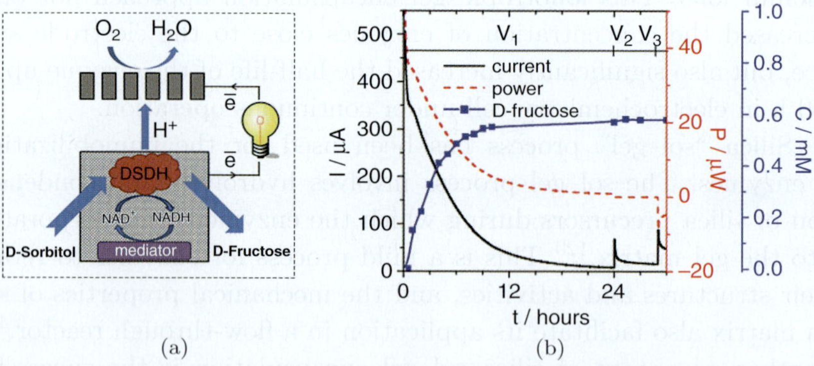

Figure 9. (a) Scheme of the flow bioreactor using D-sorbitol dehydrogenase (DSDH) immobilized bioelectrode as anode while a gas diffusion electrode (GDE) as cathode. (b) Conversion of 1 mM D-sorbitol to D-Fructose in the GDE bioreactor with simultaneous energy production. Reprinted from Ref. 67.

reaction should occur in the presence of a NAD^+ cofactor that is not mentioned in the bioelectrochemical reaction.

6.3.3.4 *Conducting polymers*

As electrochemistry is involved in enzymatic electrosynthesis, conducting polymers could provide unique advantages allowing both the protein immobilization and cofactor regeneration and wiring. Conducting biocompatible polymer materials (*e.g.* polypyrrole, poly(3,4-ethylenedioxythiophene)) have indeed been applied to this field. In a work with enzymes encapsulated in the polypyrrole matrix, the film was electrografted on a platinum electrode from pyrrole solution in the presence of GOx, resulting in a uniform polymer structure and enzyme distribution. The modified electrode was adaptable to a bioelectrochemical process for gluconic acid production.[80] The electrochemical process benefited from the oxygen regeneration at the electrode and from the enzyme protection effect from the polymer matrix, leading to an increase in the transformation ratios by a factor of two.[81]

The P450 enzymes were immobilized in conducting polymers like polypyrrole films[91] and poly(3,4-ethylenedioxythiophene) doped with polystyrene sulfonic acid (PEDOT/PSS)[90] on electrode in an active form, the conductive polymer was electrochemically wiring the enzymes for DET reaction (see Scheme 2(C)). Figure 10 reports a comparison of enzymatic production in the presence of the natural cofactor NADPH (black symbols) with the electrochemically driven biocatalysis (white symbols). Both experiments have been performed with the cytochrome P450 CYP102A1 immobilized in a polypyrrole matrix. Two different electrode materials were used, either Pt (squares) or glassy carbon (circles). This immobilization did not hinder the enzymatic activity of the proteins on glassy carbon electrode. The electrochemically driven biocatalysis was as efficient as the one fed by NADPH on this latter electrode material. The enzymatic activity determined on Pt electrodes show at the contrary a more complex behavior, largely unfavorable to electrochemically driven reactions.[91]

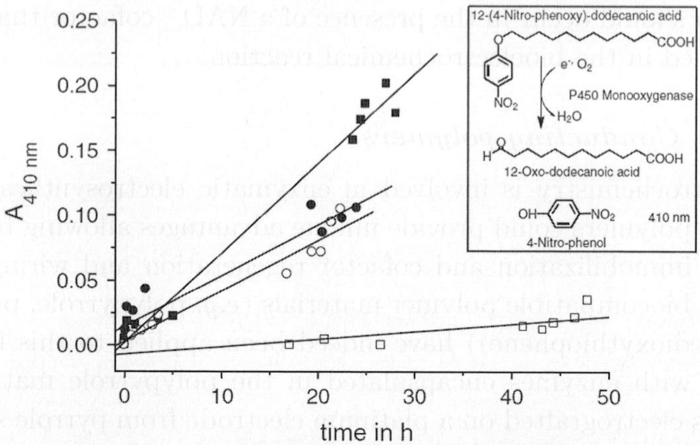

Figure 10. Product (p-nitrophenolate, detected at 410 nm) formation kinetics of P450 BM-3 QM embedded in polypyrrole on platinum (■) or glassy carbon (●) electrodes with the natural cofactor NADPH (□) or by electrochemistry (○). The insert is the scheme of corresponding reaction. Reprinted from Ref. 91.

6.4 Immobilization of Mediators

In practical electrosynthesis, the immobilization of mediators is also necessary from economic and ecological considerations. The immobilization of mediators would not only increase their reusability and simplify the product purification step, but also limit the undesirable interactions between enzymes and mediators. A huge effort has been made for the immobilization of redox mediator for flavoenzymes (especially for GOx) or NAD-dependent dehydrogenases in the field of biosensors,[13,64] or enzymatic fuel cells.[129–131] This topic was more rarely studied in the field of electrosynthesis and essentially for NADH regenerations, which concerns two molecules methyl-viologen and (2,2′-bipyridyl)(pentamethylcyclopentadienyl)-rhodium family ($[Cp^*Rh(bpy)Cl]^+$). Here, we will briefly review the immobilization of methyl-viologen and rhodium complex in relation with electrosynthesis.

6.4.1 Methyl-viologen

Methyl-viologen has been widely used as a mediator for enzymatic electroregeneration of NADH, always co-immobilized with a flavoenzyme, diaphorase,[27–29] lipoamide dehydrogenase,[22,29,30] ferredoxin NAD(P)$^+$ reductase[30,31] and AMAPORS.[32] Methyl-viologen immobilization was first achieved by membrane confinement[44,46,48,49,51] but this must lead to leaching. One strategy for more efficient immobilization of methyl viologen was bonding the molecule to co-flavoenzymes.[45,52] For example, in a dialysis membrane reactor, when the cutoff of the membrane was larger than the molecular size of methyl-viologen, this molecule was covalently bonded with the regeneration enzyme lipoamide dehydrogenase.[45] The enzyme-mediator species were also immobilized on the electrode by a Nafion coating. The methyl-viologen species were first functionalized with amino group, followed by attaching to a lipoamide dehydrogenase by *N*-ethyl-*N*′-(′*t*-dimethylaminopropyl)carbodiimide and then a stable membrane-coated electrode was obtained by dipping the electrode into a Nafion solution.[52] Another strategy was directly bonding to the electrode by polymerization[53] or covalent bonding.[54] Compared to physical entrapment under a Nafion membrane, electrochemical copolymerization of lipoamide dehydrogenase with oligomeric polyviologen on the electrodes allowed a higher production rate in lactate production.[53]

6.4.2 [Cp*Rh(bpy)Cl]$^+$

The most efficient non-enzymatic NADH regeneration catalyst is the (2,2′-bipyridyl) (pentamethylcyclopentadienyl)–rhodium family ([Cp*Rh(bpy)Cl]$^+$).[132] However, low enzymatic catalytic stability has been reported in the presence of this rhodium complex due to the interaction between enzyme surface functional groups and the rhodium species. Hildebrand *et al.* suggested that immobilization of the rhodium complex was necessary in order to separate it from enzymes.[56] A membrane reactor was reported with complete separation of the proteins from the rhodium complex modified with a large size polymer modified as shown in Figure 11(a).[56] This complete separation significantly improved the long-term stability of the system. A bioconversion from *p*-chloroacetophenone to *p*-chloro-(*R*)-phenylethanol catalyzed by *Lactobacillus brevis* alcohol

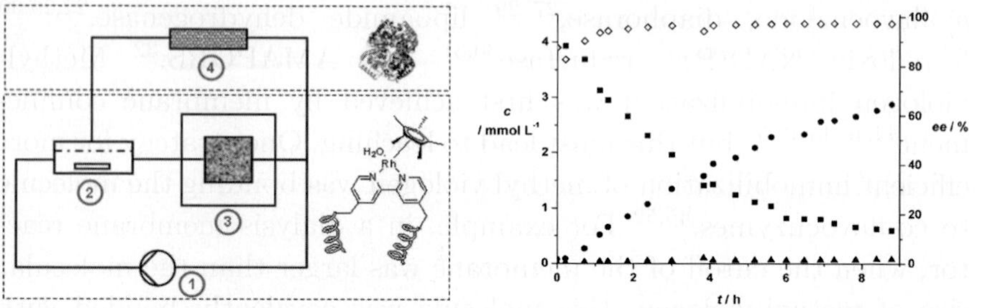

(a) (b)

Figure 11. (a) Setup of a membrane reactor with complete separation of the proteins from the large-size polymer-modified rhodium complex with setup: (1) pump, (2) membrane reactor, (3) electrochemical cell, (4) immobilized enzyme. (b) Conversion from 4 mM *p*-chloroacetophenone to *p*-chloro-(*R*)-phenylethanol by immobilized *Lactobacillus brevis* alcohol dehydrogenase with electrochemical cofactor regeneration. Reprinted from Ref. 56.

dehydrogenase (Lb-ADH) was performed in this bioreactor, a conversion rate up to 90% was achieved after 6 h (Figure 11(b)), and the activity of immobilized enzymes was completely kept (5.5 ± 1 U mg^{-1} before and 6 ± 1 U mg^{-1} after). In addition, 86% of the rhodium complex could be recovered.

Two other strategies have been successfully developed to immobilize rhodium complexes on an electrode surface in an active form. The first one is based on the electrodeposition of conducting polymers on the electrode. Chardon-Noblat *et al.* prepared polypyrrole rhodium (III) complex (2,2′-bipyridine or 1,10-phenanthroline ligand) films on an electrodes by anodic electropolymerization of pyrrole monomers. The electrocatalytic interest for hydrogen evolution was demonstrated.[133] Beley *et al.* bound a rhodium (III) bis-terpyridinepyrrole complex by electropolymerization on a reticulated vitreous carbon electrode, the long-term stability of the modified electrode was characterized by reduction of NAD^+ into 1,4-NADH. The system was then applied to the reduction of cyclohexanone to cyclohexanol in the presence of alcohol dehydrogenase.[20] In a report from Nolte *et al.*, rhodium complexes were incorporated into bilayer vesicles formed from a polymerizable ammonium surfactant by copolymerization reaction. The catalytic activity of the rhodium complexes was dependent on its position in the bilayer.[134] γ-irradiation cross-linking was also reported to obtain $[Rh^{III}(C_5Me_5)(L)C1]^+$ complex polymer-modified electrodes.[135] Even through this method provides a stable bounding, it suffered from the electron shielding effect caused by the thick porous polymer film. The electron transfer was influenced by the film structure and the position of rhodium complex inside the polymer, which reduced the efficiency significantly.

The second strategy is to take advantage of the π–π stacking effect between carbonaceous materials (graphene and CNTs) and aromatic moieties from the modified rhodium complex. This phenomenon would favor a strong adsorption of the rhodium complex on carbon support. Compared to the first route, the electron shielding effect caused by the polymer film was eliminated and the electron transfer between the rhodium complex and the electrode was facilitated.

Park *et al.* reported the synthesis of a 3D-structured graphene–Rh-complex hydrogel, on which a phenanthroline modified Rh complex was strongly immobilized on graphene through π–π interaction and π–cation interaction, the repeated usability for electrochemical NADH regeneration was demonstrated.[25] Later, Minteer *et al.* synthesized a rhodium complex with a pyrene-substituted phenanthroline ligand, which was immobilized onto multi-walled carbon nanotubes (MWCNTs) via π–π stacking interaction with a MWCNT substrate. Good catalytic response for the regeneration of NADH and reusability up to 10 cycles was achieved (each cycle lasted 30 min).[24] In both cases, the immobilization of the rhodium complex significantly increased the activity of the catalyst over time, making it applicable in NADH-dependent enzymatic catalytic systems. However, this method could be only applied on CNTs or 3D-structured graphene which has aromatic rings as electrode. In addition, π–π stacking is a reversible non-covalent bond with energy less than 10 kcal per mole, much lower than the covalent bond which is 80–100 kcal per mole.[136,137] This could limit the application for continuous long-term bioconversion. In order to take advantage of the stable covalent bonding and get rid of the electron shielding effect, up to now, some attempts have been made to covalently immobilize rhodium complex by self-assembled monolayers (SAMs) on gold electrodes, or by incorporating into sol–gel matrix on electrode by amino-modified rhodium complex.[138] However, it was found very difficult to keep the mediator in an active form after covalent bonding. So, new methods for covalent immobilization of rhodium complex need to be developed to fulfill the demand in bioconversion.

6.5 Conclusions

Enzymatic catalysis has become a central area for the synthesis/transformation of industrially important products and intermediates. They allow high regio- and stereo-selectivity for the desired reactions and operates at mild conditions. Inside this large field of research and biotechnological applications, electrochemistry is providing the possibility to regenerate efficiently the electroactive cofactor associated with redox proteins. For the community of electrochemists, electroenzymatic synthesis has a lot of similarities with enzymatic biofuel cells or biosensors. If one compares today, the quantity of publications in these different domains, it appears that electrosynthesis is much less popular than other mentioned fields. The reason is probably related to major challenges associated with bioconversion experiments, *i.e.* the need for long-term stability of the biocatalyst, its reusability and, somehow related to the previous items, the economic competitiveness.

In general, enzymes need to be immobilized in order to increase their operational stability and their recycling. In electrosynthesis, they need to be immobilized at a close distance to the electrode surface in order to favor efficient electron transfer reactions between the electrode and the redox cofactors. Membrane confinement has been successfully proposed in the past for bioelectrochemical conversion, but this approach has the major drawback that enzymes cannot be recycled after experiments. Surface grafting or encapsulation is probably the more suitable approach that needs to be investigated for further improvement of this biotechnology.

Electrochemical regeneration of the enzymatic cofactor implies either a mediator to shuttle electron from the redox center inside the protein to the electrode surface or a catalyst to facilitate the electrochemical conversion of nicotinamide adenine dinucleotide cofactors (NAD^+, NADH, $NADP^+$, NADPH). One of the most popular reactions in electroenzymatic synthesis is the electrochemical reduction of NAD^+ to NADH that involves either methyl-viologen as a mediator in the presence of a co-enzyme, or $[Cp^*Rh(bpy)Cl]^+$ as a catalyst. Methyl-viologen has the disadvantage of being highly toxic and should either be strongly immobilized on the electrode surface or simply replaced, *e.g.* by FAD, or removed from the bioelectrochemical pathway by exploring DET reactions with proteins allowing regeneration of NADH, as reported for diaphorase fragments. $[Cp^*Rh(bpy)Cl]^+$ is one of the best catalysts for this reaction. Most applications in electrosynthesis with this catalyst has been performed when using it in solution. Immobilization and separation from the proteins have been proposed and this is surely one interesting area to investigate in the future. Recyclability of both the electrochemical catalyst and proteins should be the major objective to achieve. In addition, the development of NAD analogs should also be kept in mind as it would allow to overpass the limited stability of the original biomolecule[139] that could then be immobilized on the electrode surface with the goal to build the optimal electronic pathway.[140]

Finally, DET is certainly one topic to be further investigated in the coming years for electroenzymatic synthesis. Developing stable and recyclable enzymes allowing long-term enzymatic activity and DET reactions would allow significant steps forward for major industrial developments.

List of Abbreviations

ABTS 2,2-azinobis(3-ethylbenzothiazoline-6-sulfonate)
ADH alcohol dehydrogenase
ADPy L-aminopropyl-1 methyl-4,4'-dipyridine

AMAPORS	artificial mediator accepting oxidoreductases
AOx	amino acid oxidase
bpy	bipyridine
CF	carbon felt
CNT	carbon nanotube
Cp	cyclopentadienyl
CPE	carbon paste electrode
CPO	chloroperoxidase
CPR	cytochrome P450 reductase
CYP	cytochrome P450
Cys-DSDH	cysteine-tagged D-sorbitol dehydrogenase
Cys-GatDH	cysteine-tagged galactitol dehydrogenase
Cyt	cytochrome
DDAB	didodecyldimethlammonium bromide
DET	direct electron transfer
3,4-DHB	3,4-dihydroxybenzaldehyde
D-MER	dialysis membrane electrochemical reactor
DSDH	D-sorbitol dehydrogenase
DTME	dithio-bismaleimidoethane
EPMH	4-ethylphenol methylenehydroxylase
ee	enantiomeric excess
FAD	flavin adenine dinucleotide
FDH	formate dehydrogenase
FMN	flavin mononucleotide
FNR	ferredoxin: $NADP^+$ reductase
G6PDH	glucose-6-phosphate dehydrogenase
Gat-DH	galactitol dehydrogenase
GC	glassy carbon
GDE	gas diffusion electrode
GDH	glucose dehydrogenase
GF	graphite felt
GOx	glucose oxidase
HPB	4-hydroxy-1-(3pyridyl)-1-butanone
HPR	horseradish peroxidase
LDH	L-lactate dehydrogenase
L-DOPA	L-3,4-dihydroxyphenylalanine

LipDH	lipoamide dehydrogenase
Med_{ox}	oxidized mediator
Med_{red}	reduced mediator
MDH	malate dehydrogenase
MV	methyl viologen
MWCNT	multi-walled carbon nanotube
NAD+	nicotinamide adenine dinucleotide, oxidized form
NADH	nicotinamide adenine dinucleotide, reduced form
NADP+	nicotinamide adenine dinucleotide phosphate, oxidized form
NADPH	nicotinamide adenine dinucleotide phosphate, reduced form
NP	nanoparticle
PCMH	p-cresol methylenehydroxylase
PEDOT	poly(3,4-ethylenedioxythiophene)
pFAD	polymerized FAD film
PMS	phenazine methosulfate
PSS	polystyrene sulfonic acid
PV	polyviologen
RDH	ribitol dehydrogenase
RGO	reduced graphene oxide
RVC	reticulated vitreous carbon
S–MDH	(S)-mandelate dehydrogenase
SAM	self-assembled monolayer
SCE	saturated calomel electrode
TTF	tetrathiafulvalene
TTN	total turnover number
Tyr	tyrosinase
UF-MER	ultrafiltration membrane electrochemical reactor
VAPOR	viologen accepting pyridine nucleotide oxidoreductase

References

1. K. Drauz and H. Waldmann (2007). Frontmatter, in K. Drauz and H. Waldmann (eds.), *Enzyme Catalysis in Organic Synthesis: A Comprehensive Handbook*, Wiley-VCH Verlag GmbH, Weinheim, Germany, pp. i–xxxvii. doi:10.1002/9783527618262.fmatter.

2. A. Schmid, J.S. Dordick, B. Hauer, A. Kiener, M. Wubbolts and B. Witholt (2001). Industrial biocatalysis today and tomorrow. *Nature* **409**, 258–268. doi:10.1038/35051736.
3. V. Uppada, S. Bhaduri and S.B. Noronha (2014). Cofactor regeneration — an important aspect of biocatalysis. *Current Science* **106**, 946–957.
4. T.R. Besanger and J.D. Brennan (2006). Entrapment of membrane proteins in sol-gel derived silica. *Journal of Sol-Gel Science Technology* **40**, 209–225. doi:10.1007/s10971-006-8132-1.
5. F. Hollmann, I.W.C.E. Arends and K. Buehler (2010). Biocatalytic redox reactions for organic synthesis: Nonconventional regeneration methods. *ChemCatChem* **2**, 762–782. doi:10.1002/cctc.201000069.
6. C. Kohlmann, W. Märkle and S. Lütz (2008). Electroenzymatic synthesis. *Journal of Molecular Catalysis B: Enzymatic* **51**, 57–72. doi:10.1016/j.molcatb.2007.10.001.
7. R. Wichmann and D. Vasic-Racki (2005). Cofactor regeneration at the lab Scale. *Advances in Biochemical Engineering Biotechnology* **92**, 225–260.
8. W. Andrea, G. Harald and H. Werner (2010). Regeneration of nicotinamide coenzymes: Principles and applications for the synthesis of chiral compounds. *Advances in Biochemical Engineering Biotechnology* **120**, 195–242. doi:10.1007/10.
9. V. Urbanova, G.-W. Kohring, T. Klein, Z. Wang, O. Mert, M. Emrullahoglu. *et al.* (2013). Sol-gel approaches for elaboration of polyol dehydrogenase-based bioelectrodes. *Zeitschrift für Physikalische Chemie* **227**, 667–689. doi:10.1524/zpch.2013.0324.
10. H. Wu, C. Tian, X. Song, C. Liu, D. Yang and Z. Jiang (2013). Methods for the regeneration of nicotinamide coenzymes. *Green Chemistry* **15**, 1773–1789. doi:10.1039/c3gc37129h.
11. U. Hanefeld, L. Gardossi and E. Magner (2009). Understanding enzyme immobilisation. *Chemical Society Reviews* **38**, 453–468. doi:10.1039/b711564b.
12. X. Dominguez-Benetton, S. Srikanth, Y. Satyawali, K. Vanbroekhoven and D. Pant (2013). Enzymatic electrosynthesis: An overview on the progress in enzyme-electrodes for the production of electricity, fuels and chemicals. *Journal of Microbials Biochemical Technology* **S6**, 7. doi:10.4172/1948-5948.S6-007.
13. L. Gorton and P.N. Bartlett (2008). NAD(P)-based biosensors, in *Bioelectrochemistry*, John Wiley & Sons, Ltd, pp. 157–198. doi:10.1002/9780470753842.ch4.
14. M. Rasmussen, S. Abdellaoui and S.D. Minteer (2016). Enzymatic biofuel cells: 30 years of critical advancements. *Biosensors and Bioelectronics* **76**, 91–102. doi:10.1016/j.bios.2015.06.029.
15. L. Gorton and E. Dominguez (2007). Electrochemistry of NAD(P) +/NAD(P)H, in *Encyclopedia Electrochemistry*, Wiley-VCH Verlag GmbH & Co. KGaA, Weinheim, Germany, pp. 67–143. doi:10.1002/9783527610426.bard090004.

16. P.J. Elving, C.O. Schmakel, K.S. V Santhanam and P. Zuman (1976). Nicotinamide-nad sequence: Redox processes and related behavior: Behavior and properties of intermediate and final products. *CRC Critical Reviews Analysis in Chemistry* **6**, 1–67. doi:10.1080/10408347608542689.
17. B.E. Steckhan, S. Herrmann and R. Ruppert (1990). Continuous generation of NADH from NAD+ and formate using a homogeneous catalyst with enhanced molecular weight. *Angewandte Chemie International Edition* **29**, 388–390. doi:10.1002/anie.199003881.
18. E. Steckhan (1994). Electroenzymatic synthesis. *Topics in Current Chemistry* **170**, 83–111. doi:10.1007/3-540-57729-7_3.
19. R. Wienkamp and E. Steckhan (1982). Indirect electrochemical regeneration of NADH by a bipyridinerhodium(I) complex as electron-transfer agent. *Angewandte Chemie International Edition* **21**, 782–783. doi :10.1002/anie.198207822.
20. M. Beley and J.-P. Collin (1993). Electrochemical regeneration of nicotinamide cofactor using a polypyrrole rhodium bis-terpyridine modified electrode. *Journal of Molecular Catalysis* **79**, 133–140. doi:10.1016/0304-5102(93)85096-C.
21. K. Délécouls-Servat, R. Basséguy and A. Bergel (2002). Membrane electrochemical reactor (MER): Application to NADH regeneration for ADH-catalysed synthesis. *Chemical Engineering Sciences* **57**, 4 633–4642. doi:10.1016/S0009-2509(02)00393-7.
22. K. Délécouls-Servat, R. Basséguy and A. Bergel (2002). Designing membrane electrochemical reactors for oxidoreductase-catalysed synthesis. *Bioelectrochemistry* **55**, 93–95. doi:10.1016/S1567-5394(01)00132-3.
23. F. Hildebrand and S. Lu (2007). Electroenzymatic synthesis of chiral alcohols in an aqueous — organic two-phase system. *Tetrahedron: Asymmetry.* **18**, 1187–1193. doi:10.1016/j.tetasy.2007.05.002.
24. B. Tan, D.P. Hickey, R.D. Milton, F. Giroud and S.D. Minteer (2015). Regeneration of the NADH cofactor by a rhodium complex immobilized on multi-walled carbon nanotubes. *Journal of Electrochemical Society* **162**, H102–H107. doi:10.1149/2.0111503jes.
25. J.S. Lee, S.H. Lee, J. Kim and C.B. Park (2013). Graphene–Rh-complex hydrogels for boosting redox biocatalysis. *Journal of Material Chemistry A.* **1**, 1040–1044. doi:10.1039/C2TA00358A.
26. A.A. Karyakin, O.A. Bobrova and E.E. Karyakina (1995). Electroreduction of NAD+ to enzymatically active NADH at poly(neutral red) modified electrodes. *Journal of Electroanalytical Chemistry* **399**, 179–184. doi:10.1016/0022-0728(95)04300-4.
27. Y. Kashiwagi, Y. Yanagisawa, N. Shibayama, K. Nakahara, J. Anzai and T. Osa (1997). Preparative, electroenzymatic reduction of ketones on an all components-immobilized graphite felt electrode. *Electrochimica Acta* **42**, 2267–2270. doi: 10.1016/S0013-4686(97)85509-0.
28. Y.W. Kang, C. Kang, J.S. Hong and S.E. Yun (2001). Optimization of the mediated electrocatalytic reduction of NAD+ by cyclic voltammetry and

construction of electrochemically driven enzyme bioreactor. *Biotechnology Letters* **23**, 599–604. doi:10.1023/A:1010316708080.
29. M.H. Kim and S.E. Yun (2004). Construction of an electro-enzymatic bioreactor for the production of (R)-mandelate from benzoylformate. *Biotechnology Letters* **26**, 21–26. doi:10.1023/B:BILE.0000009454.19082.03.
30. R. DiCosimo, C.-H. Wong, L. Daniels and G.M. Whitesides (1981). Enzyme-catalyzed organic synthesis: Electrochemical regeneration of NAD(P)H from NAD(P) using methyl viologen and flavoenzymes. *Journal of Organic Chemistry* **46**, 4622–4623. doi:10.1021/jo00335a078.
31. R. Yuan, S. Watanabe, S. Kuwabata and H. Yoneyama (1997). Asymmetric electroreduction of ketone and aldehyde derivatives to the corresponding alcohols using alcohol dehydrogenase as an electrocatalyst. *Journal of Organic Chemistry* **62**, 2494–2499. doi:10.1021/jo9620573.
32. M. Schulz, H. Leichmann, H. Günther and H. Simon (1995). Electromicrobial regeneration of pyridine nucleotides and other preparative redox transformations with *Clostridium thermoaceticum*. *Applied Microbiology and Biotechnology* **42**, 916–922. doi:10.1007/BF00191191.
33. A. Bergel and M. Comtat (1991). Thin-layer spectroelectrochemical study of the reversible reaction between nicotinamide adenine dinucleotide and flavin adenine dinucleotide. *Journal of Electroanalytical Chemistry and Interfacial Electrochemistry* **302**, 219–231. doi:10.1016/0022-0728(91)85042-N.
34. T. Tzedakis, K. Cheikhou, R. Jerôme, G.S. Karine and R. Olivier (2010). Electrochemical study in both classical cell and microreactors of flavin adenine dinucleotide as a redox mediator for NADH regeneration. *Electrochimica Acta* **55**, 2286–2294. doi:10.1016/j.electacta.2009.11.071.
35. Y. Ogino, K. Takagi, K. Kano and T. Ikeda (1995). Reactions between diaphorase and quinone compounds in bioelectrocatalytic redox reactions of NADH and NAD+. *Journal of Electroanalytical Chemistry* **396**, 517–524. doi:10.1016/0022-0728(95)04089-7.
36. P. Gros, C. Zaborosch, H.G. Schlegel and A. Bergel (1996). Direct electrochemistry of *Rhodococcus opacus* hydrogenase for the catalysis of NAD+ reduction. *Journal of Electroanalytical Chemistry* **405**, 189–195. doi:10.1016/0022-0728(95)04419-1.
37. K. Délécouls and P. Saint-Aguet (1999). C. Zaborosch and A. Bergel, Mechanism of the catalysis by Alcaligenes eutrophus H16 hydrogenase of direct electrochemical reduction of NAD+. *Journal of Electroanalytical Chemistry* **468**, 139–149. doi:10.1016/S0022-0728(99)00156-4.
38. J. Cantet, A. Bergel and M. Comtat (1992). Kinetics of the catalysis by the Alcaligenes eutrophus H16 Kinetics of the catalysis by the Alcaligenes eutrophus H16 hydrogenase of the electrochemical reduction of NAD+. *Journal of Molecular Catalysis* **73**, 371–380. doi:doi:10.1016/0304-5102(92)80088-X.
39. J. Cantet, A. Bergel and M. Comtat (1992). Bioelectrocatalysis of NAD+ reduction. *Journal of Electroanalytical Chemistry* **342**, 475–486. doi:10.1016/0022-0728(92)85140-X.

40. J. Cantet, A. Bergel and M. Comtat (1996). Coupling of the electroenzymatic reduction of NAD+ with a synthesis reaction. *Enzyme and Microbial Technology* **18**, 72–79. doi:10.1016/0141-0229(96)00059-2.
41. H.A. Reeve, L. Lauterbach, P.A. Ash, O. Lenz and K.A. Vincent (2012). A modular system for regeneration of NADcofactors using graphite particles modified with hydrogenase and diaphorase moieties. *Chemical Communications* **48**, 1589–1591. doi:10.1039/C1CC14826E.
42. H.A. Reeve, L. Lauterbach, O. Lenz and K.A. Vincent (2015). Enzyme-modified particles for selective biocatalytic hydrogenation by hydrogen-driven NADH recycling. *ChemCatChem.* **7**, 3480–3487. doi:10.1002/cctc.201500766.
43. S. Shipovskov, A. Bonamore, A. Boffi and E.E. Ferapontova (2015). Electrocatalytic interconversion of NADH and NAD+ by Escherichia coli flavohemoglobin. *Chemical Communications* **51**, 16096–16098. doi:10.1039/C5CC06317E.
44. M.T. Grimes and D.G. Drueckhammer (1993). Membrane-enclosed electroenzymatic catalysis with a low molecular weight electron-transfer mediator. *Journal of Organic Chemistry* **58**, 6148–6150. doi:10.1021/jo00074a056.
45. S.B. Sobolov, M.D. Leonida, A. Bartoszko-Malik, K.I. Voivodov, F. McKinney, J. Kim *et al.* (1996). Cross-linked LDH crystals for lactate synthesis coupled to electroenzymatic regeneration of NADH. *Journal of Organic Chemistry* **61**, 2125–2128. doi: 10.1021/jo951408v.
46. M.E. Zawalich and R.J. Fisher (1998). In-situ regeneration of co-enzyme NADH with an electro-enzymatic biomimetic membranes. *Proceedings of the IEEE 24th Annual Northeast Bioengineering Conference*, 85–87.
47. M.D. Leonida, S.B. Sobolov and A.J. Fry (1998). FAD-mediated enzymatic conversion of NAD+ to NADH: Application to chiral synthesis of L-lactate. *Bioorganics Medical Chemistry Letters* **8**, 2819–2824. doi:10.1016/S0960-894X(98)00519-8.
48. A.J. Fry, S.B. Sobolov, M.D. Leonida and K.I. Voivodov (1994). Electroenzymatic synthesis (regeneration of NADH coenzyme): Use of nafion ion exchange films for immobilization of enzyme and redox mediator. *Tetrahedron Letters* **35**, 5607–5610. doi:10.1016/S0040-4039(00)77258-3.
49. R.J. Fisher, J.M. Fenton and J. Iranmahboob (2000). Electro-enzymatic synthesis of lactate using electron transfer chain biomimetic membranes. *Journal of Membrane Science* **177**, 17–24. doi:http://dx.doi.org/10.1016/S0376-7388(00)00446-4.
50. K. Cheikhou and T. Tzédakis (2008). Electrochemical microreactor for chiral syntheses using the cofactor NADH. *AIChE Journal* **54**, 1365–1376. doi:10.1002/aic.11463.
51. X. Chen, J.M. Fenton, R.J. Fishera and R.A. Peattie (2004). Evaluation of in situ electroenzymatic regeneration of coenzyme NADH in packed bed membrane reactors. *Journal of Electrochemical Society* **151**, E56–E60. doi:10.1149/1.1638386.

52. K.L. Voivodov, S.B. Sobolov, M.D. Leonida and A.J. Fry (1995). Enzymatic transformation in an electrochemical reactor utilizing a redox mediator-modified enzyme electrode for NAD(H) regeneration. *Bioorganic Medicinal Chemistry Letters* **5**, 681–686. doi: 10.1016/0960-894X(95)00099-F.
53. M.D. Leonida, A.J. Fry, S.B. Sobolov and K.I. Voivodov (1996). Co-electropolymerization of a viologen oligomer and lipoamide dehydrogenase on an electronic surface. Application to cofactor regeneration. *Bioorganic Medicinal Chemistry Letters* **6**, 1663–1666. doi: 10.1016/0960-894X(96)00298-3.
54. H. Günther, A.S. Paxinos, M. Schulz, C. Van Dijk and H. Simon (1990). Direct electron transfer between carbon electrodes, immobilized mediator and an immobilized viologen-accepting pyridine nucleotide oxidoreductase. *Angewandte Chemie International Edition* **29**, 1053–1055. doi: 10.1002/anie.199010531.
55. K. Delecouls-Servat, A. Bergel and R. Basseguy (2004). Membrane electrochemical reactors (MER) for NADH regeneration in HLADH-catalysed synthesis: Comparison of effectiveness. *Bioprocess and Biosystems Engineering* **26**, 205–215. doi:10.1007/s00449-004-0356-2.
56. F. Hildebrand and S. Lütz (2009). Stable electroenzymatic processes by catalyst separation. *Chemistry* **15**, 4998–5001. doi:10.1002/chem.200900219.
57. E. Siu, K. Won and C.B. Park (2007). Electrochemical regeneration of NADH using conductive vanadia-silica xerogels. *Biotechnology Progress* **23**, 293–296. doi:10.1021/bp060247l.
58. L. Zhang, N. Vilà, T. Klein, G.-W. Kohring, I. Mazurenko, A. Walcarius *et al.* (2016). Immobilization of cysteine-tagged proteins on electrode surfaces by thiol–ene click chemistry. *ACS Applied Materials and Interfaces* **8**, 17591–17598. doi:10.1021/acsami.6b02364.
59. R.W. Coughlin, M. Aizawa, F. Bruce and M. Charles (1975). Immobilized-enzyme continuous-flow reactor incorporating continuous electrochemical regeneration of NAD. *Biotechnology and Bioengineering* **17**, 515–526. doi: 10.1002/bit.260170405.
60. J.M. Obón, P. Casanova, A. Manjón, V.M. Fernández and J.L. Iborra (1997). Stabilization of glucose dehydrogenase with polyethyleneimine in an electrochemical reactor with NAD(P)+ regeneration. *Biotechnology Progress* **13**, 557–561. doi:10.1021/bp970063u.
61. A. Radoi and D. Compag (2009). Recent advances in NADH electrochemical sensing design, *Bioelectrochemistry.* **76**, 126–134. doi:10.1016/j.bioelechem.2009.06.008.
62. Y. Dilgin, L. Gorton and G. Nisli (2007). Photoelectrocatalytic oxidation of NADH with electropolymerized toluidine blue O. *Electroanalysis* **19**, 286–293. doi:10.1002/elan.200603730.
63. D.G. Dilgin, D. Gligor, H.I. Gökçel, Z. Dursun and Y. Dilgin (2011). Glassy carbon electrode modified with poly-Neutral Red for photoelectrocatalytic oxidation of NADH. *Microchimica Acta* **173**, 469–476. doi:10.1007/s00604-011-0582-7.

64. S. Kochius, A.O. Magnusson, F. Hollmann, J. Schrader and D. Holtmann (2002). Immobilized redox mediators for electrochemical NAD(P)(+) regeneration. *Applied Microbiology and Biotechnology* **93**, 2251–2264. doi:10.1007/s00253-012-3900-z.
65. G. Hilt and E. Steckhan (1993) Transition Metal Complexes of 1,10-Phenanthroline-5,6-Dione as Efficient Mediators for the Regeneration of NAD+ in Enzymatic Synthesis. *Journal of the Chemical Society, Chemical Communications*, 1706–1707.
66. J. Komoschinski and E. Steckhan (1988). Efficient indirect electrochemical regeneration of NAD+ and NADP+ for enzymatic oxidations using iron bipyridine and phenanthroline complexes as redox catalysts. *Tetrahedron Letters* **29**, 3299–3300. doi:10.1016/0040-4039(88)85145-1.
67. I. Mazurenko, M. Etienne, G.-W. Kohring, F. Lapicque and A. Walcarius (2016). Enzymatic bioreactor for simultaneous electrosynthesis and energy production. *Electrochimica Acta* **199**, 342–348. doi:10.1016/j.electacta.2016.02.126.
68. O. Miyawaki and T. Yano (1993). Electrochemical bioreactor with immobilized glucose-6-phosphate dehydrogenase on the rotating graphite disc electrode modified with phenazine methosulfate. *Enzyme and Microbial Technology* **15**, 525–529. doi:10.1016/0141-0229(93)90087-I.
69. P. Kornberger, J. Gajdzik, H. Natter, G. Wenz, F. Giffhorn, G.W. Kohring *et al.* (2009). Modification of galactitol dehydrogenase from Rhodobacter sphaeroides D for immobilization on polycrystalline gold surfaces. *Langmuir* **25**, 12380 12386. doi:10.1021/la9010168.
70. A. Manjón, J.M. Obón, P. Casanova, V.M. Fernández and J.L. Ilborra (2002). Increased activity of glucose dehydrogenase co-immobilized with a redox mediator in a bioreactor with electrochemical NAD+ regeneration. *Biotechnology Letters* **24**, 1227–1232. doi:10.1023/A:1016270820734.
71. I. Schröder, E. Steckhan and A. Liese (2003). In situ NAD+ regeneration using 2,2′-azinobis(3-ethylbenzothiazoline-6-sulfonate) as an electron transfer mediator, *Journal of Electroanalytical Chemistry* **541**, 109–115. doi:10.1016/S0022-0728(02)01420-1.
72. Y. Kashiwagi and T. Osa (1993). Electrocatalytic oxidation of NADH on thin poly(acrylic acid) film coated graphite felt electrode coimmobilizing ferrocene and diaphorase. *Chemical Letters*, 677–680. doi:10.1246/cl.1993.677.
73. T. Osa, Y. Kashiwagi and Y. Yanagisawa (1994). Electroenzymatic oxidation of alcohols on a poly(acrylic acid)-coated graphite felt electrode immobilizing ferrocene, diaphorase and alcohol dehydrogenase. *Chemical Letters* **23**, 367–370. doi:10.1246/cl.1994.367.
74. O. Nikitina, S. Shleev, G. Gayda, O. Demkiv, M. Gonchar, L. Gorton *et al.* (2007). Bi-enzyme biosensor based on NAD+- and glutathione-dependent recombinant formaldehyde dehydrogenase and diaphorase for formaldehyde assay. *Sensors and Actuators B: Chemical* **125**, 1–9. doi:10.1016/j.snb.2007.01.025.

75. S. Itoh, H. Fukushima, M. Komatsu and Y. Ohshiro (1992). Heterocyclic o-quinones. Mediator for electrochemical oxidation of NADH. *Chemical Letters* **21**, 1583–1586. doi:10.1246/cl.1992.1583.
76. E. Steckhan, T. Arns, W.R. Heineman, G. Hilt, D. Hoormann, J. Jörissen *et al.* (2001). Environmental protection and economization of resources by electroorganic and electroenzymatic syntheses. *Chemosphere* **43**, 63–73. doi:10.1016/S0045-6535(00)00325-8.
77. D. Degenring, I. Schröder, C. Wandrey, A. Liese and L. Greiner (2004). Resolution of 1,2-diols by enzyme-catalyzed oxidation with anodic, mediated cofactor regeneration in the extractive membrane reactor: gaining insight by adaptive simulation. *Organic Process Research Development* **8**, 213–218. doi:10.1021/op034122y.
78. S. Gauer, Z. Wang, H. Otten, M. Etienne, M.J. Bjerrum, L. Lo Leggio *et al.* (2014). An l-glucitol oxidizing dehydrogenase from Bradyrhizobium japonicum USDA 110 for production of d-sorbose with enzymatic or electrochemical cofactor regeneration. *Applied Microbiology and Biotechnology* **98**, 3023–3032. doi:10.1007/s00253-013-5180-7.
79. R. Basséguy, K. Délécouls-Servat and A. Bergel (2004). Glucose oxidase catalysed oxidation of glucose in a dialysis membrane electrochemical reactor (D-MER). *Bioprocess and Biosystems Engineering* **26**, 165–168. doi:10.1007/s00449-004-0351-7.
80. R. Devaux-Basséguy, P. Gros and A. Bergel (1997). Electroenzymatic processes: A clean technology alternative for highly selective synthesis? *Journal of Chemical Technology and Biotechnology* **68**, 389–396. doi:10.1002/(SICI)1097-4660(199704)68:4<389::AID-JCTB647>3.0.CO;2-9.
81. P. Gros and A. Bergel (2005). Electrochemically enhanced biosynthesis of gluconic acid. *AIChE Journal* **51**, 989–997. doi:10.1002/aic.10411.
82. C. Bourdillon, R. Lortie and J.M. Laval (1988). Gluconic acid production by an mobilized glucose oxidase reactor with electrochemical regeneration of an artificial electron acceptor. *Biotechnology and Bioengineering* **31**, 553–558. doi:10.1002/bit.260310607.
83. B. Brielbeck, M. Frede and E. Steckhan (1994). Continuous electroenzymatic synthesis employing the electrochemical enzyme membrane reactor. *Biocatalysis and Biotransformation* **10**, 49–64. doi:10.3109/10242429409065217.
84. I. Mazurenko, W. Ghach, G.-W. Kohring, C. Despas, A. Walcarius and M. Etienne (2015). Immobilization of membrane-bounded (S)-mandelate dehydrogenase in sol–gel matrix for electroenzymatic synthesis. *Bioelectrochemistry* **104**, 65–70. doi:10.1016/j.bioelechem.2015.03.004.
85. S. Kawabata, N. Iwata and H. Yoneyama (2000). Asymmetric electrosynthesis of amino acid using an electrode modified with amino acid oxidase and electron mediator. *Chemical Letters* **29**, 110–111. doi:10.1246/cl.2000.110.
86. S. Krishnan, J.B. Schenkman and J.F. Rusling (2011). Bioelectronic delivery of electrons to cytochrome P450 enzymes. *Journal of Physical Chemistry B.* **115**, 8371–8380. doi:10.1021/jp201235m.

87. B.D. Fleming, D.L. Johnson, A.M. Bond and L.L. Martin (2006). Recent progress in cytochrome P450 enzyme electrochemistry. Expert opinion on drug metabolism and toxicology **2**, 581–589. doi:10.1517/17425255.2.4.581.
88. C.E. Immoos, J. Chou, M. Bayachou, E. Blair, J. Greaves and P.J. Farmer (2004). Electrocatalytic reductions of nitrite, nitric oxide, and nitrous oxide by thermophilic cytochrome P450 CYP119 in film-modified electrodes and an analytical comparison of its catalytic activities with myoglobin. *Journal of American Chemical Society* **126**, 4934–4942. doi:10.1021/ja038925c.
89. L.H. Mak, S.J. Sadeghi, A. Fantuzzi and G. Gilardi (2010). Control of human cytochrome P450 2E1 electrocatalytic response as a result of unique orientation on gold electrodes. *Analytical Chemistry* **82**, 5357–5362. doi:10.1021/ac101072h.
90. G. Güven (2011). Process optimization for the electroenzymatic hydroxylation of 3-phenoxytoluene catalyzed by P450 BM-3. *ECS Transactions* **35**, 9–23. doi:10.1149/1.3646484.
91. D. Holtmann, K.-M. Mangold and J. Schrader (2009). Entrapment of cytochrome P450 BM-3 in polypyrrole for electrochemically-driven biocatalysis. *Biotechnology Letters* **31**, 765–770. doi:10.1007/s10529-009-9925-4.
92. J. Lu, D. Cui, H. Li, Y. Zhang and S. Liu (2015). Cytochrome P450 bienzymes assembled on Au/chitosan/reduced graphene oxide nanosheets for electrochemically-driven drug cascade metabolism. *Electrochimica Acta* **165**, 36–44. doi:10.1016/j.electacta.2015.02.183.
93. S. Krishnan, D. Wasalathanthri, L. Zhao, J.B. Schenkman and J.F. Rusling (2011). Efficient bioelectronic actuation of the natural catalytic pathway of human metabolic cytochrome P450s. *Journal of American Chemical Society* **133**, 1459–1465. doi:10.1021/ja108637s.
94. H. Tian, S. Mu, H. Li, X. Wu and Z. Lu (2012). Electrode-driven regioselective oxidation of cinnamyl alcohol by use of thin biologically active chloroperoxidase composite films. *ChemCatChem* **4**, 1850–1855. doi:10.1002/cctc.201200172.
95. T. Krieg, S. Hüttmann, K.-M. Mangold, J. Schrader and D. Holtmann (2011). Gas diffusion electrode as novel reaction system for an electro-enzymatic process with chloroperoxidase. *Green Chemistry* **13**, 2686–2689. doi:10.1039/c1gc15391a.
96. A.E.W. Horst, K.-M. Mangold and D. Holtmann (2016). Application of gas diffusion electrodes in bioelectrochemical syntheses and energy conversion. *Biotechnology and Bioengineering* **113**, 260–267. doi:10.1002/bit.25698.
97. D. Holtmann, T. Krieg, L. Getrey and J. Schrader (2014). Electroenzymatic process to overcome enzyme instabilities. *Catalysis Communications* **51**, 82–85. doi:10.1016/j.catcom.2014.03.033.
98. L. Getrey, T. Krieg, F. Hollmann, J. Schrader and D. Holtmann (2014). Enzymatic halogenation of the phenolic monoterpenes thymol and carvacrol with chloroperoxidase. *Green Chemistry* **16**, 1104. doi:10.1039/c3gc42269k.
99. K. Min, D.-H.H. Park and Y.J. Yoo (2010). Electroenzymatic synthesis of l-DOPA. *Journal of Biotechnology* **146**, 40–44. doi:10.1016/j.jbiotec.2010.01.002.

100. K. Min, T. Kathavarayan, K. Park and Y.J. Yoo (2013). Novel strategy for enhancing productivity in l-DOPA synthesis: The electroenzymatic approach using well-dispersed l-tyrosine. *Journal of Molecular Catalysis B: Enzymatics* **90**, 87–90. doi:10.1016/j.molcatb.2013.01.027.
101. T. Krieg, A. Sydow, U. Schröder, J. Schrader and D. Holtmann (2014). Reactor concepts for bioelectrochemical syntheses and energy conversion. *Trends in Biotechnology* **32**, 645–655. doi:10.1016/j.tibtech.2014.10.004.
102. J. Cantet, A. Bergel and M. Comtat (1992). Kinetics of the catalysis by the *Alcaligenes eutrophus* H16 kinetics of the catalysis by the *Alcaligenes eutrophus*H16 hydrogenase of the electrochemical reduction of NAD+. *Journal of Molecular Catalysis* **73**, 371–380. doi:10.1016/0304-5102(92)80088-X.
103. J. Madoz-Gúrpide, J.M. Abad, J. Fernández-Recio, M. Vélez, L. Vázquez, C. Gómez-Moreno *et al.* (2000). Modulation of electroenzymatic NADPH oxidation through oriented immobilization of ferredoxin:NADP + reductase onto modified gold electrodes. *Journal of American Chemical Society* **122**, 9808–9817. doi:10.1021/ja001365m.
104. R. Szamocki, S. Reculusa, S. Ravaine, P.N. Bartlett, A. Kuhn and R. Hempelmann (2006). Tailored mesostructuring and biofunctionalization of gold for increased electroactivity. *Angewandte Chemie International Edition* **45**, 1317–1321. doi:10.1002/anie.200503292.
105. Y.B. Saint Côme, H. Lalo, Z. Wang, G.-W. Kohring, R. Hempelmann, M. Etienne *et al.* (2013). Interest of the sol-gel approach for multiscale tailoring of porous bioelectrode surfaces. *Electroanalysis* **25**, 621–629. doi:10.1002/elan.201200319.
106. F. Qu, R. Nasraoui, M. Etienne, Y.B. Saint Côme, A. Kuhn, J. Lenz *et al.* (2011). Electrogeneration of ultra-thin silica films for the functionalization of macroporous electrodes. *Electrochemistry Communications* **13**, 138–142. doi:10.1016/j.elecom.2010.11.034.
107. I. Ali, A. Gill and S. Omanovic (2012). Direct electrochemical regeneration of the enzymatic cofactor 1,4-NADH employing nano-patterned glassy carbon/Pt and glassy carbon/Ni electrodes. *Chemical Engineering Journal* **188**, 173–180. doi:10.1016/j.cej.2012.02.005.
108. I. Ali, T. Khan and S. Omanovic (2014). Direct electrochemical regeneration of the cofactor NADH on bare Ti, Ni, Co and Cd electrodes: The influence of electrode potential and electrode material. *Journal of Molecular Catalysis A: Chemistry* **387**, 86–91. doi:10.1016/j.molcata.2014.02.029.
109. Y.H. Kim and Y.J. Yoo (2009). Regeneration of the nicotinamide cofactor using a mediator-free electrochemical method with a tin oxide electrode. *Enzyme and Microbial Technology* **44**, 129–134. doi:10.1016/j.enzmictec.2008.10.019.
110. N. Ullah, I. Ali and S. Omanovic (2015). Direct electrocatalytic reduction of coenzyme NAD+ to enzymatically-active 1,4-NADH employing an iridium/ruthenium-oxide electrode. *Materials Chemistry and Physics* **149**, 413–417. doi:10.1016/j.matchemphys.2014.10.038.
111. R.L. McCreery (2008). Advanced carbon electrode materials for molecular electrochemistry. *Chemical Review* **108**, 2646–2687. doi:10.1021/cr068076m.

112. W. Zhang, S. Zhu, R. Luque, S. Han, L. Hu and G. Xu (2016). Recent development of carbon electrode materials and their bioanalytical and environmental applications. *Chemical Society Reviews* **45**, 715–752. doi:10.1039/C5CS00297D.
113. I. Švancara, K. Vytřas, J. Barek and J. Zima (2001). Carbon paste electrodes in modern electroanalysis. *Critical Reviews in Analytical Chemistry* **31**, 311–345. doi:10.1080/20014091076785.
114. B.J. Sanghavi and A.K. Srivastava (2010). Simultaneous voltammetric determination of acetaminophen, aspirin and caffeine using an in situ surfactant-modified multiwalled carbon nanotube paste electrode. *Electrochimica Acta* **55**, 8638–8648. doi:10.1016/j.electacta.2010.07.093.
115. M. Merino, L.J. Nunez-Vergara and J.A. Squella (1999). Reduction of different nitro-compounds on electroenzymatic system: Cytochrome *c* reductase-NAD(P)H modified carbon paste electrode. *Electroanalysis* **11**, 1285–1292. doi: 10.1002/(SICI)1521-4109(199911)11:17<1285::AID-ELAN1285>3.0.CO;2-T.
116. S.-H. Shuo-Han, J. Zhang, X. Gao and H. Ying (2015). Carbon nanostructure modified enzyme-catalyzed biosensor for bio-electrochemical NADH regeneration. *Journal of Material Chemistry B.* **3**, 6626–6633. doi:10.1039/C5TB00873E.
117. P. Zucca and E. Sanjust (2014). Inorganic materials as supports for covalent enzyme immobilization: Methods and mechanisms. *Molecules* **19**, 14139–14194. doi:10.3390/molecules190914139.
118. L. Cao (2006). Covalent enzyme immobilization, in *Carrier-Bound Immobilized Enzymes*, Wiley-VCH Verlag GmbH & Co. KGaA, pp. 169–316. doi:10.1002/3527607668.ch3.
119. S.Y.R. Pugh (1986). Immobilised redox enzymes and their use as catalysts for fine chemical synthesis, in *Enzyme as Catalysts Organic Synthesis*, M.P. Schneider (ed.), D. Reidel Publishing Company, pp. 217–232.
120. J. Krenkova and F. Foret (2004). Immobilized microfluidic enzymatic reactors. *Electrophoresis* **25**, 3550–3563. doi:10.1002/elps.200406096.
121. F. Hildebrand and S. Lütz (2006). Immobilisation of alcohol dehydrogenase from *Lactobacillus brevis* and its application in a plug-flow reactor. *Tetrahedron: Asymmetry* **17**, 3219–3225. doi:10.1016/j.tetasy.2006.11.013.
122. C. Ley, D. Holtmann, K.-M. Mangold and J. Schrader (2011). Immobilization of histidine-tagged proteins on electrodes. *Colloids Surfaces B Biointerfaces* **88**, 539–551. doi:10.1016/j.colsurfb.2011.07.044.
123. J.G. Vos, R.J. Forster and T.E. Keyes (2003). Formation and characterization of modified surfaces, in *Interfacial Supramolecular Assemblies*, John Wiley & Sons, Ltd, Chichester, UK, pp. 87–152. doi:10.1002/0470861517.ch4.
124. E. Nouri-Nigjeh, R. Bischoff, A. P. Bruins and H. P. Permentier (2011). Electrochemistry in the mimicry of oxidative drug metabolism by cytochrome P450s. *Current Drug Metabolism* **12**, 359–371. doi:10.2174/138920011795202929.

125. I.C.G. Thanos and H. Simon (1987). Electro-enzymic viologen-mediated stereospecific reduction of 2-enoates with free and immobilized enoate reductase on cellulose filters or modified carbon electrodes. *Journal of Biotechnology* **6**, 13–29. doi:10.1016/0168-1656(87)90042-3.
126. H.R. Luckarift, J.C. Spain, R.R. Naik and M.O. Stone (2004). Enzyme immobilization in a biomimetic silica support. *Nature Biotechnology* **22**, 211–213. doi:10.1038/nbt931.
127. M. Varničić, T. Vidaković-Koch and K. Sundmacher (2015). Gluconic acid synthesis in an electroenzymatic reactor. *Electrochimica Acta* **174**, 480–487. doi:10.1016/j.electacta.2015.05.151.
128. S. Schlager, H. Neugebauer, M. Haberbauer, G. Hinterberger and N.S. Sariciftci (2015). Direct electrochemical addressing of immobilized alcohol dehydrogenase for the heterogeneous bioelectrocatalytic reduction of butyraldehyde to butanol. *ChemCatChem* **7**, 967–971. doi:10.1002/cctc.201402932.
129. D. Membreno, L. Smith and B. Dunn (2014). Silica sol–gel chemistry: creating materials and architectures for energy generation and storage, *Journal of Sol-Gel Science and Technology*, 203–215. doi:10.1007/s10971-014-3299-3.
130. M.J. Moehlenbrock and S.D. Minteer (2008). Extended lifetime biofuel cells. *Chemical Society Reviews* **37** (2008) 1188–1196. doi:10.1039/b708013c.
131. E. Lojou (2011). Hydrogenases as catalysts for fuel cells: Strategies for efficient immobilization at electrode interfaces. *Electrochimica Acta* **56**, 10385–10397. doi:10.1016/j.electacta.2011.03.002.
132. F. Hollmann, B. Witholt and A. Schmid (2002). [Cp*Rh(bpy)(H2O)]2+: A versatile tool for efficient and non-enzymatic regeneration of nicotinamide and flavin coenzymes. *Journal of Molecular Catalysis B: Enzyme*, pp. 167–176. doi:10.1016/S1381-1177(02)00164-9.
133. S. Chardon-Noblat, S. Cosnier, A. Deronzier and N. Vlachopoulos (1993). Electrochemical properties of [(C5Me5)RhIII(L)Cl]+ complexes (L = 2,2′-bipyridine or 1,10-phenanthroline derivatives) in solution in related polypyrrolic films. Application to electrocatalytic hydrogen generation. *Journal of Electroanalytical Chemistry* **352**, 213–228. doi:10.1016/0022-0728(93)80265-J.
134. J.H. van Esch, M.A.M. Hoffmann and R.J.M. Nolte (1995). Reduction of nicotinamides, flavins, and manganese porphyrins by formate, catalyzed by membrane-bound rhodium complexes. *Journal of Organic Chemistry* **60**, 1599–1610. doi:10.1021/jo00111a018.
135. E. Höfer, E. Steckhan, B. Ramos and W.R. Heineman (1996). Polymer-modified electrodes with pendant [RhIII(C5Me5)(L) C1]+-complexes formed by y-irradiation cross-linking, *J. Electroanal. Chem.* **402**, 115–122. doi: 10.1016/0022-0728(95)04243-1.
136. L. Qin, K. Lv, Z. Shen and M. Liu (2015). Self-assembly of organic molecules into nanostructures, in *Soft Material Nanotechnology*, Wiley-VCH Verlag GmbH & Co. KGaA, pp. 21–94. doi:10.1002/9783527682157.ch02.
137. T. Schlick (2010). Molecular modeling and simulation an interdisciplinary guide, Springer-Verlag, New York, doi:10.1007/978-1-4419-6351-2.

138. A. Walcarius, R. Nasraoui, Z. Wang, F. Qu, V. Urbanova, M. Etienne *et al.* (2011). Factors affecting the electrochemical regeneration of NADH by (2,2′-bipyridyl) (pentamethylcyclopentadienyl)-rhodium complexes: Impact on their immobilization onto electrode surfaces. *Bioelectrochemistry* **82**, 46–54. doi:10.1016/j.bioelechem.2011.05.002.
139. H.J. Lee, S.H. Lee, C.B. Park and K. Won (2011). Coenzyme analogs: Excellent substitutes (not poor imitations) for electrochemical regeneration. *Chemical Communications* **47**, 12538–12540. doi:10.1039/C1CC14313A.
140. Z. Wang, M. Etienne, F. Quilès, G.-W. Kohring and A. Walcarius (2012). Durable cofactor immobilization in sol-gel bio-composite thin films for reagentless biosensors and bioreactors using dehydrogenases. *Biosensors & Bioelectronics* **32**, 111–117. doi:10.1016/j.bios.2011.11.043.

Chapter 7

Redox Hydrogels as an Efficient Strategy for Immobilization of Enzymes at Electrode Interfaces

Joshua W. Gallaway[*,‡] and Scott Calabrese Barton[†,§]

[*] *Department of Chemical Engineering, Northeastern University, 360 Huntington Avenue, Boston, MA 02115*

[†] *Department of Chemical Engineering and Materials Science, Michigan State University, 428 S. Shaw Lane, Room 3243 East Lansing, MI, USA 48824*

[‡] *j.gallaway@northeastern.edu*

[§] *scb@egr.msu.edu*

7.1 Introduction

Enzyme electrodes are formed by functionalizing an electrode surface with enzymes that have been separated from their natural organisms and purified. The catalytic functions of the enzymes are concentrated on the electrode surface in an enzyme film, resulting in high catalytic current measured per electrode area or per enzyme film volume. However, because enzyme's active sites are surrounded by insulating protein layers, only a single monolayer of enzymes is possible if direct electrical communication from enzyme to electrode is required. Entrapping the enzymes in a redox hydrogel overcomes this challenge and allows many monolayer equivalents of enzymes to

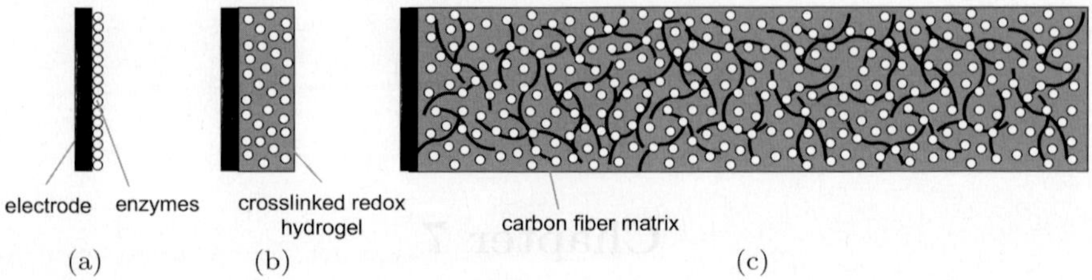

Figure 1. Enzyme-functionalized electrodes: (a) Enzyme monolayer. (b) Enzyme–redox hydrogel film. (c) Enzyme–redox hydrogel film on a conductive carbon fiber matrix.

communicate current to an electrode.[1,2] The enzymes are held in the hydrogel film electrostatically, and current is communicated to the electrode via redox complexes covalently bound to the hydrogel. Electron hopping between these redox complexes allows current to flow by Fickian diffusion of electrons from a low to high potential, driven by a potential gradient. This is shown conceptually in Figure 1, with (a) an enzyme monolayer, (b) an enzyme–redox hydrogel film, and (c) the same type of film on a conductive matrix such as carbon paper or cloth, which can further increase enzyme density per electrode area. Oxygen-reducing enzyme electrodes as in Figure 1(b) have been designed to produce extremely high volumetric current of 40 A/cm^3 at 0.7 V. This is possible through efficient design of the enzyme–redox hydrogel system.[3]

Thus, electrocatalytic films of enzymes entrapped in a redox-conductive hydrogel matrix allow the harnessing of electrical current produced by the catalytic action of the enzyme on an ambient substrate.[1,2] This current can be useful for sensing of the substrate as an analyte, as in biosensors, or as one electrode in a power-producing cell, such as a biofuel cell. The redox-conductive hydrogel structure provides an interface between the ionic-conducting aqueous phase of an ambient solution and the electron-conducting phase of the current collector. In this way, the system resembles a porous electrode model, wherein the electron-conducting phase consists of a network of redox moieties capable of electron self-exchange, rather than a continuous conductive substance.[4]

For power-producing applications, high current density and therefore high reaction rate is a significant engineering objective.

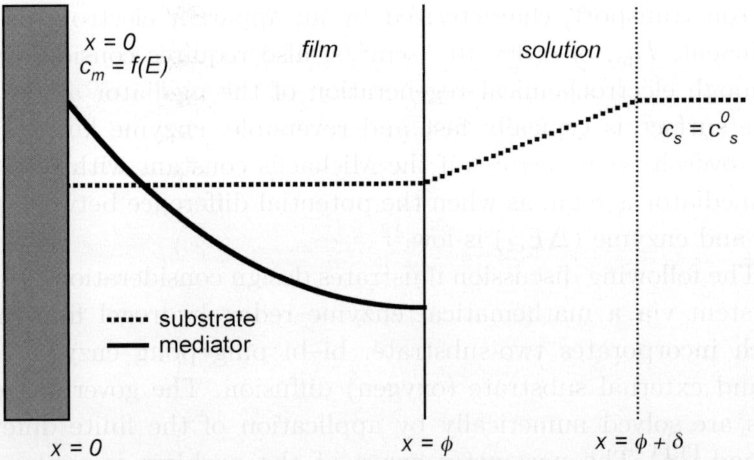

Figure 2. Sample concentration profiles for a 1D enzyme–redox hydrogel film model. The film boundary is $x = \phi$, and the diffusion layer thickness in the external electrolyte is δ. Substrate concentration (dotted line) is $c_s = f(x)$, with the bulk value c_s°. Concentration of the active form of the mediator (solid line), which is the *reduced* or electron-having form in the case of a cathode, is $c_m = f(x)$, with c_m° being the total bulk concentration of mediator sites.

Engineering such a system requires consideration of multiple processes, including reaction kinetics, electron transport, and species transport, both internal and external to the film; any combination of these may represent a performance bottleneck.[5] Mathematical treatments of such multi-dimensional systems have been presented in the literature, most notably the work of Andrieux and Saveant, who explored the complete parameter space using a second-order rate expression.[6–8] Their treatment related the various controlling processes to concentration profiles internal and external to a 1D film, such as that in Figure 2. The work of Bartlett and Pratt extended the analysis to consider Michaelis–Menten enzyme kinetics, introducing the need for numerical solution in some cases.[9]

The efficiently designed oxygen-reducing enzyme — redox hydrogel film discussed above — was based on an oxygen-reducing fungal laccase enzyme.[3,10] For such electrodes, where the film thickness ϕ is generally less than 1 μm, gradients in the substrate concentration c_s within the film are usually negligible, and only gradients external to the film ($x > \phi$) need be considered. However, limited electron transport, characterized by an apparent electron diffusion coefficient, D_{m}, of order 10^{-9} cm^2/s, also requires consideration.[11] Although electrochemical regeneration of the mediator at the electrode surface is typically fast and reversible, enzyme kinetics may limit overall reaction rates if the Michaelis constant with respect to the mediator is high, as when the potential difference between mediator and enzyme (ΔE_{et}) is low.[12]

The following discussion illustrates design considerations for such a system via a mathematical enzyme–redox hydrogel film model, which incorporates two-substrate, bi–bi ping–pong enzyme kinetics and external substrate (oxygen) diffusion. The governing equations are solved numerically by application of the finite difference method.[13,14] The parameter space of the problem is explored by both simulation and solution of the governing equations for limiting cases. Circumstances are determined under which mediator properties and/or substrate transport control electrode performance.

7.2 An Enzyme–Redox Hydrogel Model

An oxidoreductase enzyme catalyzing the transfer of electrons between two substrates, a reductant M and an oxidant S with stoichiometric coefficients ν_i, can be described by the scheme

$$\nu_{\mathrm{s}} S + E_{\mathrm{red}} \underset{k_{-1}}{\overset{k_1}{\rightleftharpoons}} \mathrm{ES} \xrightarrow{k_2} \nu_p P + E_{\mathrm{ox}}, \quad (1)$$

$$\nu_{\mathrm{m}} M_{\mathrm{red}} + E_{\mathrm{ox}} \underset{k_{-3}}{\overset{k_3}{\rightleftharpoons}} EM \xrightarrow{k_4} \nu_{\mathrm{m}} M_{\mathrm{ox}} + E_{\mathrm{red}}, \quad (2)$$

where each substrate forms a reversible complex with the relevant binding pocket of the enzyme, followed by irreversible decoupling. This scheme, in which no complex of all three species exists, is termed ping-pong.[15] The oxygen-reducing enzyme cathode is thus described, where S is oxygen, P is H_2O, and M is the reduced osmium moiety of the redox polymer; ν_s, ν_p, and ν_m are 0.25, 0.5, and 1, respectively.[12] A steady-state assumption for the enzyme and its various complexes results in the relevant rate expression

$$r = \frac{k_{\text{cat}} c_{\text{e}} c_{\text{m}} c_{\text{s}}}{c_{\text{m}} c_{\text{s}} + K_{\text{s}} c_{\text{m}} + K_{\text{m}} c_{\text{s}}}, \tag{3}$$

where c_{e} refers to the overall enzyme concentration (a constant) and c_{m} and c_{s} are the concentrations of reductant substrate, and oxidant substrate, respectively, hereafter referred to as *mediator* and *substrate*. The individual rate constants in Equations (1) and (2) are collected into a global rate constant k_{cat}, representing the maximal turnover rate of the enzyme, and K_{m} and K_{s}, the Michaelis constants for mediator and substrate, which are indications of these species' individual affinities for reaction with the enzyme. The quantity $k_{\text{cat}}/K_{\text{i}}$ is termed the bimolecular rate constant, and represents the rate of second-order kinetics between species i and the enzyme when the other substrate is present at saturated concentration, $c_i \gg K_i$. The value of $k_{\text{cat}}/K_{\text{m}}$ depends only on k_3, k_{-3}, and k_4, and is independent of the substrate kinetics. Likewise $k_{\text{cat}}/K_{\text{s}}$ is independent of mediator kinetics.

For the system of an enzyme redox hydrogel film with a working electrode at $x = 0$ and an electrolyte at the hydrogel boundary, $x = \phi$, 1D material balances on the mediator and substrate species result in the expression for each:

$$\frac{\partial c_i}{\partial t} = D_i \frac{\partial^2 c_i}{\partial x^2} - \nu_i r, \tag{4}$$

where D_i is the diffusion coefficient of species i, and migration caused by a potential field is considered negligible due to a high ionic strength. At steady state, for a system in which the mediator–enzyme reaction is one-electron and the substrate–enzyme reaction is four-electron, the resulting governing equations are

$$D_{\text{m}} \frac{\partial^2 c_{\text{m}}}{\partial x^2} = \frac{\nu_{\text{m}} k_{\text{cat}} K_{\text{m}}^{-1} c_{\text{e}} c_{\text{m}} c_{\text{s}}}{c_{\text{m}} K_{\text{m}}^{-1}(c_{\text{s}} + K_{\text{s}}) + c_{\text{s}}}, \tag{5}$$

$$D_{\text{s}} \frac{\partial^2 c_{\text{s}}}{\partial x^2} = \frac{\nu_{\text{s}} k_{\text{cat}} K_{\text{m}}^{-1} c_{\text{e}} c_{\text{m}} c_{\text{s}}}{c_{\text{m}} K_{\text{m}}^{-1}(c_{\text{s}} + K_{\text{s}}) + c_{\text{s}}}. \tag{6}$$

At the boundary $x = 0$, there is no flux of substrate, and mediator flux is controlled by the working electrode potential in the manner indicated by a Butler–Volmer expression.

$$x = 0: \quad \frac{\mathrm{d}c_{\mathrm{s}}}{\mathrm{d}x} = 0,$$

$$i = -nFD_{\mathrm{m}}\frac{\mathrm{d}c_{\mathrm{m}}}{\mathrm{d}x}$$

$$= i_0\left\{\frac{c_{\mathrm{m,ox}}}{c_{\mathrm{m}}^0}\exp\left[-\frac{E-E^0}{b}\right] - \frac{c_{\mathrm{m,red}}}{c_{\mathrm{m}}^0}\exp\left[\frac{E-E^0}{b}\right]\right\}, \quad (7)$$

where F is Faraday's constant, c_{m}^0 is the bulk mediator concentration, i_0 is the exchange current density, b is the Tafel slope, and E^0 is the redox potential of the mediator couple.

At $x = \phi$, there is no flux of mediator, and substrate flux is balanced across the film–solution interface. Here, we assume a linear concentration profile in the diffusion layer outside the film,

$$x = \phi: \quad \frac{\mathrm{d}c_{\mathrm{m}}}{\mathrm{d}x} = 0,$$

$$D_{\mathrm{s}}\left.\frac{\mathrm{d}c_{\mathrm{s}}}{\mathrm{d}x}\right|_{x=\phi-} = D(c_{\mathrm{s}}^0 - (c_{\mathrm{s}})_{\phi+})\delta^{-1}, (c_{\mathrm{s}})_{\phi-} = k(c_{\mathrm{s}})_{\phi+} \quad (8)$$

in which D is the substrate diffusion in the bulk solution, c_{s}^0 is the bulk substrate concentration, δ is the thickness of the diffusion layer between bulk solution and the film interface and k is the partition coefficient into the hydrogel film from the solution.[6]

The model is non-dimensionalized in a manner following the method of Bartlett and Pratt.[9] Dimensionless description of an entrapped enzyme electrode requires two transport parameters and an additional two parameters representing the nonlinearity of the kinetics in mediator and substrate. Defining $m = c_{\mathrm{m}}/c_{\mathrm{m}}^0$, $s = c_{\mathrm{s}}/c_{\mathrm{s}}^0$, and $\chi = x/\phi$, the governing equations become

$$\frac{\mathrm{d}^2m}{\mathrm{d}\chi^2} = \frac{\kappa^2 ms}{\gamma m(1+\mu s)+s}, \quad (9)$$

$$\frac{\mathrm{d}^2s}{\mathrm{d}\chi^2} = \frac{\gamma\eta^{-1}\kappa^2 ms}{\gamma m(1+\mu s)+s} \quad (10)$$

with the dimensionless groups defined as

$$\kappa^2 = \frac{\nu_\mathrm{m}\phi^2}{D_\mathrm{m}}\frac{k_\mathrm{cat}}{K_\mathrm{m}}c_\mathrm{e}, \quad \eta = \frac{\nu_\mathrm{m}D_\mathrm{s}K_\mathrm{s}}{\nu_\mathrm{s}D_\mathrm{m}K_\mathrm{m}}, \quad \gamma = \frac{K_\mathrm{s}}{c_\mathrm{s}^0}\frac{c_\mathrm{m}^0}{K_\mathrm{m}}, \quad \mu = \frac{c_\mathrm{s}^0}{K_\mathrm{s}}, \tag{11}$$

where κ^2 is a Thiele modulus with respect to the mediator, $\kappa^2\eta^{-1}$ the Thiele modulus with respect to substrate. The parameter μ indicates the degree of substrate saturation, and γ reflects mediator saturation relative to saturation of the substrate. An additional dimensionless parameter characterizes the substrate boundary condition

$$\sigma = \frac{\delta D_\mathrm{s} k}{\phi D}. \tag{12}$$

Oxygen diffusivity within a hydrogel is generally lower than in bulk solution, but the difference is much less than an order of magnitude.[16] Primarily because substrate concentration gradients internal to the film are generally found to be negligibly small, D and D_s are assumed to be equal in this analysis. Similarly, we assume a substrate partition coefficient, $k = 1$. The substrate boundary condition becomes

$$\chi = 1; \quad s = 1 - \sigma\left(\frac{\mathrm{d}s}{\mathrm{d}\chi}\right), \quad \frac{\mathrm{d}m}{\mathrm{d}\chi} = 0 \tag{13}$$

Discussion will be limited to instances of high overpotential between the mediator and current collector, simplifying the relevant boundary condition to

$$\chi = 0; \quad m = 1, \quad \frac{\mathrm{d}s}{\mathrm{d}\chi} = 0. \tag{14}$$

7.3 Enzyme–Redox Hydrogel Design Considerations

When considering the design of enzyme–redox hydrogel electrodes, a convenient starting point is the dimensionless parameter γ/η, which is the ratio of electron diffusion in the film to that of substrate diffusion in the film:

$$\frac{\gamma}{\eta} = \frac{D_m c_m^0}{D_s c_s^0}. \tag{15}$$

For the oxygen-reducing system discussed above, this quantity is small, on the order of 10^{-2} or less. This situation may be termed a thin-film system, in which substrate permeates the film thickness at an infinite rate, and will be present at a near-constant concentration throughout. The governing equations then simplify to Eq. (9) only, with s being a constant given by

$$s = 1 + \sigma\frac{\gamma}{\eta}\left(\frac{\mathrm{d}m}{\mathrm{d}\chi}\right)_{\chi=0}. \tag{16}$$

Therefore, the thickness of the external diffusion layer and the current density will determine the substrate concentration in the film, as quantified by $\sigma\gamma/\eta$, which is a ratio of electron diffusion in the film to that of substrate in the diffusion layer. In high-convection flow systems, which can be approximated by a rotating disk electrode (RDE) rotating at infinite rate, $\sigma\gamma/\eta \to 0$, and the substrate is present at bulk concentration throughout the system.

7.3.1 Effect of mediator diffusivity

While the parameter space of the system remains large, only a limited number of characteristics are amenable to design. For example, the diffusional transport of electrons via the mediator may be manipulated by engineering of the redox mediator hydrogel.[5,12] Figure 3 shows the effect on electrode performance as D_m is varied in a system with no limitation by the external diffusion layer ($\sigma \to 0$) and which is not impractically limited by mediator kinetics (non-vanishing γ). In all cases, two regions are present. At low D_m, current density scales with $D_\mathrm{m}^{1/2}$, as indicated by a line of slope 0.5 on a log–log plot. This relation arises because the reaction layer thickness falls within the film thickness, due to mediator depletion. The presence of this reaction layer is indicated by the condition $\kappa^2/\gamma(1+\mu) > 2$, derived by dividing the film into depleted and non-depleted zones and solving

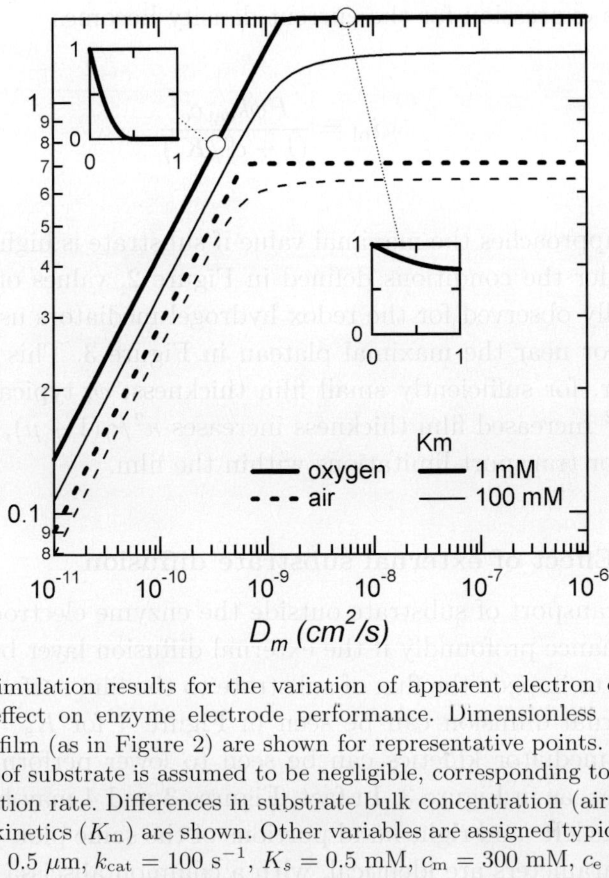

Figure 3. Simulation results for the variation of apparent electron diffusion, D_{m}, and its effect on enzyme electrode performance. Dimensionless mediator profiles in the film (as in Figure 2) are shown for representative points. External diffusion layer of substrate is assumed to be negligible, corresponding to an RDE at infinite rotation rate. Differences in substrate bulk concentration (air/oxygen) and mediator kinetics (K_{m}) are shown. Other variables are assigned typical values as follows: $\phi = 0.5\ \mu$m, $k_{\mathrm{cat}} = 100\ \mathrm{s}^{-1}$, $K_s = 0.5$ mM, $c_{\mathrm{m}} = 300$ mM, $c_{\mathrm{e}} = 5$ mM. Slope of the low-D_m region is 0.5. The transition region for each curve corresponds to $\kappa^2/\gamma(1+\mu) = 2$, indicating formation of a mediator reaction layer.

Eq. (9). A zone boundary corresponding to the film thickness marks the initial formation of a reaction layer.

As D_{m} increases, active mediator is present throughout the film. At sufficiently high values of D_{m}, mediator diffusion is no longer limiting, and current density has no dependence on D_{m}. In cases where $\gamma \gg 1$, Eq. (9) then reduces to

$$\frac{d^2 m}{d\chi^2} = \frac{\kappa^2}{\gamma(1+\mu)} \tag{17}$$

and the expression for the current density becomes

$$i_{\text{pl}} = \frac{F\phi k_{\text{cat}} c_{\text{e}}}{(1 + c_{\text{s}}^0/K_{\text{s}})} \tag{18}$$

which approaches the maximal value if substrate is highly saturated.

Under the conditions defined in Figure 2, values of $D_{\text{m}} > 10^{-9}$ (typically observed for the redox hydrogel mediators used in Ref. 3) fall on or near the maximal plateau in Figure 3. This is only true, however, for sufficiently small film thickness, ϕ, typically less than 1 μm.[12] Increased film thickness increases $\kappa^2/\gamma(1+\mu)$, and leads to mediator transport limitations within the film.

7.3.2 Effect of external substrate diffusion

Mass transport of substrate outside the enzyme electrode may limit performance profoundly if the external diffusion layer becomes large enough to impede the flux of substrate to the film surface. The effect of external diffusion can be seen in Figure 4 for $K_{\text{s}} = 0.5$ mM.[12] Again, mediator kinetics can be seen to lower performance as K_{m} increases, as in Figure 3. In fact, Figures 3 and 4 may be considered to be the left- and right-hand portions of the same plot, joining when their parameters are identical, with a common abscissa representing the ratio of electron diffusion in the film to that of substrate diffusion from the bulk, $\gamma\sigma/\eta$.[6] The dimensionless concentration profiles shown in Figures 3 and 4 represent all the possible situations for the system under examination.

For the purpose of characterizing the enzyme–redox hydrogel system, it is useful to consider the Koutecky–Levich plot generated from a system on an RDE, which aids in revealing and quantifying the bulk mass transport.[17] Two trivial cases exist, which will produce Koutecky–Levich plots with a slope of zero, meaning there is no response to additional substrate diffusion from the bulk. These cases are (a) substrate concentrations well above saturation, $\mu \gg 1$, or (b) mediator concentrations well below saturation, $\gamma\mu \ll 1$. Either of

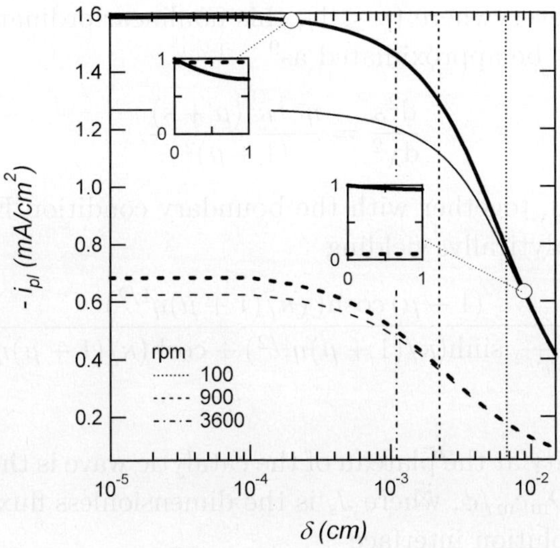

Figure 4. Simulation results for the effect of external diffusion layer δ on enzyme electrode performance. Notation for oxygen/air and mediator K_{m} are as in Figure 3, with vertical lines showing the diffusion layer thicknesses for an RDE rotating at 100, 900 and 3,600 rpm. Dimensionless concentration profiles are shown for representative points, with mediator given by the solid line and substrate by the dotted line. Simulation variables are assigned typical values as follows: $\phi = 0.5\ \mu$m, $k_{\mathrm{cat}} = 100\ \mathrm{s}^{-1}$, $K_{\mathrm{s}} = 0.5$ mM, $c_{\mathrm{m}} = 300$ mM, $c_{\mathrm{e}} = 5$ mM, $D_{\mathrm{m}} = 2 \times 10^{-9}\ \mathrm{cm}^2/\mathrm{s}$.

these conditions lead to vanishing values of γ, reducing Eq. (9) to

$$\frac{\mathrm{d}^2 m}{\mathrm{d}\chi^2} = \kappa^2 m \tag{19}$$

which is independent of substrate concentration, s.

Koutecky–Levich plots with a slope that is linearly dependent on substrate diffusivity, D, and which extrapolate to the y-intercept at infinite rotation rate, are expected in certain cases. If the kinetics are substrate limited, corresponding to $\gamma \gg 1$, this allows the material balance to be written as

$$\frac{\mathrm{d}^2 s}{\mathrm{d}\chi^2} = \frac{\eta^{-1}\kappa^2 s}{(1+\mu s)} \tag{20}$$

provided that mediator concentration does not fall to zero at any point in the film, or $\kappa^2/\gamma(1+\mu) < 2$. In cases where the substrate is unsaturated everywhere ($\mu < 1$), this nonlinear ordinary differential equation may be approximated as[9]

$$\frac{\mathrm{d}^2 s}{\mathrm{d}\chi^2} = \frac{\eta^{-1}\kappa^2(\mu + s)}{(1+\mu)^2}. \tag{21}$$

This equation, together with the boundary condition Eq. (13), may be solved analytically, yielding

$$s = \frac{(1+\mu)\cosh(\chi\kappa/(1+\mu)\eta^{1/2})}{\frac{\delta}{\phi}\frac{\kappa}{(1+\mu)\eta^{1/2}}\sinh(\kappa/(1+\mu)\eta^{1/2}) + \cosh(\kappa/(1+\mu)\eta^{1/2})} + \mu. \tag{22}$$

Current density at the plateau of the catalytic wave is then calculated as $i_{\mathrm{pl}} = FJ_{\mathrm{s}}D_{\mathrm{m}}c_{\mathrm{m}}/\phi$, where J_{s} is the dimensionless flux of substrate at the film/solution interface

$$J_{\mathrm{s}} = \frac{\eta}{\gamma}\left(\frac{\mathrm{d}s}{\mathrm{d}\chi}\right)_{\chi=1} = \frac{\frac{\eta^{1/2}\kappa}{\gamma}\sinh(\kappa/(1+\mu)\eta^{1/2})}{\frac{\delta}{\phi}\frac{\kappa}{(1+\mu)\eta^{1/2}}\sinh(\kappa/(1+\mu)\eta^{1/2}) + \cosh(\kappa/(1+\mu)\eta^{1/2})}. \tag{23}$$

This results in the expression for the current density

$$i_{\mathrm{pl}} = \frac{nF(D_{\mathrm{s}}k_{\mathrm{cat}}K_{\mathrm{s}}^{-1}c_{\mathrm{e}}c_{\mathrm{s}}^{0})^{1/2}\sinh(\Psi)}{(\delta/\phi)\Psi\sinh(\Psi) + \cosh(\Psi)};$$

$$\Psi = \frac{\phi}{D_{\mathrm{s}}^{1/2}}\left(\frac{k_{\mathrm{cat}}}{K_{\mathrm{s}}}\right)^{1/2}\frac{c_{\mathrm{e}}^{1/2}}{(1 + c_{\mathrm{s}}^{0}/K_{\mathrm{s}})} \tag{24}$$

which is independent of mediator concentration, c_{m}, and current is a kinetic function of bulk substrate concentration, c_{s}^{0}, and enzyme loading, c_{e}. Linear Koutecky–Levich plots are expected in the form of

$$\frac{1}{i_{\mathrm{pl}}} = \frac{1}{i_{\mathrm{l}}} + \frac{1}{i_{\mathrm{pl,k}}}; \quad i_{\mathrm{l}} = m_{\mathrm{l}}\omega^{1/2} \tag{25}$$

where the Levich equation for the diffusion layer at an RDE surface $\delta = 4.98D_{\mathrm{s}}^{1/3}\nu^{1/6}\omega^{-1/2}$ may be substituted, with ν being the kinematic viscosity of the external electrolyte and ω is the rotation rate

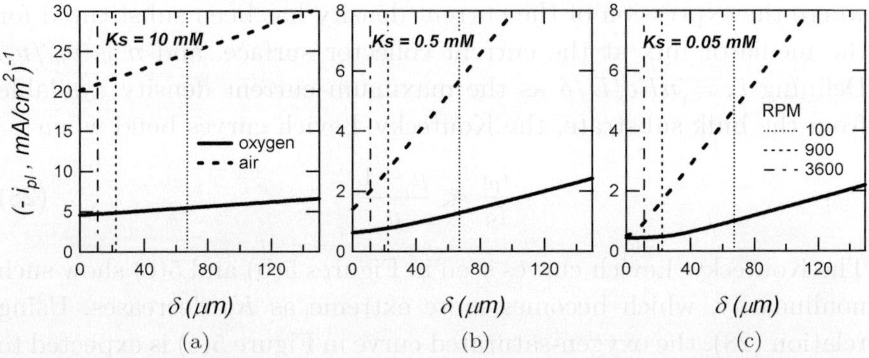

Figure 5. Simulated Koutecky–Levich plots for an enzyme electrode on an RDE, with all parameters as indicated in Figure 4. As K_s decreases, the system can be seen to saturate at high rotation rates, when diffusion layer is small.

in rpm. For the Koutecky–Levich plot, the slope and intercept — indicating the current density in the absence of external mass transport limitation — are then, respectively, given by

$$\frac{1}{m_{\mathrm{l}}} = \frac{4.98\nu^{1/6}}{nFD^{2/3}c_{\mathrm{s}}^{0}(1 + c_{\mathrm{s}}^{0}/K_{s})};$$

$$\frac{1}{i_{\mathrm{pl,k}}} = \frac{K_{\mathrm{s}}^{1/2}}{nFc_{\mathrm{s}}^{0}D_{\mathrm{s}}^{1/2}k_{\mathrm{cat}}^{1/2}c_{\mathrm{e}}^{1/2}}\coth\left(\frac{\phi}{D_{\mathrm{s}}^{1/2}}\frac{k_{\mathrm{cat}}^{1/2}}{K_{\mathrm{s}}^{1/2}}\frac{c_{\mathrm{e}}^{1/2}}{(1 + c_{\mathrm{s}}^{0})/K_{\mathrm{s}}}\right). \tag{26}$$

Such a case can be seen in Figure 5(a).

In cases where $\gamma \gg 1$ but substrate is not universally saturated, the corresponding Koutecky–Levich plots are expected to be linear at large values of σ, and bend to horizontal as $\sigma \to 0$ and the Michaelis–Menten kinetics begin to level off due to substrate saturation. This is seen in Figure 4 as the curves level off as $\delta \to 0$. Taking Eq. (9) as the relevant governing equation, with constant substrate profiles determined by Eq. (15), the current response will become insensitive to substrate when $\mu s \gg 1$. It follows that the Koutecky–Levich curve bends horizontally around the condition

$$1 \ll \mu\left(1 - \frac{\delta i_{\mathrm{pl}}}{nFc_{\mathrm{s}}^{0}D}\right), \tag{27}$$

where the expression of the current density has been substituted for the mediator flux at the current collector surface, and n is ν_m/ν_s. Defining $i_s = nFc_s^0 D/\delta$ as the maximum current density available from the bulk substrate, the Koutecky–Levich curves bend when

$$\frac{i_{\mathrm{pl}}}{i_{\mathrm{S}}} \ll \frac{\mu - 1}{\mu}. \tag{28}$$

The Koutecky–Levich curves seen in Figures 5(b) and 5(c) show such nonlinearity, which becomes more extreme as K_{s} decreases. Using relation (28), the oxygen-saturated curve in Figure 5(c) is expected to become nonlinear when the diffusion layer thickness falls to approximately 50 μm, corresponding to around 200 rpm for an RDE in aqueous electrolyte. This demonstrates that real-world oxygen-reducing enzyme electrodes operating on dissolved oxygen are likely to be limited by external oxygen diffusion.

7.3.3 Effect of mediator concentration

The preceding discussion examined D_{m}, δ and c_{s}^0 as design variables for enzyme electrode films. Another significant design variable in these systems is mediator concentration within the film, c_{m}^0. From a dimensionless analysis on the oxygen-reducing enzyme–redox hydrogel cathodes in Ref. [3], two possible causes of limitation were evident. First, mediation at potentials high enough to result in appreciable device power gave rise to kinetic limitations. Parametrically, this meant that K_{m} values for mediator–enzyme electron transfer were large and system performance therefore would benefit from increased mediator concentration. Second, a reaction layer was in evidence, indicated by the condition $\kappa^2/\gamma(1+\mu) > 2$. The solution to either of these limitations was an increase in mediator concentration c_{m}^0. Concerning the reaction layer, this would increase γ and decrease κ^2. A high enough value of c_{m}^0 would result in $\kappa^2/\gamma(1+\mu) < 2$, meaning that all enzymes entrapped in the redox hydrogel would potentially be active, and none of the film would be wasted by reaction layer formation.

Simulated current density dependence on osmium complex concentration, in which D_{m} is allowed to vary as a linear function of c_{m}^0,

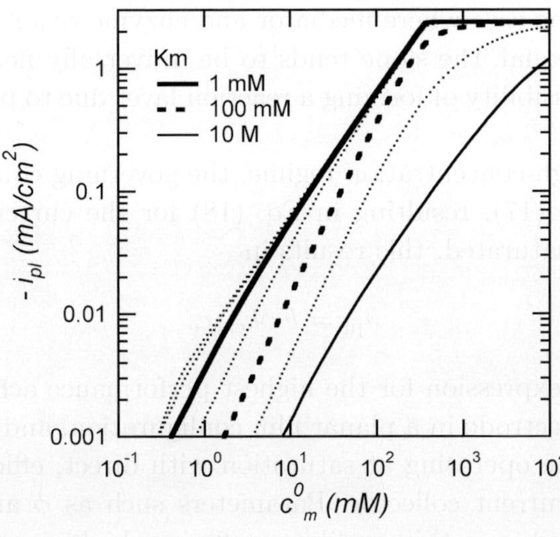

Figure 6. Simulation results for current density as affected by mediator concentration. Charge transport is defined as a function of mediator, $D_{\rm m} = 4.5 \times 10^{-9}$ $c^0_{\rm m}$ ($c^0_{\rm m}$ in M).[12] Mediator kinetics ($K_{\rm m}$) are varied by decades, with values of 1 mM, 100 mM, and 10 M marked by the lines indicated. Simulation parameters are: $\phi = 0.5$ μm, $k_{\rm cat} = 100$ s^{-1}, $K_{\rm s} = 0.1$ mM, $c_{\rm e} = 5$ mM, $\delta = 0$.

is shown in Figure 6. It can be seen that at low osmium concentrations, the current varies with a slope between 1.5 and 1 on a log–log plot. At higher concentrations, current levels off to its maximal value, as determined by substrate saturation.

In the low-concentration region, where a mediator reaction layer is generally present, the governing equations reduce to Eq. (19), as dictated by mediator kinetic limitation ($\gamma \ll 1$). The expression for the current density in this case is

$$i_{\rm pl} = F c^0_{\rm m} D_{\rm m}^{1/2} \left(\frac{k_{\rm cat}}{K_{\rm m}}\right)^{1/2} c_{\rm e}^{1/2} \tanh\left(\frac{\phi}{D_{\rm m}^{1/2}} \frac{k_{\rm cat}^{1/2}}{K_{\rm m}^{1/2}} c_{\rm e}^{1/2}\right). \quad (29)$$

The current density is controlled by mediator transport, and proportional to $c^0_{\rm m}$ $D_{\rm m}^{1/2}$, at low concentrations (slope = 1.5) and is controlled by mediator kinetics, and proportional to $c^0_{\rm m}$, at higher concentrations (slope = 1). Also, it can be seen that if $K_{\rm m}$ is unusually large, as in the case where mediator and enzyme redox potential are very nearly equal, the slope tends to be universally near 1, indicating the impossibility of forming a reaction layer due to poor mediator kinetics.

In the high-concentration regime, the governing equation will be given by Eq. (17), resulting in Eq. (18) for the current density. If substrate is saturated, this results in

$$i_{\mathrm{pl}} = F\phi k_{\mathrm{cat}} c_{\mathrm{e}} \tag{30}$$

which is the expression for the highest performance achievable from an enzyme electrode in a planar film configuration, and is equivalent to the enzyme operating at saturation with direct, efficient connection to the current collector. Parameters such as ϕ and c_{e} are, in reality, coupled, as a thinner film results in a higher enzyme concentration for constant mass loading.

7.4 Conclusion

Immobilization of enzymes in redox hydrogels allows multiple layers of enzyme to communicate current to an electrode, giving this strategy the potential to provide higher current than techniques that produce only a monolayer of enzymes. Performance of an enzyme–redox hydrogel film depends on five dimensionless variables, relating reaction kinetics and transport phenomena of mediator, enzyme, and substrate. Within this parameter space exists an optimum design to yield maximum current from the electrode, which will be desired in power-producing applications. (Other applications such as biosensing may have a different metric to assess "performance," and thus require a different set of optimal parameters.) A 1D model of a mediated, oxygen-reducing enzyme–redox hydrogel electrode was used to illustrate performance limitations under a broad range of conditions. Under typical conditions, electron diffusion, external substrate diffusion, and reaction kinetics are most likely to limit performance for this system. In one practical design situation, the concentration of redox mediator was the not a critical parameter, and it was shown that an increase in this value would increase catalytic current. The maximum possible current from an enzyme–redox hydrogel electrode is given in Eq. (30). Electrode design focuses on doing everything practically possible to obtain current approaching this value.

References

1. A. Heller (1992). Electrical connection of enzyme redox centers to electrodes. *Journal of Physical Chemistry* **96**(9), 3579–3587.
2. S.A. Calabrese Barton, J.W. Gallaway and P. Atanassov (2004). Enzymatic biofuel cells for Implantable and microscale devices. *Chemical Reviews* **104**(10), 4867–4886.
3. J.W. Gallaway and S.A. Calabrese Barton (2009). Effect of redox polymer synthesis on the performance of a mediated laccase oxygen cathode. *Journal of Electroanalytical Chemistry* **626**, 149–155.
4. J.W. Newman and C.W. Tobias (1962). Theoretical analysis of current distribution in porous electrodes. *Journal of the Electrochemical Society* **109**(8), C193–C193.
5. M. Majda (1992). Dynamics of electron transport in polymeric assemblies of redox centers. in R.W. Murray (ed.), in *Molecular Design of Electrode Surfaces*, New York: Wiley, pp. 159–206.
6. C.P. Andrieux, J.M. Dumasbouchiat and J.M. Saveant (1982). Catalysis of electrochemical reactions at redox polymer electrodes — kinetic model for stationary voltammetric techniques. *Journal of Electroanalytical Chemistry* **131**, 1–35.
7. C.P. Andrieux and J.M. Saveant (1982). Kinetics of electrochemical reactions mediated by redox polymer-films — irreversible cross-exchange reactions — formulation in terms of characteristic currents for stationary techniques. *Journal of Electroanalytical Chemistry* **134**(1), 163–166.
8. C.P. Andrieux and J.M. Saveant (1992). Catalysis at redox polymer coated electrodes. in R.W. Murray (ed.), *Molecular Design of Electrode Surfaces*, New York: Wiley pp. 207–270.
9. P.N. Bartlett and K.F.E. Pratt (1995). Theoretical treatment of diffusion and kinetics in amperometric immobilized enzyme electrodes 1. Redox mediator entrapped within the film. *Journal of Electroanalytical Chemistry* **397**(1–2), 61–78.
10. S.A. Calabrese Barton (2005). Oxygen transport in composite mediated biocathodes. *Electrochimica Acta* **50**(10), 2145–2153.
11. R.J. Forster, J.G. Vos and M.E.G. Lyons (1991). Controlling processes in the rate of charge transport through [Os(Bipy)$_2$(PVP)$_n$Cl]Cl redox polymer-modified electrodes. *Journal of the Chemical Society-Faraday Transactions* **87**(23), 3761–3767.
12. J.W. Gallaway and S.A. Calabrese Barton (2008). Kinetics of redox polymer-mediated electrodes. *Journal of the American Chemical Society* **130**, 8527–8536.
13. R.E. White (1978). Newman numerical technique for solving boundary-value problems. *Industrial & Engineering Chemistry Fundamentals* **17**(4), 367–369.

14. J.W. Newman (1968). Numerical solution of coupled ordinary differential equations. *Industrial & Engineering Chemistry Fundamentals* **7**(3), 514–517.
15. K. Kano and T. Ikeda (2000). Fundamentals and practices of mediated bioelectrocatalysis. *Analytical Sciences* **16**(10), 1013–1021.
16. S.A.M. Vanstroebiezen, F.M. Everaerts, L.J.J. Janssen and R.A. Tacken (1993). Diffusion-coefficients of oxygen, hydrogen-peroxide and glucose in a hydrogel. *Analytica Chimica Acta* **273**(1–2), 553–560.
17. A.J. Bard and L.R. Faulkner (2001). *Electrochemical Methods*, 2nd Edition. New York: Wiley.
18. J.S. Facci, R.H. Schmehl and R.W. Murray (1982). Effect of redox site concentration on the rate of electron-transport in a redox co-polymer film. *Journal of the American Chemical Society* **104**(18), 4959–4960.
19. R.J. Forster and J.G. Vos (1990). Synthesis, characterization, and properties of a series of osmium-containing and ruthenium-containing metallopolymers. *Macromolecules* **23**(20), 4372–4377.

Chapter 8

Conducting Polymer Hydrogels and Their Applications as Electrode Materials

Yu Zhao,* Lanlan Li,* Lijia Pan,*,‡ Guihua Yu†,§, and Yi Shi*,¶

*School of Electronic Science and Engineering, Collaborative Innovation Center of Advanced Microstructures, Nanjing University, Nanjing 210093, China

†Materials Science and Engineering Program and Department of Mechanical Engineering, the University of Texas at Austin, Texas 78712, USA

‡ljpan@nju.edu.cn

§ghyu@austin.utexas.edu

¶yshi@nju.edu.cn

8.1 Introduction

Nowadays, nanostructured materials are extremely important in a wide range of applications from electronics, sustainable energy, and photonics to medical and life science, because of their unique physical/chemical properties enabled by the confined dimensions of such materials.[1–4] Owing to their advantageous features, such as larger electrochemically active surface area, shortened pathways for charge/mass transport, and better accommodation of the strain within electrodes, nanostructured materials can be used as ideal electrodes. In miscellaneous nanostructured materials, hydrogels with highly cross-linked networks of polymer chains are of extensive interest because they have 3D structures at both microscale and nanometer levels.[5] Different from inorganic nanomaterials, hydrogels exhibit mechanical properties ranging from soft and weak to hard and tough, owing to their special composition with large quantity of liquid dispersed within a solid.[6] Because of its cross-linking structure and swelling nature,[7] hydrogel usually possesses a highly porous microstructure, high surface area, good compatibility with bio- or other hydrophilic molecules, and tunable mechanical properties, thus becoming an ideal material for diverse applications in energy, electronics and life science.[8,9]

Conducting polymers (CPs) are a class of functional polymers that have a delocalized *p*-system backbone along the polymeric chains. The highly conjugated polymer chain can be assigned reversible chemical, electrochemical and physical properties controlled by a doping/de-doping process. Conductive polymers usually possess highly π-conjugated polymeric chains. Typical conductive polymers include polyacetylene (PA), polyaniline (PAni), polypyrrole (PPy), polythiophene (PTh), poly(phenylenevinylene) (PPV), etc. (Figure 1(b)). The conductivity of conductive polymers could be tuned in a wide range from 10^{-10} up to 10^4 S cm^{-1} through chemical or electrochemical redox reactions which is called "doping",[10] endowing the conductive polymers with the ability to act as insulators, semiconductors or conductors by changing the dopants and/or the level of doping.[11–14] In addition, doping is a reversible process that makes the backbone of conductive polymers positively (*p*-doping) or negatively charged (*n*-doping), enabling a number of device applications such as energy storage and conversion devices, electronics, optics, medicine, actuators and composites as a viable alternative to metallic or inorganic semiconductor counterparts.[15,16] Conductive polymers can also be rationally designed and fabricated to nanostructures, thus providing new exciting features including tunable conductivities, flexibility and mixed conductive mechanism, which lowers the interfacial impedance between electrodes and electrolytes.[17]

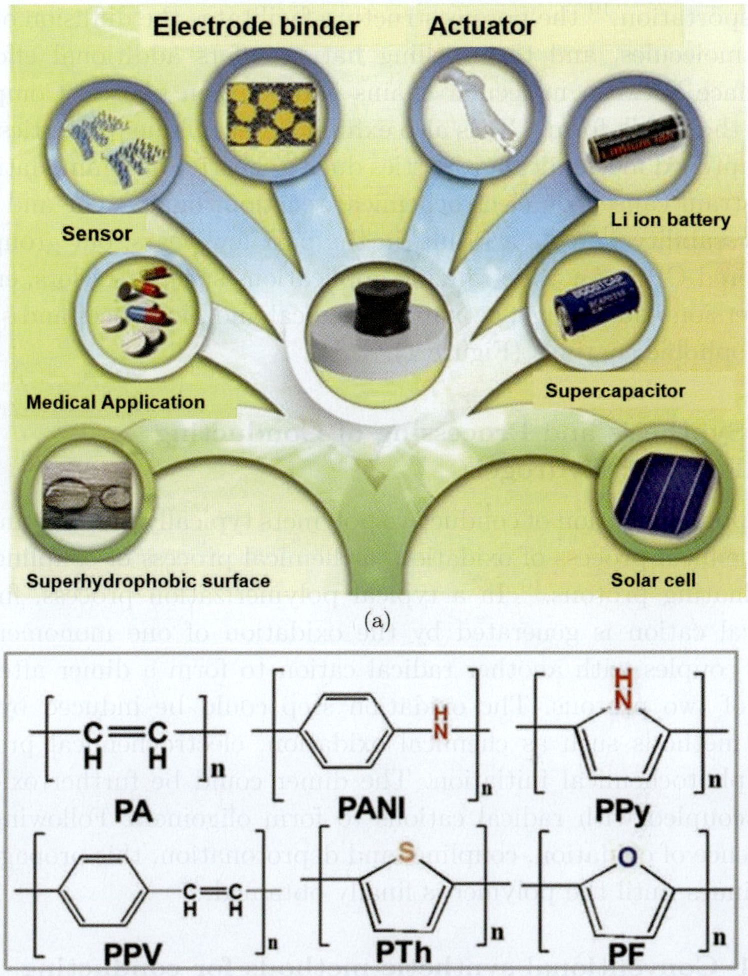

Figure 1. (a) Representative applications of CPHs. (b) The chemical structures of typical conductive polymers. Reproduced with permission from Ref. 5. Copyright 2015 Royal Society of Chemistry.

As a special class of conductive polymer materials, conducting polymer hydrogels (CPHs) synergize the advantages of hydrogels and organic conductors.[18] CPHs with 3D network structures provide an electrically conductive yet mechanically robust framework: the continuous conductive backbones can accelerate charge transportation,[19] the porous structure facilitates the diffusion of ions and molecules, and the swelling nature offers additional effective interface between molecular chains and solution phases. Compared with their bulk form, CPHs also exhibit other unique properties such as improved mechanical properties due to effective accommodation of the strain caused by electrochemical reaction, lightweight and good processability.[20,21] As a result, in the past few years, our group has explored CPHs for wide-ranging applications such as sensors, energy conversion and storage, actuators, medical and biodevices and superhydrophobic coatings (Figure 1).[22–25]

8.2 Synthesis and Processing of Conducting Polymer Hydrogels

The polymerization of conductive polymers typically involves an electrochemical process of oxidation, a chemical process of coupling and eliminating protons.[10] In a typical polymerization process, first, a radical cation is generated by the oxidation of one monomer and then couples with another radical cation to form a dimer after the loss of two protons. The oxidation step could be induced by various methods such as chemical oxidation, electrochemical process, and photochemical initiation. The dimer could be further oxidized and coupled with radical cations to form oligomers. Following the sequence of oxidation, coupling, and deprotonation, this propagation continues until the polymer is finally obtained.[26]

8.2.1 Conventional synthetic methods for conducting polymer hydrogels

Traditional synthetic routes include polymerization of monomers in an existing non-conductive hydrogel matrix, which serves as a template, copolymerization of conductive polymers with non-conductive polymers, and cross-linking conductive polymers such as poly(3,4-ethylenedioxythiophene):polystyrene sulfonate (PEDOT:PSS) by multi-valent metal ions (Fe^{3+} or Mg^{2+}).[18] The hydrogel-template method is most commonly used (Figure 2(a)). Typically, a hydrogel template matrix is first prepared and de-swollen, and then reswollen

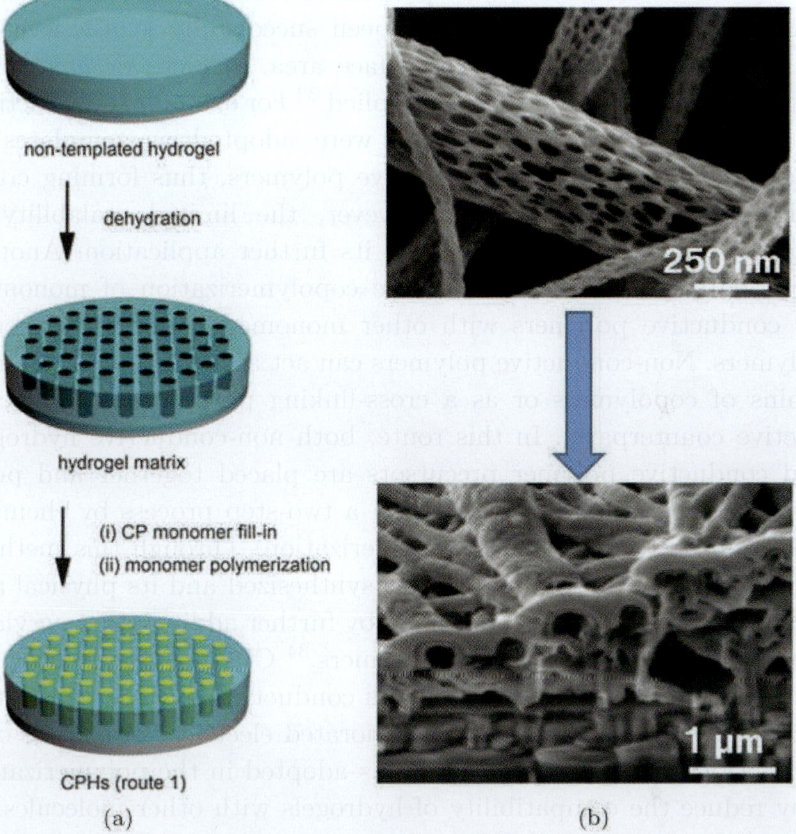

Figure 2. (a) Conventional synthetic methods for CPHs, in which monomers of conductive polymers are polymerized onto non-conductive hydrogel matrix templates. Reproduced with permission from Ref. 18. Copyright 2013 Royal Society of Chemistry. (b) SEM images of PLGA nanoscale fibers used as the template (top panel) and PEDOT hydrogels synthesized based on the PLGA template (bottom panel). Reproduced with permission from Ref. 33. Copyright 2006 WILEY-VCH Verlag GmbH & Co.

in the solution of monomers for conductive polymer synthesis, which is *in situ* electrochemical polymerization or by adding chemical oxidants. PAni or PPy–PNIPAM [poly (*N*-isopropylacrylamide)],[27] PPy–pHEMA [poly(2-hydroxyethyl methacrylate)],[28] PEDOT–alginate,[29] PEDOT–PAA (polyacrylic acid),[30] PEDOT–PAMPS [poly(2-acrylamido-2-methyl-1-propanesulfonic acid)][31] and PPy–PAAM (polyacrylamide).[32] have been successfully synthesized via this strategy. To increase the surface area, a micro or nanostructured template would be further applied.[33] For example, poly(lactide-*co*-glycolide) (PLGA) nanofibers were adopted as templates to electrochemically deposit conductive polymers, thus forming cross-linked CPHs (Figure 2(b)). However, the limited scalability of template-guided synthesis inhibits its further application. Another method to fabricate a CPH is the copolymerization of monomers for conductive polymers with other monomers for non-conductive polymers. Non-conductive polymers can act as a component of main chains of copolymers or as a cross-linking part to cross-link conductive counterparts. In this route, both non-conductive hydrogels and conductive polymer precursors are placed together and polymerized either simultaneously or in a two-step process by chemical oxidation or electrochemical polymerization. Through this method, a PPy–pHEMA hybrid hydrogel is synthesized and its physical and chemical properties could be tuned by further adding other acrylate, methacrylate and acrylamide monomers.[34] CPHs synthesized by the conventional method consist of both conductive and non-conductive components, and may lead to deteriorated electrical properties over time. Moreover, excessive metal ions adopted in the polymerization may reduce the compatibility of hydrogels with other molecules. In terms of potential scalability, the conventional synthetic methods discussed above have significant limitations to overcome.

8.2.2 Recently developed preparation routes for conducting polymer hydrogels

A new synthetic route for conductive hydrogels, which exhibit facile processability, excellent electronic properties, high electrochemical activity and high biocompatibility (Figures 3(a) and 3(b)), has recently been reported.[8] Phytic acid is adopted as the gelator and dopant in the synthetic process and conductive hydrogel networks could be directly formed. PAni and PPy hydrogels have been successfully synthesized via this method. The gelation mechanism

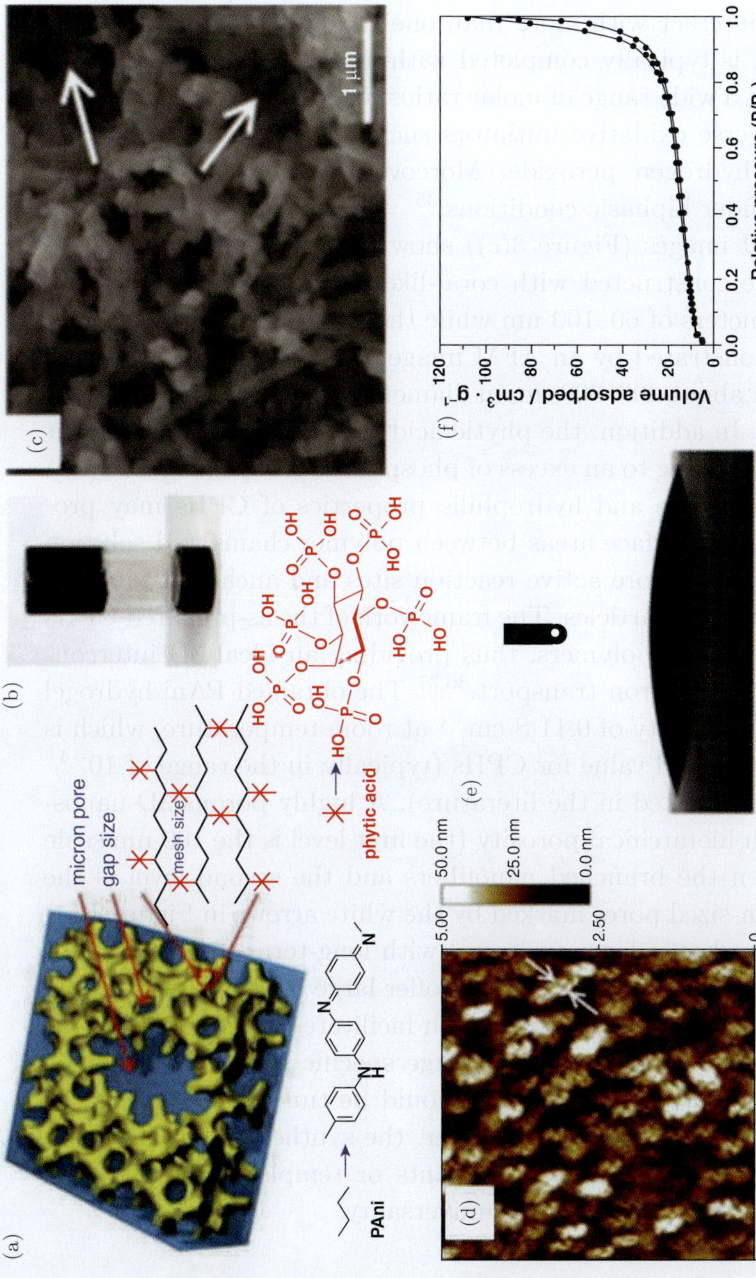

Figure 3. (a) Schematic illustration of the 3D hierarchical microstructure of the gelated PAni hydrogel where phytic acid plays the role as a dopant and a cross-linker. (b) A photograph of the PAni hydrogel. (c) A SEM image showing the interconnected network of dendritic PAni nanofibers. (d) AFM image of the PAni hydrogel. (e) Contact angle of phytic acid doped PAni. (f) Nitrogen adsorption–desorption isotherm of the dehydrated PAni hydrogel. Reproduced with permission from Ref. 8. Copyright 2012 Proceedings of the National Academy of Sciences of the United States of America.

of the CPHs could be explained by the ability of each phytic acid molecule to interact with more than one conductive polymer chain. The gelation is typically completed within several minutes and can be induced in a wide range of molar ratios of monomer to phytic acid, and with diverse oxidative initiators such as ammonium persulfate (APS) and hydrogen peroxide. Moreover, the reaction could be conducted under biphasic conditions.[35]

The SEM images (Figure 3(c)) show that the dehydrated PAni hydrogels are constructed with coral-like dendritic nanofibers with uniform diameters of 60–100 nm while the fully swollen PAni hydrogels are demonstrated by an AFM image (Figure 3(d)) to consist of nanofibers of about 200–300 nm in diameter, showing their high level of hydration. In addition, the phytic acid renders CPHs a significant hydrophilicity owing to an excess of phosphorus groups (Figure 3(e)). The swelling nature and hydrophilic properties of CPHs may provide additional interface areas between polymer chains and solution phases to support more active reaction sites and anchoring sites for active molecules or particles. The framework of the as-prepared CPHs is free of insulating polymers, thus providing an ideal 3D interconnected path for electron transport.[36,37] The obtained PAni hydrogel reaches a conductivity of 0.11 S cm^{-1} at room temperature, which is the highest reported value for CPHs (typically in the range of 10^{-4}–10^{-2} S cm^{-1} reported in the literature). A highly porous 3D nanostructure with hierarchical porosity (the first level is the 100 nm scale pores between the branched nanofibers and the second level is the bigger micron-sized pores marked by the white arrows in Figure 3(c)) can be achieved on a large-scale and with long-term stability in this synthesis. Such 3D porous structures offer large open channels on the microscale and mesoscale pores, which facilitate the diffusion of ions and molecules and also provide a large specific surface area, which is measured to be $\sim$42 m^2 g^{-1} and could be tuned from 40 m^2 g^{-1} to 100 m^2 g^{-1} (Figure 3(f)). Moreover, the synthetic route needs no external ingredients such as surfactants or templates, thus greatly enhancing its processability and universality.

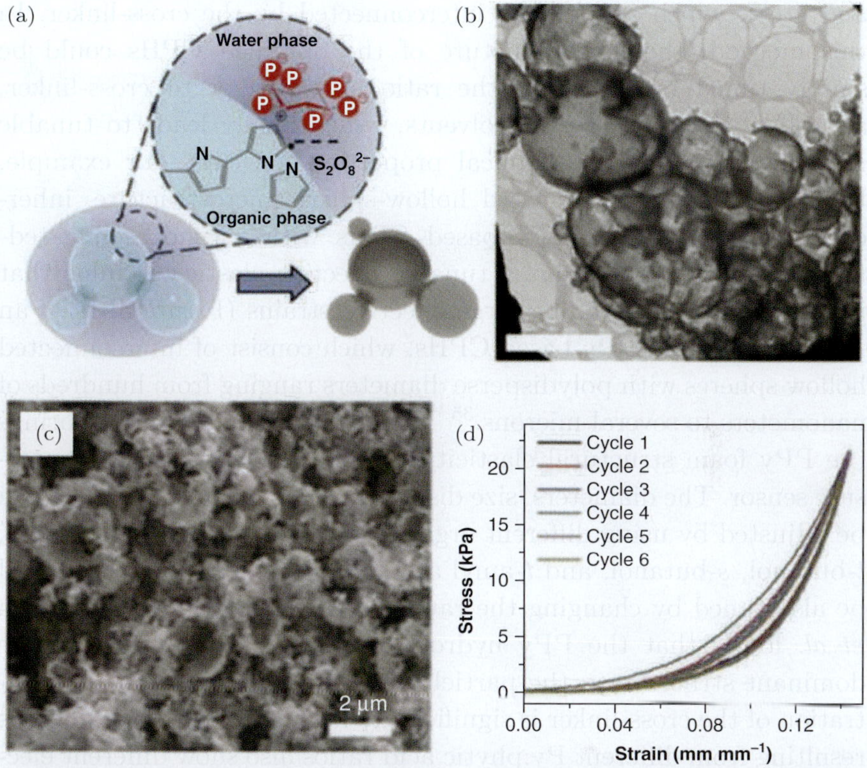

Figure 4. (a) Schematic of the structural elasticity of the hollow-sphere structured PPy. (b) TEM image of PPy showing its interconnected hollow-sphere structure, scale bar: 1 μm. (c) SEM images of the PPy hydrogel with a pyrrole:phytic acid ratio of 5:1; (d) six consecutive compression tests on the PPy thin film. Reproduced with permission from Ref. 35. Copyright 2013 Nature Publishing Group.

Adopting phytic acid as a cross-linker and a dopant, another synthetic approach based on multi-phase interfacial reactions could be applied for CPHs (Figure 4(a)).[35] In a typical synthesis, an aqueous solution of an oxidative reagent and another solution containing a mixture of the monomer, organic solvent and phytic acid are prepared and then mixed together. The monomers would polymerize at the interface, resulting in hollow sphere microstructures (Figures 4(b) and 4(c)), which are further interconnected by the cross-linker. In this method, the microstructure of the obtained CPHs could be simply tuned by adjusting the ratio of monomer to cross-linker, as well as different organic solvents, which finally leads to tunable mechanical and electrochemical properties of CPHs. For example, by forming an interconnected hollow-sphere microstructure, inherently stiff and brittle PPy based CPHs with a rigid conjugated-ring backbone can exhibit a tunable effective elastic modulus that is capable of withstanding large effective strains (Figure 4(d)). Pan *et al.* synthesized PPy-based CPHs, which consist of interconnected hollow spheres with polydisperse diameters ranging from hundreds of nanometers to several microns.[35] The hollow-sphere structure brings the PPy foam structural elasticity, enabling an ultrasensitive pressure sensor. The diameters, size dispersions, and shell thicknesses can be adjusted by using different organic solvents, including 1-butanol, *t*-butanol, *s*-butanol, and *t*-amyl alcohol. The micro structure could be also tuned by changing the ratio of monomer to phytic acid. Yu *et al.* found that the PPy hydrogel turns from the hollow-sphere dominant structure to the particle-dominant one when the concentration of the cross-linker is significantly lowered. The PPy hydrogels resulting from different Py:phytic acid ratios also show different electrochemical properties.[38]

Owing to their facile synthesis and tunable mechanical properties, the CPHs show high scalability and processability that they could be processed by scalable techniques such as ink-jet printing or spray coating, to fabricate the desired micropatterns for large arrays of electrochemical devices (Figure 5).[39] Ink-jet printing of the polymer is often hampered by the limited solubility and high viscosity of polymer solutions. This difficulty is overcome by separately depositing two distinct solutions, which contain an oxidative initiator and phytic acid, and monomers for conductive polymer synthesis, respectively, onto the substrate. The patterned PAni hydrogel is formed when two solutions are able to interact and the morphology of the printed hydrogel is found to be the same as that of the bulk synthesized hydrogel. The conductive hydrogel

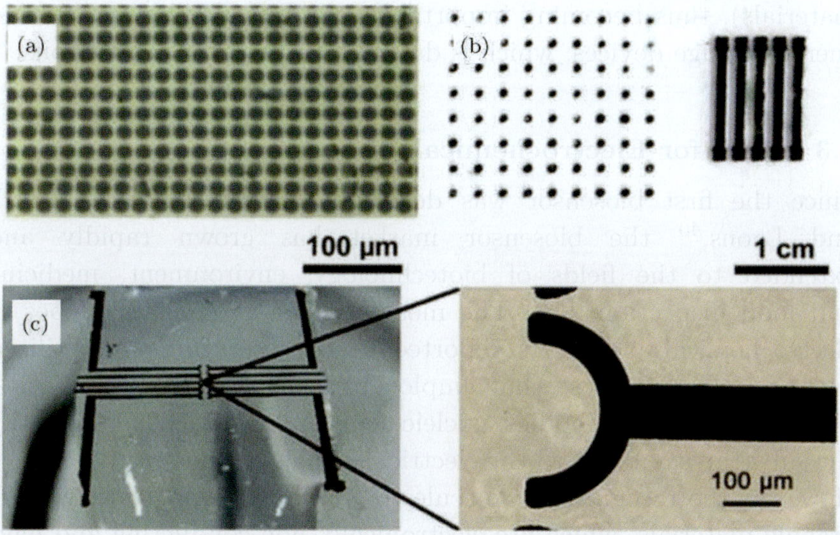

Figure 5. Micropatterned PAni hydrogel arrays fabricated by (a) ink-jet printing and (b) spray coating. (c) Photograph of a microelectrode array based on PEDOT on a flexible gel substrate. Reproduced with permission from Ref. 39. Copyright 2010 American Chemical Society.

can also be processed by spray coating to form micropatterns of millimeter size. Similar to ink-jet printing, the two solutions are alternatively deposited multiple times through poly(dimethylsiloxane) (PDMS) soft stencil masks onto a range of substrates. Through these approaches, we can fabricate large arrays of patterned CPHs, leading to highly conductive, functional microelectrodes potentially useful for supercapacitors, lithium batteries, biosensors, chemical sensors, and other bioelectrodes.

These newly developed synthetic methods for nanostructured CPHs offer the ability to rationally design and control the morphology and microstructure of conductive polymers, as well as tuning their mechanical and electrochemical properties. With these synthetically tunable chemical/physical characteristics, the newly developed CPHs possess the possibility to meet requirements in various applications ranging from functioning as active electrode materials to serving as other functional components (*e.g.* as functional binder materials), thus becoming important candidates for biosensors and energy storage devices, which is detailed in the following sections.

8.3 CPHs for Electrochemical Biosensors

Since the first biosensor was developed in the 1960s by Clark and Lyons,[40] the biosensor market has grown rapidly and extended to the fields of biotechnology, environment, medicine and food inspection.[41–45] The most popular and reliable types of electrochemical biosensors reported in the literature and utilized commercially are those that employ biological recognition species, such as enzymes, antibodies, nucleic acids and whole cells, to identify the substrates and record the electrical signals by bioelectrodes.[46–48] However, the recognition molecules or biosubstrates themselves are organic materials, which are electronically non-conducting and have distinct physical properties compared to those of inorganic electrodes. An appropriate material that wires and interfaces between the biospecies and the artificial electrode is essential for enhancing the sensing performance of electrochemical biosensors and other bioelectronic devices.[49] Moreover, the interfacial material should incorporate high hydrophilicity and biocompatibility for immobilization, as well as retain bioactivity of the biospecies.

A variety of interface materials have been developed for biosensors.[50] Among them, CPs are one of the most widely used materials because of their tunable physical properties via either the redox states or doping/dedoping states, which leads to a change of resistance, current, or electrochemical potential upon its interaction with the electrochemical enzyme.[51,52] Moreover, CPs have received intensive interest as an interface between biosubstrates and inorganic electrodes because of their hybrid conducting mechanism that combines both electron and ionic charge carriers, which favors lowered impedance between the electrode and electrolyte interface. Another candidate interfacial material for biosensors is a hydrogel, a cross-linked polymer with a 3D structure that can absorb large quantities of water without dissolving.[53] Hydrogels exhibit attractive features. For example, they are good immobilization materials for biological recognition

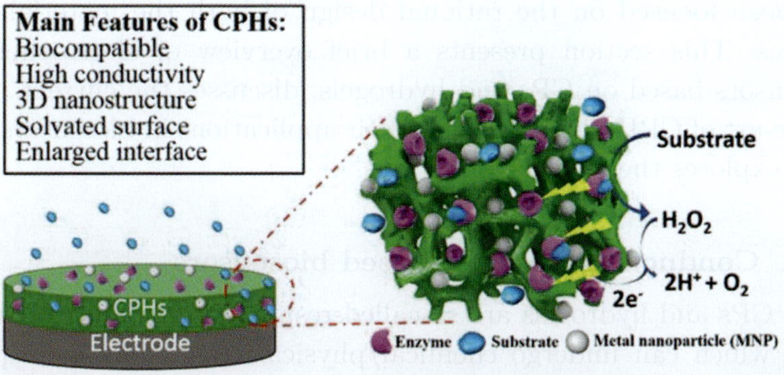

Figure 6. Schematic of a typical example of rational designed electrochemical biosensor based on CPHs and a list of the main features of CPHs. Reproduced with permission from Ref. 22. Copyright 2015 Royal Society of Chemistry.

species because they prevent leakage and inactivation of the species. Additionally, because of their high water content and organic composition that is similar to the extracellular matrix (ECM), they exhibit excellent biocompatibility for biospecies.[54–56] However, most hydrogel materials are an electron insulator, which has limited the performance of hydrogel-based biosensors.

Therefore, growing attention has been directed to the synthesis of CPHs and the fabrication of them into functional biosensors. Combining the features of both CPs and hydrogels, these CPHs exhibited great potential as advanced interface materials for biosensors. We schematically illustrated a rationally designed electrochemical biosensor based on CPHs in Figure 6. The CPHs played critical roles in enhancing the sensing performance by (1) extending the planar artificial electrode to the 3D organic matrix to increase the effective interface area, (2) providing linkages between soft and hard materials for favorable enzyme immobilization, (3) interfacing the ionic transporting phase and the electron transporting phase to lower the impedance, and (4) high-density loading with catalytic nanoparticles (NPs) to promote electron collection.[18] Based on these unique features, biosensors based on CPHs have exhibited attractive properties, such as high sensitivity, low detecting limitation, and fast response time. Until now, the development of CPH-based biosensors has been focused on the rational design of both the materials and devices. This section presents a brief overview of electrochemical biosensors based on CPs and hydrogels, discusses the current state-of-the-art of CPHs synthesis and their applications in biosensors, and then explores their future directions.

8.3.1 Conducting polymer-based biosensors

Both CPs and hydrogels are so-called responsive "intelligent" materials, which can undergo chemical/physical transitions in response to external stimuli.[57] Significant progress has been made in the application of CPs and hydrogels in electrochemical biosensors.

CPs are the cationic salts of highly conjugated polymers that have a delocalized π-system backbone along their main chains.[58] The conjugated backbone endows CPs with unusual electronic properties such as electrical conductivity, low energy optical transitions, low ionization potential, and high electron affinity.[42,59] In addition to similar electrical and optical properties to inorganic metals and semiconductors, CPs also possess mechanical flexibility and good processability. The intrinsic conductivity of CPs enables rapid electron transfer in the application of electro-analytic biosensors. Because of their unique organic functional groups and electrochemical properties, CPs are also suitable immobilization matrices for biospecies and enable a variety of immobilization techniques such as chemical and electrochemical immobilization on electrodes with variable sizes or geometries.[60] CPs have been used for electrode interface materials owning to these exciting features.[61]

In recent years, nanostructured CPs have received growing attention because their small size. Nanostructured CPs are favorable for electron transfer and biomolecule stabilization due to their increased surface area and high surface free energy. Nanostructured CPs, such as PPy, PAni and PTh, offer excellent prospects for interfacing biological recognition species and transducing electronic signals for designing novel bioelectronic devices.[44,45,62–65] For example, Tao *et al.* produced a miniature glucose sensor using PAni/enzyme nanojunctions.[66] The signal transduction mechanism of the sensor was based on the conductance change of the PAni nanofibers as a result of a redox state change induced by glucose oxidation. Due to the enlarged surface area of PAni nanofibers, the enzyme could be naturally regenerated without redox mediators and enabled the sensor to rapidly respond (<200 ms). Cai *et al.* developed an amperometric glucose biosensor based on a direct electron transfer mechanism.[67] They electrochemically entrapped glucose oxidase (GOx) onto the inner wall of PAni nanotube arrays prepared from an anodic aluminum oxide (AAO) template.[68] A pair of symmetric redox peaks was displayed in the cyclic voltammogram, and an apparent electron transfer rate constant of 5.8 ± 1.6 s^{-1} was measured. The biosensor exhibited good sensing performance with a rapid response (3 s), a low detection limit (0.3 ± 0.1 μM), a useful linear range (0.01–5.5 mM), and high sensitivity (97.18 ± 4.62 μA mM^{-1} cm^{-2}). The biosensor showed anti-interfering properties against common species, such as ascorbic acid, uric acid, and 4-acetamidophenol. To further improve sensitivity and selectivity, a variety of biosensors based on CP composites have been developed, for example, the composites of CPs and carbon nanotubes, metal particles, metal oxide particles, metal salts, chitosan, and biocompatible materials. The second component aims to increase the electric conductivity of the conducting polymers, enhancing the affinity of the electrode to a biospecies or acting as a catalyst.[52] For example, Xian *et al.* developed a glucose biosensor based on a composite of Au nanoparticles (AuNPs) and conductive PAni nanofibers.[69] The high conductivity and excellent electroactivity of the AuNPs–PAni nanocomposites were favorable for promoting the electrocatalytic activity of GOx. The glucose sensor showed a liner range from 1 μM to 0.8 mM, a detection limit of 0.5 μM, and a response time of less than 5 s.

Despite the improvements of biosensors with CPs, CPs have shortcomings such as their hydrophobicity, which denatures the immobilization of biomolecules and plays a role as a diffusion barrier for entrapped enzymes, thus degrading the performance of the biosensor.[70]

8.3.2 Hydrogel-based biosensors

Hydrogels are networks of polymers that consist mostly of water, thus resembling biological soft tissues, and have great potential for use in biomedical applications.[53,71–75] Hydrogels present a variety of unique properties, such as swelling, solvated surface, and super-permeation, which are unprecedented in either a pure liquid or a pure solid.[53,76,77] The continuous 3D microstructure across the entire macroscopic scale of a hydrogel can be used as a superb immobilization matrix for biosensors due to their high hydrophilicity,[78] their large surface area, and their good biocompatibility.[79] These features improve the stability of biomolecules and prolong the shelf-life of biosensors.[80] Water-soluble analytes and charge neutralizing ions can rapidly diffuse through a water-swollen 3D hydrogel network layer.[81]

Moreover, certain hydrogels may also show reversible swelling behavior because of their sensitivity in the physiological or biological environment and hence can exhibit modulated release in response to external stimuli such as pH, temperature, ionic strength, electric field, or specific analyte concentration gradients, as shown in Figure 7.[82–89] The responsive behavior of hydrogels can be summarized into two categories. (1) Hydrogels act as carriers and "passively" respond to external stimuli due to physicochemical interactions and associated phenomena. Such phenomena can be pH dependence, temperature dependence, ionic strength dependence, magnetic or electrical dependence, and response to thermodynamic compatibility. (2) Hydrogels exhibit a certain intelligence that confers an ability to respond to a compound, such as a biomarker or a bioanalyte in the surrounding medium, or material. These systems are decorated with functional groups that can provide artificial ligand–receptor interactions. A series of stimuli-responsive hydrogels have been developed for medical and pharmaceutical applications. Peppas's group has incorporated activated GOx into pH-sensitive cationic hydrogels.[90,91] When these hydrogels were exposed to glucose in the body, the GOx catalyzed the production of gluconic acid, thus lowering the local environment pH and thereby causing the pH-sensitive hydrogels to swell. Depending on the biomolecule, the bioanalyte incorporated

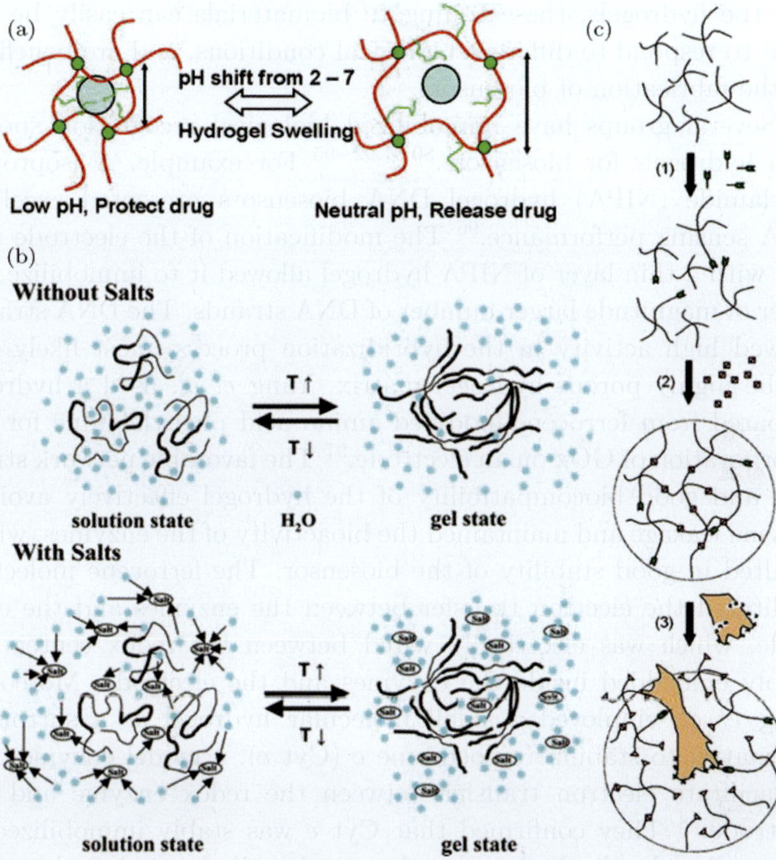

Figure 7. Schematic illustrations of: (a) pH-sensitive hydrogels: when the hydrogel is in a low pH environment, the polymer network entraps the loaded agents within the network; however, once the network is exposed to a higher pH environment, the polymer network swells and releases the drug. Reprinted with permission from Ref. 87. Copyright (2008) American Chemical Society. (b) Temperature-sensitive hydrogels: physical structures of aqueous methylcellulose with and without addition of salts at lower temperatures (solution state) and at elevated temperatures (gel state). Reprinted with permission from Ref. 88. Copyright (2004) American Chemical Society. (c) Biomolecule-sensitive hydrogels: these gels have a stimuli-responsive biomolecule directly incorporated into their polymer–hydrogel network. When these molecules are exposed to the particular agent they are sensitive to, the biomolecule will breakdown and in turn the hydrogel network will degrade and release the drug that was encapsulated within the gel. Reprinted with permission from Ref. 89. Copyright (2003) National Academy of Sciences.

into the hydrogels, these conjugate biomaterials can easily be tailored to respond to different biological conditions, and are beneficial for the fabrication of biosensors.

Several groups have immobilized biological recognition species with hydrogels for biosensors.[80,81,92–95] For example, *N*-isopropylacrylamide (NIPA) hydrogel DNA biosensors presented excellent DNA sensing performance.[96] The modification of the electrode surface with a thin layer of NIPA hydrogel allowed it to immobilize one order of magnitude larger number of DNA strands. The DNA strands showed high activity in the hybridization process, most likely due to the highly porous hydrogel matrix. Yang *et al.* used a hydrogel prepared from ferrocene-modified amino acid phenylalanine for the incorporation of GOx on an electrode.[97] The favorable network structure and good biocompatibility of the hydrogel effectively avoided enzyme leakage and maintained the bioactivity of the enzymes, which resulted in good stability of the biosensor. The ferrocene molecules facilitated the electron transfer between the enzymes and the electrode, which was electrically wired between the redox center and deeply embedded inside the enzymes and the electrode. Moreover, Yang *et al.* employed a small molecular hydrogel as a surrounding matrix to stabilize cytochrome *c* (Cyt *c*), a model enzyme, and to facilitate electron transfer between the redox enzyme and the electrode.[98] They confirmed that Cyt *c* was stably immobilized in the small molecular hydrogel and was primarily located at the cross-linking sites of the nanofibers. The Cyt *c* immobilized in the hydrogel retained its intrinsic activity toward H_2O_2.

8.3.3 Ionically cross-linked conducting polymer hydrogels and their applications in biosensors

For biosensors, many factors must be taken into account for developing novel interface materials.[99] Two main requisites for electrode materials are high conductivity for efficient charge collection and long-term biostability of the biospecies. Integrating the features of both hydrogels and organic conductors, CPHs were studied as a novel interface material for biosensors.

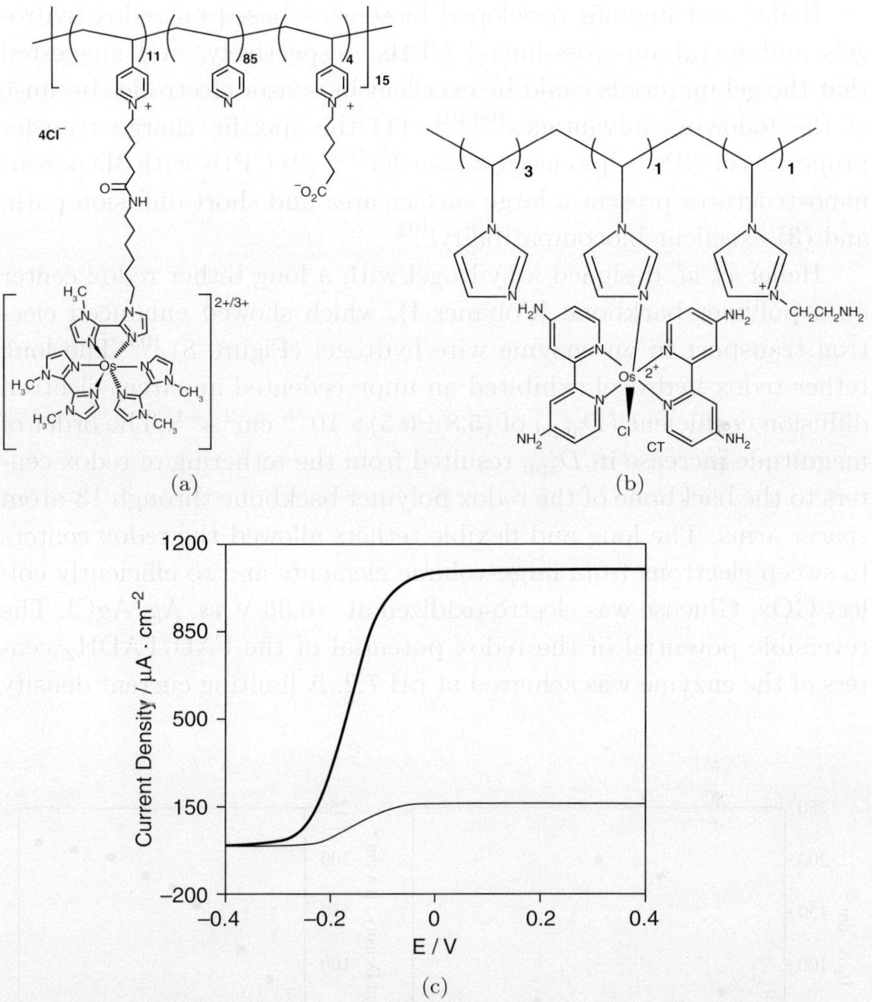

Figure 8. The enhancing effect of long tether binding redox centers on the polymer backbone on the electron transport in enzyme "wiring" hydrogels. (a) Structure of Polymer I, the chloride of PVP-[Os(N,N'-dialkylated-2,2′-biimidazole)3]2+/3+a. (b) Structure of Polymer II, PVI-[Os(4,4′-diamino-2,2′-bipyridine)2Cl]+/2+. (c) Polarization of 7-ím-diameter, 2-cm-long carbon fiber anode modified with Polymer II (fine line) and with Polymer I (heavy line). Quiescent solution, under air, 37.5°C, PBS buffer, 15 mM glucose, 1 mV/s. Reprinted with permission from Ref. 103. Copyright (2003) American Chemical Society.

Heller and Inganäs developed biosensors based on redox hydrogels and metal-ion-cross-linked CPHs, respectively, and suggested that the gel materials could be excellent biosensor electrodes because of the following advantages[100,101]: (1) the specific charge transfer properties of CPs helps electron transfer[102]; (2) CPHs with 3D porous nanostructures present a large surface area and short diffusion path; and (3) excellent biocompatibility.[103]

Heller *et al.* designed a hydrogel with a long tether redox center on a polymer backbone (Polymer I), which showed enhanced electron transport in an enzyme wire hydrogel (Figure 8).[103] The long tether redox hydrogel exhibited an unprecedented apparent electron diffusion coefficient (D_{app}) of $(5.8 \pm 0.5) \times 10^{-6}$ cm^2 s^{-1}. The order of magnitude increase in D_{app} resulted from the tethering of redox centers to the backbone of the redox polymer backbone through 13-atom spacer arms. The long and flexible tethers allowed the redox centers to sweep electrons from large-volume elements and to efficiently collect GOx. Glucose was electro-oxidized at −0.36 V *vs.* Ag/AgCl. The reversible potential of the redox potential of the FAD/FADH$_2$ centers of the enzyme was achieved at pH 7.2. A limiting current density

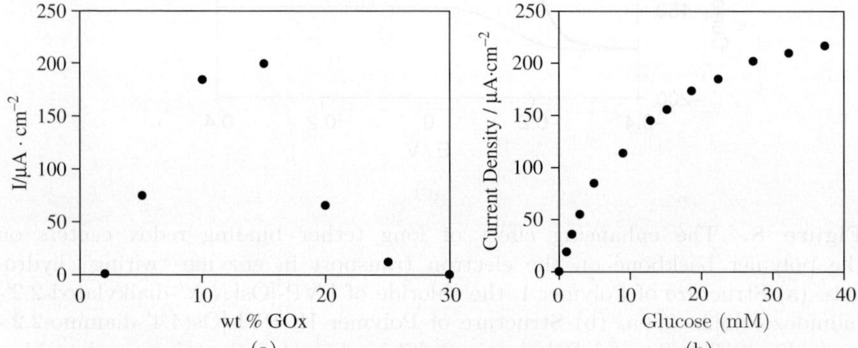

Figure 9. (a) Dependence of the current density on the GOx weight percentage with the glucose concentration maintained at 32 mM. (b) Dependence of the steady-state current density on the glucose concentration for an electrode poised at +0.3 V *vs.* Ag/AgCl. Reprinted with permission from Ref. 104. Copyright (2007) American Chemical Society.

of 1.15 mA cm^{-2} was reached at a potential as low as −0.1 V *vs.* Ag/AgCl.

Heller *et al.* also produced an electron-conducting cross-linked PAni-based redox hydrogel. The active layer was formed in one step by cross-linking a polymer acid-templated PAni with a water-soluble di-epoxide, poly(ethylene glycol) diglycidyl ether (PEGDGE).[104] GOx was immobilized in the hydrogel matrix by simultaneously cocross-linking and led to the electrical wiring of the enzyme. The cross-link-immobilized GOx allowed the electrocatalytic oxidation of glucose at a steady-state current density of 225 μA cm^{-2} at 40 mM glucose (Figure 9). Without hydration, PAni conducted through 1D bands only and such conduction required at least 1D semicrystalline PAni.[13] Because it was cross-linked and hydrated PAni behaved as a polymeric redox couple when dissolved and could form an electron-conducting redox hydrogel.[105,106] In such types of redox hydrogels, electrons diffuse through electron transferring collisions between hydrated reducible and oxidizable polymer segments. Hydration lowers the local viscosity and reduces attractive coulombic interactions, thereby increasing the segmental mobility and electron diffusion. It also increases the permeability of water-soluble ions and molecules.

Inganäs *et al.* developed a general method to synthesize PEDOT hydrogels using multi-valence metal ions such as magnesium, iron, and osmium (Os) to cross-link the poly(styrenesulfonate) (PSS) groups on PEDOT:PSS into a mesoscopic conducting network of hydrogels.[101,107–109] Inganäs *et al.* investigated a biomolecule-enhanced electrode with the electroactive polymer hydrogel PEDOT: PSS in combination with poly(4-vinylpyridine) (P4VPy) cross-linked by Os. The porous structure of the PEDOT:PSS hydrogel matrix on the surface of electrode made it feasible for analytes to easily diffuse in and out. Os played roles as both the cross-linking points of P4VPy and the mediator between the prosthetic group of the enzyme and the conducting polymer matrix. Based on the hydrogel, they demonstrated a biosensor utilizing horseradish peroxidase (HRP) to detect H_2O_2. A distinct cathodic peak at a low potential was observed on the CVs, corresponding to the reduction of H_2O_2 to H_2O catalyzed by HRP. When no H_2O_2 was present, only the capacitance changed in the PEDOT hydrogel without a peak.[110] The low redox potential was favorable for the biosensors to eliminate the interference from other electro-oxidizables. The inset graph displayed that current increased with the increased H_2O_2 concentration, and the bioelectrode started to decay at the concentration of 6 mM because this concentration was too high for HRP. The sensitivity of the biosensor was calculated to be 0.13 A M^{-1} cm^{-2}.

8.3.4 Doping acid cross-linking as a novel method to fabricate conducting polymer hydrogels and their application in biosensors

Until now, only a few strategies have been developed to synthesize CPHs because of the following challenges: (1) most of the monomers of CPs are hydrophobic; therefore, it is unknown how to render the polymer a hydrophilic and maintain the water content; and (2) it is unknown how to cross-link CPs while maintaining their conductivity and obtain good biocompatibility. Polymer hydrogels were used as templates to polymerize CPs and form composite hydrogels. However, the insulating component limited the electrical properties of the conducting polymers.[111] A water-soluble monomer (poly(3-thiopheneacetic acid)) was used to synthesize CPHs; however, the conductivity of the hydrogel was approximately 4.0–7.0 10^{-4} S cm^{-1} [112] because the polar side group affected the conjugation of the polymer main chain. Multi-valent metal ions (Fe^{3+} or Mg^{2+}) have been used to cross-link (PEDOT:PSS); however, excessive metal ions may result in osmosis and reduce the biocompatibility of the hydrogels.

To solve these crucial problems, our group developed a facile, efficient, and general method to synthesize CPHs by using a multi-functional-group acid to dope and cross-link CPs into hydrogels. The multi-functional doping acid *in situ* cross-linked the polymer and rendered the polymer sufficiently hydrophilic to retain enough water to form a hydrogel. The microstructure of the CPHs could be effectively tuned using varied reaction conditions. These CPHs can serve as a biosensor platform with attractive sensing performance such as wide linear ranges, high sensitivities, low sensing limits, and rapid response times to detect different metabolites, such as glucose, uric acid, cholesterol, and triglyceride.[23,24]

Earlier in the chapter, the gelation mechanism of PAni hydrogels is illustrated in Figure 3. We obtained PAni hydrogels with conductivity as high as 0.11 S cm^{-1}, which was higher than the typical conductivity range of 0.1–10 mS cm^{-1} of CPHs.[112,113] The hydrogel formed a continuous 3D porous nanostructure, which was constructed with dendritic nanofibers of uniform diameter (60–100 nm) (Figures 10(a)–10(c)). Such 3D interconnected foam-like structures could be more effective than wires and particles for electrochemical biosensors because of the large open channels of micro and nanoscale pores within the structures.[81] Due to the rigidity of the PAni chains and the interconnected frame structure, the 3D nanostructure matrix was stable and sustained with changes in its water content or the dedoping process against an ammonia solution. Moreover, the simple chemistry enabled us to shape CPHs into desired micropatterns by ink-jet printing or spray coating through stencil masks (Figure 5). Such micropatterning processability is highly desired for producing multiplex biosensor arrays.

The 3D hierarchically porous PAni hydrogel was employed as a glucose biosensor by immobilizing enzymes on the CPHs. The working mechanism of the biosensor was based on the redox properties of GOx, in which the enzymatic reaction of GOx and glucose was monitored via electrochemical measurement of the GOx-PAni hydrogel electrode.[114] As shown in Figures 10(d) and 10(e), the linear range of the GOx-PAni hydrogel-based biosensor to successive addition of glucose in PBS was 0.1–2.6 mM with a correlation coefficient as high as 0.9998. Upon the addition of increased glucose, the current rapidly increased within ~0.3 s with an average response time of ~1.1 s, which were comparable with the reported fastest response time of a biosensor made with a single PAni nanojunction (~0.2 s).[66] The sensitivity of the PAni hydrogel-based glucose biosensor was calculated to be 85.4 μA mM^{-1} cm^{-2}. The fast

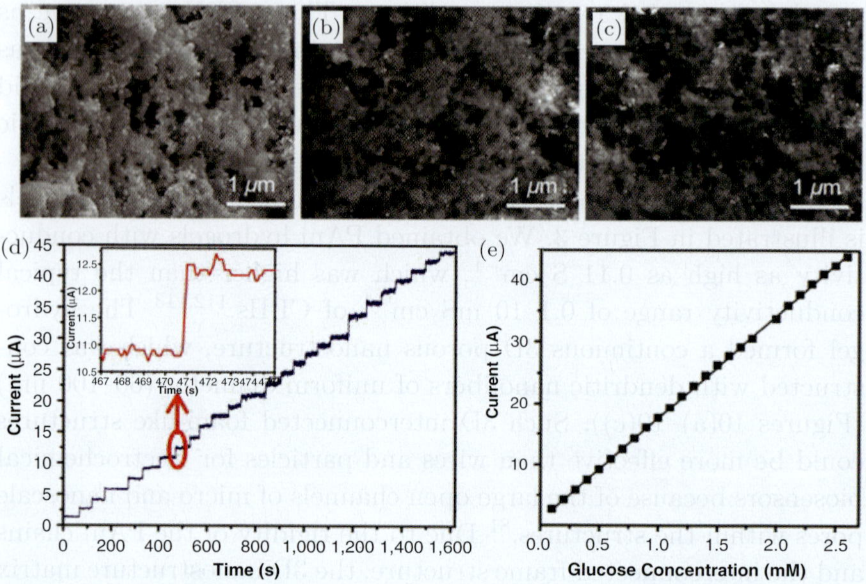

Figure 10. (a) SEM image of a piece of dehydrated hydrogel. (b) Higher magnification SEM image showing the interconnected network of dendritic PAni nanofibers. (c) TEM image showing the continuous nanostructured network of the dehydrated PAni hydrogel. The white arrows in D and E denote the micron-sized pores in PAni hydrogel. (d) Amperometric response of the GOx–PAni hydrogel-based enzymatic biosensor to successive addition of glucose in PBS (0.1 M). The Inset shows a magnification of the seventh addition of glucose. (e) The corresponding calibration plot showing a nearly linear amperometric response *vs.* glucose concentration. Reprinted with permission from Ref. 8. Copyright (2012) National Academy of Sciences of the United States of America.

response time and high sensitivity were mainly attributed to the relatively short diffusion path for molecular and electronic transport through the open channels of the hierarchical nanostructure and the continuously conductive framework of the PAni hydrogel.

The sensing performance of CPH-based biosensors can be improved by incorporating catalytic PtNPs.

For enzyme biosensors with electrochemical technology, the detection of H_2O_2 produced during the enzymatic reaction is a rapid, accurate, and efficient method.[115] However, CPH themselves do not have a catalytic effect to the electrochemical oxidization of H_2O_2 at the detection potential, thus limiting the sensing performance

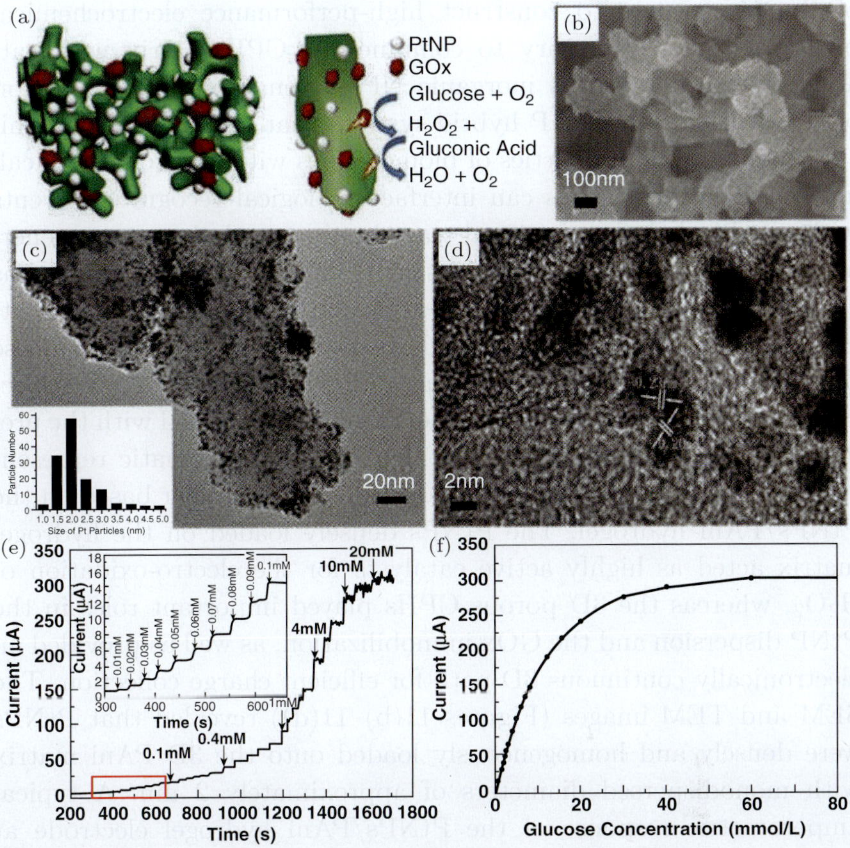

Figure 11. (a) Schematic representation of the 3D heterostructure of the PtNPs/PAni hydrogel (left) and the reaction mechanism of the glucose sensor based on the PtNPs/PAni hydrogel heterostructure (right). (b) SEM image of the PtNPs-loaded PAni hydrogel. (c) TEM image indicating the high density of PtNPs homogeneously loaded onto the PAni hydrogel matrix. The inset shows the statistical distribution of the PtNP sizes. (d) HR-TEM image of the PtNPs on the PAni hydrogel. (e) Amperometric response of the PtNPs/PAni hydrogel electrode after successive addition of glucose in 0.1 M PBS at an applied potential of 0.56 V. Inset: the magnified portion of the curve marked with red square. (f) The calibration curve for glucose concentrations from 1 mM to 80 mM. Reprinted with permission from Ref. 24. Copyright (2013) American Chemical Society.

of the biosensors. To construct high-performance electrochemical biosensors, it is necessary to combine the CPH matrix with catalytic materials such as inorganic NPs to enable electro-oxidation of H_2O_2. Biomolecule–NP hybrid systems that combine the recognition and catalytic properties of biomolecules with electronic, optical, and magnetic properties can interface biological recognition events with amplified electronic signal transduction, to design a new generation of biosensors with high sensitivity.[116] PtNPs could be used as electrochemical catalysts for the oxidation of H_2O_2 in amperometric biosensors.[117,118] We designed second generation CPH glucose biosensors based on PtNPs/PAni hydrogel heterostructured electrodes, which presented superior performance compared with the previously reported works.[23] Figure 11(a) shows a schematic representation of the reaction mechanism of a glucose biosensor based on the PtNPs/PAni hydrogel. The PtNPs densely loaded on the hydrogel matrix acted as highly active catalysts for the electro-oxidation of H_2O_2, whereas the 3D porous CPHs played important roles in the PtNP dispersion and the GOx immobilization, as well as provided an electronically continuous 3D path for efficient charge collection. The SEM and TEM images (Figures 11(b)–11(d)) revealed that PtNPs were densely and homogeneously loaded onto the 3D PAni matrix with monodispersed diameters of approximately 2 nm. A typical amperometric response of the PtNPs/PAni hydrogel electrode at 0.56 V with the successive addition of glucose is presented in Figure 11(e). The current immediately increased with increased glucose concentration, and it rapidly reached a steady state. Figure 11(f) shows that the steady-state current was linear up to the glucose concentration range of 0.01–8 mM with a correlation coefficient of 0.993. The sensitivity of this hydrogel-based sensor reached 96.1 μA mM^{-1} cm^{-2}, which was higher than previously reported values based on composites of Pt and PAni, PPy, or multi-walled carbon nanotubes (MWCNT).[119–121] The average response time of this glucose biosensor was approximately 3 s. The high sensitivity resulted from the catalytic effects of the homogeneously dispersed PtNPs on the hydrogel matrix for the oxidation of H_2O_2. The glucose sensor offered a low detection limit of 0.7 μM at a signal-to-noise ratio of

3, which was excellent compared with previous values of PAni-based glucose sensors.[122,123] Moreover, the glucose sensor showed almost no response to ascorbic acid (AA) or glutathione (GSH) and a negligible response to L-cysteine (Cys) as concentrations similar to the human physiological environment. Moreover, by using the PtNP-embedded PAni hydrogel, the expensive Pt electrode was no longer a necessary substrate for the electrochemical biosensors.

The PtNPs/PAni hydrogel sensor showed excellent performance with inexpensive glassy carbon (GC) electrodes even without using the Pt electrode. The electrocatalytic activity of the PtNPs/PAni hydrogel sensor for H_2O_2 oxidation increased up to three orders of magnitude compared with the GC + PAni electrode. The material cost (amount of Pt used) in future commercialization could be reduced by using a PtNPs/PAni hydrogel and other inexpensive substrates such as carbon/metal sheets/wires.

We developed a biosensor platform based on the PtNP/PAni hydrogel electrode, which could sense other metabolites such as uric acid, cholesterol and triglyceride. The estimation of metabolites, such as glucose, urea, cholesterol and triglycerides, in whole blood is of central importance in clinical diagnostics. The biosensor platform could be used as low-cost integration sensor kits for stringent personal control of blood metabolite concentration and avoid some high risks of several serious illnesses, for example, monitoring cholesterol and triglycerides to avoid cardiovascular diseases.[124–126] and monitoring uric acid levels to prevent gout.[127]

Our biosensors demonstrated excellent sensing performance to these metabolites with a wide linear range, high sensitivity, low sensing limit and rapid response time. The sensing mechanism of the proposed biosensor platform based on a PtNPs/CPHs/enzyme hybrid electrode is schematically illustrated in Figure 6. To enhance the immobilization of enzymes, we used a bifunctional compound, glutaraldehyde (GA), to cross-link the enzymes and the PAni hydrogel matrix through the formation of covalent C=N bonds. The SEM images of PtNPs/PAni hydrogel immobilized with enzymes are shown in Figures 12(a), 12(c) and 12(e), corresponding to the immobilization of uricase (UOx), cholesterol esterase/cholesterol oxidase

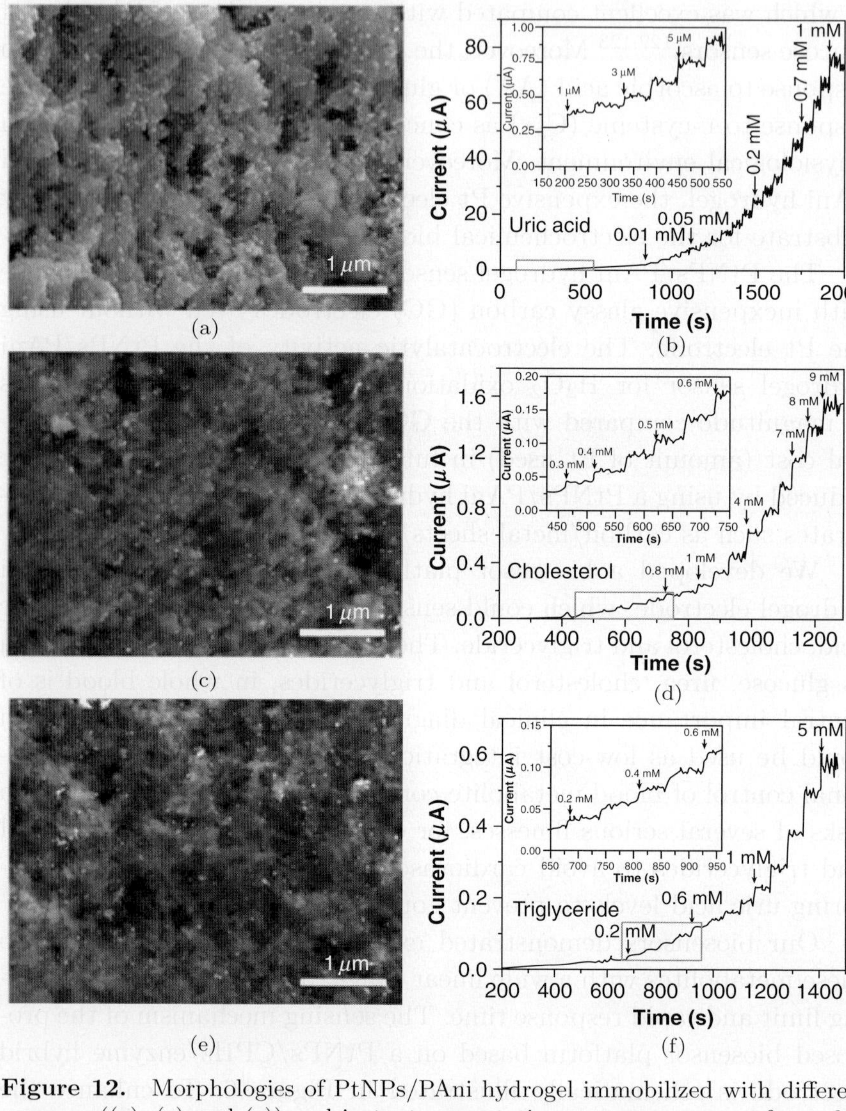

Figure 12. Morphologies of PtNPs/PAni hydrogel immobilized with different enzymes ((a), (c) and (e)) and instant current–time response curves of metabolites being successively added in PBS solutions ((b), (d) and (f)). ((a) and (b)) PtNPs/PAni hydrogel modified uric acid electrode. ((c) and (d)) PtNPs/PAni hydrogel modified cholesterol electrode. ((e) and (f)) PtNPs/PAni hydrogel modified triglyceride electrode. Reprinted with permission from. Ref. 23. Copyright (2015) American Chemical Society.

(ChEt/ChOx) and lipase/glycerol kinase/glycerol-3-phosphate oxidase (LIP/GK/GPO), respectively. Due to the good wetting of the enzyme solution on the hydrophilic hydrogel matrix, the enzymes could be homogeneously deposited with high density on the PAni 3D nanostructural thin film. The even and thin coating of enzymes on the PAni hydrogel matrix was critical for the sensing performance because of the decreased path length for molecules and electrons. Figures 12(b), 12(d) and 12(f) show the instant current–time response curves with successive additions of uric acid, cholesterol oleate and triolein into PBS solutions, respectively. The currents immediately increased upon the addition of the substrates and promptly reached a steady state with an average response time of ~3 s, which was much faster than previously reported sensors using different electrode materials.[106–108] For the detection of uric acid, our CPH-based biosensor demonstrated a wide linear range of 0.07–1 mM with a correlation coefficient of 0.994 and a lower detection limit of 1 μM ($S/N = 3$). The sensitivity of the uric acid sensor was calculated to be 322.8 μA mM^{-1} cm^{-2}, which was significantly higher than the sensitivities of the previously reported biosensors.[109,110] The linear relationship between the steady current and the concentration of the cholesterol biosensor was in the range of 0.3–9 mM, which was broader than the normal level in human serum. The correlation coefficient was calculated to be 0.997, and the sensitivity was 0.88 μA mM^{-1} cm^{-2} with a detection limit of 0.3 mM. The CPH-based sensor platform could detect triglyceride with a linear range from 0.2 to 5 mM, a correlation coefficient of 0.992, and a sensitivity of 0.57 μA mM^{-1} cm^{-2}. The detection limit of the triglyceride biosensor was calculated to be 0.2 mM. In addition, the easily oxidized species of urea, lactic acid and glucose with their normal concentration in human blood almost had no effect on the CPH-based enzyme biosensors.

8.3.5 Perspective and outlook

In portable electronics, the shape of the energy source is a particular limiting factor for the creation of practical and aesthetic devices.

The adoption of flexible and stretchable lithium-ion battery and supercapacitor would be a promising solution. Inspired by structure-derived mechanical properties and high scalability and processability, CPHs have been regarded as potential candidates for the fabrication of highly flexible and stretchable energy storage devices. To realize practical devices based on CPHs, two strategies would be future researching directions in this field. The first is to improve the mechanical strength and elasticity of CPHs by molecular modification or compositing with other materials. Although flexible supercapacitor based on PPy hydrogel prepared by interfacial synthesis has been realized, most CPHs suffer from mechanical brittleness and rigidity and may not bear large strain deformation, and large shape deformation. The introduction of polymer segments with high toughness or elasticity could potentially improve the strength and elasticity of CPHs and hybrid materials consisting of CPHs and a second reinforcing component could also modify the mechanical properties of CPHs. The second strategy is the development of printable flexible energy storage devices based on CPHs. Printing technologies which deposit electrode on flexible substrates are promising for the fabrication of flexible devices and have been booming all over the world because of their large-scale, good flexibility, and low-lost features.[128] Because CPHs have the ability to be processed by ink-jet printing or spray coating, future effort regarding printable energy-storage devices based on CPHs that can provide large-scale, cheap production processes and high flexibility is needed.

Hybridization of CPHs with other gel systems is another attractive direction which could potentially bring the hybrid hydrogel additional new features, such as stimuli-responsive property, self-healing property, and enhanced biocompatibility, thus opening the door to new functional energy storage devices such as self-healing and/or responsive devices. Though practical devices based on hybrid hydrogels have not been realized, some potentially useful materials have been synthesized and reported. For example, Yu *et al.*[27] developed a hybrid hydrogel based on PNIPAM and CPHs with a unique combination of high electrical conductivity, high thermo-responsive sensitivity, and good mechanical properties. The hybrid hydrogels cannot only maintain the unique features of CPHs, but also bring new properties such as enhanced mechanical property, stimuli-responsive and self-healing behaviors. By proper modification and device design, these hybrid hydrogels could play an increasingly important role in various fields such as energy storage, biomedical application and chemical sensors.

For the future development of biosensors based on CPHs, the combination of biosensors with other detection techniques, such as the wireless technology and implantable detection will be an important direction. Integrating with modules with ranging detecting and signal processing functions, the wearable CPHs-based biosensors may possibly be realized to be attached on the skin of human body and realize the wireless detection of desired signals without the invasive tests. To meet the requirements of precise detection with low sample quantity and simultaneous multiplex metabolic sensing, technologies including micropatterning, microfluidics, and measuring technologies should be combined to develop a novel design of CPHs biosensor devices, thus realizing miniature wearable multiplex biosensor.

8.4 Energy-Oriented Applications of Conducting Polymer Hydrogels

8.4.1 CPHs as an active material for fuel cells

Fuel cells are clean and efficient energy conversion devices that can directly convert the chemical energy of fuels to electricity through electrochemical processes. Fuel cells using hydrogen or methanol as fuels can be operated at close to room temperature and are considered promising power solutions for future electric vehicles and portable electronic devices. In fuel cells, the fuel is oxidized at the anode, and the released electrons are transferred by the driving circuit to the cathode, where O_2 is reduced.[129] The cathodic reaction is the oxygen reduction reaction (ORR), which has sluggish reaction kinetics and usually requires high loadings of Pt-based materials as electrocatalysts. Pt has the problems of scarcity, easy deactivation by CO poisoning, and the fuel crossover effect caused by undesired catalytic oxidation of leaked fuel at the cathode. The performance

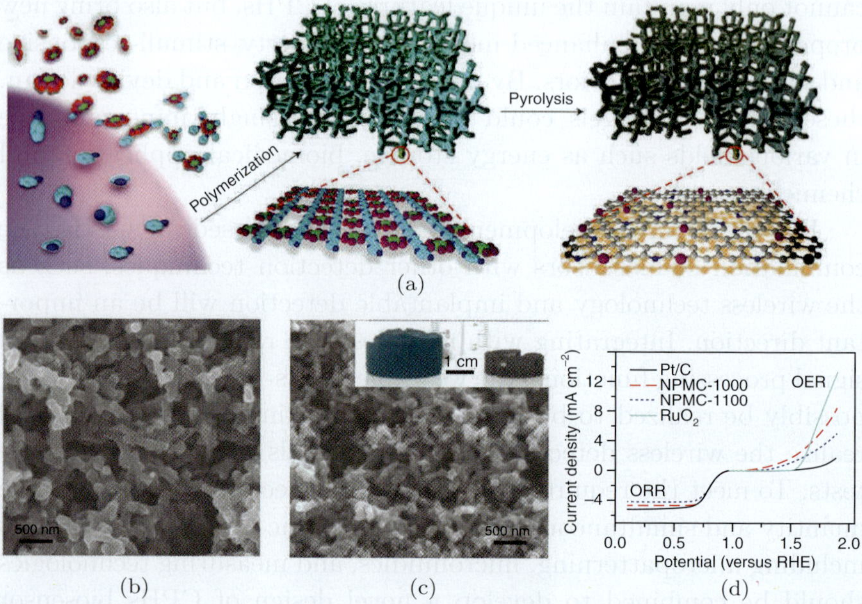

Figure 13. (a) Schematic illustration of the preparation process for the NPMC; (b),(c) SEM images of PANi CPHs; (d) LSV curves of NPMC-1000, NPMC-1100, RuO_2 and commercial Pt/C catalyst on an RDE (1,600 rpm.) in 0.1 M KOH (scan rate, 5 mV s^{-1}), showing the electrocatalytic activities towards both ORR and OER. Reprinted with permission from Ref. 131. Copyright (2015) Nature Publishing Group.

and costs of fuel cells therefore largely depend on the ORR electrocatalysts used.[130]

Dai *et al.* prepared 3D N- and P-codoped mesoporous nanocarbon (NPMC) foams, simply by pyrolysis of PANi CPHs. The resultant NPMCs show bifunctional catalytic activities towards ORR and oxygen evolution reaction (OER) which is the heart of metal–air batteries.[131] The good ORR catalytic activities for NPMCs were reflected by their onset potentials and higher currents (Figure 13).

A microbial fuel cell (MFC) or biological fuel cell is a bioelectrochemical system that drives a current by using bacteria and mimicking bacterial interactions found in nature (Figure 14).[132] As a promising wastewater treatment biotechnology and non-conventional energy resource, MFCs have attracted much attention in the past

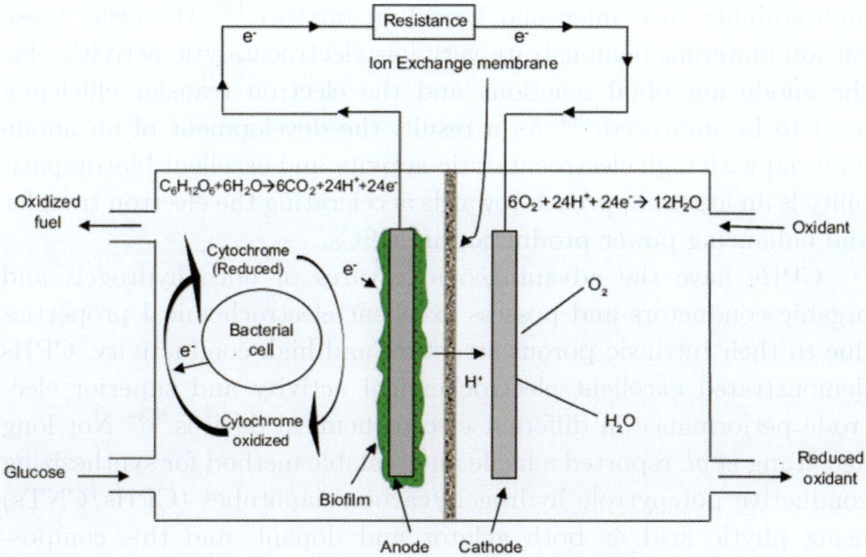

Figure 14. Schematic diagram of MFC.

decade.[133] Great efforts has been made in this area and this technology has demonstrated its potential for use in applications such as wastewater treatment, biochemical oxygen demand detection, energy production, chemical and fuel synthesis and bioremediation.[134–136] However, the commercial applications of MFCs are still far behind due to their relatively small power density. The extracellular electron transfer from bacteria to electrode, a key parameter that defines the theoretical limits of energy conversion, is currently a bottleneck in enhancing the power density of MFCs.[137] Therefore, great study should be made to facilitate the extracellular electron transfer in order to improve power generation efficiency in MFCs.

The anode material plays a crucial part in determining the power production of MFCs, because it not only influences the biofilm formation on the electrode surface, but also affects the extracellular electron transfer between the microorganisms and the electrode. To date, carbon materials, such as carbon paper, carbon cloth, and graphite felt, are used commonly as anodes in MFCs due to their excellent biocompatibility, high specific surface area, good conductivity, and high stability in a microbial inoculum mixture.[138] However, these carbon materials demonstrate very low electrocatalytic activities for the anode microbial reactions and the electron transfer efficiency need to be improved.[139] As a result, the development of an anode material with high electrocatalytic activity and excellent biocompatibility is an urgent approach towards accelerating the electron transfer and enhancing power production in MFCs.

CPHs have the advantageous features of both hydrogels and organic conductors and possess excellent electrochemical properties due to their intrinsic porous structure and high conductivity. CPHs demonstrated excellent electrochemical activity and superior electrode performance in different electrochemical devices.[8,38] Not long ago, Tang *et al.* reported a facile and scalable method for synthesizing conductive polypyrrole hydrogels/carbon nanotubes (CPHs/CNTs) using phytic acid as both gelator and dopant, and this composite was used as an anode in a dual-chamber MFC,[140] as shown in Figure 15. The high electrocatalytic activity of this material significantly reduced the interfacial charge transfer resistance and facilitated the extracellular electron transfer on the anode surface. The 3D porous structure and hydrophilicity of this composite enhanced the biofilm formation on the anode surface. CPHs/CNTs anode increased the maximum power density from 871 mW m^{-2} to 1,898 mW m^{-2} and exhibited high stability in the two-chambered MFC. These results demonstrated that the synthesis of the CPHs/CNTs composite offered an effective approach towards enhancing the power production in MFCs.

8.4.2 CPHs as electrode materials for electrochemical capacitors and lithium ion battery

Conductive polymers, such as PAni, PPy and their derivatives, have been widely investigated for electrochemical capacitor (EC) applications as pseudocapacitive materials, owing to their fast and reversible redox reactions at the electrode surface.[141] However, when used as bulk materials, conductive polymers suffer from a limited cycling stability due to structural change, which leads to the fast decay of

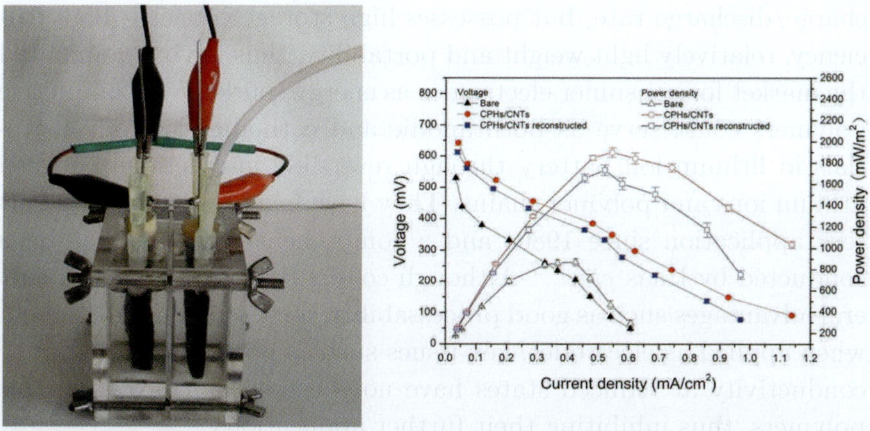

(a) (b)

Figure 15. (a) Image of the two-chambered MFC reactor; (b) Polarization and power density curves of the MFCs with the CPHs/CNTs anode and the bare graphite felt anode. Reproduced with permission from Ref. 140. Copyright 2015 Royal Society of Chemistry.

their electrochemical performance. CPHs are emerging as a promising class of materials for EC applications because their 3D hierarchical nanostructure can provide an excellent interface for electrolyte ions to access the electroactive surface of CPHs and an intrinsically conducting and robust framework that promotes the charge/mass transport, thus improving the electrochemical usage of active materials.

The 3D nanostructured PAni hydrogels have been demonstrated as high-performance EC electrodes recently.[8] The PAni hydrogel electrode shows remarkably small charge transfer resistance, suggesting favorable ion transport within the 3D continuous framework. The excellent electrochemical activity of this hydrogel-based electrodes was demonstrated by galvanostatic charge–discharge measurements in which the active PAni hydrogel materials yielded a specific capacitance of ~480 F g^{-1} at a current density of 0.2 A g^{-1}.

Lithium-ion battery is a class of rechargeable batteries in which lithium ions moves from the negative electrode to the positive electrode during discharging and back during charging. Compared to supercapacitor, lithium ion battery shows limitations on cycle life and charge/discharge rate, but possesses high storage capacity, high efficiency, relatively light weight and portability, thus having dominated the market for consumer electronics as energy sources.[142] Conductive polymers could serve as both anodic and cathodic electrode materials in lithium ion battery through reversible interactions between lithium ions and polymer chains. They have been researched for battery application since 1980s and a comprehensive review has been conducted by Haas *et al.*[17] Although conductive polymers show several advantages such as good processability, low cost, and light weight when applied as electrodes, key issues such as poor stability and low conductivity at reduced states have not been solved for conductive polymers, thus inhibiting their further applications.

CPHs has emerged as a suitable candidate due to their combination of properties such as large surface area, tunable mechanical property, high electrical conductivity, high porosity for ionic diffusion, and good compatibility with active materials.

Recently, a high-performance lithium ion battery based on anode consisting of Si NPs and PAni nanoframework was developed.[143] PAni hydrogel was *in situ* polymerized to form a bi-functional conformal coating, which bonded to the surface of Si particles via hydrogen bonding and served as a continuous 3D pathway for electronic conduction. The capacity of this Si nanoparticle–PAni composite electrode varies from 2,500 mAh g^{-1} to 1,100 mAh g^{-1}at charge/discharge rate from 0.3 to 3.0 A g^{-1} and retains ~91% after 5,000 cycles at a high current density of 6.0 A g^{-1}. The superior performance could be explained by following mechanisms. First, the porous hydrogel matrix accommodates the large volume expansion of the Si particles during lithiation process. Second, the conductive and continuous 3D framework of the CPH, as well as the conformal conductive coating surrounding Si particles, provides a good electrical connection to the particles. Third, the conducting hydrogels play multiple functions including coating, conductive binder and surface modification, thus reducing the weight fraction of the binder and improving the SEI interface between electrode and electrolyte. At last, the electrostatic interaction between positively charged PAni due to doping of phytic acid and negatively charged surface oxide could also contribute to the improved cycling performance.

8.4.3 Perspective and outlook

In portable electronics, the shape of the energy source is a particular limiting factor for the creation of practical and aesthetic devices. The adoption of flexible and stretchable lithium-ion battery and supercapacitor would be a promising solution. Inspired by structure-derived mechanical properties and high scalability and processability, CPHs have been regarded as potential candidates for the fabrication of highly flexible and stretchable energy storage devices. To realize practical devices based on CPHs, two strategies would be future research directions in this field. The first is to improve the mechanical strength and elasticity of CPHs by molecular modification or compositing with other materials. Although flexible supercapacitor based on PPy hydrogel prepared by interfacial synthesis has been realized, most CPHs suffer from mechanical brittleness and rigidity and may not bear large strain deformation, and large shape deformation. The introduction of polymer segments with high toughness or elasticity could potentially improve the strength and elasticity of CPHs and hybrid materials consisting of CPHs and a second reinforcing component could also modify the mechanical properties of CPHs. The second strategy is the development of printable flexible energy storage devices based on CPHs. Printing technologies which deposit electrode on flexible substrates are promising for the fabrication of flexible devices and has been booming all over the world because of their large-scale, good flexibility, and low-lost features. Because CPHs have the ability to be processed by ink-jet printing or spray coating, future effort regarding printable energy-storage devices based on CPHs that can provide large-scale, cheap production processes and high flexibility is needed.

8.5 Summary and Conclusion

CPHs synthesized by adopting doping acid as cross-linker exhibit advantageous features of high conductivity, large surface area, hierarchically porous structure, facile processability and mechanical flexibility, thus becoming a class of promising polymeric materials for a broad range of applications including energy storage devices and electrochemical biosensors.

CPHs have been explored as a powerful material platform for next-generation energy storage devices. Acting as active electrode material in supercapacitors, CPHs provide a robust and intrinsic conducting framework to promote the transport of charges, as well as large interfaces between the electrode and the electrolyte enabling the high rate capability. In lithium ion battery, CPHs can serve as functional binder materials to maintain mechanical and electrical integrity of electrodes and accommodate large volume change during charge/discharge processes. CPHs also represent a promising interface platform for biosensor electrodes. The hierarchical nanostructures of CPHs offer an open porous structure, a large surface area, and an organic matrix, which are favorable for enhancing the permeability of biosubstrates, increasing the interfacial area of electrodes, and immobilizing enzymes evenly and densely.

For future development, the performance of CPHs could be further improved by comprehensive utilization of advanced organic synthetic techniques, knowledge in interface chemistry, simulation/modeling studies, and state-of-the-art microscopic and spectroscopic techniques to delicate control over the electrical, electrochemical, mechanical and surface properties of CPHs, as well as studying their fundamental electrochemistry of and electrochemical dynamics. Moreover, composites based on CPHs with newly excited functionalities need to be exploited. The innovation of CPH-based devices requires interdisciplinary researches across chemistry, nanotechnology, biology, physics and electric engineering.

Acknowledgments

Lijia Pan and Yu Zhao are grateful for funding support from the Chinese National Key Fundamental Research Project (2013CB932900, 2011CB92210), the National Natural Science Foundation of China (61229401, 61076017, 41401257, 60990314), Natural Science

Foundation of Jiangsu Province of China (BK20141054) and the NCET and the PAPD Programs.

List of Abbreviations

AAO	anodic aluminum oxide
APS	ammonium persulphate
AuNPs	Au nanoparticles
CPs	conducting polymers
CPHs	conducting polymer hydrogels
ECM	extracellular matrix
FAD	flavin adenine dinucleotide
$FADH_2$	flavin adenine dinucleotide, reduced
GC	glassy carbon
GOx	glucose oxidase
GSH	glutathione
HRP	horseradish peroxidase
MFC	microbial fuel cell
NIPA	N-isopropylacrylamide
P4VPy	poly(4-vinylpyridine)
PA	polyacetylene
PAni	polyaniline
PPy	polypyrrole
PTh	polythiophene
PPV	poly(phenylenevinylene)
PNIPAM	poly(N-isopropylacrylamide)
PHEMA	poly(2-hydroxyethyl methacrylate)
PAA	polyacrylic acid
PEDOT	poly(3,4-ethylenedioxythiophene)
PAMPS	poly(2-acrylamido-2-methyl-1-propanesulfonic acid)
PAAM	polyacrylamide
PLGA	poly(lactide-co-glycolide)
PEGDGE	poly(ethylene glycol) diglycidyl ether
PSS	poly(styrenesulfonate)
PtNPs	platinum nanoparticles
UOx	uricase

References

1. A.S. Arico, P. Bruce, B. Scrosati, J.M. Tarascon, and W. van Schalkwijk (2005). Nanostructured materials for advanced energy conversion and storage devices. *Nature Materials* **4**, 366–377, doi:10.1038/nmat1368 (2005).
2. R.K. Joshi and J.J. Schneider (2012). Assembly of one dimensional inorganic nanostructures into functional 2D and 3D architectures. Synthesis, arrangement and functionality. *Chemical Society Reviews* **41**, 5285–5312, doi:10.1039/c2cs35089k.
3. D.R. Rolison, *et al.* (2009). Multifunctional 3D nanoarchitectures for energy storage and conversion. *Chemical Society Reviews* **38**, 226–252, doi:10.1039/b801151f (2009).
4. F. Bonaccorso, *et al.* 2D materials (2015). Graphene, related two-dimensional crystals, and hybrid systems for energy conversion and storage. *Science* **347**, 1246501, doi:10.1126/science.1246501.
5. Y. Shi, L. Peng, and G. Yu (2015). Nanostructured conducting polymer hydrogels for energy storage applications. *Nanoscale* **7**, 12796–12806, doi:10.1039/c5nr03403e.
6. N. Sahiner (2013). Soft and flexible hydrogel templates of different sizes and various functionalities for metal nanoparticle preparation and their use in catalysis. *Progress in Polymer Science* **38**, 1329–1356, doi:10.1016/j.progpolymsci.2013.06.004.
7. J. Kopecek (2002). Polymer chemistry: Swell gels. *Nature* **417**, 388–389, 391, doi:10.1038/417388a.
8. L. Pan, *et al.* (2012). Hierarchical nanostructured conducting polymer hydrogel with high electrochemical activity. *Proceedings National Academy of Science United States America* **109**, 9287–9292, doi:10.1073/pnas.1202636109.
9. J. Janata and M. Josowicz (2003). Conducting polymers in electronic chemical sensors. *Nature Material* **2**, 19–24, doi:10.1038/nmat768.
10. C. Li, H. Bai, and G. Shi (2009). Conducting polymer nanomaterials: electrosynthesis and applications. *Chemical Society Reviews* **38**, 2397–2409, doi:10.1039/b816681c.
11. G. Sabouraud, S. Sadki and N. Brodie (2000). The mechanisms of pyrrole electropolymerization. *Chemical Society Reviews* **29**, 283–293, doi:10.1039/a807124a.
12. A.J. Heeger (2001). Semiconducting and metallic polymers: The fourth generation of polymeric materials (Nobel Lecture). *Angewandte Chemie International Edition* **40**, 2591–2611, doi:10.1002/1521-3773(20010716)40:14<2591::aid-anie2591>3.0.co;2-0.
13. K. Lee, *et al.* (2006). Metallic transport in polyaniline. *Nature* **441**, 65–68, doi:10.1038/nature04705.
14. D. Li, J. Huang, and R.B. Kaner (2009). Polyaniline nanofibers: A unique polymer nanostructure for versatile applications. *Accounts of Chemical Research* **42**, 135–145, doi:10.1021/ar800080n.

15. H. Yoon, (2013). Current trends in sensors based on conducting polymer nanomaterials. *Nanomaterials* **3**, 524–549, doi:10.3390/nano3030524.
16. Y.-Z. Long, *et al.* (2011). Recent advances in synthesis, physical properties and applications of conducting polymer nanotubes and nanofibers. *Progress in Polymer Science* **36**, 1415–1442, doi:10.1016/j.progpolymsci.2011.04.001.
17. P. Novák, K. Müller, K.S.V. Santhanam and O. Haas (1997). Electrochemically active polymers for rechargeable batteries. *Chemical Reviews* **97**, 207–282, doi:10.1021/cr941181o.
18. Y. Zhao, B. Liu, L. Pan, and G. Yu (2013). 3D nanostructured conductive polymer hydrogels for high-performance electrochemical devices. *Energy and Environmental Science* **6**, 2856, doi:10.1039/c3ee40997j.
19. C.R. Martin (1995). Template Synthesis of Electronically Conductive Polymer Nanostructures. *Accounts of Chemical Research* **28**, 61–68, doi:10.1021/ar00050a002.
20. L. Xia, Z. Wei and M. Wan (2010). Conducting polymer nanostructures and their application in biosensors. *Journal Colloidal and Interface Science* **341**, 1–11, doi:10.1016/j.jcis.2009.09.029.
21. H.D. Tran, D. Li and R.B. Kaner (2009). One-dimensional conducting polymer nanostructures: Bulk synthesis and applications. *Advanced Materials* **21**, 1487–1499, doi:10.1002/adma.200802289.
22. L. Li, Y. Shi, L. Pan, Y. Shi and G. Yu (2015). Rational design and applications of conducting polymer hydrogels as electrochemical biosensors. *Journal of Materials Chemistry B* **3**, 2920–2930, doi:10.1039/c5tb00090d.
23. L. Li, *et al.* (2015). A nanostructured conductive hydrogels-based biosensor platform for human metabolite detection. *Nano Letters* **15**, 1146–1151, doi:10.1021/nl504217p.
24. D. Zhai, *et al.* (2013). Highly sensitive glucose sensor based on pt nanoparticle/polyaniline hydrogel heterostructures. *ACS Nano* **7**, 3540–3546, doi:10.1021/nn400482d.
25. Y. Wang, *et al.* (2014). Multifunctional superhydrophobic surfaces templated from innately microstructured hydrogel matrix. *Nano Letters* **14**, 4803–4809, doi:10.1021/nl5019782.
26. Z. Yin and Q. Zheng (2012). Controlled synthesis and energy applications of one-dimensional conducting polymer nanostructures: An overview. *Advanced Energy Materials* **2**, 179–218, doi:10.1002/aenm.201100560.
27. Y. Shi, C. Ma, L. Peng and G. Yu, (2015). Conductive "Smart" hybrid hydrogels with PNIPAM and nanostructured conductive polymers. *Advanced Functional Materials* **25**, 1219–1225, doi:10.1002/adfm.201404247.
28. G. Justin and A. Guiseppi-Elie (2009). Characterization of electroconductive blends of poly(HEMA-co-PEGMA-co-HMMA-co-SPMA) and poly(Py-co-PyBA). *Biomacromolecules* **10**, 2539–2549, doi:10.1021/bm900486d.
29. M.R. Abidian and D.C. Martin (2009). Multifunctional nanobiomaterials for neural interfaces. *Advanced Functional Materials* **19**, 573–585, doi:10.1002/adfm.200801473.

30. T. Dai, X. Qing, Y. Lu and Y. Xia (2009). Conducting hydrogels with enhanced mechanical strength. *Polymer* **50**, 5236–5241, doi:10.1016/j.polymer.2009.09.025.
31. S.K. Siddhanta and R. Gangopadhyay (2005). Conducting polymer gel: Formation of a novel semi-IPN from polyaniline and crosslinked poly(2-acrylamido-2-methyl propanesulphonicacid). *Polymer* **46**, 2993–3000, doi:10.1016/j.polymer.2005.01.084.
32. B.C. Kim, G.M. Spinks, G.G. Wallace and R. John (2000). Electroformation of conducting polymers in a hydrogel support matrix. *Polymer* **41**, 1783–1790, doi:10.1016/s0032-3861(99)00308-0.
33. M.R. Abidian, D.H. Kim and D.C. Martin (2006). Conducting-polymer nanotubes for controlled drug release. *Advanced Materials* **18**, 405–409, doi:10.1002/adma.200501726.
34. S. Brahim, D. Narinesingh and A. Guiseppi-Elie (2003). Synthesis and hydration properties of pH-sensitive p(HEMA)-based hydrogels containing 3-(trimethoxysilyl)propyl methacrylate. *Biomacromolecules* **4**, 497–503, doi:10.1021/bm020080u.
35. L. Pan, *et al.* (2014). An ultra-sensitive resistive pressure sensor based on hollow-sphere microstructure induced elasticity in conducting polymer film. *Nature Communications* **5**, 3002, doi:10.1038/ncomms4002.
36. L.Q. Mai, *et al.* (2011). Hierarchical MnMoO(4)/CoMoO(4) heterostructured nanowires with enhanced supercapacitor performance. *Nature Communications* **2**, 381, doi:10.1038/ncomms1387.
37. H. Zhang, X. Yu and P.V. Braun (2011). Three-dimensional bicontinuous ultrafast-charge and -discharge bulk battery electrodes. *Nature Nanotechnology* **6**, 277–281, doi:10.1038/nnano.2011.38.
38. Y. Shi, *et al.* (2014). Nanostructured conductive polypyrrole hydrogels as high-performance, flexible supercapacitor electrodes. *Journal of Materials Chemistry A* **2**, 6086, doi:10.1039/c4ta00484a.
39. S. Sekine, Y. Ido, T. Miyake, K. Nagamine and M. Nishizawa (2010). Conducting polymer electrodes printed on hydrogel. *Journal of American Chemistory Society* **132**, 13174–13175, doi:10.1021/ja1062357.
40. L.C. Clark and C. Lyons (1962). Electrode systems for continuous monitoring in cardiovascular surgery. *Annals of the New York Academy of Sciences* **102**, 29–45, doi:10.1111/j.1749-6632.1962.tb13623.x.
41. R.H. Farahi, A. Passian, L. Tetard and T. Thundat (2012). Critical issues in sensor science to aid food and water safety. *ACS Nano* **6**, 4548–4556, doi:10.1021/nn204999j.
42. M. Gerard (2002). Application of conducting polymers to biosensors. *Biosensors and Bioelectronics* **17**, 345–359, doi:10.1016/s0956-5663(01)00312-8.
43. J. Wang (2008). Electrochemical glucose biosensors. *Chemistry Reviews* **108**, 814–825, doi:10.1021/cr068123a.
44. P. Lin and F. Yan (2012). Organic thin-film transistors for chemical and biological sensing. *Advanced Materials* **24**, 34–51, doi:10.1002/adma.201103334.

45. S. Cosnier (1999). Biomolecule immobilization on electrode surfaces by entrapment or attachment to electrochemically polymerized films: A review. *Biosensors and Bioelectronics* **14**, 443–456, doi:10.1016/s0956-5663(99)00024-x.
46. E. Vargas, *et al.* (2013). Development of an integrated electrochemical biosensor for sucrose and its implementation in a continuous flow system for the simultaneous monitoring of sucrose, fructose and glucose. *Talanta* **105**, 93–100, doi:10.1016/j.talanta.2012.11.058.
47. M.D. Gouda, M.A. Kumar, M.S. Thakur and N.G. Karanth (2002). Enhancement of operational stability of an enzyme biosensor for glucose and sucrose using protein based stabilizing agents. *Biosensors and Bioelectronics* **17**, 503–507, doi:10.1016/s0956-5663(02)00021-0.
48. D.R. Thévenot, K. Toth, R.A. Durst and G.S. Wilson (2001). Electrochemical biosensors: Recommended definitions and classification1International Union of Pure and Applied Chemistry: Physical Chemistry Division, Commission I.7 (Biophysical Chemistry); Analytical Chemistry Division, Commission V.5 (Electroanalytical Chemistry).1. *Biosensors and Bioelectronics* **16**, 121–131, doi:10.1016/s0956-5663(01)00115-4.
49. M. Muskovich and C.J. Bettinger (2012). Biomaterials-based electronics: Polymers and interfaces for biology and medicine. *Advanced Healthcare Materials* **1**, 248–266, doi:10.1002/adhm.201200071.
50. A.A. Saei, J.E.N. Dolatabadi, P. Najafi-Marandi, A. Abhari and M. de la Guardia (2013). Electrochemical biosensors for glucose based on metal nanoparticles. *TrAC Trends in Analytical Chemistry* **42**, 216–227, doi:10.1016/j.trac.2012.09.011.
51. U. Lange, N.V. Roznyatovskaya and V.M. Mirsky (2008). Conducting polymers in chemical sensors and arrays. *Anal Chim Acta* **614**, 1–26, doi:10.1016/j.aca.2008.02.068.
52. X. Lu, W. Zhang, C. Wang, T.-C. Wen and Y. Wei (2011). One-dimensional conducting polymer nanocomposites: Synthesis, properties and applications. *Progress in Polymer Science* **36**, 671–712, doi:10.1016/j.progpolymsci.2010.07.010.
53. N. Annabi, *et al.* (2014). 25th Anniversary Article: Rational Design and Applications of Hydrogels in Regenerative Medicine. *Advanced Materials* **26**, 85–124, doi:10.1002/adma.201303233.
54. P.T. Charles, *et al.* (2004). Fabrication and characterization of 3D hydrogel microarrays to measure antigenicity and antibody functionality for biosensor applications. *Biosens Bioelectron* **20**, 753–764, doi:10.1016/j.bios.2004.04.007.
55. B.V. Slaughter, S.S. Khurshid, O.Z. Fisher, A. Khademhosseini and N.A. Peppas (2009). Hydrogels in regenerative medicine. *Advanced Materials* **21**, 3307–3329, doi:10.1002/adma.200802106.
56. H. Ding, *et al.* (2014). Biologically derived soft conducting hydrogels using heparin-doped polymer networks. *ACS Nano* **8**, 4348–4357, doi:10.1021/nn406019m.

57. B.C. Kim, G. Spinks, C.O. Too, G.G. Wallace and Y.H. Bae (2000). Preparation and characterisation of processable conducting polymer–hydrogel composites. *Reactive and Functional Polymers* **44**, 31–40, doi:10.1016/s1381-5148(99)00074-7.
58. J.D. Stenger-Smith (1998). Intrinsically electrically conducting polymers. Synthesis, characterization, and their applications. *Progress in Polymer Science* **23**, 57–79, doi:10.1016/s0079-6700(97)00024-5.
59. S. Nambiar and J.T. Yeow (2011). Conductive polymer-based sensors for biomedical applications. *Biosens Bioelectron* **26**, 1825–1832, doi:10.1016/j.bios.2010.09.046.
60. D.H. Kim, S.M. Richardson-Burns, J.L. Hendricks, C. Sequera and D.C. Martin (2007). Effect of immobilized nerve growth factor on conductive polymers: Electrical properties and cellular response. *Advanced Functional Materials* **17**, 79–86, doi:10.1002/adfm.200500594.
61. G.G. Wallace, S.E. Moulton and G.M. Clark, (2009). Applied physics: Electrode-cellular interface. *Science* **324**, 185–186, doi:10.1126/science.1168346.
62. S. Kirchmeyer and K. Reuter (2005). Scientific importance, properties and growing applications of poly(3,4-ethylenedioxythiophene). *Journal of Materials Chemistry* **15**, 2077, doi:10.1039/b417803n.
63. X. Cui (2003). In vivo studies of polypyrrole/peptide coated neural probes. *Biomaterials* **24**, 777–787, doi:10.1016/s0142-9612(02)00415-5.
64. J. Huang, S. Virji, B.H. Weiller and R.B. Kaner (2003). Polyaniline nanofibers: Facile synthesis and chemical sensors. *Journal of American Chemistry Society* **125**, 314–315, doi:10.1021/ja028371y.
65. Y. Xiao, *et al.* (2004). Electrochemical polymerization of poly (hydroxymethylated-3,4-ethylenedioxythiophene) (PEDOT-MeOH) on multichannel neural probes. *Sensors and Actuators B: Chemical* **99**, 437–443, doi:10.1016/j.snb.2003.12.067.
66. E.S. Forzani, *et al.* (2004). A conducting polymer nanojunction sensor for glucose detection. *Nano Letters* **4**, 1785–1788, doi:10.1021/nl049080l.
67. Z. Wang, S. Liu, P. Wu and C. Cai (2009). Detection of glucose based on direct electron transfer reaction of glucose oxidase immobilized on highly ordered polyaniline nanotubes. *Analytical Chemistry* **81**, 1638–1645, doi:10.1021/ac802421h.
68. T. Ruzgas, E. Csöregi, J. Emnéus, L. Gorton and G. Marko-Varga (1996). Peroxidase-modified electrodes: Fundamentals and application. *Analytica Chimica Acta* **330**, 123–138, doi:10.1016/0003-2670(96)00169-9.
69. Y. Xian, *et al.* (2006). Glucose biosensor based on Au nanoparticles-conductive polyaniline nanocomposite. *Biosens Bioelectron* **21**, 1996–2000, doi:10.1016/j.bios.2005.09.014.
70. N.K. Guimard, N. Gomez and C.E. Schmidt (2007). Conducting polymers in biomedical engineering. *Progress in Polymer Science* **32**, 876–921, doi:10.1016/j.progpolymsci.2007.05.012.
71. A. Singh and N.A. Peppas (2014). Hydrogels and scaffolds for immunomodulation. *Advanced Materials* **26**, 6530–6541, doi:10.1002/adma.201402105.

72. J.P. Gong (2014). Materials science. Materials both tough and soft. *Science* **344**, 161–162, doi:10.1126/science.1252389.
73. J.E. Elliott, M. Macdonald, J. Nie and C.N. Bowman (2004). Structure and swelling of poly(acrylic acid) hydrogels: Effect of pH, ionic strength, and dilution on the cross-linked polymer structure. *Polymer* **45**, 1503–1510, doi:10.1016/j.polymer.2003.12.040.
74. M.A. Krepker and E. Segal (2013). Dual-functionalized porous Si/hydrogel hybrid for label-free biosensing of organophosphorus compounds. *Analytical Chemistry* **85**, 7353–7360, doi:10.1021/ac4011815.
75. O.Z. Fisher, A. Khademhosseini, R. Langer and N.A. Peppas (2010). Bioinspired materials for controlling stem cell fate. *Accounts of Chemical Research* **43**, 419–428, doi:10.1021/ar900226q.
76. T. Miyata, T. Uragami and K. Nakamae (2002). Biomolecule-sensitive hydrogels. *Advanced Drug Delivery Reviews* **54**, 79–98, doi:10.1016/s0169-409x(01)00241-1.
77. A.K. Gaharwar, N.A. Peppas and A. Khademhosseini (2014). Nanocomposite hydrogels for biomedical applications. *Biotechnology Bioengineering* **111**, 441–453, doi:10.1002/bit.25160.
78. A. Guiseppi-Elie (2010). Electroconductive hydrogels: synthesis, characterization and biomedical applications. *Biomaterials* **31**, 2701–2716, doi:10.1016/j.biomaterials.2009.12.052.
79. W. Yang, H. Xue, L.R. Carr, J. Wang and S. Jiang (2011). Zwitterionic poly(carboxybetaine) hydrogels for glucose biosensors in complex media. *Biosens Bioelectron* **26**, 2454–2459, doi:10.1016/j.bios.2010.10.031.
80. R. Wang and Y. Li (2013). Hydrogel based QCM aptasensor for detection of avian influenza virus. *Biosens Bioelectron* **42**, 148–155, doi:10.1016/j.bios.2012.10.038.
81. S. Brahim, D. Narinesingh and A. Guiseppi-Elie (2002). Bio-smart hydrogels: co-joined molecular recognition and signal transduction in biosensor fabrication and drug delivery. *Biosensors and Bioelectronics* **17**, 973–981, doi:10.1016/s0956-5663(02)00089-1.
82. N.A. Peppas, Y. Huang, M. Torres-Lugo, J.H. Ward and J. Zhang (2000). Physicochemical foundations and structural design of hydrogels in medicine and biology. *Annual Reviews Biomed Eng* **2**, 9–29, doi:10.1146/annurev.bioeng.2.1.9.
83. N. Peppas (2000). Hydrogels in pharmaceutical formulations. *European Journal of Pharmaceutics and Biopharmaceutics* **50**, 27–46, doi:10.1016/s0939-6411(00)00090-4.
84. M.E. Byrne, K. Park and N.A. Peppas (2002). Molecular imprinting within hydrogels. *Advanced Drug Delivery Reviews* **54**, 149–161, doi:10.1016/s0169-409x(01)00246-0.
85. S.J. Kim, *et al.* (2006). Surprising shrinkage of expanding gels under an external load. *Nature Materials* **5**, 48–51, doi:10.1038/nmat1553.
86. M. Caldorera-Moore and N.A. Peppas (2009). Micro- and nanotechnologies for intelligent and responsive biomaterial-based medical systems. *Adv Drug Deliv Rev* **61**, 1391–1401, doi:10.1016/j.addr.2009.09.002.

87. K.M. Wood, G.M. Stone and N.A. Peppas (2008). Wheat germ agglutinin functionalized complexation hydrogels for oral insulin delivery. *Biomacromolecules* **9**, 1293–1298, doi:10.1021/bm701274p.
88. H.F. Liang, *et al.* (2004). Novel method using a temperature-sensitive polymer (methylcellulose) to thermally gel aqueous alginate as a pH-sensitive hydrogel. *Biomacromolecules* **5**, 1917–1925, doi:10.1021/bm049813w.
89. M.P. Lutolf, *et al.* (2003). Synthetic matrix metalloproteinase-sensitive hydrogels for the conduction of tissue regeneration: Engineering cell-invasion characteristics. *Proceedings of National Academy Science USA* **100**, 5413–5418, doi:10.1073/pnas.0737381100.
90. K. Podual, F.J. Doyle and N.A. Peppas (2000). Glucose-sensitivity of glucose oxidase-containing cationic copolymer hydrogels having poly(ethylene glycol) grafts. *Journal of Controlled Release* **67**, 9–17, doi:10.1016/s0168-3659(00)00195-4.
91. T.G. Farmer, Jr., T.F. Edgar and N.A. Peppas (2008). In Vivo simulations of the intravenous dynamics of submicron particles of pH-responsive cationic hydrogels in diabetic patients. *Ind Eng Chem Res* **47**, 10053–10063, doi:10.1021/ie070957b.
92. M. Toma, U. Jonas, A. Mateescu, W. Knoll and J. Dostalek (2013). Active control of SPR by thermoresponsive hydrogels for biosensor applications. *Journal of Physical Chemistry C Nanomaterials Interfaces* **117**, 11705–11712, doi:10.1021/jp400255u.
93. S. Brahim, D. Narinesingh and A. Guiseppi-Elie (2002). Polypyrrole-hydrogel composites for the construction of clinically important biosensors. *Biosensors and Bioelectronics* **17**, 53–59, doi:10.1016/s0956-5663(01)00262-7.
94. D. Ivekovic, S. Milardovic and B.S. Grabaric (2004). Palladium hexacyanoferrate hydrogel as a novel and simple enzyme immobilization matrix for amperometric biosensors. *Biosensors and Bioelectronics* **20**, 872–878, doi:10.1016/j.bios.2004.04.001.
95. X.L. Luo, J.J. Xu, Q. Zhang, G.J. Yang and H.Y. Chen (2005). Electrochemically deposited chitosan hydrogel for horseradish peroxidase immobilization through gold nanoparticles self-assembly. *Biosensors and Bioelectronics* **21**, 190–196, doi:10.1016/j.bios.2004.07.029.
96. A. Kowalczyk, *et al.* (2014). Hydrogel with chains functionalized with carboxyl groups as universal 3D platform in DNA biosensors. *Biosensors and Bioelectronics* **54**, 222–228, doi:10.1016/j.bios.2013.11.017.
97. F. Qu, Y. Zhang, A. Rasooly and M. Yang (2014). Electrochemical biosensing platform using hydrogel prepared from ferrocene modified amino acid as highly efficient immobilization matrix. *Analytical Chemistry* **86**, 973–976, doi:10.1021/ac403478z.
98. J. Zhou, C. Liao, L. Zhang, Q. Wang and Y. Tian (2014). Molecular hydrogel-stabilized enzyme with facilitated electron transfer for determination of H2O2 released from live cells. *Analytical Chemistry* **86**, 4395–4401, doi:10.1021/ac500231e.

99. T. Nyberg, O. Inganäs H. Jerregård (2002), Polymer hydrogel microelectrodes for neural communications. *Biomedical Microdevices* **4**, 43–52, doi:10.1023/a:1014219828983.
100. A. Heller (2006). Electron-conducting redox hydrogels: Design, characteristics and synthesis. *Current Opinion Chemistry Biology* **10**, 664–672, doi:10.1016/j.cbpa.2006.09.018.
101. P. Åsberg and O. Inganäs (2003). Hydrogels of a conducting conjugated polymer as 3-D enzyme electrode. *Biosensors and Bioelectronics* **19**, 199–207, doi:10.1016/s0956-5663(03)00220-3.
102. B. Guo, A. Finne-Wistrand and A.-C.Albertsson (2011). Degradable and electroactive hydrogels with tunable electrical conductivity and swelling behavior. *Chemistry of Materials* **23**, 1254–1262, doi:10.1021/cm103498s.
103. F. Mao, N. Mano and A. Heller (2003). Long tethers binding redox centers to polymer backbones enhance electron transport in enzyme "Wiring" hydrogels. *Journal of American Chemical Society* **125**, 4951–4957, doi:10.1021/ja029510e.
104. N. Mano, J.E. Yoo, J. Tarver, Y.L. Loo and A. Heller (2007). An electron-conducting cross-linked polyaniline-based redox hydrogel, formed in one step at pH 7.2, wires glucose oxidase. *Journal American Chemical Society* **129**, 7006–7007, doi:10.1021/ja071946c.
105. G. Horányi and G. Inzelt (1988). Anion-involvement in electrochemical transformations of polyaniline. A radiotracer study. *Electrochimica Acta* **33**, 947–952, doi:10.1016/0013-4686(88)80093-8.
106. S. Brahim, A.M. Wilson, D. Narinesingh, E. Iwuoha and A. Guiseppi-Elie (2003). Chemical and biological sensors based on electrochemical detection using conducting electroactive polymers. *Microchimica Acta* **143**, 123–137, doi:10.1007/s00604-003-0065-6.
107. D.P. Åsberg and O. Inganäs (2003). PEDOT/PSS hydrogel networks as 3-D enzyme electrodes. *Synthetic Metals* **137**, 1403–1404, doi:10.1016/s0379-6779(02)01060-3.
108. S. Ghosh and O. Inganäs (2000). Electrochemical characterization of poly(3,4-ethylene dioxythiophene) based conducting hydrogel networks. *Journal of The Electrochemical Society* **147**, 1872, doi:10.1149/1.1393450.
109. L. Kergoat, *et al.* (2014). Detection of glutamate and acetylcholine with organic electrochemical transistors based on conducting polymer/platinum nanoparticle composites. *Advanced Materials* **26**, 5658–5664, doi:10.1002/adma.201401608.
110. S. Ghosh and O. Inganäs (1999). Conducting polymer hydrogels as 3D electrodes: Applications for supercapacitors. *Advanced Materials* **11**, 1214-1218, doi:10.1002/(sici)1521-4095(199910)11:14<1214::aid-adma1214>3.0.co;2-3.
111. R.A. Green, S. Baek, L.A. Poole-Warren and P.J. Martens (2010). Conducting polymer-hydrogels for medical electrode applications. *Science Technology Advanced Materials* **11**, 014107, doi:10.1088/1468-6996/11/1/014107.
112. L. Chen, Kim, M. Nishino, J.P. Gong and Y. Osada, (2000). Environmental Responses of Polythiophene Hydrogels. *Macromolecules* **33**, 1232–1236, doi:10.1021/ma990923i.

113. Y. Xia and H. Zhu (2011). Polyaniline nanofiber-reinforced conducting hydrogel with unique pH-sensitivity. *Soft Matter* **7**, 9388, doi:10.1039/c1sm05890h.
114. C. Dhand, M. Das, M. Datta and B.D. Malhotra (2011). Recent advances in polyaniline based biosensors. *Biosens Bioelectron* **26**, 2811–2821, doi:10.1016/j.bios.2010.10.017.
115. R.S. Dey and C.R. Raj (2010). Development of an amperometric cholesterol biosensor based on graphene-pt nanoparticle hybrid material. *The Journal of Physical Chemistry C* **114**, 21427–21433, doi:10.1021/jp105895a.
116. S. Guo and S. Dong (2009). Biomolecule-nanoparticle hybrids for electrochemical biosensors. *TrAC Trends in Analytical Chemistry* **28**, 96–109, doi:10.1016/j.trac.2008.10.014.
117. P. Karam and L.I. Halaoui (2008). Sensing of H_2O_2 at low surface density assemblies of Pt nanoparticles in polyelectrolyte. *Analytical Chemistry* **80**, 5441–5448, doi:10.1021/ac702358d.
118. M. Yang, Y. Yang, H. Yang, G. Shen and R. Yu (2006). Layer-by-layer self-assembled multilayer films of carbon nanotubes and platinum nanoparticles with polyelectrolyte for the fabrication of biosensors. *Biomaterials* **27**, 246-255, doi:10.1016/j.biomaterials.2005.05.077.
119. H. Xue, Z. Shen and C. Li (2005). Improved selectivity and stability of glucose biosensor based on in situ electropolymerized polyaniline-polyacrylonitrile composite film. *Biosensors and Bioelectronics* **20**, 2330–2334, doi:10.1016/j.bios.2004.07.018.
120. E.M. Ekanayake, D.M. Preethichandra and K. Kaneto (2007). Polypyrrole nanotube array sensor for enhanced adsorption of glucose oxidase in glucose biosensors. *Biosensors and Bioelectronics* **23**, 107–113, doi:10.1016/j.bios.2007.03.022.
121. H. Zhong, *et al.* (2011). In situ chemo-synthesized multi-wall carbon nanotube-conductive polyaniline nanocomposites: Characterization and application for a glucose amperometric biosensor. *Talanta* **85**, 104–111, doi:10.1016/j.talanta.2011.03.040.
122. Y. Liu, X. Feng, J. Shen, J.J. Zhu and W. Hou (2008). Fabrication of a novel glucose biosensor based on a highly electroactive polystyrene/polyaniline/Au nanocomposite. *Journal of Physical Chemistry B* **112**, 9237–9242, doi:10.1021/jp801938w.
123. Y.-Y. Horng, *et al.* (2009). Direct-growth of polyaniline nanowires for enzyme-immobilization and glucose detection. *Electrochemistry Communications* **11**, 850–853, doi:10.1016/j.elecom.2009.02.010.
124. S.K. Arya, P. Pandey, S.P. Singh, M. Datta and B.D. Malhotra (2007). Dithiobissuccinimidyl propionate self assembled monolayer based cholesterol biosensor. *Analyst* **132**, 1005–1009, doi:10.1039/b707000d.
125. J. Yang, H. Lee, M. Cho, J. Nam and Y. Lee (2012). Nonenzymatic cholesterol sensor based on spontaneous deposition of platinum nanoparticles on layer-by-layer assembled CNT thin film. *Sensors and Actuators B: Chemical* **171–172**, 374–379, doi:10.1016/j.snb.2012.04.070.

126. A. Phongphut, *et al.* (2013). A disposable amperometric biosensor based on inkjet-printed Au/PEDOT-PSS nanocomposite for triglyceride determination. *Sensors and Actuators B: Chemical* **178**, 501–507, doi:10.1016/j.snb.2013.01.012.
127. Y. Zhang, *et al.* (2007). Development and analytical application of an uric acid biosensor using an uricase-immobilized eggshell membrane. *Biosens Bioelectron* **22**, 1791–1797, doi:10.1016/j.bios.2006.08.038.
128. X. Wang, *et al.* (2014). Flexible energy-storage devices: design consideration and recent progress. *Advanced Materials* **26**, 4763–4782, doi:10.1002/adma.201400910.
129. M. Winter and R.J. Brodd (2004). What are batteries, fuel cells, and supercapacitors? *Chemical Reviews* **104**, 4245–4270, doi:10.1021/cr020730k.
130. L. Yang, *et al.* (2013). A mini review on carbon-based metal-free electrocatalysts for oxygen reduction reaction. *Chinese Journal of Catalysis* **34**, 1986–1991, doi:10.1016/s1872-2067(12)60713-x.
131. J. Zhang, Z. Zhao, Z. Xia and L. Dai (2015). A metal-free bifunctional electrocatalyst for oxygen reduction and oxygen evolution reactions. *Nature Nanotechnology* **10**, 444–452, doi:10.1038/nnano.2015.48.
132. H. Liu, R. Ramnarayanan and B.E. Logan (2004). Production of electricity during wastewater treatment using a single chamber microbial fuel cell. *Environmental Science and Technology* **38**, 2281–2285, doi:10.1021/es034923g.
133. Z. Du, H. Li and T.A. Gu (2007). state of the art review on microbial fuel cells: A promising technology for wastewater treatment and bioenergy. *Biotechnology Advances* **25**, 464–482, doi:10.1016/j.biotechadv.2007.05.004.
134. Y. Huang, *et al.* (2012). Electricity generation from a floating microbial fuel cell. *Bioresources and Technology* **114**, 308–313, doi:10.1016/j.biortech.2012.02.142.
135. I.S. Chang, H. Moon, J.K. Jang and B.H. Kim (2005). Improvement of a microbial fuel cell performance as a BOD sensor using respiratory inhibitors. *Biosensors and Bioelectronics* **20**, 1856–1859, doi:10.1016/j.bios.2004.06.003.
136. K.B. Gregory and D.R. Lovley (2005). Remediation and recovery of uranium from contaminated subsurface environments with electrodes. *Environmental Science and Technology* **39**, 8943–8947, doi:10.1021/es050457e.
137. K. Rabaey, N. Boon, M. Höfte and W. Verstraete (2005). Microbial phenazine production enhances electron transfer in biofuel cells. *Environmental Science and Technology* **39**, 3401–3408, doi:10.1021/es048563o.
138. M. Zhou, M. Chi, J. Luo, H. He and T. Jin (2011). An overview of electrode materials in microbial fuel cells. *Journal of Power Sources* **196**, 4427–4435, doi:10.1016/j.jpowsour.2011.01.012.
139. Y. Qiao, C.M. Li, S.-J. Bao and Q.-L. Bao (2007) Carbon nanotube/polyaniline composite as anode material for microbial fuel cells. *Journal of Power Sources* **170**, 79–84, doi:10.1016/j.jpowsour.2007.03.048.
140. X. Tang, H. Li, Z. Du, W. Wang and H.Y. Ng (2015). Conductive polypyrrole hydrogels and carbon nanotubes composite as an anode for microbial fuel cells. *RSC Advances* **5**, 50968–50974, doi:10.1039/c5ra06064h.

141. A. Rudge, J. Davey, I. Raistrick, S. Gottesfeld and J.P. Ferraris (1994). Conducting polymers as active materials in electrochemical capacitors. *Journal of Power Sources* **47**, 89–107, doi:10.1016/0378-7753(94)80053-7.
142. A. Yoshino (2012). The birth of the lithium-ion battery, *Angewandte Chemie International Edition in English* **51**, 5798–5800, doi:10.1002/anie.201105006.
143. H. Wu, *et al.* (2013), Stable Li-ion battery anodes by in-situ polymerization of conducting hydrogel to conformally coat silicon nanoparticles. *Nature Communication* **4**, 1943, doi:10.1038/ncomms2941.

Chapter 9

Nanocarbon-Based Enzymatic Electrodes

Nicolas Brun[*,§], Mohamed Baccour[*], and Victoria Flexer[†,‡]

[*]*Institut Charles Gerhardt de Montpellier,*
UMR 5253 CNRS, ENSCM,
Université de Montpellier, France
[†]*Centro de Investigación y Desarrollo*
en Materiales Avanzados y Almacenamiento
de Energía de Jujuy-CIDMEJu
(CONICET-Universidad Nacional de Jujuy), Argentina
[‡]*Centro de Investigación y Transferencia Jujuy,*
CIT-JUJUY, Argentina
[§]*nicolas.brun@enscm.fr*

9.1 Carbon Nanotubes: Key Materials for the Design of High-performance Bioelectrodes

Due to versatility, stability, abundance, and relatively low cost, carbonaceous substrates have been extremely popular as far as catalysis, energy storage, and conversion devices are concerned. Bioelectrochemical systems did not contravene such interest, justifying why carbonaceous materials are widely addressed in this book. The diversity of carbon allotropes (Figure 1) — *e.g.* diamond, graphite, fullerene, carbon black, glassy carbon — provides a flexible toolbox for the design of functional electrodes. Among them, nanocarbons have been the focus of considerable research efforts for the

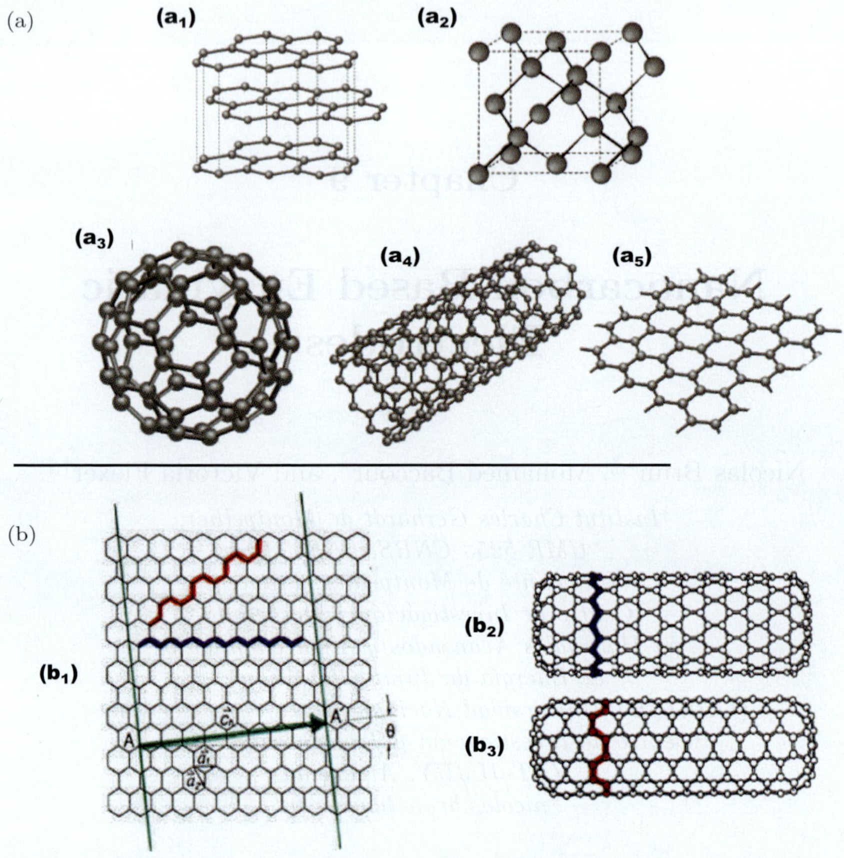

Figure 1. (a) Structure of some carbon allotropes: (a_1) graphite, (a_2) diamond, (a_3) fullerene (C60), (a_4) SWCNT and (a_5) graphene. (b) Graphene network (b_1) before and (b_2 and b_3) after being rolled-up. a_1 and a_2 are the lattice vectors. The chiral vector, $C = na_1 + ma_2$, represents a possible wrapping of the 2D graphene sheet into a tubular form. The direction perpendicular to C is the tube axis while θ is the chiral angle. Rolled-up graphene sheets in (b_2) a zig-zag and (b_3) an armchair topological structure. Reproduced with permission from Ref. 12. All rights reserved. ©IOP Publishing.

last decades, greatly boosted by the successive discoveries of nanodiamond,[1] buckminsterfullerene,[2] carbon nanotube,[3,4] and more recently single-layer graphene (SLG).[5] Like nanomaterials, nanocarbons are defined as carbon materials with at least one external dimension in the size range between 1 and 100 nm. One may distinguish between 0D (*e.g.* buckminsterfullerene), 1D (*e.g.* carbon nanotube) and 2D (*e.g.* graphene) nanocarbons, which display zero, one and two dimensions larger than 100 nm respectively. Taking advantage of an exceptional electrical conductivity, a high theoretical external specific surface area (*i.e.* 1,315 m^2 g^{-1}),[6] singular mechanical properties[7] and a relative — even though controversial[8] — biocompatibility,[9] carbon nanotubes (CNTs) have been arousing particular interest.

First evidenced by Radushkevich *et al.* in 1952,[4] but passed almost unnoticed, CNTs became popular in the early 1990s after Ijima reported the observation of "helical microtubules of graphitic carbon" while working on the synthesis of fullerene via an arc discharge evaporation method.[3] From that moment on, both electric arc discharge and chemical vapor deposition (CVD) methods have been developed to synthesize nanotubes. As most of these methods involve inorganic catalysts, CNTs usually hold metallic impurities, which should be considered when dealing with catalysis and/or electrochemistry. CNTs possess a tubular nanostructure made of one, two, or several rolled-up graphene sheets, called *walls*, yielding single-walled carbon nanotubes (SWCNTs), double-walled carbon nanotubes (DWCNTs), or multi-walled carbon nanotubes (MWCNTs), respectively. CNTs exhibit diameters ranging from 0.7 to 30 nm, depending on the number of walls, and have a high aspect ratio with a length at least 100 times longer than the diameter. The interlayer distance between graphene layers is close to the one reported in graphite, *i.e.* 3.4 Å. In an ideal representation, CNTs are closed on each extremity by two half-fullerenes. In practice, catalyst particles and/or defects are often present at one or each extremity. Both the number of walls, the purity and the geometrical way graphene sheets are rolled-up — *i.e.* armchair, zig-zag or chiral (Figure 1) — intimately influence the electronic properties of CNT, which can behave either as a metal or a semiconductor.[10–12] As the band gap decreases with the number of walls, MWCNTs display metallic behavior with a resistivity as low as 10^{-7}–10^{-8}Ωm at 300 K,[13] which is comparable to copper, gold, or single-layer graphene, but two orders of magnitude lower than graphite.[14] On the contrary, SWCNTs and even DWCNTs with small diameters can have a finite gap depending on the geometrical arrangement of graphene sheets. The use of selective metal/semiconductor separation methods may be required prior to using them in electrochemical devices.[15] Also, one should assume that, due to numerous particle contacts, the electrical resistivity of individual CNTs is by far lower than the bulk resistivity of macroscopic assemblies.[16] The same comment can be made regarding the theoretical specific surface area (SSA) calculated for individual SWCNTs (*i.e.* 1,315 m^2g^{-1}),[6] which is significantly higher than the SSA of CNTs' bundles obtained in practice.[16] These aspects should be considered when engineering electrodes and electrochemical devices such as biofuel cells.

In 1995, Allen *et al.* reported on the use of opened MWCNTs for the attachment and immobilization of small proteins.[17] The main focus of this precursory study was to exploit the tubular opened nanostructure of pretreated MWCNTs to entrap proteins. Lately, the same group proved CNTs to provide efficient electrical contact with two model redox proteins, *i.e.* cytochrome *c* and azurin, without inherent denaturation.[18] From that moment on, CNTs have been considered as key materials for the formulation of high-performance enzymatic electrodes, such as implantable fuel cells bearing high power density (Section 9.2.2). Interestingly, CNTs showed an ability to favor direct electron transfer (DET; cf. Chapter 1) with several enzymes, such as hemoglobin,[19] glucose oxidase (GOx),[20–22] multicopper oxidases (*e.g.* laccase and bilirubin oxidase),[23] catalase,[24] and even [Ni–Fe] hydrogenase.[25] Both SWCNTs and MWCNTs have been used for the design of bioelectrodes, with considerable efforts in structuration, shaping, and functionalization.

The aim of this chapter is to introduce recent advances and trends in the rapidly developing field of CNT-based enzymatic electrodes. Beyond CNTs, the usage of graphene, graphene oxide (GO) and even less popular nanocarbons will be also discussed. Some insights regarding other nanomaterials and nanocomposites will be also provided. The design and synthesis of nanocarbon-based enzymatic electrodes will be the main focus of this chapter rather than the final application. Thus, both electrodes for sensors and biofuel cells will be discussed. For a complete overview, the readers are invited to refer to few topical reviews regarding nanomaterials,[26,27] CNT[28] and graphene[29,30] based enzymatic electrodes.

9.2 Assembly, Structuration, and Shaping of CNT-based Bioelectrodes

9.2.1 From flat electrode to three-dimensional architectures

As mentioned earlier, a maximum SSA of 1,315 $m^2\ g^{-1}$ was calculated for individual SWCNTs.[6] In practice, one may assume that such value is really hard to reach as CNTs seldom occur as discrete individual nanoparticles, but most often form large bundles. Thus, the assembly of CNTs into porous functional architectures has been a challenging task. Since Allen *et al.* reported on the use of CNTs for the immobilization of proteins, various strategies have been developed to assemble CNTs and enzymes onto flat substrates (*e.g.* glassy carbon electrodes (GCE)) or even into free-standing porous architectures. Drop casting of pristine CNTs previously dispersed in organic solvents (*e.g.* DMF or THF), followed by drying and deposition of enzymatic solutions, is probably the most straightforward approach.[24,31,32] The entrapment of enzymes is usually ensured by the addition of polymers (*e.g.* poly(ethyleneimine) or Nafion), electropolymerizable monomers (*e.g.* pyrrole) or even ionic liquids.[33] Interestingly, drop casting allows a precise control over thickness and porosity of the deposits. Cosnier *et al.* could easily enhance the catalytic current density toward hydrogen oxidation of PPy/hydrogenase/MWCNTs modified electrodes by increasing the loading of CNTs and the thickness of the film (Figure 2(a)).[32] Dong *et al.*[34] proposed the layer-by-layer (LbL) deposition of MWCNTs, poly-L-lysine and laccases onto ITO electrodes to produce multi-layer structured cathodes with high power density of about 330 $\mu W\ cm^{-2}$. More recently, Kwon *et al.* employed the LbL approach for the successive deposition of poly(ethyleneimine) (PEI) and GOx onto CNTs via electrostatic interactions between positively charges PEI and negatively charged GOx.[35] Such LbL assemblies were incorporated into biofuel cells (BFC) and allowed reaching power density as high as

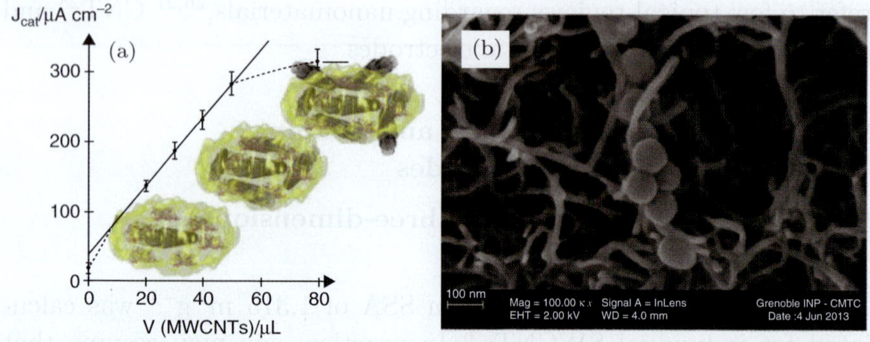

Figure 2. (a) Plot of catalytic current for hydrogen oxidation towards the amount of MWCNT deposits. Reproduced with permission from Ref. 28. © 2012, Elsevier Publishing. (b) Scanning electrochemical micrograph of a cross-section of a chitosan/MWCNTs/laccase biocathode. The agglomerates of laccases are the spherical nodes attached to the chitosan/MWCNTs matrix. Reproduced with permission from Ref. 36. © 2014, RSC Publishing.

1.3 mW cm^{-2}. The preliminary adsorption of PEI onto CNTs allowed to work in aqueous solutions and to prepare biocatalytic inks before drop casting.

Working with modified CNTs dispersible in water rather than with pristine CNTs in organic solvents opens up new perspectives in terms of structuration and shaping and avoids employing multi-step drop casting approaches. Thus, various strategies to disperse CNTs in water have been developed. As pristine CNTs are highly hydrophobic, preliminary functionalization is needed. In particular, surfactants,[19] oxidation via acid treatments,[37] or even physical adsorption and/or wrapping with (bio)polymers[35,38,39] have been developed. Among these three approaches, the use of surfactants can potentially denature biological macromolecules, while chemical oxidation can drastically impair the physical properties of CNTs (*i.e.* increase in resistivity and decrease in tensile strength). Thus, polymer-induced dispersions seem more appropriate for the preparation of CNT-based bioelectrodes and have been widely used in the last decade. Interestingly, various polysaccharides such as chitosan[39] and carboxymethyl cellulose[23] are known to adsorb onto CNTs via Van der Waals interactions. Polysaccharides, and particularly chitosan, are biocompatible and chemically modifiable for potential covalent immobilization of enzymes and cofactors through hydroxyl and/or amine groups. Martin *et al.* showed that chitosan could greatly improve the stability of MWCNTs/laccase-based biocathodes by creating a protective microenvironment and preventing the release of enzymes, while providing good oxygen diffusion (Figure 2(b)).[36] Regarding diffusion and mass transport improvement, Cooney *et al.* prepared porous scaffolds from freeze-dried chitosan-based hydrogels.[40,41] Following these studies, Atanassov *et al.* could prepare porous bioelectrodes by drop casting and freeze-drying enzymatic inks — made of butyraldehyde-modified chitosan, MWCNTs, NAD^+ cofactor, and malate dehydrogenase — onto GCE.[42] Flow-through 3D bioelectrodes were also designed by drop casting similar enzymatic inks within reticulated vitreous carbon (RVE) foams. In the same vein, Martin *et al.* used genipin as a cross-linker to synthesize chitosan/MWCNTs/laccase free-standing porous cathodes.[43] The as-prepared biocathodes could be compressed and implanted in the body of rats (cf. Section 9.2.2.). They demonstrated an *in vivo* operational lifetime of about 6 months with stable electrical output.

The use of a commercial 3D scaffold as a premade support for the deposit and assembly of CNT/enzyme mixtures has been infrequently employed to prepare hierarchical porous bioelectrodes. Apart from RVE foams as mentioned above,[42] Atanassov *et al.* proposed to use Toray® carbon paper (TP) as a matrix for selectively growing MWCNTs by the CVD technique.[44] After drop casting of GOx/PEI mixture followed by impregnation of Nafion® solution, free-standing hierarchical porous bioanodes could be designed. The highly porous 3D MWCNTs network has been demonstrated to promote the DET between the enzymes and the electrode surface. Later, the same group employed the sol–gel process to entrap GOx/SWCNTs within a silica matrix.[45] The as-synthesized gels were then coated onto TP by flow-through impregnation. Such immobilization not only stabilized enzyme activity over a period of 1 month but could also favor DET. More recently, Nishirazawa *et al.* proposed to impregnate

CNT-decorated carbon fabric strips with redox enzymes and to assemble them into flexible, laminate BFCs.[46]

Another strategy to assemble CNTs into functional free-standing architectures has been recently developed by Cosnier *et al.*[47,48] Pyrene-functionalized molecules[47] or polymers[48] were employed to cross-link MWCNTs into mechanically reinforced buckypapers via non-covalent π-stacking interactions. Interestingly, redox mediators were also included to ensure mediated electron transfer with the enzymes (cf. Section 9.3.1).

9.2.2 Miniaturized devices: Toward implantable enzymatic fuel cells

9.2.2.1 *Oriented carbon nanotubes scaffolds*

As mentioned previously, CNTs display peculiar properties due in part to their high aspect ratio. However, one should consider that, due to numerous particle contacts, the electrical resistivity of individual CNTs is by far lower than the bulk resistivity of macroscopic assemblies.[16] In order to take advantage of their intrinsic anisotropy and to preserve it at the macroscopic scale, various strategies have been developed to prepare oriented nanotube scaffolds.[49,50] In particular, few studies focused on such assemblies to design enzymatic electrodes.[51,52] For instance, Wallace *et al.*[52] grew CNTs on a quartz substrate by pyrolyzing iron(II) phthalocyanine and could obtain aligned arrays. These arrays were employed as conductive supports to prepare glucose sensors based on GOx/PPy electrodes. In the same vein, Gooding *et al.*[51] modified a gold electrode with a cysteamine self-assembled monolayer (SAM) to align shortened SWCNTs perpendicularly to the surface. Microperoxidases were then covalently attached to the free edges of the nanotubes, which were demonstrated to act as efficient wires for DET.[51] In 2010, Mano *et al.*[53] reported on the use of porous microfibers comprised of oriented CNTs to engineer high-power miniature glucose/oxygen BFC (Figure 3).

Prepared by a nanotube coagulation spinning process, such microfibers allowed reaching power density as high as 740 μW cm^{-2}, which was four-fold higher than the power density obtained for a

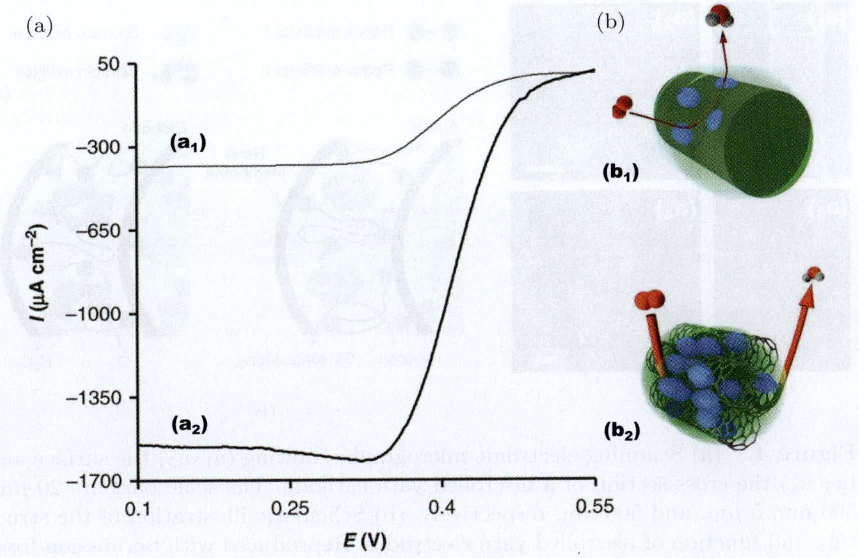

Figure 3. (a) Polarization curves and (b) schematic representation of the oxygen reduction on a (a_1 and b_1) modified carbon fiber and (a_2 and b_2) modified CNT fiber. The green shade represents the osmium-containing redox polymer and the blue spheres represent the enzymes, bilirubin oxidases. Reproduced with permission from Ref. 53. © 2010, Nature Publishing.

BFC made with conventional carbon fibers (Figure 3). Osmium-containing redox polymers were employed to ensure the electron transfer and entrap the enzymes. Nishizawa *et al.*[54] employed free-standing films of aligned and extremely long SWCNTs (about 1 mm), so-called CNTs' forests, to entrap enzymes through liquid-induced shrinkage. This approach allowed preparing binder-free electrodes, which were easily handled with tweezers and patched on flexible substrates to design miniature biodevices. Both fructose dehydrogenase-based anodes and laccase-based cathodes could operate without any redox mediator. A BFC bearing a maximum power density of 1.8 mW cm^{-2} was engineered. More recently, Kwon *et al.*[55] employed 50 μm diameter biscrolled CNT yarns as flexible porous scaffolds to entrap enzymes and redox mediators while minimizing diffusion distances (Figure 4). Complete woven textile BFC was assembled and a maximum power density of 2.18 mW cm^{-2} was obtained.

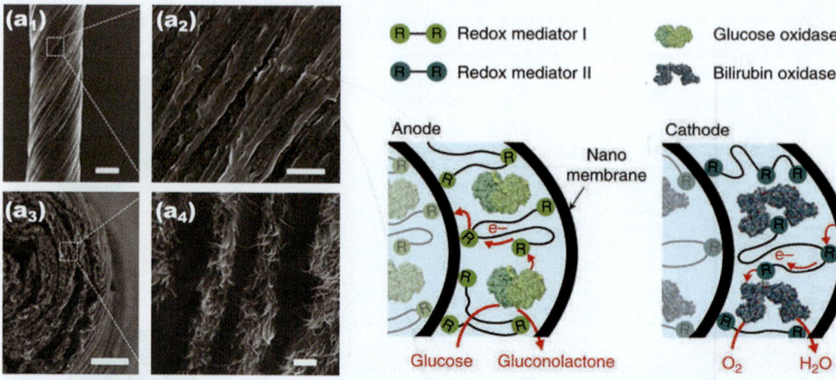

(a) (b)

Figure 4. (a) Scanning electronic micrographs showing (a_1–a_2) the surface and (a_3–a_4) the cross-section of a biscrolled yarn cathode. The scale bars are 20 μm, 500 nm, 5 μm, and 500 nm, respectively. (b) Schematic illustration of the structure and function of biscrolled yarn electrodes. Reproduced with permission from Ref. 55. © 2014, Nature Publishing.

Miniaturized devices were engineered by helically wrapping both the anodic and cathodic biscrolled yarns around a 1.2 mm diameter glass rod. Such high-performance BFCs have been considered as promising candidates for powering implantable devices.

9.2.2.2 *Densely packed CNT tangles*

Even though oriented CNT fibers and yarns allowed the design of high-performance devices, densely packed CNT tangles obtained by simple mechanical compression in the presence of enzymes have been widely employed to engineer implantable BFCs.[56,57] Unlike sophisticated bioelectrodes reported earlier (Section 9.2.2.1), compressed buckypapers are rather simple-in-preparation and favor DET between enzymes and CNTs. Mechanical compression allows reaching densely packed CNTs scaffolds in which enzymes can find a nearby conducting wire regardless of the enzyme orientation, as mentioned by Katz *et al.*[57] Such intimate contact between enzymes and CNTs favors efficient electrical wiring via DET. Taking advantage of this technology, Cosnier *et al.* engineered glucose/oxygen BFCs remaining stable for 1 month and delivering a high power density of 1 mW cm^{-2} under physiological conditions.[56] More recently, the same group reported a mixed operational/storage stability of similar BFCs over one year.[58] Interestingly, compressed buckypapers BFCs could be implanted in living creatures,[59] such as snails,[57] clams,[60] lobsters,[61] and rats.[43,62,63]

9.3 Electrical Wiring via Functionalization of Carbon Nanotubes

9.3.1 Functionalization of carbon nanotubes

Beyond assembly, structuration, and shaping of CNTs into porous bioelectrodes, considerable efforts have been devoted to functionalize them in order to ensure efficient electrical wiring with enzymes. While redox mediators are still necessary for some challenging enzymes, the immobilization of DET molecular promoters onto CNTs prior to enzyme adsorption has been leading to real breakthroughs for the design of high-performance bioelectrodes (Section 9.3.3). Various pathways have been developed to immobilize enzymes, redox mediators, and/or DET promoters, either through covalent bonding (Figures 5(a)–5(c)) or non-covalent interactions (Figure 5(d)).[28]

In the case of covalent modification, the functional groups can be directly attached to the graphitic surface via aryldiazonium reductions[64] (Figure 5(c)) or cycloadditions[65] (1,3-dipolar cycloaddition or Diels–Alder reaction). While cycloadditions have not been used much to functionalize enzymatic electrodes, the (electro)chemical grafting of aryldiazonium derivatives has been widely employed to immobilize mediators, such as ferrocene,[64] or even DET promoters.[66,67] This pathway allows for high surface coverage, thanks to the high reactivity of aryl radicals toward CNT sidewalls. However, it usually leads to complex surface modification with the formation of multilayered films composed of an oligomerized structure with possible azo bonds within the layer.[68] Covalent modifications can be also performed through the hydroxyl or carboxyl groups at the edges and/or defect sites of oxidized CNTs.[22,69,70] A preliminary oxidation via acid treatment is often necessary. However, one may consider that

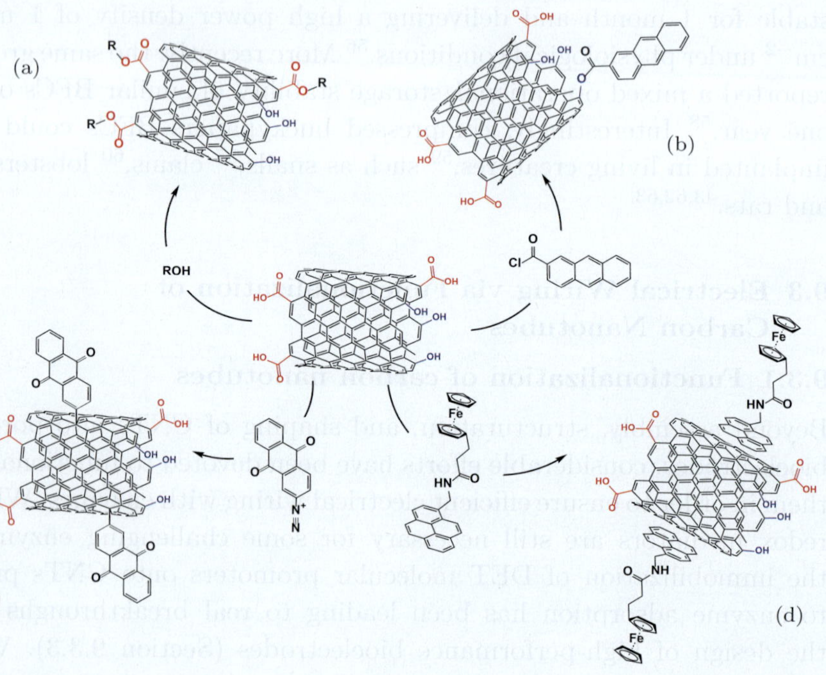

Figure 5. Main pathways to functionalize CNTs. (a)–(c) Covalent modification: (a) esterification of carboxyl groups from intrinsic or induced defects, (b) nucleophilic addition/elimination reaction between anthracene-2-carbonyl chloride and hydroxyl groups, (c) reduction of an anthraquinone-derived diazonium salt. (d) Non-covalent interactions: π-stacking of a ferrocene derivative bearing a pyrene group.

such harsh treatment can shorten the tubes and alter their properties. Further modifications can involve esterification (Figure 5(a)), coupling of carboxyl groups with amines[22,71] or even nucleophilic addition/elimination reaction between acyl chlorides and hydroxyl groups[70] (Figure 5(b)). This pathway enabled the specific functionalization of the edges of SWCNTs (see Section 9.3.3).[22,69]

As the covalent modification of graphitic materials, *e.g.* CNTs and graphene, generates a full rehybridization of the derivatized carbon atoms from sp^2 to sp^3, such modification can lead to a reduction in the mobility of the charge carriers within the material and to a significant decrease in its intrinsic conductivity.[72] To avoid such destructive rehybridization, Dai *et al.*[73] proposed the utilization of non-covalent interactions between a pyrene derivative (*i.e.* 1-pyrenebutanoic acid, succinimidyl ester) and CNT sidewalls to immobilize proteins. More recently, Cosnier *et al.* employed the strong interaction between CNT sidewalls and π-extended planar molecules, such as pyrene and porphyrin, to modify CNTs with small molecules before enzyme immobilization.[74–79] The double functionalization of MWCNTs with two different pyrene derivatives was also reported.[79] Both ruthenium complexes, employed as NADH oxidation electrocatalysts, and NAD-dependent glucose dehydrogenases could be immobilized. The as-synthesized electrodes showed high current densities toward glucose oxidation.[79]

Another noteworthy aspect of the pyrene approach concerns surface coverage. As compared with covalent immobilization by the diazonium salt approach, the π-stacking of ferrocene derivatives bearing pyrene afforded a surface coverage density within the same order of magnitude.[64] However, the optimization of such non-covalent immobilizations requires anticipating the solubility of the pyrene derivative within the electrolyte in order to avoid leaching.

9.3.2 Immobilization of redox mediators

For some challenging enzymes, for instance when the active site is deeply embedded, redox mediators are still necessary. Few pathways to immobilize molecular redox mediators, such as ferrocene[64] or ABTS[47] (*i.e.* 2,2′-azino-bis(3-ethylbenzothiazoline-6-sulfonic acid)), were introduced above (see Section 9.3.1). However, most of the bioelectrochemical systems requiring mediated electron transfers (MET) involve macromolecular redox mediators such as osmium-based hydrogels (cf. Chapters 5 and 7)[53] or electropolymerizable monomers bearing redox entities such as pyrrolic viologen derivatives.[32] Both imply long-range non-covalent interactions with CNTs and offer a strong host matrix for the efficient entrapment of enzymes.

9.3.3 Molecular functionalization of carbon nanotubes to favor direct electron transfer

As mentioned earlier (Section 9.1), CNTs showed ability to favor efficient DET with several enzymes.[19,20,22–25] One method to favor

DET consists in providing a highly conductive matrix that closely surrounds the enzymes, for instance via entrapment in porous scaffolds or intimate contact in compressed buckypapers (Section 9.2.2.2). The shortening of CNTs, providing disorder and porosity, was also reported as an efficient strategy to promote DET, in particular in the case of hydrogenases.[80] Even though enzymes are randomly oriented, most of the active sites can find a nearby conductive surface to ensure efficient electrical wiring.

Beyond porosity and mechanical compression, DET can be favored by appropriate molecular functionalization. In particular, Willner *et al.*[22] could covalently link FAD cofactors at the edges of SWCNTs and specifically adsorb the apoenzyme, apo-GOx, *i.e.* GOx without its cofactor (Figure 6). Surface-standing nanotubes with different lengths (from 20 nm to 150 nm) could be covalently grafted onto flat gold electrodes via amide bonds (Figure 6(a)). The authors observed that the electrocatalytic current depended linearly on the length of the SWCNTs and could conclude that the electron-transfer barrier was lower for shorter nanotubes. Similar experiments were performed by Gooding *et al.*[81] The authors covalently attached the whole *holoenzyme* — *i.e.* GOx together with its FAD cofactor — at

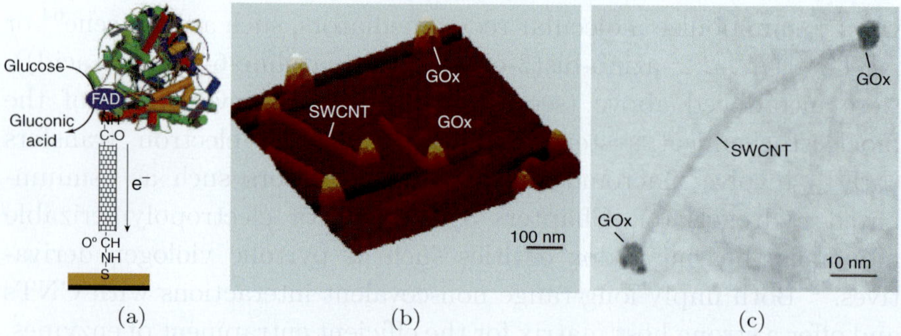

Figure 6. (a) Schematic illustration of the assembly of SWCNTs modified at its end with FAD units allowing for specific adsorption of apo-GOx units at the edges. (b) Atomic force microscopy topographic image of SWCNTs reconstituted at their ends with GOx units. (c) High-resolution transmission electronic micrograph of an SWCNT modified at its ends with GOx units. Reproduced with permission from Ref. 22. © 2004, Wiley-VCH Publishing.

the edges of the aligned SWCNTs. However, the rate constant for electron transfer was significantly smaller than the one reached with the apo-GOx reconstituted over the immobilized FAD.

Apart from biological cofactors, synthetic molecules have been also immobilized onto CNTs to favor the specific orientation of immobilized enzymes. In particular, the molecular manipulation of the orientation of blue multi-copper oxidases — *e.g.* laccases and bilirubin oxidase (BOD) — immobilized onto CNTs has been widely studied.[67,70,75–78,82,83] These oxidases present two distinct copper-containing clusters, located at each side of the enzyme: a trinuclear copper active site (TNC), at which O_2 is reduced to two H_2O molecules, and a mononuclear copper site (T1) from which electrons are transferred to the TNC site. Interestingly, the T1 site is located nearby a wide hydrophobic pocket. In 2007, Armstrong *et al.* took advantage of this hydrophobic pocket to orientate and immobilize laccase onto anthracene-modified graphite surface.[84] As assumed by the authors, each grafted anthracene molecule provided a hydrophobic conjugated system with the right dimension to fit in the hydrophobic pocket. Such "plugin socket' attachment" favored fast DET transfer. Later, this approach was extended to CNTs via covalent attachment of anthracene[70,82] or anthraquinone.[83] Cosnier *et al.* also developed non-covalent pathways to functionalize CNTs with anthraquinone derivatives through π–π interactions with pyrene groups.[76] Apart from anthracene and anthraquinone, pyrene (Figure 7),[75] porphyrins,[77] and adamantane[78] were also demonstrated to orientate oxidases and favor DET. In particular, adamantane-modified pyrene derivative demonstrated higher efficiency than its analogous anthraquinone derivative for the electrical wiring of laccase onto MWCNTs.[78] Cosnier *et al.* could also target which copper site — either T1 or TNC — was involved in the DET between BOD and functionalized MWCNTs. While carboxyl groups favored DET via the T1 site, amino groups favored DET via the TNC active site at lower potential.[67] The diazonium functionalization pathway had also a significant influence over the morphology and porosity of the CNT film, impacting the oxygen diffusion through the as-synthesized air-breathing biocathode.

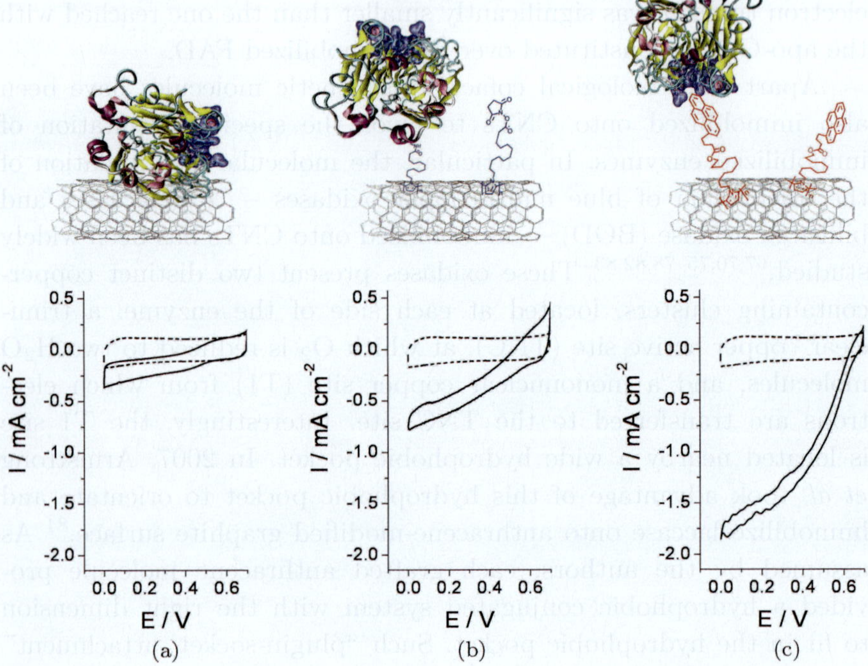

Figure 7. Cyclic voltammograms of laccase-functionalized CNT electrodes under argon (dashed lines) and under oxygen (solid lines) for (a) MWCNT, (b) MWCNT/polypyrrole-NHS and (c) MWCNT/poly(pyrrole–pyrene). Reproduced with permission from Ref. 75. 2013, Royal Society of Chemistry.

Beyond GOx and blue multi-copper oxidases, the orientation of (NiFe) hydrogenases at the electrochemical interface of graphitic materials has been shown to be closely dependent on the chemical end groups.[80] The covalent grafting of DET molecular promoters onto CNTs was also reported for (NiFe) hydrogenases.[66,85] In particular, Lojou *et al.* showed that the covalent functionalization of MWCNTs with carboxyl groups could efficiently favor DET and minimize the reversible deactivation of hydrogenases at high overpotential.[66] This last feature was not observed with pristine MWCNTs. The authors took advantage of these promising properties to design a membraneless air-breathing hydrogen biofuel cell.

In the last decade, considerable efforts have been devoted to assemble and functionalize CNTs into advanced porous bioelectrodes. CNTs opened new perspectives in the design and engineering of high-performance bioelectrochemical devices. In particular, real breakthroughs were made in the challenging field of implantable BFCs. With the emergence of additional singular nanomaterials such as graphene (see below) the "toolbox" has been getting bigger, thus making further achievements conceivable.

9.4 Beyond Carbon Nanotubes

9.4.1 "The rise of graphene"

Following the example of CNTs, graphene displays peculiar properties, among which a low resistivity of about 10^{-8} Ωm at 300 K,[86] an ultrahigh charge carrier mobility[86,87] and a high theoretical SSA of 2630 m^2g^{-1}, surpassing that of graphite (10 m^2g^{-1}) and SWCNTs (1,315 m^2g^{-1}).[88] While a SWCNT can behave either as a metal or a semiconductor, an unrolled SWCNT, *i.e.* a single graphene nanosheet, is considered as a zero-band-gap semiconductor or a zero-overlap semimetal.[89] Like CNTs, the electronic structure of a graphene nanosheet closely depends on its thickness.[86] While a double-layer graphene still behaves as a zero-band-gap semimetal, from three layers the bands start notably overlapping. In 2006, Partoens *et al.*[90] could estimate the number of layers from which graphene should be rather considered as a 3D thin film of graphite. According to the authors, above 10 layers of graphene, the electronic band structure is really close to the one observed in bulk graphite. Thus, one may distinguish between single-layer graphene, double-layer graphene, few-layer graphene (from three to nine layers) and graphite nanosheets (more than nine layers).

Aside from the number of layers, the preparation method and the presence of defects and/or metallic impurities drastically influence the physical properties of graphene. Single-layer graphene (SLG) was first evidenced by Novoselov *et al.*[5] from the repeated peeling of bulk graphite with adhesive tape. Since then, several exfoliation methods have been developed and applied to graphite including ball milling,[91] chemical oxidation,[92] intercalation compounds, liquid phase,[93] and electrochemical routes.[94] In practice, the most convenient method to obtain single graphene-like nanosheets involves the chemical oxidation and subsequent exfoliation of flake graphite.[92] As-synthesized GO and its derivatives have been widely employed for the fabrication of enzymatic electrodes. However, considering the fact that GO and pristine graphene present significant structural differences,[92] a clear distinction will be made in this chapter. For more information about GO-based bioelectrodes, the readers are invited to refer to the dedicated section (Section 9.4.2).

Exfoliated pristine graphene has been of particular interest for the fabrication of enzymatic electrodes.[29,30] The strategies employed to modify and assemble graphene nanosheets into functional bioelectrodes have been rather similar to the ones developed for other graphitic nanocarbons, such as CNTs (Section 9.2). Chen *et al.* used exfoliated nanographene (synthesized by a ball milling method) mixed with chitosan to prepare tyrosinase-based biosensors for the detection of bisphenol A.[91] Interestingly, the authors compared the performances of their nanographene-based biosensors with their analogs made of MWCNTs. Faster response, better repeatability, lower background current, and limit of detection were reached with nanographene and were attributed to its larger SSA and potential LbL self-assembly with enzymes. More recently, graphene nanosheets produced through a sonochemical-assisted solvent exfoliation method[93] were modified via non-covalent π–π interactions with a pyrene derivative (*i.e.* 1-pyrenebutanoic acid, succinimidyl ester).[95] Drawing their inspiration from analogos works performed on CNTs,[73,75] the authors immobilized GOx on the basal plane of the as-modified graphene nanosheets and demonstrated efficient electron transfer kinetics at the electrode surface. In the same vein, Schuhmann *et al.* designed and developed acrylate polymers modified with osmium complexes for the sonochemical-assisted exfoliation of graphite in water.[96] Due to the presence of osmium complexes, these acrylate polymers could not only stabilize pristine graphene nanosheets in aqueous media, but could also ensure mediated electron transfer with cellobiose dehydrogenases. Such functional polymers seem very promising and present a great potential for the design of graphene-based electrodes for enzymatic electrosynthesis and biofuel cells.

Apart from exfoliation methods, graphene nanosheets can be directly grown on metal supports via the CVD of hydrocarbons such as methane.[97] Even though CVD allows producing large-area and high-quality pristine graphene nanosheets,[97] their handling and processability remain difficult. For this reason, exfoliation methods have been preferred to design graphene-based porous frameworks. Alternatively, a solution to engineer 3D open pore architectures via CVD lies in the use of a premade porous scaffold as growth substrate. For example, Chen *et al.* reported on the growth of pristine graphene onto commercial nickel foams.[98] After chemical etching of the metal substrate and soaking in a SWCNTs' dispersion, graphene/SWCNTs nanocomposite foams were obtained. Both 3D GOx-based anodes and laccase-based cathodes were prepared and assembled into a high-performance glucose/oxygen BFC, delivering a maximum power density of 2.27 mW cm^{-2}.[98] The authors claimed unprecedented electron transfer kinetics at the anode and pointed out a probable synergistic effect by combining graphene and SWCNTs.

9.4.2 Graphene oxide as a convenient alternative to pristine graphene

As mentioned earlier, the most convenient method to obtain and handle SLG involves the chemical oxidation and subsequent exfoliation of graphite to yield dispersible single-layer GO.[92] One the one hand, the presence of oxygenated groups makes GO easily dispersible as single sheets in few polar solvents including water and allows assembling them into 3D porous frameworks.[99,100] On the other hand, due to the presence of oxygenated functions, such as carboxyl and epoxy groups, GO is chemically reactive and easy to functionalize.[92,101] This feature has been of particular interest for the immobilization of enzymes.[102–105] Last but not least, the disruption of the π-conjugated system makes GO an electrically insulating material with a resistivity as high as 50 Ωm at 300 K.[106] Even though GO was demonstrated to facilitate the DET of several heme-containing metalloproteins, such as cytochrome *c*, horseradish peroxidase,[107] and hemoglobin (Hb),[108] its reduction to reduced-graphene oxide (rGO) may be mandatory when electrochemical devices are concerned.[29,30] Whatever the reduction method employed, *i.e.* chemical (yielding CrGO), electrochemical (yielding ErGO) or thermal (yielding TrGO), defect-free reduced-graphene oxides are almost impossible to obtain[92] and their electrical properties barely meet the ones of pristine graphite.[106] While graphene and rGO are often confused, one may assume that these two nanomaterials present significant differences in terms of structure and properties.[92] Nevertheless, like CNTs and GO, CrGO,[83,109–111] ErGO,[112,113] and TrGO[114–116] have been reported for the direct electrical wiring of various enzymes, such as GOx,[111,114] laccase,[83] and BOD.[113]

Following the example of graphitic carbons, both cycloadditions[117] and aryldiazonium salt reduction[83,113] have been applied to rGO for the covalent immobilization of enzymes,[113] molecular DET promoters[83] (*e.g.* anthraquinone) and redox mediators[117] (*e.g.* ferrocene). The non-covalent immobilization of prefunctionalized enzymes was also reported by Liu *et al.*[118] The authors developed a LbL methodology to direct the self-assembly of graphene nanosheets and pyrene-modified GOx via π–π stacking interactions. Such multi-layered enzymatic electrodes exhibited good stability and high sensitivity towards glucose. Aside from those more conventional approaches, some judicious works[111,119] proposed to take advantage of the self-polymerization of a reducing agent, *i.e.* dopamine, to stabilize the resulting CrGO while providing functional groups for the covalent immobilization of GOx. The assembly of rGO with other nanomaterials, such as MWCNTs[83,120] or metal nanoparticles,[109,115,116] has been also demonstrated to facilitate DET and/or to prevent rGO from restacking, thereby increasing the overall porosity. In particular, Salimi *et al.*[120] showed that employing rGO nanosheets as a substrate for MWCNTs/laccase-based bioelectrodes significantly improved electron transfer kinetics.

Since first evidenced by Novoselov and Geim,[5] graphene has been nourishing considerable expectations for the fabrication of advanced materials and cutting-edge devices. Considering bioelectrochemical devices, in particular enzymatic fuel cells, the most significant achievements have been made when combining graphene-like nanosheets with CNTs.[98,120] However, one may assume that the exceptional potential of graphene-like materials has not been fully exploited yet.[29] The development of novel strategies to disperse pristine graphene in water could help breaking down some barriers.[121] The emergence of graphene quantum dots,[122] a new class of 0D nanocarbons made of graphene sheets smaller than 100 nm, might also open new perspectives in the field of bioelectrochemical devices.[123]

9.4.3 Alternative nanocarbons

Beyond the two "top billing" materials, *i.e.* CNTs and graphene-like nanosheets, less popular and/or emerging nanocarbons were also used for the synthesis of enzymatic electrodes. For example, Dong *et al.* employed single-walled carbon nanohorns (SWCNHs) to design BFCs.[124,125] Ijima *et al.* could produce SWCNHs by carbon dioxide laser ablation of graphite at room temperature without any metal catalyst,[126] making them a cost-effective, non-toxic/non-irritant,[127] and highly pure (*i.e.* metal-free) alternative to CNTs. Like SWCNTs, nanohorns are made of graphene-like sp^2 hybridized carbon atoms. Unlike SWCNTs, nanohorns are tubules of small aspect-ratio with a closed conical tip, a diameter of 2–3 nm and length of about 50 nm. They were shown to assemble into radial nano-aggregates usually compared to "*dahlia* flower type structure".[126,128] Dong *et al.* took advantage of these nano-assemblies to prepare SWCNH-based enzymatic electrodes[124] and engineer miniature glucose/air BFCs.[125] Interestingly, the authors demonstrated the possibility to power these BFCs with four different soft drinks and could reach a maximum power density of about 240 μW cm^{-2}. Even though this value is still far below the ones reported for the most competitive CNT-based BFCs (see Section 9.2.2), few studies have been reported and

SWCNHs remain promising candidates to design bioelectrochemical devices.

More recently, Lojou *et al.* employed herringbone carbon nanofibers (HCNFs) as supports for the enzymatic oxidation of hydrogen.[129] Usually, carbon nanofibers have a diameter of about 100–200 nm and a length of few micrometers.[130,131] While the core is either hollow or amorphous, the outer part is made of graphene layers stacked obliquely with respect to the fiber axis, yielding conical nanostructures.[131] Unlike SWCNHs, HCNFs are typically grown by CVD using nickel or nickel–copper catalysts, which may cause the presence of metallic impurities trapped within the core.[131] According to Lojou *et al.*[129] CNFs present the advantage of displaying numerous edge sites potentially favorable toward enzyme adsorption. Despite mass transport limitations, the authors demonstrated the efficient immobilization and electrical wiring of (NiFe) hydrogenases.[129] In the same vein, Rivas *et al.* proposed to use bamboo-like MWCNTs benefiting from the presence of many transverse edge-like defects along the walls.[132]

Beyond pristine sp^2 hybridized carbons, other nanocarbon allotropes were also employed for the design of enzymatic electrodes. For example, Cosnier *et al.*[133] combined nanodiamonds with SWCNTs to prepare 3D composite scaffolds for the immobilization of GOx. While the authors proposed different approaches to assemble SWCNTs and nanodiamonds, the LbL deposition technique turned out to be the most promising and allowed the preparation of highly competitive glucose biosensors.[133] Interestingly, SWCNTs/nanodiamonds layered assemblies showed better performances than their analogous without nanodiamonds. This work, together with the one reported earlier about graphene/SWCNTs electrodes[98] (Section 9.4.1), reminds the vast potential of nanocarbon composites and should encourage the design and development of novel nanoassemblies. More recently, Gao *et al.* proposed to employ carbon nanodots (CNDs)[134] for the design of enzymatic biosensors[135] and BFCs.[136] Like nanodiamonds, CNDs are quasi-spherical nanoparticles with small diameters ranging from 1 to 10 nm and rather high resistivity as compared with CNTs and graphene. The main difference between these two 0D nanocarbons lies in their structural order and hybridization. While CNDs have been described as paracrystalline carbons composed of sub-nanometric polyaromatic domains (*i.e.* sp^2-hybridized carbon atoms) surrounded by amorphous coatings, nanodiamonds can exhibit a quasi-perfect crystalline structure composed of sp^3-hydridized carbon atoms. Interestingly, while nanodiamonds are usually synthesized from the detonation of TNT/RDX (*i.e.* 2,4,6-trinitrotoluene/cyclotrimethylenetrinitramine) mixtures in oxygen-deficient explosive chambers,[137] CNDs can be prepared in rather soft conditions from a large variety of potential precursors.[134,138] For example, Gao *et al.* used CNDs derived from candle soot for the immobilization and direct electrical wiring of both GOx and BOD.[136] The authors could fabricate a CND-based glucose/air BFC with a limited maximum power density of about 40 μW cm^{-2}.[136] One may assume that the high resistivity of CNDs was detrimental to the BFC's performance. However, following the example of Cosnier *et al.* with nanodiamonds,[133] CNDs could be of particular interest if combined with CNTs or graphene nanosheets.

The diversity of allotropes, the multiplicity of potential precursors, and their peculiar intrinsic properties make carbon nanomaterials almost essential for the design of enzymatic electrodes. In particular, few studies revealed the vast potential of carbon-based nanocomposites, which should lead to real breakthroughs in the near future. As mentioned earlier (Section 9.4.2), noble metal nanoparticles combined with rGO have also been demonstrated to facilitate DET with several enzymes. Beyond nanocarbons and carbon–carbon nanocomposites, promising perspectives should arise from metal nanoparticles and other emerging non-carbonaceous nanomaterials. The next section of this chapter aims at providing a quick insight into such non-carbonaceous nanomaterials and their potential combination with nanocarbons.

9.4.4 Non-carbonaceous nanomaterials and nanocomposites

Besides nanocarbons, noble metal nanoparticles (NPs) supported onto[139–142] or assembled into 3D porous scaffolds[143,144] have been arousing particular interest for the preparation of functional bioelectrodes. Like CNTs, metallic NPs benefit from their small diameter, typically few nanometers, to potentially decrease the electron transfer distance with enzymes' active sites, therefore favoring DET. In particular, Eychmüller *et al.* synthesized highly porous aerogels from the assembly of palladium and/or platinum nanoparticles, which were used to prepare both GOx-based mediated anodes[144] and BOD-based mediator-free cathodes.[143] The authors engineered a membraneless glucose/oxygen BFC[143] with a maximum power density of 20 μW cm^{-2} far below the ones reported for the most competitive nanocarbon-based devices (see Sections 9.2.2 and 9.4.1). In the same vein, Willner *et al.* proposed the electropolymerization of functionalized metal nanoparticles with modified enzymes to synthesize cross-linked composites directly onto flat electrodes.[145,146] In particular, the authors proposed the assembly of glucose dehydrogenase and gold nanoparticles to construct a functional bioanode.[146] When combined with a BOD/SWCNTs cathode, a glucose/oxygen BFC with a maximum power density of about 30 μW cm^{-2} was reported.

Despite such innovative approaches, the design of metallic NPs-based bioelectrodes has not encountered the same success as CNTs or graphene-based ones. Even though it might rest on some conjectural considerations, one may assume that employing noble metals rather than nanocarbons is probably less cost-effective, less versatile, and even less sustainable. In this context, a new trend has been to combine metallic NPs with nanocarbons, either graphene-like nanosheets[109,115,116] or CNTs,[132,147–149] to provide more efficient electrical contacts and/or larger SSA while restricting to the minimum the amount of metal. Recently, Le Goff *et al.*[149] reported on the homogeneous coverage of MWCNTs with β-cyclodextrin-modified gold NPs to maximize DET with pyrene-modified laccases (Figure 8). Such nanostructured composites allowed reaching high catalytic current densities at low overpotential toward dioxygen reduction, while narrowing electron transfers pathways (Figure 8(a)). The authors suggested a cooperative and beneficial combination of MWCNTs and β-cyclodextrin-modified gold NPs.

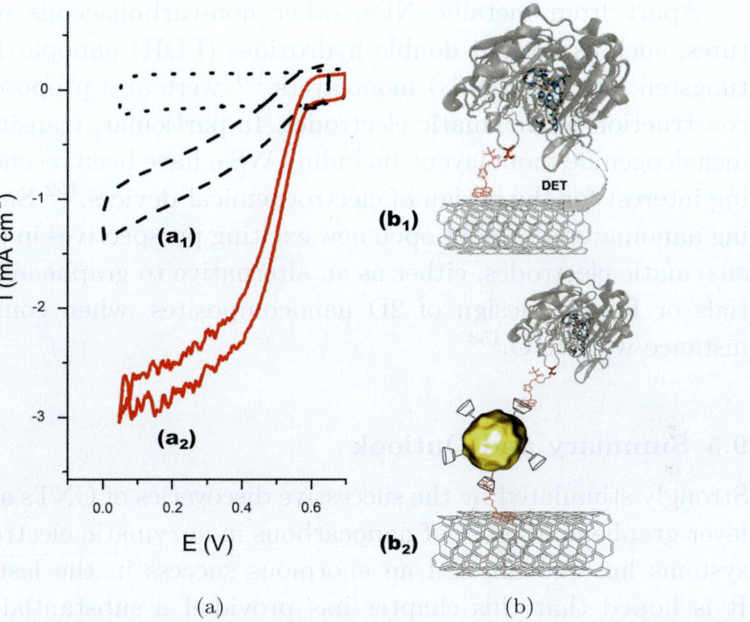

(a) (b)

Figure 8. (a) Cyclic voltammograms recorded under oxygen of pyrene-modified laccases immobilized onto (a_1) MWCNTs (dashed line) and (a_2) MWCNTs/β-cyclodextrin modified gold nanoparticles (solid line). The cyclic voltammogram recorded under argon of pyrene-modified laccases immobilized onto MWCNTs (dotted line) is presented for comparison. (b) Schemes of pyrene-modified laccases immobilized and oriented onto (b_1) MWCNTs and (b_2) MWCNTs/β-cyclodextrin-modified gold nanoparticles. Reproduced with permission from Ref. 149. © 2016, ACS Publishing.

Interestingly, metallic nanoparticles can be also beneficial when dealing with specific redox mediators or cofactors (cf. Chapter 1). In particular, gold, platinum, or even nickel NPs are known to be efficient catalysts toward the electrooxidation of the nicotinamide adenine dinucleotide cofactor in its reduced form (NADH). This feature is of particular interest for the design of bioelectrochemical devices involving NAD^+-dependent dehydrogenases. For example, Minteer *et al.*[132,147,148] could significantly improve the overall performances of ethanol/oxygen BFCs via the immobilization of NAD^+-dependent alcohol dehydrogenases onto gold NPs/MWCNTs nanocomposite anodes.

Apart from metallic NPs, other non-carbonaceous nanostructures, such as layered double hydroxides (LDH) nanoparticles[150] or tungsten disulfide (WS_2) monolayers,[151] were also proposed for the construction of enzymatic electrodes. In particular, transition-metal dichalcogenide monolayers, including WS_2, have been recently receiving interest for the design of electrochemical devices.[152] Such emerging nanomaterials might open new exciting perspectives in the field of enzymatic electrodes, either as an alternative to graphene-like materials or for the design of 2D nanocomposites, when combined for instance with rGO.[153]

9.5 Summary and Outlook

Strongly stimulated by the successive discoveries of CNTs and single-layer graphene, the use of nanocarbons in enzymatic electrochemical systems has encountered an enormous success in the last decades. It is hoped that this chapter has provided a substantial overview for the readers of the recent advances and trends regarding this field of research. A particular focus has been devoted to CNTs. Taking advantage of exceptional electronic and mechanical properties together with a singular nanostructure, CNTs have been widely employed as functional building blocks for the preparation of enzymatic electrodes. Considerable efforts have been devoted to assemble them into 3D architectures in order to stabilize and entrap a large amount of enzymes while ensuring efficient electrical wirings. Selective and versatile functionalization methods have been also developed to anchor enzymes, control their specific orientation and favor the DET with their active sites. Meanwhile, the pursuit of miniaturization has been significantly accelerated through the use of CNTs, which led to real breakthroughs in the challenging field of implantable biofuel cells.

Aside from CNTs, graphene has been nourishing considerable expectations for the fabrication of advanced materials and cutting-edge devices. So far, the exceptional potential of graphene has not been fully exploited when dealing with enzymatic electrochemical systems. The most significant achievements have been made through the combination of graphene with CNTs, proving the vast potential of carbon nanocomposites. These results should encourage the design and development of innovative nanoassemblies. With the emergence of singular 0D carbons (*e.g.* graphene quantum dots and carbon nanodots) together with other promising non-carbonaceous nanomaterials (*e.g.* transition-metal dichalcogenide monolayers), the "toolbox" has been getting bigger, thus making further achievements conceivable in a near future.

List of Abbreviations

ABTS	2,2′-azino-bis(3-ethylbenzothiazoline-6-sulfonic acid)
BFC	biofuel cell
BOD	bilirubin oxidase
CND	carbon nanodot
CNF	carbon nanofibers
CNT	carbon nanotube
CrGO	chemically reduced-graphene oxide
CVD	chemical vapor deposition
DET	direct electron transfer
DMF	N,N-diméthylformamide
DWCNT	double-walled carbon nanotube
e.g.	*exempli gratia*
ErGO	electrochemically reduced-graphene oxide
FAD	flavin adenine dinucleotide
GCE	glassy carbon electrode
GO	graphene oxide
GOx	glucose oxidase
HCNF	herringbone carbon nanofibers
i.e.	*id est*
ITO	indium tin oxide
LbL	layer-by-layer
LDH	layer double hydroxide
MET	mediated electron transfer
MWCNT	multi-walled carbon nanotube
NAD	nicotinamide adenine dinucleotide

NAD^+/NADH	oxidized and reduced state (with two electrons and one proton) of NAD
NP	nanoparticle
PEI	poly(ethyleneimine)
PPy	polypyrrole
RDX	cyclotrimethylenetrinitramine
rGO	reduced-graphene oxide
RVE	reticulated vitreous carbon
SAM	self-assembled monolayer
SLG	single-layer graphene
SSA	specific surface area
SWCNH	single-walled carbon nanohorn
SWCNT	single-walled carbon nanotube
THF	tetrahydrofuran
TNC	trinuclear copper
TNT	2,4,6-trinitrotoluene
TP	Toray® carbon paper
TrGO	thermally reduced-graphene oxide

References

1. V.V. Danilenko (2004). On the history of the discovery of nanodiamond synthesis. *Physics of the Solid State* **46**(4), 595–599.
2. H.W. Kroto, J.R. Heath, S.C. O'Brien, R.F. Curl and R.E. Smalley (1985). C60: Buckminsterfullerene. *Nature* **318**, 162–163.
3. S. Ijima. Helical microtubules of graphitic carbon (1991). *Nature* **354**, 56–58.
4. L.V. Radushkevich and V.M. Lukyanovich (1952). O strukture ugleroda, obrazujucegosja pri termiceskom razlozenii okisi ugleroda na zeleznom kontakte. *Zurn. Fisic. Chim.* **26**, 88–95.
5. K.S. Novoselov, A.K. Geim, S.V. Morozov, D. Jiang, Y. Zhang, S.V. Dubonos, I.V. Grigorieva and A.A. Firsov (2004). Electric field effect in atomically thin carbon films. *Science* **306**(5696), 666–669.
6. A. Peigney, C. Laurent, E. Flahaut, R.R. Bacsa and A. Rousset (2001). Specific surface area of carbon nanotubes and bundles of carbon nanotubes. *Carbon* **39**(4), 507–514.
7. M.F. Yu, O. Lourie, M.J. Dyer, K. Moloni, T.F. Kelly and R.S. Ruoff (2000). Strength and breaking mechanism of multiwalled carbon nanotubes under tensile load. *Science* **287**(5453), 637–640.
8. A. Magrez, S. Kasas, V. Salicio, N. Pasquier, J.W. Seo, M. Celio, S. Catsicas, B. Schwaller and L. Forro (2006). Cellular toxicity of carbon-based nanomaterials. *Nano Letters* **6**(6), 1121–1125.

9. S.K. Smart, A.I. Cassady, G.Q. Lu and D.J. Martin (2006). The biocompatibility of carbon nanotubes. *Carbon* **44**(6), 1034–1047.
10. R. Saito, G. Dresselhaus and M.S. Dresselhaus (2000). Trigonal warping effect of carbon nanotubes. *Physical Review* B **61**(4), 2981–2990.
11. J.C. Charlier, X. Blase and S. Roche (2007). Electronic and transport properties of nanotubes. *Reviews of Modern Physics* **79**(2), 677–732.
12. M. Scarselli, P. Castrucci, M. De Crescenzi (2012). Electronic and optoelectronic nano-devices based on carbon nanotubes. *Journal of Physics-Condensed Matter* **24**(31), 313202.
13. T.W. Ebbesen, H.J. Lezec, H. Hiura, J.W. Bennett, H.F. Ghaemi and T. Thio (1996). Electrical conductivity of individual carbon nanotubes. *Nature* **382**(6586), 54–56.
14. H.O. Pierson (1993). Handbook of Carbon, Graphite, Diamond and Fullerenes: Properties, Processing and Applications. Noyes Publications: Park Ridge, New Jersey, *U.S.A.*, p 61.
15. H. Liu, Y. Feng, T. Tanaka, Y. Urabe and H. Kataura (2010). Diameter-selective metal/semiconductor separation of single-wall carbon nanotubes by agarose gel. *Journal of Physical Chemistry* C **114**(20), 9270–9276.
16. B. Marinho, M. Ghislandi, E. Tkalya, C.E. Koning and G. de With (2012). Electrical conductivity of compacts of graphene, multi-wall carbon nanotubes, carbon black, and graphite powder. *Powder Technology* **221**, 351–358.
17. S.C. Tsang, J.J. Davis, M.L.H. Green, H. Allen, O. Hill, Y.C. Leung and P.J. Sadler (1995). Immobilization of small proteins in carbon nanotubes — high-resolution transmission electron-microscopy study and catalytic activity. *Journal of the Chemical Society-Chemical Communications* **17**, 1803–1804.
18. J.J. Davis, R.J. Coles, H.A.O. Hill (1997). Protein electrochemistry at carbon nanotube electrodes. *Journal of Electroanalytical Chemistry* **440**(1–2), 279–282.
19. C.X. Cai and J. Chen (2004). Direct electron transfer and bioelectrocatalysis of hemoglobin at a carbon nanotube electrode. *Analytical Biochemistry* **325**(2), 285–292.
20. A. Guiseppi-Elie, C.H. Lei and R.H. Baughman (2002). Direct electron transfer of glucose oxidase on carbon nanotubes. *Nanotechnology* **13**(5), 559–564.
21. C.X. Cai and J. Chen (2004). Direct electron transfer of glucose oxidase promoted by carbon nanotubes. *Analytical Biochemistry* **332**(1), 75–83.
22. F. Patolsky, Y. Weizmann and I. Willner (2004). Long-range electrical contacting of redox enzymes by SWCNT connectors. *Angewandte Chemie-International Edition* **43**(16), 2113–2117.
23. W. Zheng, Q.F. Li, L. Su, Y.M. Yan, J. Zhang and L.Q. Mao (2006). Direct electrochemistry of multi-copper oxidases at carbon nanotubes non-covalently functionalized with cellulose derivatives. *Electroanalysis* **18**(6), 587–594.

24. L. Wang, J.X. Wang and F.M. Zhou (2004). Direct electrochemistry of catalase at a gold electrode modified with single-wall carbon nanotubes. *Electroanalysis* **16**(8), 627–632.
25. M.A. Alonso-Lomillo, O. Rudiger, A. Maroto-Valiente, M. Velez, I. Rodriguez-Ramos, F.J. Munoz, V.M. Fernandez and A.L. De Lacey (2007). Hydrogenase-coated carbon nanotubes for efficient H-2 oxidation. *Nano Letters* **7**(6), 1603–1608.
26. A. Walcarius, S.D. Minteer, J. Wang, Y.H. Lin and A. Merkoci (2013). Nanomaterials for bio-functionalized electrodes: recent trends. *Journal of Materials Chemistry* B **1**(38), 4878–4908.
27. D. Wen and A. Eychmuller (2016). Enzymatic biofuel cells on porous nanostructures. *Small* **12**(34), 4649–4661.
28. M. Holzinger, A. Le Goff and S. Cosnier (2012). Carbon nanotube/enzyme biofuel cells. *Electrochimica* Acta **82**, 179–190.
29. J. Filip and J. Tkac (2014). Is graphene worth using in biofuel cells? *Electrochimica Acta* **136**, 340–354.
30. A. Karimi, A. Othman, A. Uzunoglu, L. Stanciu and S. Andreescu (2015). Graphene based enzymatic bioelectrodes and biofuel cells. *Nanoscale* **7**(16), 6909–6923.
31. Y.M. Yan, W. Zheng, L. Su and L.Q. Mao (2006). Carbon-nanotube-based glucose/O-2 biofuel cells. *Advanced Materials* **18**(19), 2639–2643.
32. J. Baur, A. Le Goff, S. Dementin, M. Holzinger, M. Rousset and S. Cosnier (2011). Three-dimensional carbon nanotube-polypyrrole-[NiFe] hydrogenase electrodes for the efficient electrocatalytic oxidation of H-2. *International Journal of Hydrogen Energy* **36**(19), 12096–12101.
33. L.F. Sun, X.H. Zhang, W.Y. Wang and J.H. Chen (2015). Carbon nanotube-ionic liquid composite gel based high-performance bioanode for glucose/O-2 biofuel cells. *Analytical Methods* **7**(12), 5060–5066.
34. L. Deng, L. Shang, Y.Z. Wang, T. Wang, H.J. Chen and S.J. Dong (2008). Multilayer structured carbon nanotubes/poly-L-lysine/laccase composite cathode for glucose/O-2 biofuel cell. *Electrochemistry Communications* **10**(7), 1012–1015.
35. K.H. Hyun, S.W. Han, W.G. Koh and Y. Kwon (2015). Fabrication of biofuel cell containing enzyme catalyst immobilized by layer-by-layer method. *Journal of Power Sources* **286**, 197–203.
36. S. El Ichi, A. Zebda, A. Laaroussi, N. Reverdy-Bruas, D. Chaussy, M.N. Belgacem, P. Cinquin and D.K. Martin (2014). Chitosan improves stability of carbon nanotube biocathodes for glucose biofuel cells. *Chemical Communications* **50**(93), 14535–14538.
37. Y.C. Tsai, S.C. Li and S.W. Liao (2006). Electrodeposition of polypyrrole-multiwalled carbon nanotube-glucose oxidase nanobiocomposite film for the detection of glucose. *Biosensors & Bioelectronics* **22**(4), 495–500.
38. J. Wang, M. Musameh and Y.H. Lin (2003). Solubilization of carbon nanotubes by nafion toward the preparation of amperometric biosensors. *Journal of the American Chemical Society* **125**(9), 2408–2409.

39. M.G. Zhang, A. Smith and W. Gorski (2004). Carbon nanotube-chitosan system for electrochemical sensing based on dehydrogenase enzymes. *Analytical Chemistry* **76**(17), 5045–5050.
40. M.J. Cooney, C. Lau, M. Windmeisser, B.Y. Liaw, T. Klotzbach and S.D. Minteer (2008). Design of chitosan gel pore structure: towards enzyme catalyzed flow-through electrodes. *Journal of Materials Chemistry* **18**(6), 667–674.
41. C. Lau, M.J. Cooney and P. Atanassov (2008). Conductive macroporous composite chitosan-carbon nanotube scaffolds. *Langmuir* **24**(13), 7004–7010.
42. R.A. Rincon, C. Lau, K.E. Garcia and P. Atanassov (2011). Flow-through 3D biofuel cell anode for NAD(+)-dependent enzymes. *Electrochimica Acta* **56**(5), 2503–2509.
43. S. El Ichi, A. Zebda, J.P. Alcaraz, A. Laaroussi, F. Boucher, J. Boutonnat, N. Reverdy-Bruas, D. Chaussy, M.N. Belgacem, P. Cinquin and D.K. Martin (2015). Bioelectrodes modified with chitosan for long-term energy supply from the body. *Energy & Environmental Science* **8**(3), 1017–1026.
44. D. Ivnitski, B. Branch, P. Atanassov and C. Apblett (2006). Glucose oxidase anode for biofuel cell based on direct electron transfer. *Electrochemistry Communications* **8**(8), 1204–1210.
45. D. Ivnitski, K. Artyushkova, R.A. Rincon, P. Atanassov, H.R. Luckarift and G.R. Johnson (2008). Entrapment of enzymes and carbon nanotubes in biologically synthesized silica: Glucose oxidase-catalyzed direct electron transfer. *Small* **4**(3), 357–364.
46. T. Miyake, K. Haneda, S. Yoshino and M. Nishizawa (2013). Flexible, layered biofuel cells. *Biosensors & Bioelectronics* **40**(1), 45–49.
47. M. Bourourou, K. Elouarzaki, M. Holzinger, C. Agnes, A. Le Goff, N. Reverdy-Bruas, D. Chaussy, M. Party, A. Maaref and S. Cosnier (2014). Freestanding redox buckypaper electrodes from multi-wall carbon nanotubes for bioelectrocatalytic oxygen reduction via mediated electron transfer. *Chemical Science* **5**(7), 2885–2888.
48. S. Cosnier, R. Haddad, D. Moatsou, R.K. O'Reilly (2015). Biofunctionalizable flexible bucky paper by combination of multi-walled carbon nanotubes and polynorbornene-pyrene — Application to the bioelectrocatalytic reduction of oxygen. *Carbon* **93**, 713–718.
49. B. Vigolo, A. Penicaud, C. Coulon, C. Sauder, R. Pailler, C. Journet, P. Bernier and P. Poulin (2000). Macroscopic fibers and ribbons of oriented carbon nanotubes. *Science* **290**(5495), 1331–1334.
50. M. Zhang, S.L. Fang, A.A. Zakhidov, S.B. Lee, A.E. Aliev, C.D. Williams, K.R. Atkinson and R.H. Baughman (2005). Strong, transparent, multifunctional, carbon nanotube sheets. *Science* **309**(5738), 1215–1219.
51. J.J. Gooding, R. Wibowo, J.Q. Liu, W.R. Yang, D. Losic, S. Orbons, F.J. Mearns, J.G. Shapter and D.B. Hibbert (2003). Protein electrochemistry using aligned carbon nanotube arrays. *Journal of the American Chemical Society* **125**(30), 9006–9007.

52. M. Gao, L. Dai and G.G. Wallace. Glucose sensors based on glucose-oxidase-containing polypyrrole/aligned carbon nanotube coaxial nanowire electrodes. *Synthetic Metals* (2003) **137**(1–3), 1393–1394.
53. F. Gao, L. Viry, M. Maugey, P. Poulin and N. Mano (2010). Engineering hybrid nanotube wires for high-power biofuel cells. *Nature Communications* **1**.
54. T. Miyake, S. Yoshino, T. Yamada, K. Hata and M. Nishizawa (2011). Self-Regulating Enzyme-Nanotube Ensemble Films and Their Application as Flexible Electrodes for Biofuel Cells. *Journal of the American Chemical Society* **133**(13), 5129–5134.
55. C.H. Kwon, S.H. Lee, Y.B. Choi, J.A. Lee, S.H. Kim, H.H. Kim, G.M. Spinks, G.G. Wallace, M.D. Lima, M.E. Kozlov, R.H. Baughman and S.J. Kim (2014). High-power biofuel cell textiles from woven biscrolled carbon nanotube yarns. *Nature Communications* **5**.
56. A. Zebda, C. Gondran, A. Le Goff, M. Holzinger, P. Cinquin and S. Cosnier (2011). Mediatorless high-power glucose biofuel cells based on compressed carbon nanotube-enzyme electrodes. *Nature Communications* **2**.
57. L. Halamkova, J. Halamek, V. Bocharova, A. Szczupak, L. Alfonta and E. Katz (2012). Implanted biofuel cell operating in a living snail. *Journal of the American Chemical Society* **134**(11), 5040–5043.
58. B. Reuillard, C. Abreu, N. Lalaoui, A. Le Goff, M. Holzinger, O. Ondel, F. Buret and S. Cosnier (2015). One-year stability for a glucose/oxygen biofuel cell combined with pH reactivation of the laccase/carbon nanotube biocathode. *Bioelectrochemistry* **106**, 73–76.
59. E. Katz and K. MacVittie (2013). Implanted biofuel cells operating in vivo - methods, applications and perspectives - feature article. *Energy & Environmental Science* **6**(10), 2791–2803.
60. A. Szczupak, J. Halamek, L. Halamkova, V. Bocharova, L. Alfonta and E. Katz (2012). Living battery — biofuel cells operating in vivo in clams. *Energy & Environmental Science* **5**(10), 8891–8895.
61. K. MacVittie, J. Halamek, L. Halamkova, M. Southcott, W.D. Jemison, R. Lobeld and E. Katz (2013). From "cyborg" lobsters to a pacemaker powered by implantable biofuel cells. *Energy & Environmental Science* **6**(1), 81–86.
62. J.A. Castorena-Gonzalez, C. Foote, K. MacVittie, J. Halamek, L. Halamkova, L.A. Martinez-Lemus and E. Katz (2013). Biofuel cell operating in vivo in rat. *Electroanalysis* **25**(7), 1579–1584.
63. A. Zebda, S. Cosnier, J.P. Alcaraz, M. Holzinger, A. Le Goff, C. Gondran, F. Boucher, F. Giroud, K. Gorgy, H. Lamraoui and P. Cinquin (2013). Single glucose biofuel cells implanted in rats power electronic devices. *Scientific Reports* **3**.
64. A. Le Goff, F. Moggia, N. Debou, P. Jegou, V. Artero, M. Fontecave, B. Jousselme and S. Palacin (2010). Facile and tunable functionalization of carbon nanotube electrodes with ferrocene by covalent coupling and pi-stacking interactions and their relevance to glucose bio-sensing. *Journal of Electroanalytical Chemistry* **641**(1–2), 57–63.

65. A. Callegari, S. Cosnier, M. Marcaccio, D. Paolucci, F. Paolucci, V. Georgakilas, N. Tagmatarchis, E. Vazquez and M. Prato (2004). Functionalised single wall carbon nanotubes/polypyrrole composites for the preparation of amperometric glucose biosensors. *Journal of Materials Chemistry* **14**(5), 807–810.
66. N. Lalaoui, A. de Poulpiquet, R. Haddad, A. Le Goff, M. Holzinger, S. Gounel, M. Mermoux, P. Infossi, N. Mano, E. Lojou and S. Cosnier (2015). A membraneless air-breathing hydrogen biofuel cell based on direct wiring of thermostable enzymes on carbon nanotube electrodes. *Chemical Communications* **51**(35), 7447–7450.
67. N. Lalaoui, M. Holzinger, A. Le Goff and S. Cosnier (2016). Diazonium functionalisation of carbon nanotubes for specific orientation of multicopper oxidases: Controlling electron entry points and oxygen diffusion to the enzyme. *Chemistry-a European Journal* **22**(30), 10494–10500.
68. S. Mahouche-Chergui, S. Gam-Derouich, C. Mangeney and M.M. Chehimi (2011). Aryl diazonium salts: a new class of coupling agents for bonding polymers, biomacromolecules and nanoparticles to surfaces. *Chemical Society Reviews* **40**(7), 4143–4166.
69. E. Nazaruk, K. Sadowska, J.F. Biernat, J. Rogalski, G. Ginalska and R. Bilewicz (2010). Enzymatic electrodes nanostructured with functionalized carbon nanotubes for biofuel cell applications. *Analytical and Bioanalytical Chemistry* **398**(4), 1651–1660.
70. M.T. Meredith, M. Minson, D. Hickey, K. Artyushkova, D.T. Glatzhofer and S.D. Minteer (2011). Anthracene-modified multi-walled carbon nanotubes as direct electron transfer scaffolds for enzymatic oxygen reduction. *Acs Catalysis* **1**(12), 1683–1690.
71. J.J. Davis, K.S. Coleman, B.R. Azamian, C.B. Bagshaw, M.L.H. Green (2003). Chemical and biochemical sensing with modified single walled carbon nanotubes. *Chemistry-a European Journal* **9**(16), 3732–3739.
72. E. Bekyarova, S. Sarkar, F.H. Wang, M.E. Itkis, I. Kalinina, X.J. Tian and R.C. Haddon (2013). Effect of covalent chemistry on the electronic structure and properties of carbon nanotubes and graphene. *Accounts of Chemical Research* **46**(1), 65–76.
73. R.J. Chen, Y.G. Zhang, D.W. Wang and H.J. Dai (2001). Noncovalent sidewall functionalization of single-walled carbon nanotubes for protein immobilization. *Journal of the American Chemical Society* **123**(16), 3838–3839.
74. R. Haddad, M. Holzinger, R. Villalonga, A. Neumann, J. Roots, A. Maaref and S. Cosnier (2011). Pyrene-adamantane-beta-cyclodextrin: An efficient host-guest system for the biofunctionalization of SWCNT electrodes. *Carbon* **49**(7), 2571–2578.
75. N. Lalaoui, K. Elouarzaki, A. Le Goff, M. Holzinger and S. Cosnier (2013). Efficient direct oxygen reduction by laccases attached and oriented on pyrene-functionalized polypyrrole/carbon nanotube electrodes. *Chemical Communications* **49**(81), 9281–9283.

76. M. Bourourou, K. Elouarzaki, N. Lalaoui, C. Agnes, A. Le Goff, M. Holzinger, A. Maaref and S. Cosnier (2013). Supramolecular Immobilization of Laccase on Carbon Nanotube Electrodes Functionalized with (Methylpyrenylaminomethyl)anthraquinone for Direct Electron Reduction of Oxygen. *Chemistry-a European Journal* **19**(28), 9371–9375.
77. N. Lalaoui, A. Le Goff, M. Holzinger and S. Cosnier (2015). Fully oriented bilirubin oxidase on porphyrin-functionalized carbon nanotube electrodes for electrocatalytic oxygen reduction. *Chemistry-a European Journal* **21**(47), 16868–16873.
78. N. Lalaoui, R. David, H. Jamet, M. Holzinger, A. Le Goff and S. Cosnier (2016). Hosting Adamantane in the Substrate Pocket of Laccase: Direct Bioelectrocatalytic Reduction of O-2 on Functionalized Carbon Nanotubes. *Acs Catalysis* **6**(7), 4259–4264.
79. B. Reuillard, A. Le Goff and S. Cosnier (2014). Non-covalent double functionalization of carbon nanotubes with a NADH oxidation Ru(II)-based molecular catalyst and a NAD-dependent glucose dehydrogenase. *Chemical Communications* **50**(79), 11731–11734.
80. E. Lojou, X. Luo, M. Brugna, N. Candoni, S. Dementin, M.T. Giudici-Orticoni (2008). Biocatalysts for fuel cells: efficient hydrogenase orientation for H(2) oxidation at electrodes modified with carbon nanotubes. *Journal of Biological Inorganic Chemistry* **13**(7), 1157–1167.
81. J.Q. Liu, A. Chou, W. Rahmat, M.N. Paddon-Row and J.J. Gooding (2005). Achieving direct electrical connection to glucose oxidase using aligned single walled carbon nanotube arrays. *Electroanalysis* **17**(1), 38–46.
82. M. Sosna, L. Stoica, E. Wright, J.D. Kilburn, W. Schuhmann and P.N. Bartlett (2012). Mass transport controlled oxygen reduction at anthraquinone modified 3D-CNT electrodes with immobilized Trametes hirsuta laccase. *Physical Chemistry Chemical Physics* **14**(34), 11882–11885.
83. N. Lalaoui, A. Le Goff, M. Holzinger, M. Mermoux and S. Cosnier (2015). Wiring Laccase on Covalently Modified Graphene: Carbon Nanotube Assemblies for the Direct Bio-electrocatalytic Reduction of Oxygen. *Chemistry-a European Journal* **21**(8), 3198–3201.
84. C.F. Blanford, R.S. Heath and F.A. Armstrong (2007). A stable electrode for high-potential, electrocatalytic O-2 reduction based on rational attachment of a blue copper oxidase to a graphite surface. *Chemical Communications* **17**, 1710–1712.
85. A. De Poulpiquet, A. Ciaccafava, K. Szot, B. Pillain, P. Infossi, M. Guiral, M. Opallo, M.T. Giudici-Orticoni and E. Lojou (2013). Exploring Properties of a Hyperthermophilic Membrane-Bound Hydrogenase at Carbon Nanotube Modified Electrodes for a Powerful H2/O2 Biofuel Cell. *Electroanalysis* **25**(3), 685–695.
86. X.Y. Fang, X.X. Yu, H.M. Zheng, H.B. Jin, L. Wang and M.S. Cao (2015). Temperature- and thickness-dependent electrical conductivity of few-layer graphene and graphene nanosheets. *Physics Letters A* **379**(37), 2245–2251.

87. S.V. Morozov, K.S. Novoselov, M.I. Katsnelson, F. Schedin, D.C. Elias, J.A. Jaszczak and A.K. Geim (2008). Giant intrinsic carrier mobilities in graphene and its bilayer. *Physical Review Letters* **100**(1).
88. D.A.C. Brownson and C.E. Banks (2010). Graphene electrochemistry: an overview of potential applications. *Analyst* **135**(11), 2768–2778.
89. A.K. Geim and K.S. Novoselov (2007). The rise of graphene. *Nature Materials* **6**(3), 183–191.
90. B. Partoens and F.M. Peeters (2006). From graphene to graphite: Electronic structure around the K point. *Physical Review B* **74**, (7).
91. L.D. Wu, D.H. Deng, J. Jin, X.B. Lu and J.P. Chen (2012). Nano-graphene-based tyrosinase biosensor for rapid detection of bisphenol A. *Biosensors & Bioelectronics* **35**(1), 193–199.
92. D.R. Dreyer, S. Park, C.W. Bielawski and R.S. Ruoff (2010). The chemistry of graphene oxide. *Chemical Society Reviews* **39**(1), 228–240.
93. J.S.Y. Chia, M.T.T. Tan, P.S. Khiew, J.K. Chin, H. Lee, D.C.S. Bien and C.W. Siong (2014). A novel one step synthesis of graphene via sonochemical-assisted solvent exfoliation approach for electrochemical sensing application. *Chemical Engineering Journal* **249**, 270–278.
94. K.P. Loh, Q.L. Bao, P.K. Ang and J.X. Yang (2010). The chemistry of graphene. *Journal of Materials Chemistry* **20**(12), 2277–2289.
95. J.S.Y. Chia, M.T.T. Tan, P.S. Khiew, J.K. Chin and C.W. Siong (2015). A bio-electrochemical sensing platform for glucose based on irreversible, non-covalent pi-pi functionalization of graphene produced via a novel, green synthesis method. *Sensors and Actuators B-Chemical* **210**, 558–565.
96. Z.Y. Sun, J. Vivekananthan, D.A. Guschin, X. Huang, V. Kuznetsov, P. Ebbinghaus, A. Sarfraz, M. Muhler and W. Schuhmann (2014). High-concentration graphene dispersions with minimal stabilizer: A scaffold for enzyme immobilization for glucose oxidation. *Chemistry-a European Journal* **20**(19), 5752–5761.
97. X.S. Li, W.W. Cai, J.H. An, S. Kim, J. Nah, D.X. Yang, R. Piner, A. Velamakanni, I. Jung, E. Tutuc, S.K. Banerjee, L. Colombo and R.S. Ruoff (2009). Large-area synthesis of high-quality and uniform graphene films on copper foils. *Science* **324**(5932), 1312–1314.
98. K.P. Prasad, Y. Chen and P. Chen (2014). Three-dimensional graphene-carbon nanotube hybrid for high-performance enzymatic biofuel cells. *Acs Applied Materials & Interfaces* **6**(5), 3387–3393.
99. Y.X. Xu, K.X. Sheng, C. Li and G.Q. Shi (2010). Self-assembled graphene hydrogel via a one-step hydrothermal process. *Acs Nano* **4**(7), 4324–4330.
100. Z.S. Wu, A. Winter, L. Chen, Y. Sun, A. Turchanin, X.L. Feng and K. Mullen (2012). Three-Dimensional Nitrogen and Boron Co-doped Graphene for High-Performance All-Solid-State Supercapacitors. *Advanced Materials* **24**(37), 5130–5135.
101. H.F. Yang, C.S. Shan, F.H. Li, D.X. Han, Q.X. Zhang and L. Niu (2009). Covalent functionalization of polydisperse chemically-converted graphene sheets with amine-terminated ionic liquid. *Chemical Communications* (26), 3880–3882.

102. Y. Liu, D.S. Yu, C. Zeng, Z.C. Miao and L.M. Dai (2010). Biocompatible Graphene Oxide-Based Glucose Biosensors. *Langmuir* **26**(9), 6158–6160.
103. L.Y. Zhou, Y.J. Jiang, J. Gao, X.Q. Zhao, L. Ma and Q.L. Zhou (2012). Oriented immobilization of glucose oxidase on graphene oxide. *Biochemical Engineering Journal* **69**, 28–31.
104. Y.Y. Jiang, Q.X. Zhang, F.H. Li and L. Niu (2012). Glucose oxidase and graphene bionanocomposite bridged by ionic liquid unit for glucose biosensing application. *Sensors and Actuators B-Chemical* **161**(1), 728–733.
105. M.A. Ali, W. Hong, S. Oren, Q.G. Wang, Y.F. Wang, H.W. Jiang and L. Dong (2016). Tunable bioelectrodes with wrinkled-ridged graphene oxide surfaces for electrochemical nitrate sensors. *Rsc Advances* **6**(71), 67184–67195.
106. S. Stankovich, D.A. Dikin, R.D. Piner, K.A. Kohlhaas, A. Kleinhammes, Y. Jia, Y. Wu, S.T. Nguyen and R.S. Ruoff (2007). Synthesis of graphene-based nanosheets via chemical reduction of exfoliated graphite oxide. *Carbon* **45**(7), 1558–1565.
107. X.L. Zuo, S.J. He, D. Li, C. Peng, Q. Huang, S.P. Song and C.H. Fan (2010). Graphene oxide-facilitated electron transfer of metalloproteins at electrode surfaces. *Langmuir* **26**(3), 1936–1939.
108. W. Sun, S.X. Gong, F. Shi, L.L. Cao, L.Y. Ling, W.Z. Zheng and W.C. Wang (2014). Direct electrochemistry and electrocatalysis of hemoglobin in graphene oxide and ionic liquid composite film. *Materials Science & Engineering C-Materials for Biological Applications* **40**, 235–241.
109. K.J. Huang, D.J. Niu, X. Liu, Z.W. Wu, Y. Fan, Y.F. Chang and Y.Y. Wu (2011). Direct electrochemistry of catalase at amine-functionalized graphene/gold nanoparticles composite film for hydrogen peroxide sensor. *Electrochimica Acta* **56**(7), 2947–2953.
110. M.B. Gholivand and M. Khodadadian (2014). Amperometric cholesterol biosensor based on the direct electrochemistry of cholesterol oxidase and catalase on a graphene/ionic liquid-modified glassy carbon electrode. *Biosensors & Bioelectronics* **53**, 472–478.
111. H.P. Liu, C.A. Tao, Z.H. Hu, S.D. Zhang, J.F. Wang and Y.G. Zhan (2014). An electrochemical glucose biosensor based on graphene composites: use of dopamine as reducing monomer and as site for covalent immobilization of enzyme. *Rsc Advances* **4**(82), 43624–43629.
112. B. Liang, X.S. Guo, L. Fang, Y.C. Hu, G. Yang, Q. Zhu, J.W. Wei and X.S. Ye (2015). Study of direct electron transfer and enzyme activity of glucose oxidase on graphene surface. *Electrochemistry Communications* **50**, 1–5.
113. C. Di Bari, A. Goni-Urtiaga, M. Pita, S. Shleev, M.D. Toscano, R. Sainz, A.L. De Lacey (2016). Fabrication of high surface area graphene electrodes with high performance towards enzymatic oxygen reduction. *Electrochimica Acta* **191**, 500–509.
114. X.H. Kang, J. Wang, H. Wu, I.A. Aksay, J. Liu and Y.H. Lin (2009). Glucose Oxidase-graphene-chitosan modified electrode for direct electrochemistry and glucose sensing. *Biosensors & Bioelectronics* **25**(4), 901–905.

115. T.T. Baby, S.S.J. Aravind, T. Arockiadoss, R.B. Rakhi and S. Ramaprabhu (2010). Metal decorated graphene nanosheets as immobilization matrix for amperometric glucose biosensor. *Sensors and Actuators B-Chemical* **145**(1), 71–77.
116. O. Parlak, A. Tiwari, A.P.F. Turner and A. Tiwari (2013). Template-directed hierarchical self-assembly of graphene based hybrid structure for electrochemical biosensing. *Biosensors & Bioelectronics* **49**, 53–62.
117. A. Rabti, C.C. Mayorga-Martinez, L. Baptista-Pires, N. Raouafi and A. Merkoci (2016). Ferrocene-functionalized graphene electrode for biosensing applications. *Analytica Chimica Acta* **926**, 28–35.
118. J.Q. Liu, N. Kong, A.H. Li, X. Luo, L. Cui, R. Wang and S.Y. Feng (2013). Graphene bridged enzyme electrodes for glucose biosensing application. *Analyst* **138**(9), 2567–2575.
119. N. Huang, S. Zhang, L.Q. Yang, M.L. Liu, H.T. Li, Y.Y. Zhang and S.Z. Yao (2015). Multifunctional Electrochemical Platforms Based on the Michael Addition/Schiff Base Reaction of Polydopamine Modified Reduced Graphene Oxide: Construction and Application. *Acs Applied Materials & Interfaces* **7**(32), 17935–17946.
120. A. Navaee and A. Salimi (2015). Graphene-supported pyrene-functionalized amino-carbon nanotube: A novel hybrid architecture of laccase immobilization as effective bioelectrocatalyst for oxygen reduction reaction. *Journal of Materials Chemistry* A **3**(14), 7623–7630.
121. G. Bepete, E. Anglaret, L. Ortolani, V. Morandi, K. Huang, A. Pénicaud and C. Drummond (2016). Surfactant-free single-layer graphene in water. *Nature Chemistry.*
122. J.H. Shen, Y.H. Zhu, X.L. Yang and C.Z. Li (2012). Graphene quantum dots: emergent nanolights for bioimaging, sensors, catalysis and photovoltaic devices. *Chemical Communications* **48**(31), 3686–3699.
123. H. Razmi, R. Mohammad-Rezaei (2013). Graphene quantum dots as a new substrate for immobilization and direct electrochemistry of glucose oxidase: Application to sensitive glucose determination. *Biosensors & Bioelectronics* **41**, 498–504.
124. D. Wen, L. Deng, M. Zhou, S.J. Guo, L. Shang, G.B. Xu and S.J. Dong (2010). A biofuel cell with a single-walled carbon nanohorn-based bioanode operating at physiological condition. *Biosensors & Bioelectronics* **25**(6), 1544–1547.
125. D. Wen, X.L. Xu and S.J. Dong (2011). A single-walled carbon nanohorn-based miniature glucose/air biofuel cell for harvesting energy from soft drinks. *Energy & Environmental Science* **4**(4), 1358–1363.
126. S. Iijima, M. Yudasaka, R. Yamada, S. Bandow, K. Suenaga, F. Kokai and K. Takahashi (1999). Nano-aggregates of single-walled graphitic carbon nanohorns. *Chemical Physics Letters* **309**, (3–4), 165–170.
127. J. Miyawaki, M. Yudasaka, T. Azami, Y. Kubo and S. Iijima (2008). Toxicity of single-walled carbon nanohorns. *Acs Nano* **2**(2), 213–226.

128. K. Murata, K. Kaneko, F. Kokai, K. Takahashi, M. Yudasaka and S. Iijima (2000). Pore structure of single-wall carbon nanohorn aggregates. *Chemical Physics Letters* **331**(1), 14–20.
129. A. de Poulpiquet, H. Marques-Knopf, V. Wernert, M.T. Giudici-Orticoni, R. Gadiou and E. Lojou (2014). Carbon nanofiber mesoporous films: efficient platforms for bio-hydrogen oxidation in biofuel cells. *Physical Chemistry Chemical Physics* **16**(4), 1366–1378.
130. Y.A. Zhu, Z.J. Sui, T.J. Zhao, Y.C. Dai, Z.M. Cheng and W.K. Yuan (2005). Modeling of fishbone-type carbon nanofibers: A theoretical study. *Carbon* **43**(8), 1694–1699.
131. D. Luxembourg, X. Py, A. Didion, R. Gadiou, C. Vix-Guterl and G. Flamant (2007). Chemical activations of herringbone-type nanofibers. *Microporous and Mesoporous Materials* **98**(1–3), 123–131.
132. M. Eguilaz, R. Villalonga, J.M. Pingarron, N.F. Ferreyra and G.A. Rivas (2015). Functionalization of bamboo-like carbon nanotubes with 3-mercaptophenylboronic acid-modified gold nanoparticles for the development of a hybrid glucose enzyme electrochemical biosensor. *Sensors and Actuators B-Chemical* **216**, 629–637.
133. M. Singh, M. Holzinger, O. Biloivan and S. Cosnier (2013). 3D-nanostructured scaffold electrodes based on single-walled carbon nanotubes and nanodiamonds for high performance biosensors. *Carbon* **61**, 349–356.
134. H.P. Liu, T. Ye and C.D. Mao (2007). Fluorescent carbon nanoparticles derived from candle soot. *Angewandte Chemie-International Edition* **46**(34), 6473–6475.
135. M.L. Sheng, Y. Gao, J.Y. Sun and F. Gao (2014). Carbon nanodots-chitosan composite film: A platform for protein immobilization, direct electrochemistry and bioelectrocatalysis. *Biosensors & Bioelectronics* **58**, 351–358.
136. M. Zhao, Y. Gao, J.Y. Sun and F. Gao (2015). Mediator less Glucose Biosensor and Direct Electron Transfer Type Glucose/Air Biofuel Cell Enabled with Carbon Nanodots. *Analytical Chemistry* **87**(5), 2615–2622.
137. V.N. Mochalin, O. Shenderova, D. Ho and Y. Gogotsi (2012). The properties and applications of nanodiamonds. *Nature Nanotechnology* **7**(1), 11–23.
138. H.T. Li, Z.H. Kang, Y. Liu and S.T. Lee (2012). Carbon nanodots: synthesis, properties and applications. *Journal of Materials Chemistry* **22**(46), 24230–24253.
139. C. Gutierrez-Sanchez, M. Pita, C. Vaz-Dominguez, S. Shleev, A.L. De Lacey (2012). Gold Nanoparticles as Electronic Bridges for Laccase-Based Biocathodes. *Journal of the American Chemical Society* **134**(41), 17212–17220.
140. Y.H. Wang, L. Ge, C. Ma, Q.K. Kong, M. Yan, S.G. Ge and J.H. Yu (2014). Self-Powered and Sensitive DNA Detection in a Three-Dimensional Origami-Based Biofuel Cell Based on a Porous Pt-Paper Cathode. *Chemistry-a European Journal* **20**(39), 12453–12462.
141. Y.H. Wang, L. Ge, P.P. Wang, M. Yan, J.H. Yu and S.G. Ge (2014). A three-dimensional origami-based immuno-biofuel cell for self-powered, low-cost, and sensitive point-of-care testing. *Chemical Communications* **50**(16), 1947–1949.

142. R.A. Blaik, E. Lan, Y. Huang and B. Dunn (2016). Gold-coated M13 bacteriophage as a template for glucose oxidase biofuel cells with direct electron transfer. *Acs Nano* **10**(1), 324–332.
143. D. Wen, W. Liu, A.K. Herrmann and A. Eychmuller (2014). A membraneless glucose/O-2 biofuel cell based on Pd aerogels. *Chemistry-a European Journal* **20**(15), 4380–4385.
144. D. Wen, A.K. Herrmann, L. Borchardt, F. Simon, W. Liu, S. Kaskel and A. Eychmuller (2014). Controlling the growth of palladium aerogels with high-performance toward bioelectrocatalytic oxidation of glucose. *Journal of the American Chemical Society* **136**(7), 2727–2730.
145. Y.M. Yan, I. Baravik, R. Tel-Vered and I. Willner (2009). An ethanol/O-2 biofuel cell based on an electropolymerized bilirubin oxidase/Pt nanoparticle bioelectrocatalytic O-2-reduction cathode. *Advanced Materials* **21**(42), 4275–4279.
146. O. Yehezkeli, R. Tel-Vered, S. Reichlin and I. Willner (2011). Nano-engineered Flavin-Dependent Glucose Dehydrogenase/Gold Nanoparticle-Modified Electrodes for Glucose Sensing and Biofuel Cell Applications. *Acs Nano* **5**(3), 2385–2391.
147. S.A. Neto, T.S. Almeida, L.M. Palma, S.D. Minteer, A.R. de Andrade (2014). Hybrid nanocatalysts containing enzymes and metallic nanoparticles for ethanol/O-2 biofuel cell. *Journal of Power Sources* **259**, 25–32.
148. S.A. Neto, R.D. Milton, L.B. Crepaldi, D.P. Hickey, A.R. de Andrade and S.D. Minteer (2015). Co-immobilization of gold nanoparticles with glucose oxidase to improve bioelectrocatalytic glucose oxidation. *Journal of Power Sources* **285**, 493–498.
149. N. Lalaoui, P. Rousselot-Pailley, V. Robert, Y. Mekmouche, R. Villalonga, M. Holzinger, S. Cosnier, T. Tron, A. Le Goff (2016). Direct electron transfer between a site-specific pyrene-modified laccase and carbon nanotube/gold nanoparticle supramolecular assemblies for bioelectrocatalytic dioxygen reduction. *Acs Catalysis* **6**(3), 1894–1900.
150. A. Zebda, S. Tingry, C. Innocent, S. Cosnier, C. Forano and C. Mousty (2011). Hybrid layered double hydroxides-polypyrrole composites for construction of glucose/O-2 biofuel cell. *Electrochimica Acta* **56**(28), 10378–10384.
151. O. Parlak, P. Seshadri, I. Lundstrom, A.P.F. Turner and A. Tiwari (2014). Two-Dimensional Gold-Tungsten Disulphide Bio-Interface for High-Throughput Electrocatalytic Nano-Bioreactors. *Advanced Materials Interfaces* **1**(6).
152. M. Chhowalla, H.S. Shin, G. Eda, L.J. Li, K.P. Loh and H. Zhang (2013). The chemistry of two-dimensional layered transition metal dichalcogenide nanosheets. *Nature Chemistry* **5**(4), 263–275.
153. J. Yang, D. Voiry, S.J. Ahn, D. Kang, A.Y. Kim, M. Chhowalla and H.S. Shin (2013). Two-dimensional hybrid nanosheets of tungsten disulfide and reduced graphene oxide as catalysts for enhanced hydrogen evolution. *Angewandte Chemie-International Edition* **52**(51), 13751–13754.

Chapter 10

Carbonaceous Electrodes Featuring Tunable Mesopores for Use as Enzyme Electrodes

Seiya Tsujimura
Division of Materials Science,
Faculty of Pure and Applied Sciences,
University of Tsukuba, 1-1-1 Tennodai,
Tsukuba, Ibaraki 305-8573, Japan
seiya@ims.tsukuba.ac.jp

10.1 Introduction: Designing Porous Carbon Electrodes

An enzymatic biofuel cell (EBFC) is a type of fuel cell in which a purified enzyme is used as the electrocatalyst to convert the chemical energy of the oxidation reaction of reducing substances (such as sugars and alcohols) directly into electricity under ambient conditions: 10–40°C, normal atmospheric pressure, and a moderate pH range of 5–8. At the anode, the electrons captured from the reducing fuels by specific enzymes are transferred to the electrode through the direct or mediated electron-transfer pathway. The electrons transferred against external resistance are passed to O_2 molecules with the aid of Cu-containing enzymes to form water at the cathode. EBFCs minimally comprise anodic and cathodic enzymes and the electrodes, and do not require a case to seal the cell components or a separator between the anode and the cathode.

This simple structure facilitates straightforward miniaturization of EBFCs. Enzyme-catalyzed electrode reactions are classified into two types: direct0 electron transfer (DET), which involves direct transfer of electrons between enzymes and electrodes; and mediated electron transfer (MET), in which mediators transfer the electron between enzymes and electrodes. EBFCs are expected to become a ubiquitous power device in the field of communications and in medical (health-care) applications: EBFCs can power *in vivo* bioelectronics such as biosensors (*e.g.* self-driven blood-glucose sensors) or pacemakers implanted in the living body, and EBFCs are also expected to power mobile communication electronic devices and microscale machines. However, when compared with the application of conventional small power generators, the practical use of EBFCs is hindered by major barriers related to low durability and low output power density. Nevertheless, the highly favorable properties and the potential applications of EBFCs have stimulated intense interest in the basic study and development of EBFCs. In recent years, numerous review articles on EBFCs have been published,[1–8] which can serve as a useful supplement to this review.

Two critical shortcomings impede the application of EBFCs, short lifetime and low power density, and these are related to the instability of enzymes on the electrode surface and a large barrier to overcome for electron transfer between the enzyme active site and electrode surface,[2,9–11] and to the limited amount of electroactive enzyme that can be loaded on the electrode surface. To improve EBFC performance, 3D electrode materials and protein-engineered redox enzymes have been developed, and numerous attempts have been made to immobilize enzymes by using various nanostructured materials, including mesoporous media, nanoparticles, and nanofibers.[7,12] The large specific surface area of nanostructured materials is expected to enable high enzyme loading. Moreover, enzyme stabilization in nanostructured materials has been reported to extend the lifetime of enzymes.[13–15] For EBFC applications, the multi-dimensionality of the pore structure (hierarchical pore structure) provides both mesopores for enzyme immobilization and macropores to support the mass transport of fuels. This chapter reviews the most recent developments in porous carbon electrodes featuring a controlled pore size for EBFC applications.

10.2 Nanostructured Carbon-based Electrodes

Recently, numerous nanostructured carbon materials, including carbon nanotubes (CNTs), graphene, and carbon black, have become readily available for biofuel-cell studies, and this has allowed remarkable progress in enzyme-electrode technologies. These nanostructured materials are typically deposited on a current collector such as carbon, Ti, Au, or stainless steel, with the material being selected based on considering the potential window, the corrosion of the substrates, and the adhesiveness of the carbon materials on the substrate. Normally, to stably deposit carbon materials on the current collector, a polymer such as polyvinylidene difluoride (PVDF; poly-1,1-difluoroethene), polytetrafluoroethylene (PTFE; poly(1,1,2,2-tetrafluoroethylene)), or styrene–butadiene rubber (SBR) is used as a binder. This chapter focuses on carbon blacks and mesoporous carbons as the carbon materials and describes the use of these materials for EBFCs.

10.2.1 Carbon blacks

Carbon blacks, for example Vulcan XC-72 (Cabot, USA) and Ketjen black (KB; Lion Specialty Chemicals Co., Ltd, Japan), have already been used in the electrochemical fields of fuel cells, batteries, and capacitors as a catalyst support or conduction assistant. KB provides high electron conductivity with smaller amounts of loading as compared to other carbon blacks. KB features a primary particle diameter of 40 nm and a high specific surface area of 800–1,200 m^2 g^{-1}. The KB-modified electrode is fabricated by cast-coating or printing a KB ink on a substrate (current collector); the KB ink is typically prepared by mixing a KB powder with a binder by using a mortar and pestle, planetary centrifugal mixer, or ultrasonicator. For example, KB was mixed with 20 wt.% PVDF in *N*-methylpyrrolidone (NMP), and after homogenization, the dispersion was drop casted onto a carbon substrate; the resulting electrodes were then dried to evaporate the NMP.[16]

KB-modified electrodes have been used for DET and MET reaction platforms to realize an increase in current density. The first reported use of a KB-modified electrode for DET-type bioelectrocatalysis was for fructose oxidation by D-fructose dehydrogenase (FDH) from acetic acid bacteria.[16] The DET-type catalytic current density of FDH on a flat electrode was previously reported to be as low as a few μA cm^{-2}, which could be adequate for use in sensor applications but would be challenging to exploit in EBFC applications.[17] When FDH was adsorbed on a KB-modified electrode prepared according to the aforementioned method by dipping the electrode in an FDH solution, the modified electrode produced 10 mA cm^{-2} of fructose-oxidation current density per unit geometric surface area.[16] Moreover, a KB-modified carbon-felt electrode was used in a batch-type coulometric D-fructose biosensor based on the DET reaction of FDH.[18] The electric charge gained from D-fructose oxidation as determined through the biocoulometric measurement agreed closely with the theoretical value corresponding to the D-fructose amount in the 1–100-mM range with a sample volume of 1 μL.

KB-based electrodes were also used for developing a high-performance biocathode, in which CueO from *Escherichia coli* was used as the O_2-reduction biocatalyst.[19] The current density reached a value as high as 12 mA cm^{-2} at 1 atm O_2, 25°C, and 10,000 rpm on the CueO-modified KB electrode in the absence of redox mediators. The CueO-modified KB-based cathode was considerably superior to the smooth CueO cathode based on highly oriented pyrolytic graphite (HOPG) in terms of current density, electrode kinetics, pH tolerance, and thermal stability.

Conversely, Mano *et al.* co-immobilized a redox polymer and glucose oxidase (GOx) on a KB-based electrode. A current density of up to 5.1 mA cm^{-2} was achieved on the KB-modified electrode, which was 60% higher than that on a glassy carbon (GC) electrode.[20] KB features a large Brunauer–Emmett–Teller (BET) specific surface area, but the wide pore-size distribution of the KB-modified electrode suggested that a portion of its surface was inefficiently used for enzyme loading and electrochemical reactions. Moreover, the hydrogel formed in the mesopores located on the surface of the KB layer could readily prevent the mass transfer of glucose and also of GOx and the redox polymer.

The large active surface area of KB can be chemically modified. Nishizawa *et al.* studied the co-immobilization of oxidized nicotinamide adenine dinucleotide (NAD^+) with glucose dehydrogenase (GDH) on a chemically oxidized KB electrode for use as a biofuel-cell anode.[21] The GDH/NAD^+/KB electrode allowed the electrochemical oxidation of glucose at around 0 V *vs.* Ag|AgCl. Moreover, a biofuel cell was constructed with a KB cathode on which bilirubin oxidase (BOD) was immobilized, and it exhibited a maximal power density of 52 μW cm^{-2} at 0.3 V.

KB can also be applied to the air-diffusion-type biocathode, which is an enzymatic cathode that uses gas-phase O_2; the cathode consists of a hydrophobic gas-diffusion layer and a hydrophilic catalytic layer and forms a triphase interface composed of a liquid electrolyte, the solid electrode, and the air phase. The cell performance operating in a liquid is limited by the mass transfer of O_2 toward the cathode because the concentration of the O_2 that is dissolved in an air-saturated electrolyte solution is no more than 0.25 mM and the diffusion coefficient of O_2 is lower in the aqueous phase than in the gaseous phase. The liquid electrolyte offers an environment (moderate temperature and pH) for the biocatalyst reaction to occur and serves as an ion-transport medium from the anode to the cathode. To improve the mass transport of O_2 in the gas phase, it is crucial to prevent the electrode surface from flooding the electrolyte solution. The first air-diffusion-type biocathode was fabricated by coating a mixture of KB and PTFE binder to control the hydrophobicity/hydrophilicity on Toray carbon paper (CP), and CueO from *E. coli* was used as the O_2-reduction catalyst.[22] The steady-state catalytic current density was measured to be as high as 20 mA cm^{-2} in the passive mode under 1 atm without a fan or pump. The modified electrode was prepared as follows. CP or a porous carbon substrate was used as a gas-permeable current collector. The KB and PTFE powders in 2-propanal were mixed using a tip-type homogenizer or a mortar and pestle, and the slurries were applied on the surface of the CP and dried in a drying oven to evaporate the solvent. Lastly, the KB–CP electrodes were immersed in the enzyme solution to adsorb the enzyme on the KB materials, which generated the enzyme-adsorbed KB–CP electrode. The electrode was extremely hydrophobic and thus a surfactant such as Triton X-100 was added to the enzyme solution in order to increase the penetration of the enzyme through the carbon layer.[23]

Shitanda *et al.* printed a flexible and high-power biofuel cell on a Japanese paper substrate by using KB inks and a screen printer (0.12 mW cm^{-2} (at 0.4 V)).[24] The current density of the gas-diffusion-type O_2-reduction cathode was enhanced by adjusting the hydrophobicity of the porous-carbon electrodes by using a hydrophobic PTFE binder to control the hydrophilic/hydrophobic property. The carbon ink for the anode was fabricated with an SBR binder, which was dispersed in water with carboxylmethyl cellulose to make the carbon layer hydrophilic.

10.2.2 Mesoporous carbon

As noted in the preceding section, carbon black is a high-surface-area material, and it is commercially available and inexpensive; however, controlling the hierarchical pore structures of carbon black is challenging, which could lower the usage efficiency of the pore structure.

10.2.2.1 *Mesoporous silica-templated carbon*

The encapsulation of enzymes and other proteins in inorganic host materials such as ordered mesoporous silica structures (*e.g.* MCM-48, MCM-41 and SBA-15) has attracted considerable attention with regard to stabilization of the enzyme structure.[14] However, most porous silica-based materials are electronic semiconductors (insulators from the viewpoint of electrochemical applications). The mesoporous silica structure can be successfully inherited to prepare carbons featuring mesopores in various ordering (ordered mesoporous carbons), as described in previous reviews with a focus on silica-templated carbonization.[25,26] The pores in the template silica are replicated in the carbons through either impregnation or chemical vapor infiltration of a carbon precursor, and this is followed by the carbonization and removal of the templates, mostly using HF. Highly ordered mesoporous carbons based on the mesoporous silica-templated carbon have been receiving a substantial amount attention because of their extremely high surface area, defined pore size, and electronic conductivity.

Guo *et al.* reported a hydrophilic SBA-15-templated carbon prepared from glucose and poly(vinyl alcohol) as a carbon precursor.[27] The mesopore diameter was 4 nm, which was too small for encapsulating GOx: the enzyme features a dimensional size of 5–10 nm. Direct electron exchange between an electrode and the enzyme's cofactor flavin adenine dinucleotide (FAD), which is buried in a cleft in GOx, occurs extremely slowly. However, an ordered porous surface that is highly hydrophilic, such as the one used here, can enhance the electron transfer between FAD and an electrode. The carbon electrode on which GOx was immobilized showed clear bioelectrocatalysis of glucose oxidation starting at around −0.5 V *vs.* Ag|AgCl, which was attributed to the redox reaction of FAD/$FADH_2$. A previous study suggested that CNTs could promote the DET to GOx by protruding into the enzyme and reducing the electron-transfer distance to the buried flavin, but direct electrocatalysis was not achieved at a reasonable overpotential.[28]

You *et al.* also reported the effect of carbon structure on the electrochemistry of immobilized GOx on 2D CMK-3 and 3D CMK-1-like carbons, which were prepared by using the mesoporous silica FDU-5 as a template. FAD-related redox signals were observed on both electrodes and the electrochemical signal on the 3D electrode was more stable than that on 2D electrode. However, the glucose-oxidation catalytic current based on DET from FAD to the electrode was not observed.[29]

Zhou *et al.* reported a glucose/O_2 biofuel cell prepared with mesoporous carbon by using SBA-15 as the template and sucrose as the carbon precursor.[30,31] Meldola's Blue (MDB), a redox catalyst for NADH oxidation, was modified on the carbon surface, and

NAD-GDH was immobilized on the MDB-modified carbon by cross-linking with BSA. MDB was immobilized in the mesopores, and the enzyme, which was considerably larger than the mesopores, might have capped the mesopores. The open-circuit voltage of the biofuel cell was 0.82 V and 38.7 μW cm^{-2} at 0.54 V, which is low considering the specific surface area.

In conclusion, porous silica-templated carbon can create highly ordered mesopores, but the pore diameter is considerably lower than the diameter of enzyme molecules. Thus, the material is not suitable for encapsulating enzymes within mesopores in order to increase enzyme stability or enzyme loading; however, the nanostructured surface might enhance electron transfer by shortening the distance between the enzyme active site and the surface edge.

10.2.2.2 *Carbon gels*

Carbon gels (CGs) are synthesized using a resorcinol–formaldehyde (RF) sol–gel process, which is followed by either freeze-drying and carbonization (carbon cryogel, CCG) or CO_2–supercritical drying (carbon aerogel, CAG)[32–34]; CGs possess several critical and useful properties, including controllable pore sizes, high porosity, high surface area, and high electrical conductivity, all of which markedly depend on the synthesis and processing conditions.[35] Notably, CGs harboring large pores (larger than enzyme molecules) can be synthesized, and CGs have been investigated as electrode materials for the construction of bioelectrochemical devices, including biofuel cells and biosensors, to improve the electron path and the enzyme immobilization process. Previous studies have reported that the mean pore size of materials used for enzyme encapsulation must be at least four to five times larger than the mean molecular diameter of enzymes in order to guarantee unrestricted access.[36,37] Therefore, CGs featuring large pores must be prepared in order to enable effective enzyme loading. The mean diameter of the pore formed in a CG can be increased by lowering the concentration of the CO_3^{2-} catalyst (C) solution at a given ratio of resorcinol to water (R/W) and by decreasing the R/W ratio at a given R/C ratio in the process of gel synthesis. Previously, the average pore radius in the 5–40-nm range was controlled by changing the molar ratio of resorcinol to the catalyst.[38] However, CGs featuring a mean pore diameter of >40 nm could not be readily prepared with a high degree of reproducibility because the extremely low catalyst concentration (Na_2CO_3) could be altered by certain contaminants. CCGs harboring comparatively larger pores (14–108 nm) were synthesized using a sol–gel process from resorcinol and formaldehyde in carbonate buffer solutions by adjusting the total concentrations and pH of the catalyst components; the concentration of Na_2CO_3 in the sol–gel process was increased to stabilize the pH (buffer action), and the catalyst (CO_3^{2-}) concentration was controlled by adjusting the pH.[39]

FDH adsorbed on a CCG electrode exhibited large catalytic currents for fructose oxidation in the absence of redox mediators.[38] The catalytic current depended on the pore size, and the catalytic current density was very low when the CCG pore was smaller than the FDH molecule. Conversely, under the experimental conditions used, comparatively wider pores resulted in a higher catalytic current density. Thus, the enzymes caged within the carbon mesopores might be suitably positioned to communicate with the electrode directly.

A previous study demonstrated that the particle size of carbon could also be a critical factor determining the performance of an electrode: the measured catalytic current depended on the mass transfer of the substrate not only from the bulk solution to the surface of the carbon particle, but also inside the carbon particle.[40] The porous enzyme electrode harboring mesopores exhibited a high bioelectrocatalytic current when the substrate was sufficiently supplied to the FDH present inside the CG.

Flexer *et al.* presented a DET reaction of soluble pyrroloquinoline quinone (PQQ)-GDH from *Acinetobacter calcoaceticus* on CCG electrodes.[41] The response current depended on the glucose concentration, and the highest glucose-oxidation catalytic current density was 1 mA cm^{-2}, which was markedly higher than that on other materials.

CGs were also successfully used for an O_2-reducing biocathode, in which multi-copper oxidases were used as electrocatalysts.[42–45]

Diffusion-controlled O_2 reduction into water was archived without mediators by using enzymes that were physically adsorbed on CAG particles; the enzymes used were laccase from *Trametes* sp., BOD from *Myrothecium verrucaria*, and recombinant CueO from *E. coli*. The current density was predominantly controlled by the diffusion of dissolved O_2 in rotating-disk-electrode experiments and reached a value as high as 10 mA cm^{-2} at 1 atm O_2, 25°C, and 8,000 rpm on the laccase-adsorbed electrode. The laccase-adsorbed CAG-based biocathode showed stable O_2 reduction current in an air-saturated buffer for at least 10 days under a continuous-flow system; by contrast, the activity of laccase dissolved in a buffer solution decreased to half the initial activity during incubation for 5 days. Another study reported the bioelectrocatalytic reactions of a newly identified thermophilic BOD from *Bacillus pumilus* that was adsorbed on a CCG electrode in the presence and absence of a mediator.[46] BOD adsorbed on a KB electrode also showed DET-type electrochemical O_2-reduction activity, but the electrochemical response on the CCG electrode was more stable than that on the KB electrode, which could be attributed to the encapsulation of BOD by the mesoporous carbon wall.

10.2.2.3 *MgO-templated carbon*

CG technology can be effectively used for producing a desired porous carbon featuring a controlled pore size; however, the production process is time-consuming and requires a large amount of organic solvents, which is unsuitable for factory-scale mass production. A promising alternative porous material could be magnesium oxide (MgO)-templated mesoporous carbon (MgOC).[47] One of the notable advantages of MgOC is the simplicity of its production procedure as compared with the procedures reported previously for mesoporous carbons such as CCGs and CAGs. MgOC is fabricated as follows: A mixture of a carbon precursor, such as poly(vinyl alcohol), with the template MgO particles is heated in a furnace, and then the template MgO is removed by washing the carbon with a diluted acid such as HCl or H_2SO_4 to obtain the porous carbon materials.

MgOC featuring an average pore diameter of 38 nm and a BET specific surface area of 580 $m^2 g^{-1}$ was used as the electrode material for a glucose-oxidizing anode.[48] The MgOC was modified on a GC electrode by using the cast-drop technique. Anodic catalysts composed of a redox polymer, GOx, and a cross-linker were applied to the MgOC-modified electrodes. The glucose-oxidation current density on MgOC was 11.3 mA cm^{-2} at 0.5 V, which was 5.3-fold higher than that on GC.

FDH also shows improved current-production efficiency and stability when MgOC featuring an adequate mesopore size is used.[49] When the MgOC pores were substantially larger than FDH molecules (15–20 nm), a sufficient amount of FDH was adsorbed within the mesopores on and even inside the MgOC structure. By contrast, when the MgOC pore was comparable in size to FDH molecules, the catalytic current depended only on the amount of enzyme adsorbed in the mesopores formed at the surface of the carbon particles; however, FDH showed enhanced thermal stability when encaged in carbon mesopores that were comparable in size to the enzyme molecules.

10.3 Hierarchically Structured Carbon Electrodes

To increase enzyme usage efficiency, an electrode material is required that concurrently meets the criteria of a large surface area for efficient bioelectrocatalytic reactions (structure-controlled mesopores) and a rapid mass transport of reactants (macropores). Moreover, given that the electron-exchange reaction is increased near the electrode surface, the maximal catalytic current density would be suppressed as a result of substrate consumption. To successfully satisfy the requirements of both a large specific surface area for enzyme loading and the rapid mass transport of fuel, a nano/microhierarchically structured carbon electrode would be necessary: If pores are considerably larger than the enzyme molecules, leaching and low loading can occur, but if the pores are small, enzyme penetration will be prevented. Distinct pore sizes can be obtained by mixing materials harboring pores of different scales, and for constructing hierarchically structured carbon electrodes, two strategies could be considered: in one strategy, the macroporous structure is constructed first and then the mesopores are formed on the surface of the macroporous material, whereas in the other strategy, the macroporous structure is built upon mesoporous carbon materials.

10.3.1 Mesoporous carbon formed on macrostructured carbon

Macroporous structures can be obtained by using a carbon-fiber-based carbon texture such as carbon cloth, carbon felt, CP, or carbon foam, all of which are templated carbons prepared using silica or polymer particles as the template. The mesopores can be formed on the macroporous structure by (1) depositing mesoporous carbon, (2) growing CNTs though chemical vapor deposition (CVD), and (3) forming pores on the substrate through activation treatment performed using CO_2 or H_2O with metal catalysts.

Barton *et al.* reported a CNT-modified CP or carbon-fiber electrode that was prepared using CVD, which led to a 100-fold increase in surface area and a 10-fold increase in current density (22 mA cm^{-2} of glucose current at 37.5°C); CNT modification is a promising technique for fabricating a tailored hierarchical structure, but the process requires special equipment and might be impractical for mass production.[50,51] Another study reported efficient O_2 reduction catalyzed by BOD by using a redox hydrogel on a hierarchically structured carbon material obtained by growing carbon fibers on highly porous reticulated vitreous carbon.[52] Moreover, Lau *et al.* developed chitosan scaffolds doped with CNTs as 3D hieratically structured electrodes.[53] The porous and interconnected structures were formed through the process of thermally induced phase separation followed by freeze-drying applied to an aqueous solution of chitosan acetic acid. Multi-walled CNTs provided electrical conductivity throughout the chitosan structure. Flexer *et al.* reported a new meso-macropore carbon material in which the macropores were prepared from an Si template and the mesopores were introduced by using a porous carbon precursor; the 3D macrocellular carbonaceous foam electrode modified with 1,734 μg cm^{-2} hydrogel generated 18 mA cm^{-2} at 37°C and 2,000 rpm.[54]

10.3.2 Carbon particles deposited using electrophoretic deposition

Macropores can be formed using the electrophoretic deposition (EPD) method, in which carbon particles dispersed in a solvent with a binder are forced to migrate toward an electrode by applying an electric field (DC 10–200 V). The MgOC particles are collected on the electrode and form a compact macroporous deposit. The macropore size (morphology) and the carbon-layer thickness can be controlled by tuning the applied voltage and the distance between the electrodes, the carbon concentration, and the electrophoresis time. The macropore structure can be controlled by altering the applied voltage and the distance between the two electrodes. For example, in a previous study, at low applied voltage, no pores larger than 10 μm were observed. To create a glucose bioanode, the porous-carbon-modified electrode was further coated with a biocatalytic hydrogel composed of a conductive redox polymer, deglycosylated FAD-dependent GDH (*d*-FAD-GDH), and a cross-linker.[55] The glucose-oxidation current density was 100 mA cm^{-2} at 25°C and pH 7 with a hydrogel loading of 1.0 mg cm^{-2}. For the same hydrogel composition and loading, the current density on the MgOC-modified electrode was >30 times higher than that on a flat carbon electrode. In the case of MgOC applied (using the drop casting technique) on a GC electrode, which lacked macropores larger than 10 μm, the maximal current was <40 mA cm^{-2}, even at 9,000 rpm. The maximal current was ascribed to the limitations of biocatalyst and glucose transport in the carbon-layer structure on the GC support. Furthermore, the stability of the hydrogel electrode was enhanced through the use of mesoporous carbon materials: >95% of the initial catalytic current remained after storage for 220 days in phosphate buffer at 4°C, and 80% of the activity was observed after 7 days of continuous operation at 25°C. The electrochemical response of FAD-GDH in the hydrogel on the GC electrode was diminished at 37°C after 2 days of continuous operation in 0.1 M phosphate buffer at pH 7, but the response stability was successfully improved by immobilizing the hydrogel on the porous carbon structure by encapsulating the hydrogel within mesopores.[56] A glucose anode based on a GOx-hydrogel consisting of an Os-complex-tethered polymer produced a glucose-oxidation current as high as 60 mA cm^{-2} at a loading of 1,000 μg cm^{-2} at 37°C.[57]

Lastly, a macro-meso-hierarchical structure was shown to enable efficient BOD-catalyzed electrochemical O_2 reduction.[58] BOD was adsorbed on an MgOC electrode together with the redox mediator 2,2′-azinobis(3-ethylbenzothiazolin-6-sulfonate) (ABTS). The ABTS molecules adsorbed on the surface of MgOC exhibited extremely stable redox behavior without disproportionation. The ABTS adsorbed on the electrode surface accelerated the DET and MET rates by improving the orientation of BOD on the carbon surface and/or by enhancing enzyme loading (which favored electron transfer), and by mediating the electron transfer from the electrode to the electroactive site of BOD. The maximal catalytic current was doubled in the presence of ABTS across a wide pH range.

10.4 Outlook

This chapter reviews recent developments in porous carbon electrodes featuring a controlled pore size for use in EBFCs to overcome the two critical weak points of EBFC, a short lifetime and low power density. The large surface area of mesoporous carbon materials can increase the loading of enzyme in the vicinity of electrode surface and heterogeneous electron-transfer efficiency for MET and DET reactions, and the macropores can enable rapid penetration of enzymes into the porous carbon layer during enzyme-electrode fabrication and efficient fuel mass transfer during the bioelectrocatalytic reaction. Combining an efficient electron-transfer technology developed by modifying the microscopic interface between enzymes and electrodes with a hierarchically structured carbon material can help in achieving considerably increased and stable current output with the use of a small amount of enzyme, and thus contribute to a practical advance in EBFC technology.

List of Abbreviations

ABTS	2,2′-azino-bis(3-ethylbenzothiazoline-6-sulfonate)
BET	Brunauer–Emmett–Teller
BOD	bilirubin oxidase

BSA	bovine serum albumin
CNT	carbon nanotube
CG	carbon gel
CAG	carbon aerogel
CCG	carbon cryogel
CP	carbon paper
CVD	chemical vapor deposition
DET	direct electron transfer
EBFC	enzymatic biofuel cell
e.g.	*exempli gratia*
EPD	electrophoretic deposition
FAD	flavin adenine dinucleotide
FDH	D-fructose dehydrogenase
GC	glassy carbon electrode
GDH	glucose dehydrogenase
GOx	glucose oxidase
HOPG	highly oriented pyrolytic graphite
KB	Ketjen black
MDB	Meldola's blue
MET	mediated electron transfer
MgOC	magnesium oxide-templated mesoporous carbon
NAD	nicotinamide adenine dinucleotide
NAD^+/NADH	oxidized and reduced state (with two electrons and one proton) of NAD
NMP	N-methylpyrrolidone
PQQ	pyrroloquinoline quinone
PTFE	poly(1,1,2,2-tetrafluoroethylene)
PVDF	poly-1,1-difluoroethene
SBR	styrene-butadiene rubber

References

1. S. B. Calabrese, J. Gallaway and P. Atanassov (2004). Enzymatic biofuel cells for implantable and microscale devices. *Chemical Reviews* **104**(10), 4867–4886.
2. J.A. Cracknell, K.A. Vincent and F.A. Armstrong (2008). Enzymes as working or inspirational electrocatalysts for fuel cells and electrolysis. *Chemical Reviews* **108**(7), 2439–2461.

3. G.T.R. Palmore and G.M. Whitesides (1994). Microbial and enzymatic biofuel cells. In *Enzymatic Conversion of Biomass for Fuels Production*, ACS Symposium Series, Vol. 566, pp. 271–290.
4. M.H. Osman, A.A. Shah and F.C. Walsh (2011). Recent progress and continuing challenges in bio-fuel cells. Part I: Enzymatic cells. *Biosensors and Bioelectronics* **26**(7), 3087–3102.
5. D. Leech, P. Kavanagh and W. Schuhmann (2012). Enzymatic fuel cells: Recent progress. *Electrochimica Acta*, **84**, 223–234.
6. S.D. Minteer, P. Atanassov, H.R. Luckarift and G.R. Johnson (2012). New materials for biological fuel cells. *Materials Today* **15**(4), 166–173.
7. A. De Poulpiquet, A. Ciaccafava and E. Lojou (2014). New trends in enzyme immobilization at nanostructured interfaces for efficient electrocatalysis in biofuel cells. *Electrochimica Acta* **126**, 104–114.
8. M. Holzinger, A. Le Goff and S. Cosnier (2012). Carbon nanotube/enzyme biofuel cells. *Electrochimica Acta* **82**, 179–190.
9. R.A. Marcus and N. Sutin (1985). Electron transfers in chemistry and biology. *Biochimica et Biophysica Acta (BBA)-Reviews on Bioenergetics* **811**(3), 265–322.
10. H.B. Gray and J.R. Winkler (2005). Long-range electron transfer. *Proceedings of the National Academy of Sciences of the United States of America* **102**(10), 3534–3539.
11. C. Léger and P. Bertrand (2008). Direct electrochemistry of redox enzymes as a tool for mechanistic studies. *Chemical Reviews* **108**(7), 2379–2438.
12. J. Kim, H. Jia and P. Wang (2006). Challenges in biocatalysis for enzyme-based biofuel cells. *Biotechnology Advances* **24**(3), 296–308.
13. C.H. Lee, T.S. Lin and C.Y. Mou (2009). Mesoporous materials for encapsulating enzymes. *Nano Today* **4**(2), 165–179.
14. M. Hartmann (2005). Ordered mesoporous materials for bioadsorption and biocatalysis. *Chemistry of Materials* **17**(12), 4577–4593.
15. H. Takahashi, B. Li, T. Sasaki, C. Miyazaki, T. Kajino and S. Inagaki (2000). Catalytic activity in inorganic solvents and stability of immobilized enzymes depend on the pore size and surface characteristics of mesoporous silica. *Chemistry of Materials* **12**(11), 3301–3305.
16. Y. Kamitaka, S. Tsujimura and K. Kano (2007). High current density bioelectrolysis of D-fructose at fructose dehydrogenase-adsorbed and Ketjen black-modified electrodes without a mediator. *Chemistry Letters* **36**(2), 218–219.
17. T. Ikeda, F. Matsushita and M. Senda (1991). Amperometric fructose sensor based on direct bioelectrocatalysis. *Biosensors and Bioelectronics* **6**(4), 299–304.
18. S. Tsujimura, A. Nishina, Y. Kamitaka and K. Kano (2009). Coulometric D-fructose biosensor based on direct electron transfer using D-fructose dehydrogenase. *Analytical Chemistry* **81**(22), 9383–9387.
19. S. Tsujimura, Y. Miura and K. Kano (2008). CueO-immobilized porous carbon electrode exhibiting improved performance of electrochemical reduction of dioxygen to water. *Electrochimica Acta* **53**(18), 5716–5720.

20. E. Suraniti, S. Vivès, S. Tsujimura and N. Mano (2013). Designing thin films of redox hydrogel for highly efficient enzymatic anodes. *Journal of The Electrochemical Society* **160**(6), G79–G82.
21. T. Miyake, M. Oike, S Yoshino, Y. Yatagawa, K. Haneda, H. Kaji and M. Nishizawa (2009). Biofuel cell anode: NAD^+/glucose dehydrogenase-coimmobilized ketjenblack electrode. *Chemical Physics Letters* **480**(1), 123–126.
22. R. Kontani, S. Tsujimura and K. Kano (2009). Air diffusion biocathode with CueO as electrocatalyst adsorbed on carbon particle-modified electrodes. *Bioelectrochemistry* **76**(1), 10–13.
23. I. Asano, Y. Hamano, S. Tsujimura, O. Shirai and K. Kano (2012). Improved performance of gas-diffusion biocathode for oxygen reduction. *Electrochemistry* **80**(5), 324–326.
24. I. Shitanda, S. Kato, Y. Hoshi, M. Itagaki and S. Tsujimura (2013). Flexible and high-performance paper-based biofuel cells using printed porous carbon electrodes. *Chemical Communications* **49**(94), 11110–11112.
25. R. Ryoo, S.H. Joo, M. Kruk and M. Jaroniec (2001). Ordered mesoporous carbons. *Advanced Materials* **13**(9), 677–681.
26. J. Lee, S. Han and T. Hyeon (2004). Synthesis of new nanoporous carbon materials using nanostructured silica materials as templates. *Journal of Materials Chemistry* **14**(4), 478–486.
27. C.X. Guo, F.P. Hu, X.W. Lou and C.M. Li (2010). High-performance biofuel cell made with hydrophilic mesoporous carbon as electrode material. *Journal of Power Sources* **195**(13), 4090–4097.
28. D. Ivnitski, B. Branch, P. Atanassov and C. Apblett (2006). Glucose oxidase anode for biofuel cell based on direct electron transfer. *Electrochemistry Communications* **8**(8), 1204–1210.
29. C. You, X. Xu, B. Tian, J. Kong, D. Zhao and B. Liu (2009). Electrochemistry and biosensing of glucose oxidase based on mesoporous carbons with different spatially ordered dimensions. *Talanta* **78**(3), 705–710.
30. M. Zhou, L. Deng, D. Wen, L. Shang, L. Jin and S. Dong (2009). Highly ordered mesoporous carbons-based glucose/O_2 biofuel cell. *Biosensors and Bioelectronics* **24**(9), 2904–2908.
31. S. Jun, S.H. Joo, R. Ryoo, M. Kruk, M. Jaroniec, Z. Liu, T. Ohsuna and O. Terasaki (2000). Synthesis of new, nanoporous carbon with hexagonally ordered mesostructured. *Journal of the American Chemical Society* **122**(43), 10712–10713.
32. R.W. Pekala (1989). Organic aerogels from the polycondensation of resorcinol with formaldehyde. *Journal of Materials Science* **24**(9), 3221–3227.
33. H. Tamon, H. Ishizaka, T. Yamamoto and T. Suzuki (1999). Preparation of mesoporous carbon by freeze drying. *Carbon* **37**(12), 2049–2055.
34. T. Yamamoto, T. Nishimura, T. Suzuki and H. Tamon (2001). Control of mesoporosity of carbon gels prepared by sol–gel polycondensation and freeze drying. *Journal of Non-Crystalline Solids* **288**(1), 46–55.

35. S.A. Al-Muhtaseb and J.A. Ritter (2003). Preparation and properties of resorcinol–formaldehyde organic and carbon gels. *Advanced Materials* **15**(2), 101–114.
36. Y.J. Wang, T.C. Wu and C.L. Chiang (1998). Effect of pore size distribution on enzyme immobilization in porous supports. *AIChE Journal* **35**(9), 1551–1554.
37. J.A. Bosley and J.C. Clayton (1994). Blueprint for a lipase support: Use of hydrophobic controlled-pore glasses as model systems. *Biotechnology and Bioengineering* **43**(10), 934–938.
38. S. Tsujimura, A. Nishina, Y. Hamano, K. Kano and S. Shiraishi (2010). Electrochemical reaction of fructose dehydrogenase on carbon cryogel electrodes with controlled pore sizes. *Electrochemistry Communications* **12**(3), 446–449.
39. Y. Hamano, S. Tsujimura, O. Shirai and K. Kano (2014). Control of the pore size distribution of carbon cryogels by pH adjustments of catalyst solutions. *Materials Letters* **128**, 191–194.
40. Y. Hamano, S. Tsujimura, O. Shirai and K. Kano (2012). Micro-cubic monolithic carbon cryogel electrode for direct electron transfer reaction of fructose dehydrogenase. *Bioelectrochemistry* **88**(2012), 114–117.
41. V. Flexer, F. Durand, S. Tsujimura and N. Mano (2011). Efficient direct electron transfer of PQQ-glucose dehydrogenase on carbon cryogel electrodes at neutral pH. *Analytical Chemistry* **83**(14), 5721–5727.
42. S. Tsujimura, Y. Kamitaka and K. Kano (2007). Diffusion-controlled oxygen reduction on multi-copper oxidase-adsorbed carbon aerogel electrodes without mediator. *Fuel Cells* **7**(6), 463–469.
43. Y. Miura, S. Tsujimura, S. Kurose, Y. Kamitaka, K. Kataoka, T. Sakurai and K. Kano (2009). Direct electrochemistry of CueO and its mutants at residues to and near type I Cu for oxygen-reducing biocathode. *Fuel Cells* **9**(1), 70–78.
44. Y. Kamitaka, S. Tsujimura, N. Setoyama, T. Kajino and K. Kano (2007). Fructose/dioxygen biofuel cell based on direct electron transfer-type bioelectrocatalysis. *Physical Chemistry Chemical Physics* **9**(15), 1793–1801.
45. S. Tsujimura, M. Asahi, M. Goda-Tsutsumi, O. Shirai, K. Kano and K. Miyazaki (2013). Direct electron transfer to a metagenome-derived laccase fused to affinity tags near the electroactive copper site. *Physical Chemistry Chemical Physics* **15**(47), 20585–20589.
46. S. Tsujimura, E. Suraniti, F. Durand and N. Mano (2014). Oxygen reduction reactions of the thermostable bilirubin oxidase from *Bacillus pumilus* on mesoporous carbon-cryogel electrodes. *Electrochimica Acta* **117**, 263–267.
47. T. Morishita, T. Tsumura, M. Toyoda, J. Przepiórski, A.W. Morawski, H. Konno and M. Inagaki (2010). A review of the control of pore structure in MgO-templated nanoporous carbons. *Carbon* **48**(10), 2690–2707.
48. K. Murata, W. Akatsuka and S. Tsujimura (2014). Bioelectrocatalytic oxidation of glucose on MgO-templated mesoporous carbon-modified electrode. *Chemistry Letters* **43**(6), 928–930.

49. H. Funabashi, K. Murata and S. Tsujimura (2015). Effect of pore size of MgO-templated carbon on the direct electrochemistry of D-fructose dehydrogenase. *Electrochemistry* **83**(5), 372–375.
50. S. B. Calabrese, Y. Sun, B. Chandra, S. White and J. Hone (2007). Mediated enzyme electrodes with combined micro- and nanoscale supports. *Electrochemistry and Solid-State Letters* **10**(5), B96–B100.
51. H. Wen, V. Nallathambi, D. Chakraborty and S. B. Calabrese (2011). Carbon fiber microelectrodes modified with carbon nanotubes as a new support for immobilization of glucose oxidase. *Microchimica Acta* **175**(3–4), 283–289.
52. S.J. Little, S.F. Ralph, N. Mano, J. Chen and G.G. Wallace (2011). A novel enzymatic bioelectrode system combining a redox hydrogel with a carbon NanoWeb. *Chemical Communications* **47**(31), 8886–8888.
53. C. Lau, M.J. Cooney and P. Atanassov (2008). Conductive macroporous composite chitosan–carbon nanotube scaffolds. *Langmuir* **24**(13), 7004–7010.
54. V. Flexer, N. Brun, M. Destribats, R. Backov and N. Mano (2013). A novel three-dimensional macrocellular carbonaceous biofuel cell. *Physical Chemistry Chemical Physics* **15**(17), 6437–6445.
55. S. Tsujimura, K. Murata and W. Akatsuka (2014). Exceptionally high glucose current of a hierarchically structured porous carbon electrode with "wired" flavin adenine dinucleotide-dependent glucose dehydrogenase. *Journal of the American Chemical Society* **136**(41), 14432–14437.
56. A. Suzuki and S. Tsujimura (2016). Long-term continuous operation of FAD-dependent glucose dehydrogenase hydrogel-modified electrode at 37°C. *Chemistry Letters* **45**(4), 484–486.
57. A. Suzuki, K. Murata, N. Mano and S. Tsujimura (2016). Redox hydrogel of glucose oxidase on MgO-templated carbon electrode. *Bulletin of the Chemical Society of Japan* **89**(1), 24–26.
58. S. Tsujimura and K. Murata (2015). Electrochemical oxygen reduction catalyzed by bilirubin oxidase with the aid of 2,2′-azinobis(3-ethylbenzothiazolin-6-sulfonate) on a MgO-templated carbon electrode. *Electrochimica Acta* **180**, 555–559.

Section III

Electrodes for Microbial Electrochemical Systems

Chapter 11

Materials and Their Surface Modification for Use as Anode in Microbial Bioelectrochemical Systems

Kun Guo, Antonin Prévoteau, Sunil A. Patil, and Korneel Rabaey*

Center for Microbial Ecology and Technology, Ghent University, Coupure Links 653, B-9000 Ghent, Belgium

**korneel.rabaey@ugent.be*

11.1 Introduction

Microbial electrocatalysis refers to the use of microorganisms as catalysts to drive oxidation and/or reduction reactions at electrodes.[1] The processes developed on this principle are known as microbial electrochemical technologies (METs), the technology itself a microbial bioelectrochemical system (BES).[2] In the past decades, the potential applications of METs have been extended from power production (microbial fuel cells (MFCs)),[3] H_2 production (microbial electrolysis cells (MECs)),[4] water desalination (microbial desalination cells (MDCs)),[5] to valuable chemicals production from CO_2 (microbial electrosynthesis (MES)).[6] However, the practical applications of METs are still limited by low current densities and high capital costs.[7]

In the BES, one or both electrode reactions (anodic and cathodic) are catalyzed by microorganisms. Electrodes, the core part of BESs, act not only as the bacteria carrier but also as an electron acceptor or donor. In this context, the material type and surface properties of the electrodes dramatically impact the performance of BESs. Figure 1 represents the impact of electrode surface chemistry, topography, and configuration on the microbial electrocatalysis in BESs.

The materials of electrodes also determine the cost of BESs to a large extent.[9] Thus, finding cheap and high-performance electrode materials have been among the major focus of the BES researchers worldwide. The ideal materials for electrodes in BESs should possess the following properties: (1) good biocompatibility, (2) high specific and accessible surface area, (3) strong chemical stability (including corrosion resistance), (4) fast electron transfer at microbe–electrode interface, (5) high conductivity, (6) sufficient mechanical strength, (7) good processability and scalability, (8) low environmental impact (manufacturing footprint, recyclability), and (9) low cost.[8] However, none of the commercial electrode materials could meet all these requirements. Consequently, numerous studies have been carried out to modify the surface or the structure of commercial electrode materials to improve the performance of BESs. Moreover, researchers have also designed their own electrodes for specific purpose in BESs.

In this chapter, first, the commercial anode materials that are widely used for microbial electrocatalysis are reviewed, and their pros and cons are presented. Second, the anode surface modification methods that have been reported thus far are summarized and discussed. Third, purposely built anode materials for BESs are briefly presented. Finally, a brief discussion on the future research opportunities and directions on anodic materials for improving microbial electrocatalysis is provided.

11.2 Commercial Anode Materials

Anode materials used in BESs can be divided into two types: carbon-based and metal-based. They are commercially available in various configurations. Table 1 summarizes the most commonly used anode

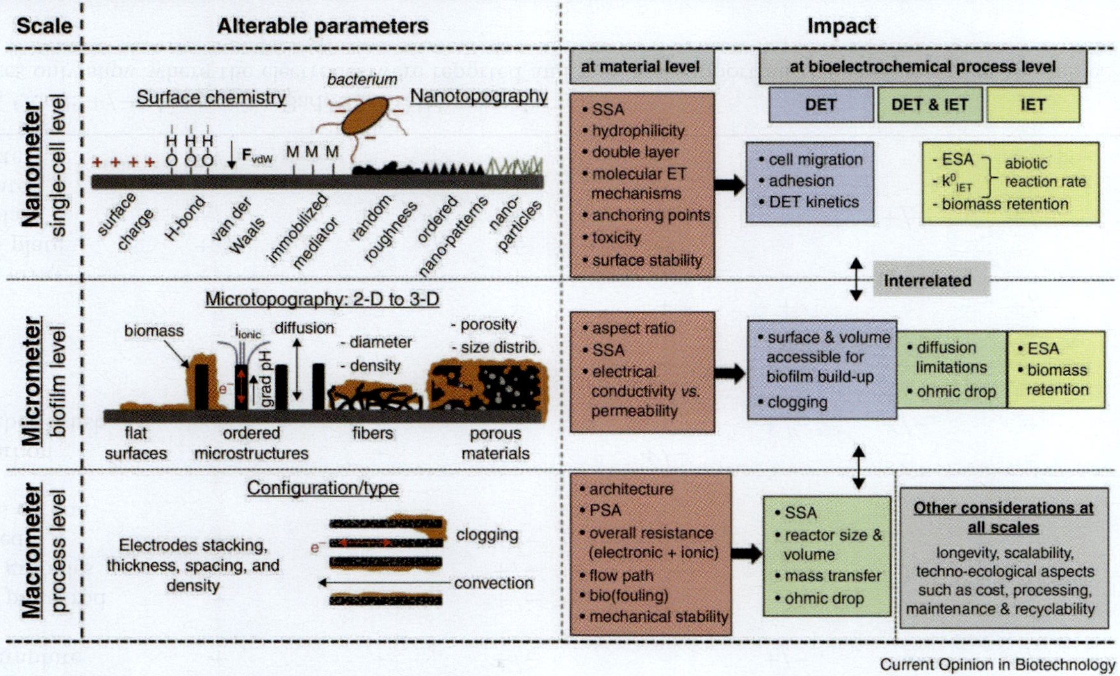

Figure 1. An overview of impact of material surfaces (chemistry and topography) on the electrode characteristics and microbial electrocatalysis in bioelectrochemical systems. 2D/3D; direct electron transfer (DET); indirect electron transfer (IET); heterogeneous electron transfer rate constant for abiotic reaction; projected surface area (PSA) [it is the largest 2-D area obtained from the projection of a planar surface or a 3D material; estimated with respect to the plane covering X and Y axes of the material]; specific surface area (SSA) [it is the total surface area of electrode per unit of projected surface area or electrode volume (for 3D architecture; often BET measurements are considered to estimate SSA)]; electroactive surface area (ESA) [it is the actual surface available for electrochemical reactions (*i.e.* solvated, electrically connected area); cyclic voltammetry for capacitance measurement can be used to estimate ESA].[8] (Reprinted from Ref. 15 with permission from Elsevier).

Table 1. A comparison of different commercial anode materials and configurations used in BESs.

Anode material	Biocompatibility	Chemical stability	Electrical conductivity	Mechanical strength	Specific surface area	Price superiority	Scale-up potential	References[a]
Carbon paper	+	+	+/−	−	−	−	−	20
Carbon cloth	+	+	+/−	+/−	+/−	−	+/−	21
Carbon mesh	+	+	+/−	+/−	+/−	+/−	+/−	22
Carbon/graphite felt	+	+	+/−	−	+/−	−	+/−	22
Graphite plate/rod	+	+	+/−	+/−	−	−	−	23
Graphite granules	+	+	+/−	+/−	+/−	+/−	−	24
Reticulated vitreous carbon (RVC)	+	+	+/−	−	+	−	−	19
Glassy carbon	+	+	+/−	+/−	−	−	−	25
Carbon fiber brush	+	+	+	+	+/−	+/−	+	26
SS plate	+/−	+/−	+	+	−	+/−	+/−	13
SS mesh	−	+/−	+	+	+/−	+	+	2
SS felt	−	+/−	+	+	+/−	+/−	+	11
SS foam	+/−	+/−	+	+	+	−	+/−	18
Gold line array	+/−	+	+	+	−	−	−	27
Titanium plate	+/−	+	+	+	−	−	−	28
Copper plate	+/−	−	+	+	−	+/−	+/−	29
Nickel plate	−	−	+	+	−	−	−	29
Silver plate	+/−	−	+	+	−	−	−	29

Notes: +, Good; +/−, Average; −, Bad; SS, Stainless steel.
[a]References only show where the electrodes were reported and may not support all the assessments in the table.

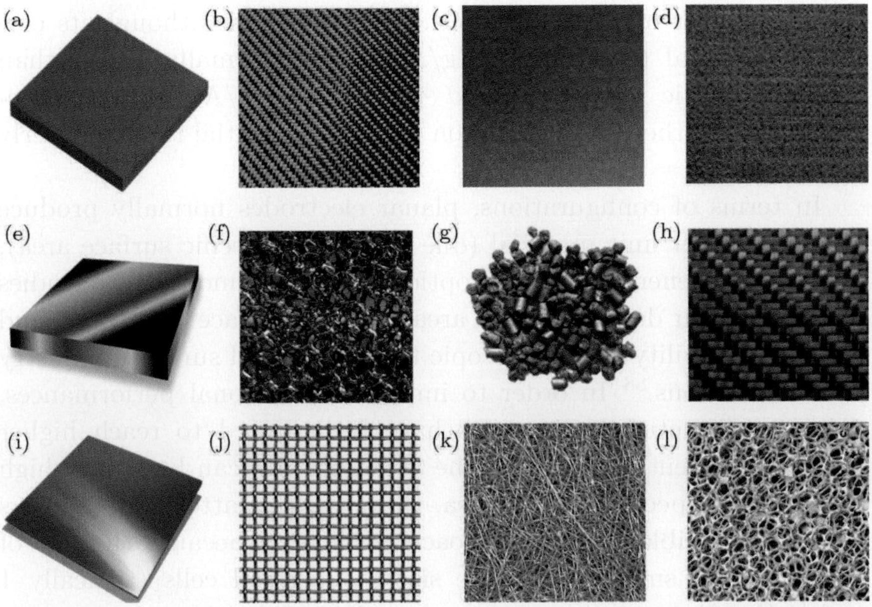

Figure 2. Pictures of the commercially available and commonly used anode materials in BESs. (a) graphite plate, (b) carbon cloth, (c) carbon paper, (d) carbon felt, (e) glassy carbon, (f) RVC, (g) graphite granules, (h) carbon mesh, (i) SS plate, (j) SS mesh, (k) SS fiber felt, (l) SS foam.

materials in BESs and provides a comparative overview of their properties. Pictures of the most commonly used carbon and stainless steel (SS) based anodes are shown in Figure 2.

Carbon-based materials have been extensively utilized due to their good biocompatibility and chemical stability.[9] However, relatively low conductivity and weak mechanical strength for most porous configurations of these materials prohibit their usage at large scales unless a metal current collector scaffold is included. Among all the metal electrodes proposed, SS present the best scale-up potential because of its low cost, high conductivity (even for porous configurations), chemical and mechanical stability, and easy processability. However, the passive layer on SS surface makes it less biocompatible and also hinders the electron transfer between electrode and bacteria. Consequently, the current density achieved on bare SS electrodes are normally lower than carbonaceous electrodes unless the system is operated under marine conditions.[10–13] Moreover, although its corrosion potential (+0.5 V *vs.* Ag/AgCl)[14] is normally higher than the BES anodic potential (max. +0.2 V *vs.* Ag/AgCl),[15] SS electrode also has the risk of corrosion when its potential is not properly controlled.[16]

In terms of configurations, planar electrodes normally produce less current per unit material (due to limited specific surface area), but they are generally the best options for most fundamental studies because of their defined surface area, uniform surface properties, and easy processability for microscopic techniques and surface chemistry characterizations.[88] In order to improve operational performances, many 3D structured electrodes have been tested to reach higher projected current densities.[89] The 3D electrodes can have very high porosity and specific surface area. However, the latter is not always entirely accessible to the electroactive bacteria because the size of some pores is smaller than the size of bacterial cells (typically 1 μm).[17] Certainly when wastewater or other solids containing streams are used as a substrate feed pore sizes below the millimeter level quickly become clogged. Among all the mentioned 3D electrodes, the macroporous SS/carbon foam and SS mesh could avoid the biofilm clogging issue.[18,19] However, the high price of SS and carbon foam makes them prohibitive for scaling up. SS mesh is much cheaper than SS/carbon foam and it can be stacked into multi-layers to further improve the specific surface area,[10] so it holds great promise for the scalable and high-performance anodes for BESs.

11.3 Anode Surface Modification Strategies

Electrode surface properties such as surface chemistry and topography dramatically influence the electroactive biofilm formation and microbial extracellular electron transfer (EET).[8] Therefore, electrode surface modification has been used as an efficient way to improve the anode performance and understand the associated phenomena. In the past decades, numerous investigations on anode surface modifications led to a considerable increase of current and power densities at least at laboratory scale, although no single modification has been applied widely by other research groups than the originating one thus far. The improvement of the microbial anode performances can be attributed to one or a combination of the following factors: (i) increase of surface area, (ii) facilitation of EET, (iii) enhancement of material conductivity, and (iv) improvement of surface biocompatibility. The strategies that are most commonly used to modify anode surfaces are classified into surface functionalization and thin film coating.

Surface functionalization introduces chemical functional groups to the electrode surface. The introduced functional groups impact the anodic biofilm formation rate and extent, and possibly the EET kinetics. It has been demonstrated that positively charged surface and hydrophilic surface favor thick electroactive biofilm formation and the associated current density.[30,31] This is likely due to the fact that electroactive bacteria are normally negatively charged and hydrophilic surface are more approachable for them. Table 2 summarizes the anode surface functionalization methods and their impact on the microbial anode performance.

Deposition of a thin-film on the surface is another efficient way to introduce additional properties and functionalities to electrode materials. Thin-film coating not only modifies the electrode surface chemistry but usually also changes the surface topography. Based on the coating materials, thin-film coating can be further divided into carbon(nanomaterials) coating, metal/metal oxide coating, and conductive polymer coating. Table 3 lists the different thin-film coating methods and their impact on microbial anode performance.

Since operational conditions (microbial inoculum, growth medium, temperature, convection, reactor design, electrode potential, etc.) and current/ power density calculation methods are often different in each study, it is complex to objectively compare microbial anode performances. Therefore, Tables 2 and 3 provide the relative improvement *vs.* the control due to the surface modification. It is very important to note that comparison of anodic performances should always been carried out in terms of maximal current density (at a specific potential, or, better, by providing an accurate cyclic

Table 2. A summary of anode surface functionalization methods and their impact on microbial anode performance.

Modification method	Anode material	Major change of surface	Performance enhancement	References
Ammonia treatment	Carbon cloth	Increase N/C ratio Introduce amine groups	Start-up time[a]: −50% Power density: +20%	21
	Carbon mesh		Power density[b]: +28%	22
Heat treatment	Carbon mesh	Decrease O/C ratio	Power density[b]: +15%	22
	Carbon brush	Increase N/C ratios Decrease C–O composition Increase surface area	Power density: +25%	26
	SS felt	Generate α-Fe_2O_3 layer	Current density: +7 times	32
Flame treatment	SS felt	Generate α-Fe_2O_3 layer	Current density: +16.5 times	11
Electrochemical oxidation	Carbon felt	Generate –COOH groups	Current density: +40%	33
	Carbon plate	Introduce O-containing groups	Kinetic activity[c]: +7 times	34
	Carbon cloth	Increase surface roughness Introduce N functional groups	Power density: +14.2%	35
H_2SO_4 treatment	Carbon brush	Render the surface hydrophilic Increase surface area	Power density: +8%	26
	SS mesh	Remove surface oxides Increase surface roughness	Current density: +11 times	10
H_2SO_4 treatment + Heat treatment	Carbon brush	Increase N/C ratio Decrease C–O composition Increase surface area	Power density: +34%	26
HNO_3 treatment	Carbon felt	Introduce O-containing groups	Power density: +58%	36

Ethylenediamine treatment	Carbon felt	Generate N-containing groups	Power density: +25%	36
Plasma treatment (N_2/O_2/Air plasma)	Glassy carbon	Improve hydrophilicity Removes surface contamination	Start-up time: –50%	37
Plasma treatment (N^+ ion implantation)	Carbon paper	Increase C–N bonds Increase roughness and hydrophilicity	Start-up time: –50% Power density: +2.5 times	38
Surfactant treatment (SDS)	Graphite/CNTs paste	Introduce–SO_4^- groups Improve hydrophilicity	Power density: +20%	39
Surfactant treatment (CTAB)	Carbon felt	Introduce–$N^+(CH_3)_3$ groups Improve hydrophilicity	Start-up time: –36% Current density: +24%	40
AQDS absorption	Carbon plate	Introduce EET mediators	Kinetic activity[c]: +216.8 times	34
	Graphite disk		Kinetic activity[c]: +70%	41
NQ absorption	Graphite disk	Introduce EET mediators	Kinetic activity[c]: +50%	41
Grafting NR by amidation reaction	Carbon felt	Covalent binding of NR Facilitate EET	Current density: +10 times	42

(*Continued*)

Table 2. (*Continued*)

Modification method	Anode material	Major change of surface	Performance enhancement	References
Grafting NR by reduction of NR diazonium salts	Carbon felt	Covalent binding of NR Facilitate EET	Current density: +3.6 times	43
Electrochemical reduction of aryl diazonium salts	Graphite plate	Introduce negatively charged phenylacetic acid group	Power density: –40%	44
		Introduce positively charged benzylphosphonium groups	Power density: +105%	
Chemical reduction of aryl diazonium salts	Carbon cloth	Introduce –$N(CH_3)_2$ groups Increase N/C ratios	Power density: +24%	45

Notes: "+" and "–" mean, "increased by " and "decreased by", respectively. SDS, sodium dodecyl sulfate; CTAB, cetyltrimethylammonium bromide; AQDS, anthraquinone-2,6-disulfonate; NQ, naphthoquinone; NR, neutral red.
[a]Start-up time refers to the initial start of the current production resulting from the microbial electrocatalysis.
[b]The control is the acetone-treated carbon mesh as untreated carbon mesh did not startup.
[c]Kinetic activity refers to the electrocatalytic activity obtained from Tafel plots.

Table 3. A summary of anode surface coating methods and their impact on microbial anode performance.

Coating materials	Modification method	Anode	Improvement of performance	References
Carbon nanoparticles				
Carbon nanotubes (CNTs)	Electrophoretic deposition	Carbon paper	Current density: +10 times	46
	Chemical vapor deposition (CVD)	RVC	Current density: +3 times	19
		Carbon felt	Power density: +7 times	47
		SS mesh	Power density: +936 times	48
	Binding by Nafion	Carbon cloth	Power density: +120%	49
		SS felt	Power density: +1599 times	50
	Dip-coating	Carbon cloth	Power density: +2.5 times	51
	Painting	Glassy carbon	Current density: +82 times	52
Carbon nanostructures	Flame synthesis	SS mesh	Power density: +60 times	12
Carbon black	Binder-free dip-coating	SS mesh	Current density: +611 times	53
Graphene	Electrophoretic deposition	Carbon cloth	Power density: +2.7 times	54
	In situ formation via electrolysis	Graphite plate	Current density: +1.5 times	55
	Binding by Nafion	SS felt	Power density: +2677 times	50
Activated carbon	Binding by Nafion	SS felt	Power density: +699 times	50
Metal/Metal oxides				
Au nanoparticles	Sputter coating + thermal annealing	Graphite disk	Current density: +20 times	56
Pd nanoparticles			Current density: +1.5 times	
Fe/Ferric oxide	Electrodeposition of Fe + air oxidation	Graphite plate	Power density: +17.2 times	57

(*Continued*)

Table 3. (*Continued*)

Coating materials	Modification method	Anode	Improvement of performance	References
Pt	Electron beam evaporation	Carbon paper	Power density: +5.4 times	58
Ferric oxide	CVD	Carbon paper	Power density: + 2.75 times	59
Iron oxide	Thermal decomposition of $Fe(CO)_5$	Carbon felt	Power density: +2.6 times	60
MnO_2	Electrodeposition	Carbon felt	Power density: +24.7%	61
RuO_2	Electrodeposition	Carbon felt	Power density: +17 times	62
Conductive polymers				
PANI	Electro polymerization	Carbon cloth	Start-up time: −30% Current density: +2.7 times	63
PPy	Electro polymerization	Carbon cloth	Power density: +7 times	64
PANI	Chemical polymerization	Carbon felt	Power density: +27.4%	65
PAOA			Power density: +23.8%	
Composite materials				
Graphene/Au	Paste painting	Carbon paper	Power density: +4 times	66
Graphite/Fe_3O_4	Paste painting	Graphite disk	Kinetic activity: +110%	41
Graphite/Fe_3O_4/Ni^+	Paste painting		Kinetic activity: +70%	
CNTs/MnO_2	Binding by Nafion	Graphite plate	Current density: +3 times Power density: +11 times	67
CNTs/SnO_2	Binding by PTFE	Glassy carbon	Power density: +2.1 times	68
PANI/Vanadate	Binding by PTFE	Graphite plate	Power density: +5.5 times	69

PANI/CNTs	Electropolymerization of PANI + Electrophoretic deposition of CNTs	Carbon felt	Power density: +3.4 times	70
PANI/Gaphene	Layer-by-layer dip-coating of graphene + soaking in PANI solution	Carbon cloth	Power density: +3 times	71
PPy/CNTs	Gelating by phytic acid	Carbon felt	Power density: +1.2 times	72
	Binding by Nafion	Carbon paper	Current density: +2 times	73
PEI/CNTs	Layer-by-layer dip-coating	Carbon paper	Power density: +20%	74
PEI/graphene	Layer-by-layer dip-coating	Carbon paper	Power density: +51%	75
PEI-AQDS	Dip coating of PEI + Grafting AQDS by amidation reaction	Carbon felt	Current density: +59 times	76
Graphene/SnO_2	Binding by PTFE	Carbon cloth	Power density: +4.8 times	77
Osmium redox polymer	Adsorption on surface	Graphite rod	Current density: +4 times Start-up time: −30–40%	78

Notes: "+" and "−" mean "increased by" and "decreased by", respectively. PANI, Polyaniline; Ppy, Polypyrrole; PAOA, poly(aniline-co-o-aminophenol); PEI, Polyethylenimine"; PTFE, Polytetrafluoroethylene."

voltammogram (CV) recorded during substrate turnover conditions) rather than power density. The latter is indeed also dependent on cathodic performance and ohmic drop across the electrochemical system, which therefore must be characterized separately. Moreover, more careful and comprehensive surface characterizations need to be done before making the conclusion that which surface physiochemical property changes can be attributed to the performance improvement.

Although all the aforementioned surface modification strategies (Tables 2 and 3) have successfully improved the current density of microbial anodes, few are suitable for practical applications. For examples, CNT coating appears to be an excellent way to enhance the specific surface area and current density, but the high cost and possible release of CNTs from the electrode hinder its use for larger-scale electrode surface modification in BESs. Besides the cost, efficiency of the method may also limit its applicability. For example, grafting NR onto carbon surface via amidation reaction has very low efficiency and requires large amount of organic solvent.[42] Moreover, the electrochemical stability of the electrode after modification is also very crucial. For instance, the flame oxidation of SS felt drastically improves biocompatibility and maximal current density, but the modified anodes could also be more susceptible to corrosion if their operational potential exceeds a certain threshold.[11]

Surface modification ideally targets the most prominent disadvantage of a specific material to be used as a microbial electrode. For example, SS shows excellent properties for scaling-up but the current densities achieved on unmodified SS electrodes are generally much lower than carbon electrodes. However, adequate surface modifications can mitigate this issue and allow SS-based anodes to even outcompete carbon-based anodes. For example, flame oxidized SS felt produced a maximal current density of 1.92 mA/cm^2 at −0.2 V *vs.* Ag/AgCl, which was 16.5 times higher than the untreated SS.[11] Coating of SS mesh with carbon black increased the current density from 0.002 to 1.53 mA/cm^2.[53] Both SS felt and SS mesh are cheap, scalable, and highly conductive materials. Surface modifications via cheap and scalable processes successfully render their surfaces biocompatible. Therefore, modified SS felt and SS mesh appear to be the most promising anodes currently available for larger-scale BESs.[90]

11.4 Purposely Built 3D Electrodes

Besides the use of the aforementioned commercially available electrodes, several 3D structured electrodes have been designed and fabricated by researchers, and high projected current densities have been achieved in BESs. Generally, three approaches have been used to build such architectures, namely fabrication of carbon fibers networks, carbonization of natural products, and coating of macroporous scaffolds. Figure 3 shows the pictures of representative anodes purposely built for BESs.

The first type comprises the carbon fiber-based materials such as carbon fiber brushes and electrospun carbon fiber mats. Unlike the commercial fiber-based materials such as carbon paper, carbon cloth and carbon felt, they present more-organized structures and better conductivities. The carbon fiber brush was fabricated by winding carbon fibers into a twisted core consisting two titanium wires.[25] The brush electrodes had a high surface-to-volume ratio (71.7 cm^2/cm^3) and the metal wire current collector considerably reduced the resistance. Consequently, the power density achieved with a MFC was 1.4 times higher with the fiber brush as anode (0.143 mW/cm^2) than with carbon paper. However, the clumping of the carbon fibers hindered the accessibility of the bacteria to the fiber surface and the substrate mass transfer in the brush inner layers. The carbon fiber mat was fabricated by an electrospinning process followed by carbonization.[81] The high porosity, high mechanical strength, and high conductivity of this material resulted in a current density of 3 mA/cm^2 (20 mA/cm^3 with respect to electrode volume).

The second type of purposely built electrodes is the carbonized sponge-like electrode. This includes carbonized loofah sponges,[79] carbonized cardboard,[82] carbonized mushroom and corn stem,[84] carbonized kenaf stem,[85,86] and carbonized pomelo peel.[87] These kinds of materials present similarity with the commercial carbon foam such as RVC, but are generally much cheaper. Thanks to the macroscale

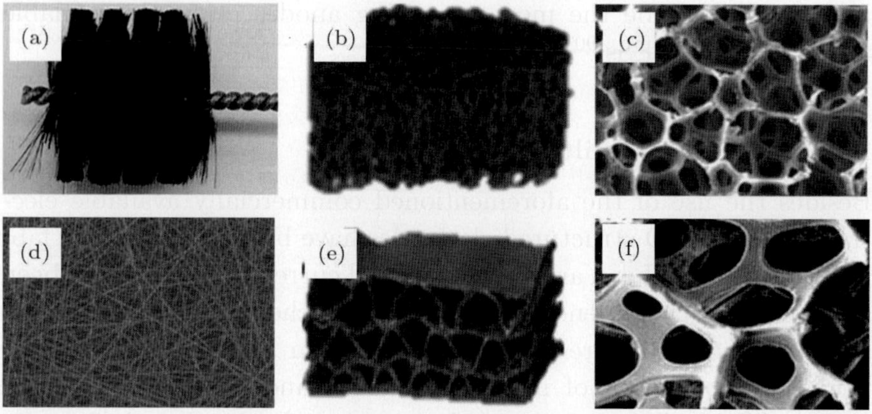

Figure 3. Images of the purposely-built anode materials in BESs. (a) Carbon fiber brush, (b) carbonized loofah sponge, (c) CNTs-coated sponge, (d) carbon fiber mat, (e) carbonized cardboard, (f) graphene-sponge.

Sources: (a) reprinted with permission from Ref. 25. Copyright (2007) American Chemical Society.
(b) reprinted with permission from Ref. 79. Copyright (2013) American Chemical Society.
(c) reprinted with permission from Ref. 80. Copyright (2011), Royal Society of Chemistry.
(d) reprinted with permission from Ref. 81. Copyright (2011), Royal Society of Chemistry.
(e) reprinted with permission from Ref. 82. Copyright (2012), Royal Society of Chemistry.
(f) reprinted with permission from Ref. 83. Copyright (2012) American Chemical Society.

holes (good mass transfer) coupled with high specific surface area (high biofilm volume), the current densities achieved with these electrodes are extremely high. For instance, the carbonized cardboard electrode arranged in multi-layers produced the highest current density, 7 mA/cm^2. Similar to the commercial carbon foam, the big disadvantage of this material is the low mechanical strength and relatively low conductivity, which somehow limits its scalability.

The third type of purposely built electrodes comprise carbon nanoparticles (CNPs) coated sponge electrodes such as CNT-coated sponge and graphene foam. The CNT-coated sponge was fabricated by coating of a ~200 nm thick CNTs layer on a plain sponge via a dipping and drying process.[80] This electrode showed good flexibility and electronic conductivity (~1 S/cm). Due to the macroporous structure and microporous CNTs layer, a maximum current density of 2.1 mA/cm^2 (10.6 mA/cm^3 of electrode volume) was achieved. The graphene foam was made by CVD of graphene on nickel foam, the latter being subsequently etched away by with HCl (3 M) solution.[83] The intrinsically high mechanical strength and conductivity of graphene was consequently imparted to the 3D structure. However, the high hydrophobicity of graphene made this material unfavorable for bacteria adhesion. Thus, the authors decorated the graphene foam with hydrophilic PANI through *in situ* polymerization. As a result, the power density achieved on the PANI-graphene foam (0.077 mW/cm^2) was four times higher than that of the carbon cloth.

11.5 Conclusions

Over the past two decades, numerous works have been carried out on anode materials and their surface modifications to improve the performance of BESs. In terms of anode materials, carbon-based, metal-based, and composite materials have been tested. It seems that composite materials with a metal backbone and carbon coating layer hold great promise for scalable anodes for BESs. Regarding configurations, macroporous 3D anodes outperformed microporous and planar anodes for current production, while planar, smooth anodes are mostly suited for fundamental studies due to their defined surface area, uniform surface properties and easy biofilm visualizations by microscopic techniques. Surface modification is an efficient way to improve the performance of the commercially available anodes. Most of the surface functionalization and thin-film coating methods are summarized in this chapter. However, some of them are not suitable for practical application because of the high cost and complexity. In future, more knowledge needs to be gained on the interaction between the electrode surface and bacterial cell membrane, as well as on the electron transfer mechanisms across electroactive biofilms. Moreover, more cost-effective surface modification methods need to be developed. Reducing the electrode cost and getting the desirable electrode performance will ultimately determine whether BESs can be up-scaled and are suitable for practical applications.

List of Abbreviations

AQDS	anthraquinone-2,6-disulfonate
BES	bioelectrochemical system
CNTs	carbon nanotubes
CTAB	cetyltrimethylammonium bromide
CV	cyclic voltammogram
CVD	chemical vapor deposition
EET	extracellular electron transfer
HNQ	2-hydroxy-l,4-naphthoquinone
MDC	microbial desalination cell
MEC	microbial electrolysis cell
MES	microbial electrosynthesis
MET	microbial electrochemical technology
MFC	microbial fuel cell
NQ	naphthoquinone
NR	neutral red
PANI	polyaniline
PAOA	poly(aniline-*co*-*o*-aminophenol)
PET	polyethylenimine
Ppy	polypyrrole
PTFE	polytetrafluoroethylene
RVC	reticulated vitreous carbon
SAMs	self-assembled monolayers
SDS	sodium dodecyl sulfate
SS	stainless steel

References

1. K. Rabaey, J. Rodriguez, L.L. Blackall, J. Keller, P. Gross, D. Batstone, W. Verstraete and K.H. Nealson. (2007). Microbial ecology meets electrochemistry: Electricity-driven and driving communities. *ISME Journal* **1**, 9–18.
2. B.E. Logan and K. Rabaey (2012). Conversion of wastes into bioelectricity and chemicals by using microbial electrochemical technologies. *Science* **337**, 686–690.

3. B.E. Logan, B. Hamelers, R.A. Rozendal, U. Schrorder, J. Keller, S. Freguia, P. Aelterman, W. Verstraete and K. Rabaey (2006). Microbial fuel cells: Methodology and technology. *Environmental Science & Technology* **40**, 5181–5192.
4. B.E. Logan, D. Call, S. Cheng, H.V.M. Hamelers, T.H.J.A. Sleutels, A.W. Jeremiasse and R.A. Rozendal (2008). Microbial electrolysis cells for high yield hydrogen gas production from organic matter. *Environmental Science & Technology* **42**, 8630–8640.
5. M. Mehanna, T. Saito, J.L. Yan, M. Hickner, X.X. Cao, X. Huang and B.E. Logan (2010). Using microbial desalination cells to reduce water salinity prior to reverse osmosis. *Energy & Environmental Science* **3**, 1114–1120.
6. K. Rabaey and R.A. Rozendal (2010). Microbial electrosynthesis — revisiting the electrical route for microbial production. *Nature Reviews Microbiology* **8**, 706–716.
7. R.A. Rozendal, H.V.M. Hamelers, K. Rabaey, J. Keller and C.J.N. Buisman (2008). Towards practical implementation of bioelectrochemical wastewater treatment. *Trends in Biotechnology* **26**, 450–459.
8. K. Guo, A. Prévoteau, S.A. Patil and K. Rabaey (2015). Engineering electrodes for microbial electrocatalysis. *Current Opinion in Biotechnology* **33**, 149–156.
9. M.H. Zhou, M.L. Chi, J.M. Luo, H.H. He and T. Jin (2011). An overview of electrode materials in microbial fuel cells. *Journal of Power Sources* **196**, 4427–4435.
10. S. Zheng, F. Yang, S. Chen, L. Liu, Q. Xiong, T. Yu, F. Zhao, U Schröder and H. Hou (2015). Binder-free carbon black/stainless steel mesh composite electrode for high-performance anode in microbial fuel cells. *Journal of Power Sources* **284**, 252–257.
11. K. Guo, B.C. Donose, A.H. Soeriyadi, A. Prévoteau, S.A. Patil, S. Freguia, J.J. Gooding and K. Rabaey (2014). Flame oxidation of stainless steel felt enhances anodic biofilm formation and current output in bioelectrochemical systems. *Environmental Science & Technology* **48**, 7151–7156.
12. J.L. Lamp, J.S. Guest, S. Naha, K.A. Radavich, N.G. Love, M.W. Ellis and I.K. Puri (2011). Flame synthesis of carbon nanostructures on stainless steel anodes for use in microbial fuel cells. *Journal of Power Sources* **196**, 5829–5834.
13. D. Pocaznoi, A. Calmet, L. Etcheverry, B. Erable and A. Bergel (2012). Stainless steel is a promising electrode material for anodes of microbial fuel cells. *Energy & Environmental Science* **5**, 9645–9652.
14. K. Guo, D. Hidalgo, T. Tommasi and K. Rabaey (2016). Pyrolytic carbon-coated stainless steel felt as a high-performance anode for bioelectrochemical systems. *Bioresource Technology* **211**, 664–668.
15. B.E. Logan (2008). *Microbial Fuel Cells*. New Jersey, Wiley.
16. P. Ledezma, B.C. Donose, S. Freguia and J. Keller (2015). Oxidised stainless steel: A very effective electrode material for microbial fuel cell bioanodes but at high risk of corrosion. *Electrochimica Acta* **158**, 356–360.

17. X. Xie, C. Criddle and Y. Cui (2015). Design and fabrication of bioelectrodes for microbial bioelectrochemical systems. *Energy & Environmental Science* **8**, 3418–3441.
18. S.F. Ketep, A. Bergel, A. Calmet and B. Erable (2014). Stainless steel foam increases the current produced by microbial bioanodes in bioelectrochemical systems. *Energy & Environmental Science* **7**, 1633–1637.
19. V. Flexer, J. Chen, B.C. Donose, P. Sherrell, G.G. Wallace and J. Keller (2013). The nanostructure of three-dimensional scaffolds enhances the current density of microbial bioelectrochemical systems. *Energy & Environmental Science* **6**, 1291–1298.
20. R. Arechederra and S.D. Minteer (2008). Organelle-based biofuel cells: Immobilized mitochondria on carbon paper electrodes. *Electrochimica Acta* **53**, 6698–6703.
21. S.A. Cheng and B.E. Logan (2007). Ammonia treatment of carbon cloth anodes to enhance power generation of microbial fuel cells. *Electrochemistry Communications* **9**, 492–496.
22. X. Wang, S.A. Cheng, Y.J. Feng, M.D. Merrill, T. Saito and B.E. Logan (2009). Use of carbon mesh anodes and the effect of different pretreatment methods on power production in microbial fuel cells. *Environmental Science & Technology* **43**, 6870–6874.
23. S.K. Chaudhuri and D.R. Lovley (2003). Electricity generation by direct oxidation of glucose in mediatorless microbial fuel cells. *Nature Biotechnology* **21**, 1229–1232.
24. S.J. You, Q.L. Zhao, J.N. Zhang, J.Q. Jiang, C.L. Wan, M.A. Du and S.Q. Zhao (2007). A graphite-granule membrane-less tubular air-cathode microbial fuel cell for power generation under continuously operational conditions. *Journal of Power Sources* **173**, 172–177.
25. B.E. Logan, S. Cheng, V. Watson and G. Estadt (2007). Graphite fiber brush anodes for increased power production in air-cathode microbial fuel cells. *Environmental Science & Technology* **41**, 3341–3346.
26. Y.J. Feng, Q. Yang, X. Wang and B.E. Logan (2010). Treatment of carbon fiber brush anodes for improving power generation in air-cathode microbial fuel cells. *Journal of Power Sources* **195**, 1841–1844.
27. Y. Liu, H. Kim, R. Franklin and D.R. Bond (2010). Gold line array electrodes increase substrate affinity and current density of electricity-producing G. sulfurreducens biofilms. *Energy & Environmental Science* **3**, 1782–1788.
28. A. ter Heijne, H.V.M. Hamelers, M. Saakes and C.J.N. Buisman (2008). Performance of non-porous graphite and titanium-based anodes in microbial fuel cells. *Electrochimica Acta* **53**, 5697–5703.
29. A. Baudler, I. Schmidt, M. Langner, A. Greiner and U. Schroder (2015). Does it have to be carbon? Metal anodes in microbial fuel cells and related bioelectrochemical systems. *Energy & Environmental Science* **8**, 2048–2055.
30. K. Guo, S. Freguia, P.G. Dennis, X. Chen, B.C. Donose, J. Keller, J.J. Gooding and K. Rabaey (2013). Effects of surface charge and hydrophobicity on anodic biofilm formation, community composition, and current generation

in bioelectrochemical systems. *Environmental Science & Technology* **47**, 7563–7570.

31. C. Santoro, S. Babanova, K. Artyushkova, J.A. Cornejo, L. Ista, O. Bretschger, E. Marsili, P. Atanassov and A.J. Schuler (2015). Influence of anode surface chemistry on microbial fuel cell operation. *Bioelectrochemistry* **106**, 141–149.
32. K. Guo, A.H. Soeriyadi, H.J. Feng, A. Prévoteau, S.A. Patil, J.J. Gooding and K. Rabaey (2015). Heat-treated stainless steel felt as scalable anode material for bioelectrochemical systems. *Bioresource Technology* **195**, 46–50.
33. X.H. Tang, K. Guo, H.R. Li, Z.W. Du and J.L. Tian (2011). Electrochemical treatment of graphite to enhance electron transfer from bacteria to electrodes. *Bioresource Technology* **102**, 3558–3560.
34. D.A. Lowy and L.M. Tender (2008). Harvesting energy from the marine sediment-water interface III. Kinetic activity of quinone- and antimony-based anode materials. *Journal of Power Sources* **185**, 70–75.
35. J. Liu, J.F. Liu, W.H. He, Y.P. Qu, N.Q. Ren and Y.J. Feng (2014). Enhanced electricity generation for microbial fuel cell by using electrochemical oxidation to modify carbon cloth anode. *Journal of Power Sources* **265**, 391–396.
36. N.W. Zhu, X. Chen, T. Zhang, P.X. Wu, P. Li and J.H. Wu (2011). Improved performance of membrane free single-chamber air-cathode microbial fuel cells with nitric acid and ethylenediamine surface modified activated carbon fiber felt anodes. *Bioresource Technology* **102**, 422–426.
37. V. Flexer, M. Marque, B.C. Donose, B. Virdis and J. Keller (2013). Plasma treatment of electrodes significantly enhances the development of anodic electrochemically active biofilms. *Electrochimica Acta* **108**, 566–574.
38. Y.R. He, X. Xiao, W.W. Li, G.P. Sheng, F.F. Yan, H.Q. Yu, H. Yuan and L.J. Wu (2012). Enhanced electricity production from microbial fuel cells with plasma-modified carbon paper anode. *Physical Chemistry Chemical Physics* **14**, 9966–9971.
39. Y.C. Song, D.S. Kim, J.H. Woo, B. Subha, S.H. Jang and S. Sivakumar (2015). Effect of surface modification of anode with surfactant on the performance of microbial fuel cell. *International Journal of Energy Research* **39**, 860–868.
40. K. Guo, A.H. Soeriyadi, S.A. Patil, A. Prévoteau, S. Freguia, J.J. Gooding and K. Rabaey (2014). Surfactant treatment of carbon felt enhances anodic microbial electrocatalysis in bioelectrochemical systems. *Electrochemistry Communications* **39**, 1–4.
41. D.A. Lowy, L.M. Tender, J.G. Zeikus, D.H. Park and D.R. Lovley (2006). Harvesting energy from the marine sediment-water interface II — Kinetic activity of anode materials. *Biosensors & Bioelectronics* **21**, 2058–2063.
42. D.H. Park and J.G. Zeikus (2002). Impact of electrode composition on electricity generation in a single-compartment fuel cell using Shewanella putrefaciens. *Applied Microbiology and Biotechnology* **59**, 58–61.
43. K. Guo, X. Chen, S. Freguia and B.C. Donose (2013). Spontaneous modification of carbon surface with neutral red from its diazonium salts for bioelectrochemical systems. *Biosensors and Bioelectronics* **47**, 184–189.

44. M. Picot, L. Lapinsonniere, M. Rothballer and F. Barriere (2011). Graphite anode surface modification with controlled reduction of specific aryl diazonium salts for improved microbial fuel cells power output. *Biosensors & Bioelectronics* **28**, 181–188.
45. T. Saito, M. Mehanna, X. Wang, R.D. Cusick, Y.J. Feng, M.A. Hickner and B.E. Logan (2011). Effect of nitrogen addition on the performance of microbial fuel cell anodes. *Bioresource Technology* **102**, 395–398.
46. X.W. Liu, J.J. Chen, Y.X. Huang, X.F. Sun, G.P. Sheng, D.B. Li, L. Xiong, Y.Y. Zhang, F. Zhao and H.Q. Yu (2014). Experimental and theoretical demonstrations for the mechanism behind enhanced microbial electron transfer by CNT network. *Scientific Reports* **4**, 3732.
47. Y. Zhao, K. Watanabe and K. Hashimoto (2011). Hierarchical micro/nano structures of carbon composites as anodes for microbial fuel cells. *Physical Chemistry Chemical Physics* **13**, 15016–15021.
48. C. Erbay, X. Pu, W. Choi, M.J. Choi, Y. Ryu, H.J. Hou, F.R. Lin, P. de Figueiredo, C. Yu and A. Han (2015). Control of geometrical properties of carbon nanotube electrodes towards high-performance microbial fuel cells. *Journal of Power Sources* **280**, 347–354.
49. S.Q. Ci, Z.H. Wen, J.H. Chen and Z. He (2012). Decorating anode with bamboo-like nitrogen-doped carbon nanotubes for microbial fuel cells. *Electrochemistry Communications* **14**, 71–74.
50. J. Hou, Z. Liu, S. Yang and Y. Zhou (2014). Three-dimensional macroporous anodes based on stainless steel fiber felt for high-performance microbial fuel cells. *Journal of Power Sources* **258**, 204–209.
51. H.Y. Tsai, C.C. Wu, C.Y. Lee and E.P. Shih (2009). Microbial fuel cell performance of multiwall carbon nanotubes on carbon cloth as electrodes. *Journal of Power Sources* **194**, 199–205.
52. L. Peng, S.J. You and J.Y. Wang (2010). Carbon nanotubes as electrode modifier promoting direct electron transfer from Shewanella oneidensis. *Biosensors & Bioelectronics* **25**, 1248–1251.
53. S.Q. Zheng, F.F. Yang, S.L. Chen, L. Liu, Q. Xiong, T. Yu, F. Zhao, U. Schroder and H.Q. Hou (2015). Binder-free carbon black/stainless steel mesh composite electrode for high-performance anode in microbial fuel cells. *Journal of Power Sources* **284**, 252–257.
54. J. Liu, Y. Qiao, C.X. Guo, S. Lim, H. Song and C.M. Li (2012). Graphene/carbon cloth anode for high-performance mediatorless microbial fuel cells. *Bioresource Technology* **114**, 275–280.
55. J.H. Tang, S.S. Chen, Y. Yuan, X.X. Cai and S.G. Zhou (2015). In situ formation of graphene layers on graphite surfaces for efficient anodes of microbial fuel cells. *Biosensors & Bioelectronics* **71**, 387–395.
56. Y.Z. Fan, S.T. Xu, R. Schaller, J. Jiao, F. Chaplen and H. Liu (2011). Nanoparticle decorated anodes for enhanced current generation in microbial electrochemical cells. *Biosensors & Bioelectronics* **26**, 1908–1912.
57. Y.B. Fu, Q. Xu, X.R. Zai, Y.Y. Liu and Z.K. Lu (2014). Low electrical potential anode modified with Fe/ferric oxide and its application in marine benthic

microbial fuel cell with higher voltage and power output. *Applied Surface Science* **289**, 472–477.

58. T. Sharma, A.L.M. Reddy, T.S. Chandra and S. Ramaprabhu (2008). High power density from Pt thin film electrodes based microbial fuel cell. *Journal of Nanoscience and Nanotechnology* **8**, 4132–4134.
59. J.R. Kim, B. Min and B.E. Logan (2005). Evaluation of procedures to acclimate a microbial fuel cell for electricity production. *Applied Microbiology and Biotechnology* **68**, 23–30.
60. P. Wang, H.R. Li and Z.W. Du (2013). Deposition of iron on graphite felts by thermal decomposition of Fe(CO)(5) for anodic modification of microbial fuel cells. *International Journal of Electrochemical Science* **8**, 4712–4722.
61. C.Y. Zhang, P. Liang, Y. Jiang and X. Huang (2015). Enhanced power generation of microbial fuel cell using manganese dioxide-coated anode in flow-through mode. *Journal of Power Sources* **273**, 580–583.
62. Z.S. Lv, D.H. Xie, X.J. Yue, C.H. Feng and C.H. Wei (2012). Ruthenium oxide-coated carbon felt electrode: A highly active anode for microbial fuel cell applications. *Journal of Power Sources* **210**, 26–31.
63. B. Lai, X.H. Tang, H.R. Li, Z.W. Du, X.W. Liu and Q. Zhang (2011). Power production enhancement with a polyaniline modified anode in microbial fuel cells. *Biosensors & Bioelectronics* **28**, 373–377.
64. D. Jiang and B. Li (2009). Novel electrode materials to enhance the bacterial adhesion and increase the power generation in microbial fuel cells (MFCs). *Water Science and Technology* **59**, 557–563.
65. C. Li, L.B. Zhang, L.L. Ding, H.Q. Ren and H. Cui (2011). Effect of conductive polymers coated anode on the performance of microbial fuel cells (MFCs) and its biodiversity analysis. *Biosensors & Bioelectronics* **26**, 4169–4176.
66. C.E. Zhao, P.P. Gai, R.B. Song, J.R. Zhang and J.J. Zhu (2015). Graphene/Au composites as an anode modifier for improving electricity generation in Shewanella-inoculated microbial fuel cells. *Analytical Methods* **7**, 4640–4644.
67. Y.B. Fu, J. Yu, Y.L. Zhang and Y. Meng (2014). Graphite coated with manganese oxide/multiwall carbon nanotubes composites as anodes in marine benthic microbial fuel cells. *Applied Surface Science* **317**, 84–89.
68. A. Mehdinia, E. Ziaei and A. Jabbari (2014). Multi-walled carbon nanotube/SnO_2 nanocomposite: A novel anode material for microbial fuel cells. *Electrochimica Acta* **130**, 512–518.
69. Y. Fu, Z. Zhao, J. Liu, K. Li, Q. Xu and S. Zhang (2011). Sulfonated polyaniline/vanadate composite as anode material and its electrochemical property in microbial fuel cells on ocean floor. *Science China Chemistry* **54**, 844–849.
70. H.F. Cui, L. Du, P.B. Guo, B. Zhu and J.H.T. Luong (2015). Controlled modification of carbon nanotubes and polyaniline on macroporous graphite felt for high-performance microbial fuel cell anode. *Journal of Power Sources* **283**, 46–53.
71. J.X. Hou, Z.L. Liu and P.Y. Zhang (2013). A new method for fabrication of graphene/polyaniline nanocomplex modified microbial fuel cell anodes. *Journal of Power Sources* **224**, 139–144.

72. X.H. Tang, H.R. Li, Z.W. Du, W.D. Wang and H.Y. Ng (2015). Conductive polypyrrole hydrogels and carbon nanotubes composite as an anode for microbial fuel cells. *RSC Advances* **5**, 50968–50974.
73. Y.J. Zou, C.L. Xiang, L.N. Yang, L.X. Sun, F. Xu and Z. Cao (2008). A mediatorless microbial fuel cell using polypyrrole coated carbon nanotubes composite as anode material. *International Journal of Hydrogen Energy* **33**, 4856–4862.
74. J.J. Sun, H.Z. Zhao, Q.Z. Yang, J. Song and A. Xue (2010). A novel layer-by-layer self-assembled carbon nanotube-based anode: Preparation, characterization, and application in microbial fuel cell. *Electrochimica Acta* **55**, 3041–3047.
75. W. Guo, Y.R. Cui, H. Song and J.H. Sun (2014). Layer-by-layer construction of graphene-based microbial fuel cell for improved power generation and methyl orange removal. *Bioprocess and Biosystems Engineering* **37**, 1749–1758.
76. M. Adachi, T. Shimomura, M. Komatsu, H. Yakuwa and A. Miya (2008). A novel mediator-polymer-modified anode for microbial fuel cells. *Chemical Communications* **7**, 2055–2057.
77. A. Mehdinia, E. Ziaei and A. Jabbari (2014). Facile microwave-assisted synthesized reduced graphene oxide/tin oxide nanocomposite and using as anode material of microbial fuel cell to improve power generation. *International Journal of Hydrogen Energy* **39**, 10724–10730.
78. S.A. Patil, K. Hasan, D. Leech, C. Hagerhall and L. Gorton (2012). Improved microbial electrocatalysis with osmium polymer modified electrodes. *Chemical Communications* **48**, 10183–10185.
79. Y. Yuan, S.G. Zhou, Y. Liu and J.H. Tang (2013). Nanostructured macroporous bioanode based on polyaniline-modified natural loofah sponge for high-performance microbial fuel cells. *Environmental Science & Technology* **47**, 14525–14532.
80. X. Xie, M. Ye, L.B. Hu, N. Liu, J.R. McDonough, W. Chen, H.N. Alshareef, C.S. Criddle and Y. Cui (2012). Carbon nanotube-coated macroporous sponge for microbial fuel cell electrodes. *Energy & Environmental Science* **5**, 5265–5270.
81. S. Chen, H. Hou, F. Harnisch, S.A. Patil, A.A. Carmona-Martinez, S. Agarwal, Y. Zhang, S. Sinha-Ray, A.L. Yarin, A. Greiner and U. Schröder (2011). Electrospun and solution blown three-dimensional carbon fiber nonwovens for application as electrodes in microbial fuel cells. *Energy & Environmental Science* **4**, 1417–1421.
82. S.L. Chen, G.H. He, Q. Liu, F. Harnisch, Y. Zhou, Y. Chen, M. Hanif, S.Q. Wang, X.W. Peng, H.Q. Hou and U. Schröder (2012). Layered corrugated electrode macrostructures boost microbial bioelectrocatalysis. *Energy & Environmental Science* **5**, 9769–9772.
83. Y.C. Yong, X.C. Dong, M.B. Chan-Park, H. Song and P. Chen (2012). Macroporous and monolithic anode based on polyaniline hybridized three-dimensional graphene for high-performance microbial fuel cells. *ACS Nano* **6**, 2394–2400.

84. R. Karthikeyan, B. Wang, J. Xuan, J.W.C. Wong, P.K.H. Lee and M.K.H. Leung (2015). Interfacial electron transfer and bioelectrocatalysis of carbonized plant material as effective anode of microbial fuel cell. *Electrochimica Acta* **157**, 314–323.
85. S.L. Chen, G.H. He, X.W. Hu, M.Y. Xie, S.Q. Wang, D.J. Zeng, H.Q. Hou and U. Schroder (2012). A three-dimensionally ordered macroporous carbon derived from a natural resource as anode for microbial bioelectrochemical systems. *Chemsuschem* **5**, 1059–1063.
86. J.H. Tang, Y. Yuan, T. Liu and S.G. Zhou (2015). High-capacity carbon-coated titanium dioxide core-shell nanoparticles modified three dimensional anodes for improved energy output in microbial fuel cells. *Journal of Power Sources* **274**, 170–176.
87. S.L. Chen, Q. Liu, G.H. He, Y. Zhou, M. Hanif, X.W. Peng, S.Q. Wang and H.Q. Hou (2012). Reticulated carbon foam derived from a sponge-like natural product as a high-performance anode in microbial fuel cells. *Journal of Materials Chemistry* **22**, 18609–18613.
88. X. Zhang, J. Philips, H. Roume, K. Guo, K. Rabaey and A. Prévoteau (2017). Rapid and quantitative assessment of redox conduction across electroactive biofilms by using double potential step chronoamperometry. *ChemElectroChem.* doi:10.1002/celc.201600853
89. A. Baudler, M. Langner, C. Rohr, A. Greiner and U. Schröder (2017). Metalpolymer hybrid architectures as novel anode platform for microbial electrochemical technologies, *ChemSusChem* **10**, 253–257.
90. K. Guo, A. Prévoteau and K. Rabaey (2017). A novel tubular microbial electrolysis cell for high rate hydrogen production, *Journal of Power Sources*, in press, http://dx.doi.org/10.1016/j. jpowsour.2017.03.029.

Chapter 12

Electrodes for Cathodic Microbial Electrosynthesis Processes: Key Developments and Criteria for Effective Research and Implementation

Ludovic Jourdin* and David Strik

Sub-department of Environmental Technology, Wageningen UniversityAxis- Z, Bornse Weilanden 9, 6708 WG Wageningen, The Netherlands

**ludovic.jourdin@wur.nl*

12.1 Introduction on Microbial Electrosynthesis

12.1.1 The biocathode: Assessing the key-element

Global concerns over fossil resource depletion, greenhouse gas emissions, and our society's high energy and commodity chemical demands have led to a rapidly growing demand for new, sustainable form of fuels and chemicals.[1] One promising direction has been the production of chemicals from microbial electrosynthesis (MES). MES is a novel biocathode-driven process, which relies on the capability of electroactive microorganisms to derive electrons from solid-state electrode to catalyze the reduction of carbon dioxide or organics to generate high-value extracellular multi-carbon chemicals,[2–5] as depicted in Figure 1. An anodic oxidation reaction provides electrons to which

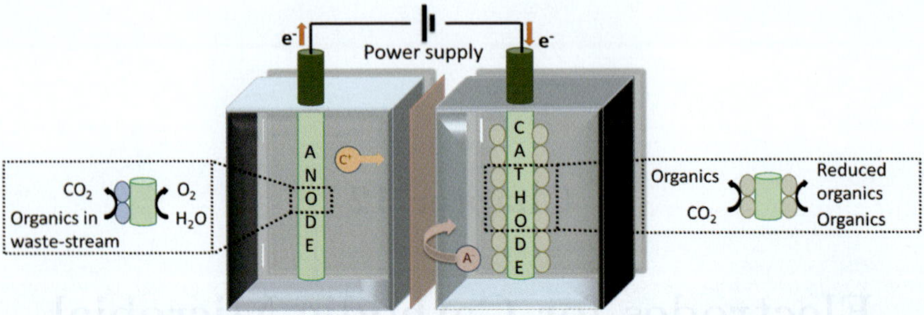

Figure 1. Microbial electrosynthesis scheme, with a cation-exchange membrane, a biocathode for CO_2 or organics reduction to organics and two examples of possible anodic reactions. Intermittent renewable energy (*e.g.* wind or solar) could be envisioned to deliver the additional power required to drive the bioproduction of high energy-density chemicals (*e.g.* off-peak energy storage). Modified from Ref. 6.

external energy is supplied to make the cathode reaction(s) thermodynamically feasible and to cover the remaining internal resistances of the system. Several anodic reactions could be envisioned, such as abiotic water oxidation or microbial oxidation of organics from waste stream (*e.g.* wastewater), with each their pros and cons, as previously discussed.[6] MES is one of the emerging processes within the field of bioelectrochemical systems (BES). Other microbial cathodic processes within BES include biocatalyzed oxygen reduction within microbial fuels cells to produce electricity, or nitrate reduction to produce clean water,[7] for instance.

MES allows converting and storing electrical energy into stable, high-energy-density liquid fuels or commodities that are easily transportable. In this perspective, MES has been envisioned as an interesting option to capture and increase the value of harvested intermittent renewable electrical energy from solar and wind.[8] MES can become a new chemical producing platform technology within a biobased economy and can be integrated in existing and new biorefineries.

The biocathode in MES is the key element since it warrants the actual production of chemicals by the cathodic microbial processes. The biocathode consists of a current collector, the actual electrode, the biocatalyst(s), and the catholyte medium. The overall MES design (including specific cathode design) and the operation methodology determine the conditions and the performance of the biocathode. There is a wide variety of MES studies that report on performances and link them to explanations on specific biocathode properties. Evidently, there are still knowledge gaps on fundamental biocathode working principles. Due to this variety, knowledge gaps, and incomplete data generation or analysis, it is not always possible to compare all studies in detail. Nevertheless, within the last few years, several breakthrough developments on MES led toward a high-rate process.

The objective of this chapter is to provide an overview on key developments on electrodes for cathodic microbial processes and set the criteria for effective research and implementation of the technology. Hereby, we distinguish between commercially available electrodes (studies that used commercially available materials as electrodes), surface-modified electrodes (studies that use modified commercially available materials), and purposely built electrodes (studies that used novel materials to on purpose build electrodes for microbial electrosynthesis). We provide an overview with criteria that should be taken into consideration when selecting electrode materials toward the implementation of cathodic microbial production processes, as illustrated in Figure 2.

12.1.2 Advantages of electro-biocatalysts

To date, most MES studies have focused on CO_2 reduction (see Table 1 for a complete overview). Creating novel approaches for the direct transformation of CO_2 into chemicals is attractive. In addition of producing valuable chemicals by recycling CO_2 as the key carbon source, CO_2-based technologies present several advantages such as independence of arable land, substrate for microorganisms, mitigation of the ever-increasing CO_2 emissions, and nearly unlimited substrate availability.[5] Several CO_2-based technologies are currently at the research and development stage for the production of fuels and chemicals, such as chemical, photochemical, electrochemical, biological, reforming and inorganic transformations.[11,12] However, most of these CO_2 mitigation and conversion techniques have major drawbacks as they require the use of expensive catalysts, large

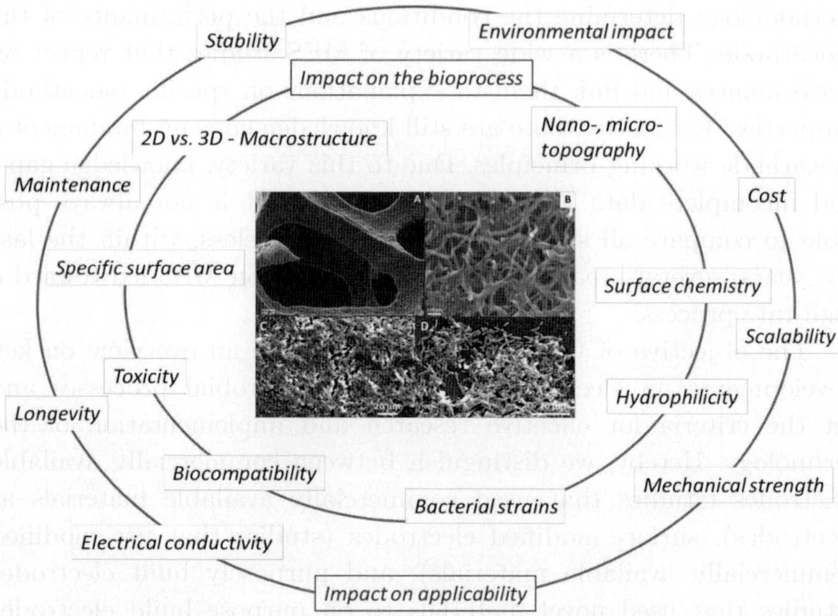

Figure 2. Schematic of parameters to consider when selecting and designing new electrode materials for cathodic microbial processes. Scanning electron microscopy images modified from Refs. 5, 9 and 10 with permission from the Royal Society of Chemistry, The American Chemical Society, and Wiley-VCH Verlag.

amounts of electricity (energy intense processing steps), extremely large surface and volumes, and/or high consumption of hydrogen.[6,11] MES of organics from carbon dioxide has been recently put forward as a promising strategy over these current techniques.[11,13] MES has the advantages to use cheap and readily available (bio)catalysts, to work in ambient conditions (temperature and pressure) to potentially obtain a larger spectrum of products directly from CO_2[13] using the complex metabolisms of microorganisms and potentially prevent the handling of gases, which are hazardous and costly. Nutrients must still be provided to start-up MES and possibly warrant long-term operation. MES is a versatile technology with regard to substrates and products. Moreover, pure CO_2 may not be necessary and MES could use different sources of CO_2 as substrate, such as biogas,[14] without any costly pretreatment steps.[6]

Table 1. Key performance indicators of most MES from CO_2 studies reported to date, in chronological order of publication.

Material category	Microbial inoculum	Cathode material	Reactor design (operation)	$E_{cathode}$ (V vs. SHE)	Current density ($A\ m^{-2}$) vs. psa	Current density ($A\ m^{-2}$) vs. tsa	Acetate production rate $g\ m^{-2}_{psa}\ day^{-1}$	Acetate production rate $g\ m^{-2}_{tsa}\ day^{-1}$	Acetate production rate $g\ m^{-3}_{electrode}\ day^{-1}$	Acetate production rate $g\ L^{-1}_{catholyte}\ day^{-1}$	Max acetate titer ($g\ L^{-1}$)	Electron recovery into acetate (%)	Microbial community analysis	Main objective	Reference
Commercial	S. ovata	Graphite rods	H-type reactor (continuous)	-0.40	-0.21	-0.21	1.30	1.30	n.d.	0.042	0.28	86 ± 21	S. ovata	Microbial electrosynthesis proof of concept using pure cultures	Nevin et al. 2010
	C. ljungdahlii	Graphite rods	H-type reactor (continuous)	-0.40	-0.03	-0.03	0.14	0.14	n.d.	0.005	n.d.	72	C. ljungdahlii		Nevin et al. 2011
	Brewery WW sludge	Graphite granules	H-type reactor (fed-batch)	-0.59	n.d.	n.d.	n.d.	1.59	1180.00	0.236	1.71	57	60.3 % Acetobacterium + Sulfurospirillum spp. and unclassified Rhodobacteraceae and sphingobacteriales	Microbial electrosynthesis proof of conept using mixed microbial culture	Marshall et al. 2012
Surface modification	S. ovata	Carbon cloth	H-type reactor (batch)	-0.40	-0.07	n.d.	0.35	n.d	n.d.	0.008	n.d.	76 ± 14	S. ovata	Effect of carbon cloth surface modifications	Zhang et al. 2013
		CC-chitosan		-0.40	-0.48	n.d.	2.70	n.d	n.d.	0.064	0.59	86 ± 12			
		CC-cyanuric acid		-0.40	-0.45	n.d.	2.42	n.d	n.d.	0.057	0.77	81 ± 16			
		CC-3-Aminopropyltriethoxysilane		-0.40	-0.21	n.d.	1.12	n.d	n.d.	0.026	0.30	82 ± 11			
		CC-Polyaniline		-0.40	-0.19	n.d.	1.06	n.d	n.d.	0.025	0.18	85 ± 7			
		CC-Melanine		-0.40	-0.07	n.d.	0.37	n.d	n.d.	0.009	n.d.	80 ± 15			
		CC-Ammonia		-0.40	-0.06	n.d.	0.33	n.d	n.d.	0.008	n.d.	82 ± 8			
		CC-Au		-0.40	-0.39	n.d.	2.14	n.d	n.d.	0.050	0.35	83 ± 14			
		CC-Pd		-0.40	-0.32	n.d.	1.66	n.d	n.d.	0.039	0.53	79 ± 16			
		CC-Ni		-0.40	-0.30	n.d.	1.60	n.d	n.d.	0.038	0.24	80 ± 15			
		CNT-cotton		-0.40	-0.22	n.d.	1.20	n.d	n.d.	0.028	0.30	83 ± 10			
		CNT-polyester		-0.40	-0.21	n.d.	1.13	n.d	n.d.	0.027	0.15	82 ± 8			
Commercial	S. ovata	Graphite plate	H-type reactor (fed-batch)	n.d.	0.49	0.49	2.94	2.94	n.d.	0.054	0.21	90	S. ovata	Sulfide anodic oxidation to drive MES	Gong et al. 2013
	Activated sludge from sewage treatment plant	Carbon felt	H-type reactor	-0.75	-3.37	n.d.	0.00	n.d.	n.d.	0.000	0.00	0	Undefined	Effect of applied cathode potentials on product spectrum (CH_4, acetate, and H_2)	Jiang et al. 2013
				-0.85	-9.29	n.d.	16.39	n.d.	n.d.	0.335	0.33	28.4			
				-0.95	n.d.	n.d.	19.39	n.d.	n.d.	0.396	0.40	15			
	Enriched brewery WW sludge	Graphite granules	H-type reactor (fed-batch)	-0.59	n.d.	n.d.	n.d.	6.98	5162.50	1.033	10.50	69	Acetobacterium spp. dominant + Sulfurospirillum spp. + unclassified Rhodobacteraceae	Long-term operation (over 150 days)	Marshall et al. 2013
	Enriched WWTP sludge	Carbon felt	H-type reactor (fed-batch)	-0.70	-3.44	n.d.	11.63	n.d.	n.d.	0.149	n.d.	53.6	Acetobacterium woodii	Effect of applied cathode potentials on acetate production rate and electron recovery	Su et al. 2013
				-0.90	n.d.	n.d.	71.89	n.d	n.d.	0.924	n.d.	89.5			
Surface modification	S. ovata	Graphite stick	H-type reactor (batch)	-0.40	n.d.	n.d.	1.45	1.45	n.d.	0.029	n.d.	n.d.	S. ovata	Effect of Ni nanostructures	Nie et al. 2013
		Ni-nanowire-network-coated graphite stick		-0.40	-0.69	-0.69	3.33	3.33	n.d.	0.067	0.24	82 ± 14			
		Ni-nanoparticle-coated graphite stick		-0.40	-0.15	-0.15	0.90	0.90	n.d.	0.018	0.12	84 ± 9			

(*Continued*)

Table 1. (*Continued*)

Material category	Microbial inoculum	Cathode material	Reactor design (operation)	$E_{cathode}$ (V vs. SHE)	Current density (A m^{-2}) vs. psa	Current density (A m^{-2}) vs. tsa	Acetate production rate g m$^{-2}_{psa}$ day^{-1}	Acetate production rate g m$^{-2}_{tsa}$ day^{-1}	Acetate production rate g m$^{-3}_{electrode}$ day^{-1}	Acetate production rate g L$^{-1}_{catholyte}$ day^{-1}	Max acetate titer (g L^{-1})	Electron recovery into acetate (%)	Microbial community analysis	Main objective	Reference
Commercial	Sediment from a bog subsurface	Carbon fiber rod	H-type reactor (fed-batch)	-0.40	-0.03	n.d.	0.05	n.d.	n.d.	0.000	1.90	n.d.	*Trichococcus palustris* DSM 9172 dominant + Oscillibacter sp. + Clostridium sp.	Pre-enrichment (glucose-based) method investigation to improve start-up of MES from CO_2 to roganics (acetate, propionate, butyrate, ethanol, butanol)	Zaybak et al. 2013
	Seed-sludge from anaerobic digester	Porous graphite felt	Tubular two-chamber reactor	-0.40	n.d.	n.d.	n.d.	n.d.	n.d.	n.d	n.d.	1	*Geobacter psychrophilus, Acidovorax ebreus, Diaphorobacter oryzae, Pedobacter duraquae,* and *Prolixibacter bellariivorans*	Cobalt recovery with simultaneous methane and acetate production	Huang et al. 2014
Purposely-built	Mixed natural & engineered sludge	Graphite rod	H-type reactor (fed-batch)	-0.85	-1.76	-1.76	5.75	5.75	n.d.	0.004	n.d.	44 ± 22	Undefined	Purposely-built electrode for improved MES - directly grown carbon nanotube on three-dimensional reticulated vitreous carbon.	Jourdin et al. 2014
		Unmodified RVC		-0.85	-3.70	-0.28	0.00	0.00	0.00	0.000	0.00	0			
		NanoWeb-RVC		-0.85	-37.00	-2.91	191.75	14.75	38645.00	0.025	1.65	70 ± 11			
	Enriched brewery WW sludge	Graphite granules	H-type reactor (fed-batch)	-0.80	n.d.	n.d.	n.d.	n.d.	n.d.	3.100	n.d.	n.d.	*Acetobacterium* dominated	Effect of pH on H_2 and acetate production	LaBelle et al. 2014
		Graphite rods		-0.60	-0.92	-0.92	8.56	8.56	n.d.	0.885	n.d.	40			
	Mesophilic WW anaerobic sludge	Graphite felt	H-type reactor and flat-plate reactor (fed-batch)	-1.10	-2.80	n.d.	10.10	n.d.	n.d.	n.d	1.40	65	*Acetobacterium spp.* + *Acetoanaerobium spp*	Bacterial composition shift during methane and acetate production	Xafenias et al. 2014
	Enriched carboxydotrophic mixed culture from a syngas fermenting lab-scale reactor	Carbon cloth	H-type reactor (Fed-batch)	-0.8	-5 to -20	n.d.	10.57	n.d.	n.d.	0.040	0.44	n.d.	Inoculum dominated by *Clostridium spp.*	Effect of the inoculum source - MES of acetate, butyrate, ethanol and butanol from CO_2	Ganigue et al. 2015
				-0.8	-5 to -20	n.d.	n.d.	n.d.	n.d.	n.d	1.62	n.d.			
Commercial	Anaerobic digester / retention basin	Graphite granules	H-type reactor (continuous)	-0.60	n.d.	n.d.	n.d.	n.d.	n.d.	0.029	n.d.	28.9 ± 6.1	Undefined	Continuous acetate production and effect oh pH	Batlle-Vilanova et al. 2015
	Primary clarifier effluent from wastewater treatment plant	Stainless steel mesh	Membrane-less single chamber reactor	-0.50	n.d.	n.d.	0.68	n.d.	n.d.	0.007	0.00	35	*Sporomusa spp.* Dominant	Methanogenesis inhibition for selective acetate production by MES using alamethicin	Zhu et al. 2015
		Carbon-coated graphite block			n.d.	n.d.	0.89	n.d.	n.d.	0.007	0.05	n.d.	Undefined		
	S. ovata	Graphite stick	H-type reactor	-0.65	-0.71	-0.71	4.07	4.07	n.d.	0.133	n.d.	82 ± 19	*S. ovata*	Simplifying MES reactor design - membrane-less reactor	Giddings et al. 2015
				-0.74	-1.70	-1.70	9.68	9.68		0.316		89 ± 12			
		Circular graphite disk	Membrane-less single chamber reactor (continuous)	-0.66	-0.17	-0.17	1.65	1.66		0.020		105 ± 5			
				-0.81	-0.48	-0.48	2.12	3.12		0.025		54 ± 10			
	CO_2-H_2 enriched mixed culture	Graphite plate	H-type reactor (batch)	-0.60	-0.35	-0.35	0.56	0.56	n.d.	0.1047	2.1	24	Undefined	MES at alkaline pH (10) coupled to bioanode	Annie Modestra et al. 2015
	Carboxydotrophic microorganisms mixed with WWTP sludge	Carbon-felt-Stainless steel assembly	H-type reactor (Fed-batch)	-0.90	-5.00	n.d.	39.83	n.d.	n.d.	0.080	0.62	40-50	Undefined	Pure culture vs. mixed microbial culture MES	Bajracharya et al. 2015
	Clostridium ljungdahlii			-0.70	-1.00	n.d.	18.88	n.d.	n.d.	0.142	0.60	40	*Clostridium ljungdahlii*		

(*Continued*)

Table 1. (*Continued*)

Material category	Microbial inoculum	Cathode material	Reactor design (operation)	$E_{cathode}$ (V vs. SHE)	Current density (A m^{-2}) vs. psa	Current density (A m^{-2}) vs. tsa	Acetate production rate g m$^{-2}_{psa}$ day^{-1}	Acetate production rate g m$^{-2}_{tsa}$ day^{-1}	Acetate production rate g m$^{-3}_{electrode}$ day^{-1}	Acetate production rate g L$^{-1}_{catholyte}$ day^{-1}	Max acetate titer (g L^{-1})	Electron recovery into acetate (%)	Microbial community analysis	Main objective	Reference
Commercial	Enriched culture (labscale anode and algae UASB sludge)	Carbon felt	H-type reactor (Fed-batch)	-1.260	-5	n.d.	19.00	n.d.	n.d.	0.060	1.29	58 ± 5	*Methanobacterium* 47% + *Acetobacterium* 16.7% in the biofilm	Pre-enrichment method investigation for acetate-selective MES (methanogens inhibition)	Patil et al. 2015
	Granular activated sludge from an anaerobic digester	VITO-CoRE-PL (VC-IS, plastic inert support)	Membrane-less single chamber reactor (Fed batch)	-0.400	-0.117	n.d.	9.49	n.d.	n.d.	0.057	4.13	10-20	Undefined	Electrode material investigation	Mohanakrishna et al. 2015
		Graphite rod 47 days		-0.400	-0.05	n.d.	3.45	n.d.	n.d.	0.021	1.52	30-35			
	Pre-enriched mixed microbial community from the effluent of a bioanode and anaerobic digester	Carbon felt with a stainless steel frame current collector	Flat plate three-chamber reactor (batch)	-1.140	-5.00	n.d.	20.40	n.d.	6399.01	0.581	13.50	60.8	Inoculum enriched of *clostridiales.*	Integrated production, extraction, and concentration of acetic acid from CO_2 using a three-chamber reactor design	Gildemyn et al. 2015
	Biological sludge from treatment plant	Stainless steel plate	H-type reactor	-0.360	0.00	0.00	0.00	0.00	0.00	0.000	0.00	0	Undefined	H_2 delivered by water reduction on cathode vs. via aquarium sparger (H_2 gas bottle) for CO_2+H_2 conversion to acetate	Blanchet et al. 2015
				-0.660	-1.50	-1.50	4.96	4.96	n.d.	0.035	0.24	53			
	Wild type *S. ovata*	graphite stick	H-type reactor (Batch)	-0.690	-0.32	-0.32	1.97	1.97	n.d.	0.007	0.00	85.3 ± 8.3	Wild type *S. ovata*	Pre-enrichment method investigation for acetate production enhancement by MES from CO2 - adaptive laboratory evolution approach by subjecting / growing pure culture of *S. ovata* to methanol.	Tremblay et al. 2015
	Methanol adapted *S. ovata*				-2.16	-2.16	12.78	12.78	n.d.	0.046	0.00	94.7 ± 2.1	Methanol adapted *S. ovata*		
Purposely-built	Enriched Mixed natural & engineered sludge	EPD-3D 45 ppi	H-type reactor (fed-batch)	-0.850	-102.00	-3.80	685.00	25.37	66469.40	0.373	11.00	100 ± 4	*Acetoanaerobium, hydrogenophaga,* and *methanobacteriales* dominant	Carbon nanotube deposition on reticulated vitreous carbon led to high rate / high selectivity MES to acetate from CO_2 + electron transfer mechanisms investigation (biologically-induced H_2)	Jourdin et al. 2015 and Jourdin et al. 2016 a
	Enriched Mixed natural & engineered sludge	EPD-3D 45 ppi	H-type reactor (fed-batch)	-0.850	-102.00	-3.80	685.00	25.37	66469.40	0.373	11.00	100 ± 4	*Acetoanaerobium, hydrogenophaga,* and *methanobacteriales* dominant	Bringing high-rate, CO_2-based MES closer to practical implementation through improved electrode design (pore diameters) and operating condition (pH, Eapplied, CO_2 source)	Jourdin et al. 2016 b
		EPD-3D 10 ppi		-0.850	-35.00	-6.30	247.00	42.36	25839.60	0.134	n.d.	100			
		EPD-3D 60 ppi		-0.850	-93.00	-2.66	620.00	17.70	69065.40	0.288	n.d.	100			
Commercial	Pre-enriched mixed community from MES and MEC	Graphite felt + SS plate current collector	Flat-plate reactor	-0.9 to -1.0	-2.26	n.d.	8.21	n.d.	544.84	0.642	0.24 - 0.19	50-60	Undefined	Microbial rechargeable battery - Energy storage and recovery through acetate	Molenaar et al. 2016

Note: Orange cells refer to values deducted/calculated from figures and/or information from materials and methods sections, but not directly referred to/mentioned in the articles. n.d. non-deductible from information given in the articles (blue cells).

MES of organics from CO_2 remains a nascent concept, with the first proof of concept reported less than a decade ago.[15] Earlier work related to this MES concept focused on conversion of organics to organics with external supplied electron shuttles or positioned electrodes in bioreactors, which is called electro-fermentation[2] (Table 2). With the global need to produce chemicals from CO_2 instead of from organics, MES research increased tremendously in the last decade. In that time frame, significant performance enhancement and improved understanding of the underlying mechanisms have been achieved, toward bringing MES closer to a practically feasible technology (Table 1). Either pure cultures or mixed microbial consortia have been used to demonstrate the process at laboratory scale (Table 1). The use of mixed microbial cultures is attractive because they are more tolerant to environmental stress and fluctuation, are readily available in large quantities, and have thus far showed higher production rates over long-term operation.[5,9,16] Pure cultures could theoretically lead toward higher product specificity, with preventing side reactions such as methanogenesis, even though to date, similar electron recoveries into the final product(s) have been reported on both pure and mixed cultures (Table 1).

12.1.3 Microbial Electrosynthesis: An emerging platform technology

MES is a versatile concept producing a wide variety of chemicals while integrated in existing or future biorefineries. To date, mainly acetate has been sustainably produced in these systems, using solid-state electrode and CO_2 as sole electron and carbon sources. Acetate has many potentially viable applications either as a useful end-product by itself or as a platform for further biological or chemical transformations.[14,17,18] Relatively high acetate titer, in the order of 11 g L^{-1}, have been achieved to date,[5,9,16] with realistic potential to reach higher, which makes it more favorable for potential downstream processing. Acetate concentrations as high as 44 g L^{-1} have been achieved in bioreactors pressurized with H_2 and CO_2 gas using pure culture of *woodi.*[19] Integrated MES acetate production at a relatively

Table 2. Key performance indicators of some MES from organics studies reported to date, in chronological order of publication. Orange cells refer to values deducted/calculated from figures and/or information from materials and methods sections, but not directly referred to/mentioned in the articles. n.d. non-deductible from information given in the articles (blue cells).

Material category	Microbial inoculum	Cathode material	Reactor design (operation)	$E_{cathode}$ (V vs. SHE)	Current density (A m^{-2})		Production rate				Max product titer (g L^{-1})	Electron recovery into product (%)	Microbial community analysis	Main objective	Reference
					vs. psa	vs. tsa	g m^{-2}_{psa} day^{-1}	g m^{-2}_{tsa} day^{-1}	g $m^{-3}_{electrode}$ day^{-1}	g $L^{-1}_{catholyte}$ day^{-1}					
Commercial	*Brevibacteriumflavum* No. 2247	Platinum, carbon or stainless steel	H-type reactor (batch)	n.d.	-0.0025	-0.0025	n.d.	n.d.	n.d.	20.4	51	n.d.	*Brevibacteriumflavum* No. 2247	Current supply through a mediator increased glutamic acid yield during glucose fermentation	Hongo and Iwahara 1979 a and b
	Clostridium acetobutylicum	Unspecified	H-type reactor (batch)	-2.5	n.d.	n.d.	n.d.	n.d.	n.d.	n.d.	1-7 - 7.2	n.d.	*Clostridium acetobutylicum*	Electron flow shift in *clostridium acetobutylicum* fermentation by electrochemically introduced reducing equivalent, with methyl viologen	Kim and Kim 1988
	Propionbacterium freudenreichii	Platinum wire	H-type reactor (batch)	-0.224 (AQDS)	-0.25	-0.25	2.156	2.156	n.d.	0.082	0.08	n.d.	*Propionbacterium freudenreichii*	The product of glucose fermentation shifted from acetate to propionate through a mediated current supply	Emde and Schink 1990
				-0.39 (CoS)	-0.5	-0.5	2.082	2.082	n.d.	0.079	0.09	n.d.			
	Actinobacillus succinogenes	finely woven graphite felt	H-type reactor (batch)	n.d.	n.d.	-0.002 to -0.011	n.d.	1.81	n.d	8.496	2.12	n.d.	*Actinobacillus succinogenes*	Fumarate to succinate conversion using neutral red	Park and Zeikus 1999
	Actinobacillus succinogenes	finely woven graphite felt	H-type reactor (batch)	n.d.	n.d.	-0.002 to -0.005	n.d.	0.57	n.d	2.700	5.40	n.d.	*Actinobacillus succinogenes*	Fumarate to succinate conversion using neutral red	Park et al. 1999
	Trichosporon capitatum	Grphite fine-woven felt	H-type reactor (batch)	n.d.	n.d.	n.d.	3.41	0.018	1136.64	0.085	2.2	n.d.	*Trichosporon capitatum*	6-bromo-2-tetraol production by yeast was stimulated by a mediated current suply	Shin et al. 2001
	Geobacter sulfurreducens	Unpolished graphite rod	H-type reactor (batch)	-0.3	-0.12	-0.12	11.96	53.80	n.d.	0.157	0.71	50	*Geobacter sulfurreducens*	Mediator-less fumarate to succinate conversion	Gregory et al. 2004
	Mixed culture sludge	Graphite felt	Flat-plate reactor (batch)	-0.55	-1.33	n.d.	0.02	n.d.	0.35	0.00006	0.621	12	Undefined	Acetate to ethanol conversion using methyl viologen	Steinbusch 2010
	Clostridium tyrobutyricum	Graphite felt	H-type reactor (batch)	-0.2	-0.69	n.d.	280	n.d.	n.d.	6.720	8.8	n.d.	*Clostridium tyrobutyricum*	Electron flow shift in *clostridium tyrobutyricum* sucrose fermentation towards enhanced butyrate production using neutral red (methyl viologen and AQDS also studied)	Choi et al. 2012
	Sewage sludge fermenter	Graphite plate	Flat-plate reactor (continuous)	n.d.	-3.2	-3.2	n.d. (large spectrum of products formed)				1.5	n.d.	*Citrobacter, pectinatus, Clostridium* species	Dynamics of cathode-associated microbial communities and metabolite profiles in a glycerol-fed bioelectrochemical system. Glycerol to 1,3-propanediol and ethanol initially, then fermentation + chain elongation	Denis et al. 2013
Purposely-built	Sulfate-reducing bacteria rich mixed culture	80-20% carbon-(PTFE) mix with 316L grade stainless steel mesh	Membrane-less single chamber reactor (batch)	-0.85	-15 from -100 to -10	n.d. (large spectrum of products formed)					< 0.1	n.d. n.d.	Undefined	Microbial electrosynthesis of alcohols and acetone from acetic and butyric acids reduction	Sharma et al. 2013
Commercial	*Geobacter sulfurreducens*	Stainless steel plate	0.45 L membrane-less single chamber reactor (batch)	-0.4	-16	-16	n.d.	n.d.	n.d.	n.d.	n.d.	n.d.	*Geobacter sulfurreducens*	Fumarate reduction to succinate coupled to electrochemical reduction of CO_2 catalysed by *Geobacter sulfurreducens* for glycerol production	Soussan et al. 2013
				-0.4	-20 to -30	-20 to -30	n.d.	n.d.	n.d.	n.d.	n.d.	n.d.			
			1.85 L membrane-less single chamber reactor (batch)	-0.2	-6 to -8	-6 to -8	n.d.	n.d.	n.d.	n.d.	n.d.	n.d.			

(*Continued*)

Table 2. (*Continued*)

Material category	Microbial inoculum	Cathode material	Reactor design (operation)	$E_{cathode}$ (V vs. SHE)	Current density (A m^{-2}) vs. psa	Current density (A m^{-2}) vs. tsa	Production rate g m$^{-2}_{psa}$ day^{-1}	Production rate g m$^{-2}_{tsa}$ day^{-1}	Production rate g m$^{-3}_{electrode}$ day^{-1}	Production rate g L$^{-1}_{catholyte}$ day^{-1}	Max product titer (g L^{-1})	Electron recovery into product (%)	Microbial community analysis	Main objective	Reference
Commercial	*Shewanella oneidensis*	Graphite	H-type reactor (batch) - 1 μM DSSN+	-0.3	-0.16	n.d.	n.d.	n.d.	n.d.	n.d.	n.d.	>100	*Shewanella oneidensis*	A lipid membrane intercalating conjugated oligoelectrolyte enables electrode driven fumarate to succinate in Shewanella.	Thomas et al. 2013
			H-type reactor (batch) - 10 μM DSSN+		-1.5	n.d.	n.d.	n.d.	n.d.	n.d.	n.d.	≈100			
	Anaerobic fixed film reactor	Graphite felt	Flat-plate reactor (continuous)	-0.9	-1.8 ± 0.6	n.d.	4.66	n.d.	n.d.	0.185	0.74	45	Undefined	Bioelectrochemical production of caproate and caprylate from acetate by mixed cultures	Van Eerten-Jansen et al. 2013
	Anaerobic glucose fed pH-stat reactor	Graphite plate	Flat-plate reactor (continuous)	-0.6	-0.6	n.d.	n.d.	n.d.	n.d.	0.913	0.15	21.2	Undefined	Carbon and electron fluxes during the electricity driven 1,3- propanediol biosynthesis from glycerol	Zhou et al. 2013
				-0.9	-7.3	n.d.	n.d.	n.d.	n.d.	2.739	0.30	47.9			
	Clostridium pasteurianum DSM 525	Graphite felt	H-type reactor (batch) - Glucose to butanol	0.045	-0.14	n.d.	59.28	n.d.	n.d.	2.134	0.97	< 1	*Clostridium pasteurianum DSM 525*	Mediator-less electricity-driven metabolic shift through direct electron uptake by electroactive heterotroph *Clostridium pasteurianum* (electro-fermentation)	Choi et al. 2014
			Glycerol to 1,3-PD		-0.83	n.d.	63.42	n.d.	n.d.	2.283	7.23				
	A. ferrooxidans			0.045	-0.26	-0.26	n.d.	n.d.	n.d.	n.d.	n.d.	n.d.	*A. ferrooxidans*		
	T. denitrificans	Coin-shaped graphite connected to a graphite rod	Membrane-less single chamber reactor (batch)	-0.5	-0.0145	-0.0145	n.d.	n.d.	n.d.	n.d.	n.d.	n.d.	*T. denitrificans*	Microbial electroreduction: screening for new cathodic biocatalysts	De Campos Rodrigues et al. 2014
	S. denitrificans			-0.5	-0.013	-0.013	n.d.	n.d.	n.d.	n.d.	n.d.	n.d.	*S. denitrificans*		
	D. orientis			-0.5	-0.0203	-0.0203	n.d.	n.d.	n.d.	n.d.	n.d.	n.d.	*D. orientis*		
	D. piger			-0.5	-0.00779	-0.00779	n.d.	n.d.	n.d.	n.d.	n.d.	n.d.	*D. piger*		
	Enriched mixed EA-SRB, TERI-MS-003 consortium	Activated carbon fabric fastened to a stainless steel (SS) mesh	Membrane-less single chamber reactor (batch)	-0.65	n.d. (large spectrum of products formed)						< 0.6	n.d.	Undefined	Influence of headspace composition on product diversity by sulfate reducing bacteria biocathode for conversion of volatile fatty acids (acetate and butyrate) to reduced organic compounds	Sharma et al. 2014
	Escherichia coli	Graphite felt coupon	H-type reactor (batch)	-0.45	-0.085	n.d. (large spectrum of products formed)						n.d.	*Escherichia coli*	Neutral red-mediated microbial electrosynthesis by *Escherichia coli* (glucose fermentation), *Klebsiella pneumoniae* (glycerol fermentation), and *Zymomonas mobilis* (glucose fermentation)	Harrington et al. 2015
	Klebsiella pneumoniae				-0.068								*Klebsiella pneumoniae*		
	Zymomonas mobilis				-0.043								*Zymomonas mobilis*		
	Anaerobic digestion effluent from wastewater treatment plant	Carbon brush	H-type reactor (fed-batch)	-0.25	-0.64	n.d. (large spectrum of products formed)					< 0.7	n.d.	Undefined	Bioelectrochemical reduction of volatile fatty acids in anaerobic digestion effluent for the production of biofuels	Kondaveeti et al. 2015
				-0.44	-0.85										
				-0.64	-1.04										
				-0.31	-0.69										
				-0.49	-0.93										
				-0.69	-1.47										

low titer and extraction using a three-chamber reactor design has also been investigated.[20] Agler *et al.*[18] described the "carboxylate platform" as a range of processes for the production of high-value compounds (*e.g.* alcohols, esters, and alkanes) from short-chain carboxylic acids (two to four carbon atoms), such as acetate. Those carboxylates can be transformed by separate postprocessing steps (*e.g.* chemical, biochemical, electrochemical and thermochemical steps) for the production of usable fuels and industrial solvents.[18] Liu *et al.*[17] recently described a biological acetate postprocessing step.[17] They presented a new direct solar-powered process for the production of acetate from CO_2 coupled with the synthesis of complex organic molecules from acetate, such as PHB biopolymer, isoprenoid and *n*-butanol, using genetically engineered *Escherichia coli* in a separate vessel.[17] Hu *et al.* also very recently described a new, continuous flow, gas-to-lipid bioconversion scheme.[21] They effectively converted CO_2 and H_2 to acetate in a first-stage anaerobic bubble column bioreactor (H_2 dark fermentation), and further converted acetate to lipids by *Yarrowia lipolytica* in a second-stage aerobic bioreactor.[21] Acetate is also widely used as a carbon substrate for industrial and environmental biological processes such as biological phosphorous removal and denitrification. In this context, MES could be envisioned as an option for the on-site production of organics for such processes at a waste-water treatment plant, *e.g.* using the CO_2 fraction from biogas and electric energy from methane combustion. However, considerable further process development of all these options is required to identify the most valuable utilization of acetate, or other organics, generated by MES.

Recently, a study has demonstrated the simultaneous conversion of CO_2 into a mixture of products comprising acetate, butyrate, ethanol, and butanol using a mixed culture previously enriched for syngas fermentation. Few MES studies have also focused on the reduction of organics, such as acetate and fumarate (Table 2). The main objective and challenge lay in the conversion of organic waste streams or feedstock into valuable organic compounds, with high selectivity.

12.2 Electrodes for Microbial Electrosynthesis

12.2.1 Why modifying the electrode (surface)?

Optimizing and scaling MES to practical applications relies on performance improvements while maintaining low investment and operation costs. Enhancement of bacterial attachment, biofilm development, electron transfer rate at the cathode surface (microorganism–electrode interaction), and chemical production/ conversion rates require optimization of several key elements, particularly improved cathode materials, selective microbial consortia, and efficient reactor designs.[6]

Despite the large number of reports on new electrode materials and surface modification strategies for the improvement of anodic processes (Chapter 8),[22,23] there is hardly any work reported on new electrode materials for microbial cathodes. This is not surprising, since there is quite a general agreement that bacteria–electrode electron transfer processes are most probably following different routes during an anodic or cathodic process.[2,4] Therefore, it is not straightforward to predict that a given electrode material yielding good results as a microbial anode will perform equally well as a cathode.

The extracellular electron transfer (EET) mechanisms play an important role in determining the electrode material requirements.[23] The opposite may also be true, with the choice of a specific material and its surface chemistry impacting the mechanisms of electron exchange. This renders the material selection for biocathode processes much more challenging. Indeed, to date, researchers have mainly focused on gathering abundant mechanistic information on the anode processes while information on the reverse process, *i.e.* electron flow from electrodes (cathodes) to microorganisms, is rather limited. Elucidating the electron transfer mechanisms involved in MES could lead to major breakthroughs in the effort to increase MES performance.[24] Some studies have recently been published toward shedding light on MES mechanisms.[10,25,26] It has also been suggested that electrode materials should be tailored to each organism or biocatalyst, depending on their different demands.[27] Common traits is that the electrode material must be scalable, highly conductive, and cost-effective.

12.2.2 Key developments on commercial electrodes

Table 1 summarizes most of the MES of organics studies reported to date. It can be seen that most studies have used commercially available carbonaceous electrode materials, due to their good electrical conductivity and chemical stability. These include graphite granules and rods, carbon paper, carbon cloth and graphite felt.

12.2.2.1 *Acetogens are able to reduce CO_2 within an attached biofilm on graphite rods*

Graphite rods were first used by Nevin *et al.*[15,28] to prove the concept of a biocathode-driven CO_2 reduction to acetate using pure cultures of acetogenic microorganisms, such as *Sporous ovata.* In their studies, they first grew the pure cultures with H_2 as the electron donor (H_2–CO_2 [80:20]). The culture was then introduced in a cathode compartment with a gas mixture containing hydrogen (N_2–CO_2–H_2) as an additional electron donor to speed up the growth of a biofilm on unpolished graphite rods. The gas feed was then switched to N_2–CO_2 (80:20) and mainly acetate was produced as well as small amounts of 2-oxybutyrate with an electron recovery in these products of over 85%, when using *Sporomusa ovata.* A potential of −400 mV (*vs.* standard hydrogen electrode) was applied to the cathode. The cathode biofilms showed relatively long-term stability as after periods of more than 3 months, their capacity for electron consumption and acetate production remained the same. However, the biofilms of acetogens remained thin, suggesting that the microorganisms gain only small amounts of energy from the reduction of CO_2. Hence, actual bacterial growth on electrons from electrode and CO_2 only as energy and carbon source was not verified. They demonstrated that a relatively wide range of acetogenic microorganisms were capable of electrosynthesis.[28] Gram-negative (*Sporomusa* species) as well as Gram-positive (*Clostridium* species) bacteria were shown capable of establishing electrical connections with electrodes. In previous studies, Gram-negative species such as Geobacter[29–32] and Anaeromyxobacter[33] have shown this capacity but Gram-positive species were only shown with the capacity of giving electrons to an electrode. *S. ovata* showed the most efficient of the pure cultures tested by Nevin *et al.*,[28] though the performance remained low, with current density up to -0.21 A m^{-2} and acetate production rate of 1.3 g m^{-2} day^{-1}, when normalized to total (as well as projected) surface area (Table 1). Due to the cathode potential applied (–0.4 V *vs.* SHE), direct electron transfer (DET) from the electrode to the microorganisms was suggested in this study. DET is defined as direct electron transfer from the cathode to the biocatalysts without involvement of added or produced electron shuttles like hydrogen. It was highlighted that DET would hold most promise for microbial cathode applications.[34] It would decrease overpotentials (energy and thus operational costs savings) and potentially simplify the product recovery. It was also argued that, from a biological metabolism perspective, wider range of products could be obtained by DET as compared to obtained from hydrogen-mediated electron transfer. However, to date, these two studies from Nevin *et al.*[15,28] have been the only ones reporting MES at cathode potentials higher than the H_2 evolution redox potentials. Nevertheless, this study does not report the pH during the production, which may change the potentials, which does thermodynamically allow H_2 evolution. Nevertheless, DET is certainly a possible mechanism while several enzyme-cathode based studies do show CO_2 reduction reactions to organics like, *e.g.* formate.[35] In addition, for DET to hold promise for the production of same value products, much higher production rates should be attained, at least in the order of what has now been obtained at lower potentials and high selectivity[9,14] (Table 1).

12.2.2.2 *Mixed cultures reduce CO_2 on graphite granules while losing electrons to H_2*

Subsequent studies by Marshall *et al.*[16,36] demonstrated for the first time the ability of mixed cultures to perform MES of acetate from CO_2, while using graphite granules as cathode material. They reported simultaneous production of methane, acetate, and hydrogen using a mixed microbial community originating from brewery wastewater anaerobic sludge selected on granular graphite cathode poised at −0.59 V *vs.* SHE.[36] When methanogenesis was inhibited following the addition of a specific chemical inhibitor, 2-bromoethanesulfonate, acetate production rate reached up to 0.236 g $L_{catholyte}^{-1}$ day^{-1} and hydrogen production up to 11 mmol L^{-1} day^{-1} (57% electron recovery into acetate). A maximum titer of about 1.71 g L^{-1} of acetate was achieved in this study. They further demonstrated improved performance over long-term operation (over 150 days) of the established acetogenic biocathode aforementioned.[16] Acetate production rates reached a maximum of 1.03 g $L_{catholyte}^{-1}$ day^{-1} with a high titer of 10.5 g L^{-1} accumulated over 20 days (69% electron recovery into acetate). Hydrogen was also found to be simultaneously produced at high rates, reaching 100 mmol L^{-1} day^{-1}. In both studies, the active electrosynthetic microbiome revealed a similar community structure, with *Acetobacterium* spp. found to be dominant, which were speculated primarily responsible for the CO_2 reduction to acetate in these MESs. Also prevalent were *Sulfurospirillum* spp. and an unclassified *Rhodobacteraceae*, although their roles were not clear. Both DET and H_2-mediated electron transfer from the cathode to the acetate producers were speculated. It could be hypothesized from this study that the electrode material packing (granules) may have not been optimal for biofilm development over the whole surface of the granules. This could explain the "waste" of most of the hydrogen, not converted to acetate. Findings by Jourdin *et al.*[14] highlighted that an optimal compromise between total surface area per volume unit available for biofilm development and effective mass transfer between the bulk liquid and the electrode and biofilm surface has to be found for achieving optimal volumetric productivity and high selectivity.

12.2.2.3 *Cathode potential affects product spectrum, production rates, and selectivity on carbon electrodes*

Since then, several other studies have also reported biocathode concomitant production of acetate and hydrogen by mixed microbial consortia on commercially available carbonaceous electrode materials.[13,20,37–47] Jiang *et al.* discussed that the metabolic pathway and end-products spectrum was dependent on the applied cathode potentials, while using carbon felt as an electrode material. They reported sole methane and H_2 production at set potentials between −0.65 and −0.75 V *vs.* SHE while acetate was also produced at potentials more negative than −0.75 V *vs.* SHE. It was also highlighted that cathode potential was a critical factor that affected MES performance with acetate production rate increasing from 12 g m^{-2} day^{-1} (normalized to projected surface area) and 53.6% electron recovery into acetate at −0.7 V *vs.* SHE to 72 g m^{-2} day^{-1} and 89.5% electron recovery at −0.9 V *vs.* SHE.[38] In another study, decreasing the cathode potential from −0.85 V to −1.1 V *vs.* SHE doubled the acetate production rate while still achieving 99% electron recovery in the form of acetate at a current density of around −200 A m^{-2} normalized to projected surface area.[14] These current densities, which led to a high acetate production rate of up to 1,330 g m^{-2} day^{-1}, were obtained using a purposely built electrode material, which is discussed in more detail below[14] (see Section 12.2.4).

12.2.2.4 *Proton availability showed to affect MES performance on graphite granules*

The pH was also shown to influence MES rates.[14,41,43] Lowering the pH from near neutral to 5, as shown by LaBelle *et al.*,[41] did not affect acetate production while concomitant H_2 production significantly increased. Further decrease of the pH below 5 was found to be detrimental to acetate production, while H_2 production rates further increased.[41] This latter study, from the same group than Marshall *et al.*[16,36] studies discussed above, used the same operational setup and conditions, with packed graphite granules as cathode material. Therefore, similar conclusions as drawn above could be suggested, with likely a non-optimized total surface area for biofilm development per volume ratio and mass transfer limitation primarily responsible for non-assimilation to acetate of most of the H_2. Controlling the pH slightly acidic (ca. 5.8) was shown to enhance acetate production in another study using graphite granules, as either a direct effect of the low pH or due to the indirect increase of substrate availability.[43] Higher proton availability was shown in another study to significantly increase the acetate production rate, with pH 5.2 found to be optimal (40% increase from pH 6.7), and 100% electron recovery into acetate, using purposely built electrodes.[14] In this latter study, it was also shown that pH lower than 4.5 was detrimental to the acetate-producing microorganisms.

12.2.2.5 *Alternative reactor designs and anodic reactions tested using graphite plate cathode: Successful proof of concepts*

Another study by Giddings *et al.*[44] demonstrated the feasibility of using a membraneless reactor design for MES of acetate from CO_2 coupled with a direct current power source, instead of a potentiostat, to supply the energy for electron delivery. Circular graphite disk and pure culture of *S. ovata* were used in that study. Relatively low current density and production rates were reported (Table 1), with decrease in electron recovery in the form of acetate from $105 \pm 5\%$ to $54 \pm 10\%$, from -0.66 V to -0.81 V *vs.* SHE. However, eliminating the need for a membrane could have a positive impact on the economic and technical feasibility of large-scale MES reactor implementation. Sulfide-oxidizing bioanode (abiotic oxidation of sulfide to sulfur, further oxidized to sulfate by *Desulfobulbus propionicus*) was also successfully used as an electron donor for microbial electrosynthesis.[40]

12.2.2.6 *Novel methanogenesis inhibition approaches tested on carbon felt and block*

In most studies using mixed microbial consortia, methanogens were showed to thrive in these reductive cathodic environments if their metabolisms for CH_4 production were not inhibited. BES for primary methane production from CO_2 have also been investigated.[48–53] However, in the frame of organics production, methane is regarded as an undesired product, whose formation redirects the carbon flow and ultimately decreases the carbon and electron recoveries into the targeted product (*e.g.* acetate). Therefore, preventing methane production in these systems is of paramount importance.[6] Most often, a chemical inhibitor such as sodium 2-bromoethanesulfonate (NaBES) was added to lab-scale experiments. NaBES has been shown to be a specific inhibitor toward methanogens while non-methanogenic organisms in pure cultures as well as in more complex environments, even at high NaBES concentrations, were not affected.[54–61] Adding external chemicals is suitable for lab-scale fundamental research. However, it will become more problematic for larger-scale applications, from both technical and economic points of view. Patil *et al.* investigated a selective enrichment approach via several culture transfers with H_2:CO_2 headspace conditions.[45] The approach was based on inhibiting methanogenesis using NaBES in the start-up of the enrichment phase (in serum bottle) followed by quick culture transfer and dilution to other serum bottles after acetate production was observed. The obtained culture was then transferred to MES reactors, using carbon felt as cathode, and controlled at a fixed current of -5 A m^{-2}. While methanogenesis did not occur initially, it was shown that this enrichment strategy was not successful to eliminate methanogens on the long term.[45] Zhu *et al.*[62] reported on the peptide alamethicin to effectively suppress methanogenic activity in mixed culture MES systems. The authors report higher effectiveness than NaBES at two orders of magnitude of lower concentration. However, it was shown to also inhibit the growth and activity of Gram-negative bacteria with thinner cell wall. Further research will need to investigate other means of effective methanogenesis inhibition in MES reactors.

12.2.2.7 *Membrane assemblies allow extraction of acetate*

One of the main challenges in large-scale chemical production is the product separation/recovery, which is often regarded as one of the major cost.[18] For instance, industrial fermentation products separation can reach up to 50–60% of the total plant costs.[63,64] Low-cost technologies are being investigated. On-line liquid/liquid extraction using membranes was reported as one possibility to remove acetate from engineered systems (*e.g.* fermenters).[65] Membrane electrolysis (ME) was also shown to be successful for the extraction of carboxylates from cathode to anode compartment, combined with biphasic esterification (BE) for the production of fine chemicals[64]; and a similar concept could be coupled to MES of acetate from CO_2. Finally, Gildemyn *et al.*[20] reported on an integrated production, extraction and concentration of acetic acid from MES from CO_2, using a three-chamber two-membrane reactor system. Carbon felt with a stainless steel (SS) current collector and applied current control (-5 A m^{-2} *vs.* projected surface area) was used. This resulted in a fairly low cathode potential of -1.14 V *vs.* SHE and an electron recovery of 60.8% in the form of acetate. Like couple of other studies that reported fairly high acetate concentration (10–11 g L^{-1}) in the cathode compartment of two-chamber MES systems,[9,16] about 13.5 g L^{-1} could be concentrated in the extraction compartment of this system. However, the use of a cation exchange membrane between the anode and the extraction compartments led to a significant loss of acetic acid in the anode compartment, with concentration above 9 g L^{-1}. In this system, the protons necessary for the reduction reaction cannot be provided from the anode reaction, hence a significant fraction of the CO_2 would also have to be used to buffer the catholyte, which cannot be used for production purposes. Though a reactor system with two membranes would significantly increase the resistance across the system and the energy need, the benefit of this system would be to recover an acidified stream for further upgrading.

12.2.2.8 *Further diversification of products is possible with mixed consortia*

Recently, a study has also shown the simultaneous conversion of CO_2 into a mixture of products composed of acetate, butyrate, ethanol, and butanol, using a mixed microbial consortia enriched from syngas fermentation, on carbon cloth polarized at -0.8 V *vs.* SHE.[13] This study highlighted the impact the source of inoculum have on the resulting product spectrum. The higher market value and demand of some of these organics (*e.g.* butanol) compared to acetate makes their direct production from CO_2 interesting. Nevertheless, high product specificity and production rates are regarded as key parameters that could enable large-scale applicability of such a technology. The MES concept from CO_2 or organics to high-value chemicals does allow a widening of the product spectrum. Though, here, electron transfer mechanisms have to be elucidated since these studies can also be explained by a combination of a MES process and an *in situ* further microbial conversion. Pure and synthetic consortium work can reveal how these MES systems with mixed microbial consortia do work. Further work should also focus on the selective production and/or selective extraction from MES of higher carbon-content organics, at high volumetric production rates. Otherwise, a mixture of chemicals is produced which is not necessarily useful.

Finally, commercially available metal materials (SS plate or mesh) have also been investigated for microbial biocathode processes.[25,62,66] However, SS biocompatibility has been argued, with no cathodic biofilm formation reported and no biological catalytic activity recorded on SS electrodes.[25] Blanchet *et al.*[25] investigated CO_2 conversion to acetate occurring via cells in suspension with hydrogen either delivered by the SS electrode or directly from a gas cylinder.[25] The cathodic process could therefore be regarded as an hydrogen dark fermentation. This process is indeed very relevant and also deserves much attention but is out of the scope of this chapter. Others have used a combination of both carbonaceous and metal components, *e.g.* carbon felt or activated carbon with SS as current collector.[20,46,67]

In summary, commercial electrodes (readily available) have been used for the development of our understanding of microbial electrosynthesis, by investigating parameters such as methanogenesis inhibition,[45,62] reactor designs,[20,44] online product extraction, or the effects of pH,[14,41,43] applied potential and applied current,[14,20,37] or source of inoculum,[13] on product(s) formation and production performance (electron consumption and production rates and efficiencies), as discussed above. However, the actual effect of the electrode material on MES performance has barely been discussed. Moreover, as seen in Table 1, the reported performance of bioelectrosynthesis processes remain largely ineffective for scaling MES to practical application when those commercially available carbonaceous or metal electrode materials are used. Alternatives have been investigated.

12.2.3 Key developments on surface modifications

Modifying commercial electrode surfaces have been shown to affect performance of bioelectrochemical reactions such as oxygen-reducing biocathodes[68] and bioanodes.[69] To the best of our knowledge, only a very limited number of studies have proposed a number of modified electrode materials for the improvement of MES processes. Lovley *et al.*[3] tested several treatments/modifications of carbon surfaces following strategies tested for bioanode electrodes (Figure 3(a)),[70,71] for the same reasons as discussed above (Chapter 8). Several studies from the same group have already been extensively discussed above. Similarly, a pure culture of *S. ovata* was used for the conversion of CO_2 to acetate. Using different treatments, they altered the surface chemistry of carbon cloth by the immobilization of positively charged molecules, or by treating it with metal nanoparticles, with the aim of enhancing microbe–electrode interactions. Moreover, they tried depositing carbon nanotubes (CNTs) on non-conductive textile composites. The best results were obtained by cathode functionalization with chitosan and cyanuric chloride, conferring positive surface charge to the carbon cloth, with six- to sevenfold higher production rates (ca. 0.02 mM cm^{-2} day^{-1}) and highest current density of 0.0475 mA cm^{-2} reached with a chitosan modified electrode. However, not all surface treatments that increase the positive charge were found to be successful. Rates obtained on melamine and ammonia-treated electrode were found similar than those on unmodified carbon cloth, suggesting that surface charge alone might not be sufficient to enhance bacterial attachment and in turn biofilm development, electrode–microbe electron transfer rate and maximal achievable production rate. It was concluded that other features of the cathode modifications designed to provide a positive charge may be critical.[71] The physical deposition of nanoparticles of metals, Au, Pd, or Ni, onto the surface of carbon cloth also led to increased acetate production rates by factors of 4.5–6 *vs.* the untreated control. While metal nanoparticles have been reported to be biocompatible and to have

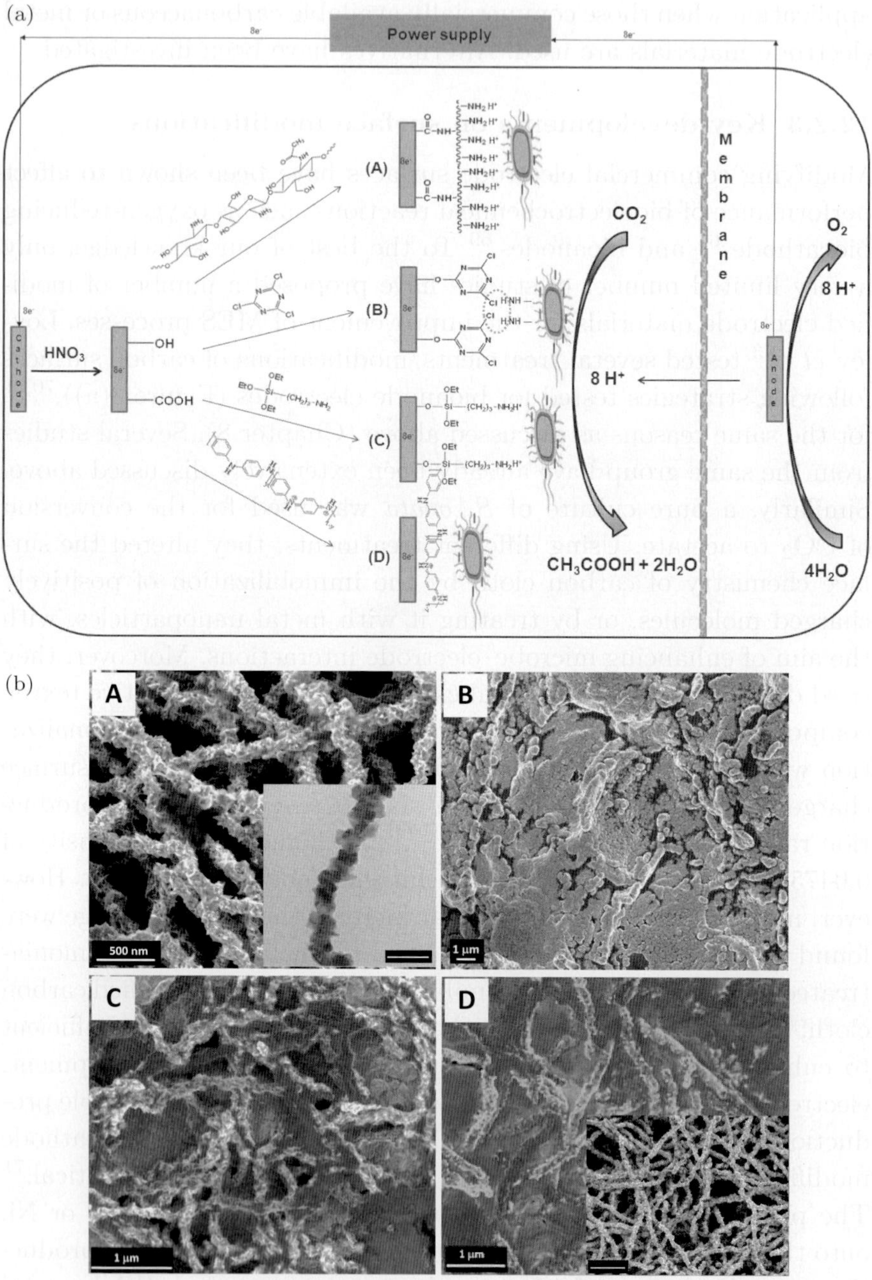
(a)
Power supply
8e⁻
Membrane
CO2
O2
8 H+
HNO3
Cathode
Anode
—OH
—COOH
(A)
(B)
(C)
(D)
8 H+
CH3COOH + 2H2O
4H2O
(b)
A
B
C
D
500 nm
1 µm
1 µm
1 µm

high active surface area and chemical stability, as discussed by the authors, it was not shown in this study whether or not the increase in performance could solely be explained by an increase in abiotic hydrogen production rate due to the superior catalytic activity of these metals over carbon. Biofilm formation was also not discussed.

A novel cathode using nickel nanowires anchored to graphite (Figure 3(b)) was also developed and tested in a similar setup than above, using *S. ovata* for the reduction of CO_2 to acetate.[70] Around 2.3-fold higher rate than obtained on an untreated graphite electrode was obtained. The porous nickel-nanowire network was argued to increase the interfacial area and enhance interactions between the cathode surface and the microbial biofilm.[70]

Most of these material modification strategies proved to be successful in increasing MES rates over carbon cloth or graphite control. However, performance remained insufficient for potential practical implementation of MES process and other approaches have been investigated.

12.2.4 Key developments on purposely built electrodes

Maximal current densities obtainable on bioanodes were shown to be limited by the transfer of substrates and products to and from the electrode surface at flat and rough electrodes.[72,73] To remedy this issue, a novel approach was recently proposed, which consists of using 3D and purposefully synthesized materials as bioanode electrodes (Chapter 8). We refer here to porous 3D materials, where microorganisms can develop in the whole volume of the electrode, as opposed to rough or dense fibrous materials. For example, on graphite felt a biofilm is only developed in the outer layers of the fiber mat, but not 2–3 mm from the surface in the deepest core of

←

Figure 3. Examples of surface modifications carried out for MES from CO_2. (a) Carbon cloth coated with chitosan (A), cyanuric acid (B), 3-aminopropyltriethoxysilane (C), and with PANi (D), reproduced from Ref. 71 with permission from The Royal Society of Chemistry. (b) Nickel-nanowire-network-coated graphite stick, reproduced from Ref. 70 with permission from The Royal Society of Chemistry.

the electrode. In the last five years, several authors reported current generation enhancement in MFC mode using such 3D electrodes as bioanode material.[6,74–80]

A similar approach was envisioned by Jourdin *et al.*[5,9,10,14] for biocathode electrode material. Novel electrode structures benefiting from the advantages of 3D reticulated vitreous carbon (RVC) and CNT was synthesized. RVC is an open-pore foam material of honeycomb structure. RVC has a high electrical conductivity, strong chemical and heat resistance, minimal reactivity over a wide range of conditions, and a high surface area to volume ratio, which are advantages for BES applications.[5] CNTs-based electrode materials have also become extremely attractive for application in BES. CNTs have large aspect ratios, high surface area, excellent mechanical strength and chemical stability, as well as high electrical conductivity along their length.[5,81]

In the first study, multi-walled carbon nanotubes (MWCNT) were grown directly on the surface of RVC by a chemical vapor deposition technique (CVD) (Figure 4(a)). The resulting electrode was called NanoWeb-RVC and was tested as biocathode material for MES of acetate from CO_2 with mixed cultures.[5] Current density and acetate production rate up to –37 A m^{-2} and 192 g m^{-2} day^{-1} were obtained, respectively. This corresponds to a significant increase when compared to other studies using flat, rough, or dense fibrous electrodes (Table 1, Figure 5(a)). The results demonstrated that (i) the high surface area to volume ratio of the macroporous RVC maximizes the available biofilm area while ensuring effective mass transfer to and from the biofilm and (ii) the carbon nanostructure enhances the bacteria–electrode interaction, biofilm development (Figure 4(a)), microbial extracellular electron transfer, and acetate production rate.[5,6]

In a following study, electrophoretic deposition (EPD), a more suitable technique for industrial scale production, was exploited to form MWCNT coatings onto RVC to generate a new biocompatible, highly conductive, 3D hierarchical porous structure, called EPD-3D (Figure 4(b)).[9] The effective mixed microbial consortium enriched from NanoWeb-RVC reactors was successfully transferred to EPD-3D reactors (very fast start-up) and demonstrated further performance

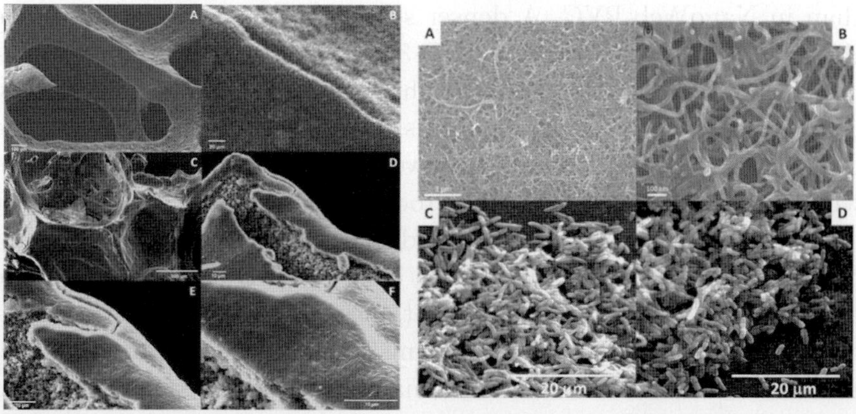

(a) (b)

Figure 4. Examples of purposely built electrodes used for MES from CO_2. (a) MWCNTs directly grown on RVC by CVD (NanoWeb-RVC; A,B) and subsequent electroactive biofilm grown 140 days after inoculation (C–F), reproduced from Ref. 5 with permission from The Royal Society of Chemistry. (b) MWCNT deposited on RVC by electrophoretic deposition (EPD-3D; A, B), and subsequent biofilm grown 63 days after inoculation (C, D), reproduced from Ref. 9 with permission from The American Chemical Society.

enhancement, reaching a biocathode current density of -102 A m^{-2} and acetate production rate of 685 g m^{-2} day^{-1} normalized to projected surface area (Figure 5(a)). Those are equivalent to high volumetric performance of ca. -9960 A $m^{-3}_{electrode}$ and 66 $kg_{acetate}$ $m^{-3}_{electrode}$ day^{-1} and CO_2 capture of 98 kg m^{-3} day^{-1}, when normalized to the electrode volume. That study showed that industrial production process standards, such as fermentation-type volumetric production rate and product yield, can be reached by MES of organics from CO_2, though there are still several issues to be overcome for the actual bioelectrochemical system (BES) technology scale-up and practical implementation.[9,10] NanoWeb-RVC and EPD-3D benefit from the same macroporous structure. The main difference between both electrodes was discussed to be the morphology and surface chemistry of the nanostructure. The MWCNT in EPD-3D was observed to be thinner in diameter as those grown by CVD (9.5 nm *vs.* 60 nm) and were also functionalized by undergoing a strong chemical oxidation prior to the electrophoretic deposition step.[9] Moreover, the MWCNT layer in EPD-3D was much denser than in NanoWeb-RVC. A denser structure was argued to improve the electrical conductivity of EPD-3D and to increase the probability of the CNTs to be at the right position to better electrically communicate with microorganisms.[9] Toward this hypothesis, the SEM images showed a remarkably well-developed, continuous biofilm with entangled, multi-layered microorganisms (Figure 4(b)). Further electron flow investigation confirmed that the high electron consumption and MES performance reached in these EPD-3D systems can be attributed to sustained bioelectrocatalytic activity through the development of a uniform biofilm covering the 3D structure of the EPD-3D.[9,10] It demonstrated that the main electron transfer mechanism occurred through biologically induced H_2 production, likely through the biological synthesis of metal copper particles, without H_2 able to escape from the biofilm ($99 \pm 1\%$ electron recovery into acetate).[10]

Another study demonstrated that to achieve such high productivities in practice, the 3D macroporous electrode materials need to also be optimized to achieve a good balance between total surface area per volume available for biofilm formation and effective mass transfer between the bulk liquid and the electrode/biofilm surface.[14] Three different EPD-3D electrode pore sizes (2.54, 0.56, and 0.42 mm average pore diameter) were investigated. Using highly open, macroporous RVC electrodes with pore sizes of about 0.56 mm diameter was found to be optimal in this respect, in this reactor configuration and given forced mass-transfer conditions.[6,14] However, this porosity influence is highly dependent on the actual flow rate through the electrode material. A better compromise between mass transfer and total surface area per volume unit available for biofilm development could then be found in a forced flow-through system, in which case the pore size could possibly be smaller to achieve the maximal performance.

12.3 Criteria for Microbial Electrosynthesis

MES is still a technology at its early stage of research and development. As seen above, most of the studies have used commercial carbonaceous electrode materials while only a handful of studies have investigated surface modifications and purposely built electrodes, which led to performance enhancement.

12.3.1 MES studies must report the full spectrum of performance indicators

Among the studies carried out on commercial carbon electrodes, barely any have analyzed and discussed the actual effect of the cathode material on microbial enrichment, biofilm coverage, coulombic efficiency, and product formation or substrate consumption. After review of most of the MES studies (Tables 1 and 2), it becomes obvious that it is challenging to compare data and extract further trends. For instance, it can be seen that a wide range of electron recovery into the final product was obtained when using carbon felt as cathode material, from around 15%[37] to 90%.[38] The main issue that renders cross-comparison of data between studies barely impossible until now is the non-existence of a logical data representation framework for MES processes, as previously discussed by Patil *et al.*[82] Indeed, many valuable information are generally missing in each publication, as seen in Tables 1 and 2, which makes comparison impossible. We recommend the aforementioned review,[82] which lists the key production parameters and other operational parameters that should systemically be reported, to allow viable cross-comparison of data. We would like to stress here that operational parameters such as pH and cathode potentials have now been showed to affect MES performance significantly, as discussed previously, and should systematically be reported. Essential performance indicators are production rate, product titer and yield, coulombic efficiency, and electron recovery as well as energy efficiency. It was discussed that production rates should be systematically reported in three different ways: normalized to total surface area of electrode (g m^{-2} day^{-1}), per unit working chamber or total working or empty bed volume (g m^{-3} day^{-1}), and per projected surface area of electrode or per unit membrane exchange area (g m^{-2} day^{-1}).[82] In light of the latest research findings, it is believed that a fourth way of reporting performance should be consistently reported, especially when using

3D or fibrous/porous electrodes, which is the production rate normalized to electrode volume (g $m^{-3}_{electrode}$ day^{-1}). Many studies have been and are being carried out in non-optimized reactor designs (*e.g.* H-type cell with non-optimized electrode volume/catholyte volume ratio), which renders performance normalized per catholyte volume much lower than what they would be in optimized systems, though they should still be reported. Normalizing and reporting rates per electrode volume is important when it comes to bringing the production process to large-scale application. Though it can be argued that a certain liquid volume is necessary for substantial production, the electrode-to-liquid volumes ratio has not yet been investigated to date, when using 3D or porous electrodes. Production rate normalized to electrode volume represents the maximal rate achievable of a theoretical cathode chamber that would be filled with the 3D/porous electrode of the same volume and is of significant importance. To date, the highest MES performance have been obtained on 3D electrode consisting of a uniformly grown or deposited CNT film over reticulated vitreous carbon, as previously discussed (Section 12.2.4).[5,9,10,14] It should also be stressed here that true 3D electrode as NanoWeb-RVC[5] and EPD-3D,[9] with relatively large pores (in the order of few hundred micrometers to millimeter scale) makes the intraporous liquid volume significant, as compared to when using fibrous materials such as carbon felt (pore diameter of a few micrometers).[14] A comparative study of different porous materials in an optimized and forced flow-through systems remains to be undertaken. Additionally, reporting performance per electrode volume across all studies (Tables 1 and 2) would allow drawing conclusions about potential mass transfer limitation using different electrode materials.

Reporting those four ways of normalizing production rates together would allow to fully characterize the material performance for a given reaction. Consequently, we strongly recommend researchers in the field to report the dimension in the three directions (*i.e.* length, width and thickness) of their electrodes, regardless of being a two (flat or rough) or 3D material and the production (and possibly substrate consumption) rates normalized to electrode volume (especially when using 3D/porous electrodes). Pore diameters of 3D/porous electrodes and current density per electrode volume should also be reported.

To date, most groups have reported current density and production rates normalized to projected surface area of electrode rather than the total surface area. This way of normalization is relevant and relates to system geometry and could determine the size of a bioelectrochemical reactor for engineering applications. However, to assess porous and 3D electrode material performance (*e.g.* carbon felt or cloth as used by many), projected performance do not mean anything unless the thickness (and porosity characteristics) of the electrode is also reported. Of two electrodes of same projected surface area and same porosity (same material), higher performance (current density and production rates) normalized to projected surface area will be obtained on the thicker electrode than on the thinner one (assuming no mass transfer limitation). However, performance normalized to total surface area (or by electrode volume) would be identical, hence no conclusion on the material performance should be drawn from normalization by projected surface area, unless thickness and porosity are reported. Unfortunately, these crucial parameters are almost never reported, neither are the performance normalized to total surface area, as seen in Tables 1 and 2, and Figure 5(c). An attempt to plot production rates normalized to projected surface area (psa), total surface area (tsa), catholyte volume and electrode volume and electron recoveries according to cathode potential, regrouping most of the MES studies to date, was made and can be seen in Figure 5.

As generally discussed above, it can be seen that a wide range of electron recovery into the final product has been reported, with as high recoveries obtained at lower cathode potentials than in more positive ones (Figure 5(e)). It has been argued that the main electron transfer mechanism occurring at lower potentials could be via hydrogen, which obviously does not prevent high electron recoveries into the target product to be reached. Additionally, as high recoveries have been recorded when using mixed microbial consortia than when using pure cultures (Table 1). Moreover, it should be noted that higher production rates when normalized to total surface area were obtained when using 3D hierarchical purposely built electrodes

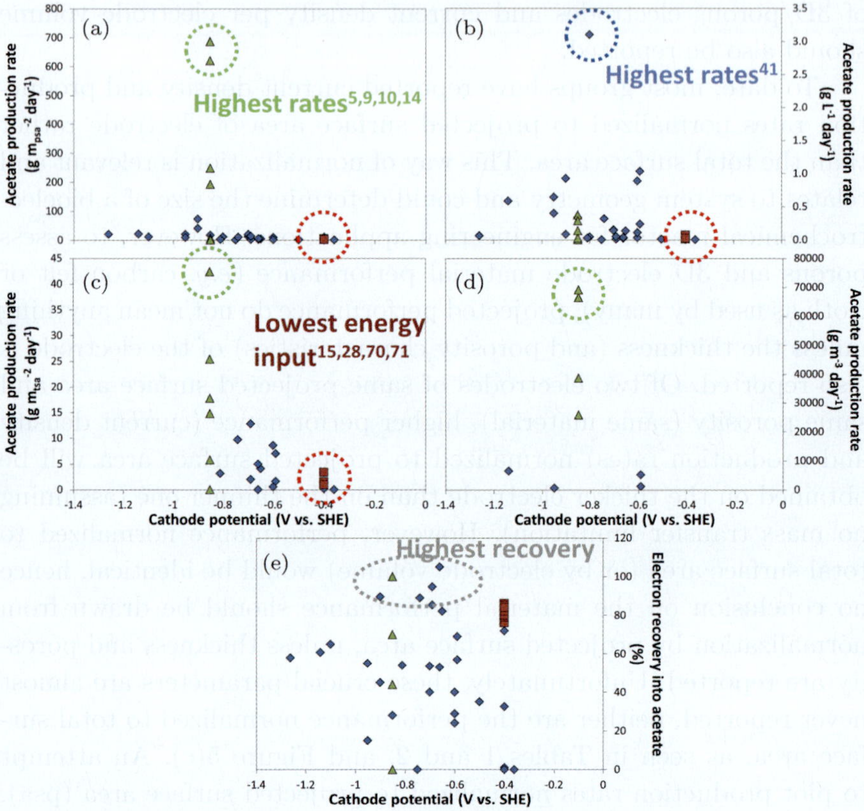

Figure 5. Compilation of all the reported and calculable acetate production rates from MES from CO_2 literature, to date, normalized to (a) projected surface area (psa), (b) catholyte volume, (c) total surface area (tsa), (d) electrode volume, and (e) electron recovery, according to the cathode potential. Studies that used commercially available electrodes (blue diamond), surface-modified electrodes (red square[70,71]), and purposely built electrodes (green triangle[5,9,10,14]).

than when flat, rough, or dense fibrous commercial electrodes were used, at similar cathode potentials (Figure 5(c)).[5] The general trend is an increase in performance with a decrease in cathode potential (*i.e.* increase in cathode overpotentials), when normalized to total surface area (Figure 5(c)). Figure 5(a) highlights what has been discussed above, with normalization to projected surface area not the optimal performance indicator on its own. Figure 5(d) stresses that the mention of electrode thickness is largely lacking, with only a few studies reporting them, when using either graphite granules,[16,36] carbon felt,[20,83] or purposely built electrodes (NanoWeb-RVC and EPD).[5,9,10,14] Due to the limited number of studies, a general trend is hardly extractable from this Figure 5(d). Nevertheless, it tends to confirm that a good balance between total surface area easily accessible for biofilm formation and a more effective mass convection between the bulk liquid and the electrode surface, such as achieved when using NanoWeb-RVC and EPD-3D compared to when using dense carbon felt and packed graphite granules, is favorable for achieving high production rates. However, this requires to be confirmed in identical operational conditions and forced-flow through systems.

Finally, when relevant, *i.e.* when optimized reactor designs are used, we urge the authors to report the energy input of their systems as well, *i.e.* the cell voltage between the anode and the cathode, which is particularly useful when referring to practical implementation of the technology.

12.3.2 The impact of the electrode (surface) properties must be further investigated to allow a mechanistic understanding

Initially, biofilms in MES reactors were pregrown on a H_2 gas mixture.[15,28] Others used fructose, methanol, or other measures to start-up via a heterotrophic growth phase.[84] Several other studies reported low electron recoveries and a not fully covered biofilm on the electrode.[85] Researchers were puzzled on the reasons for the difficult start-up as was previously observed for bioanodes. Currently, biocathodes are started-up with (enriched) inoculum obtained from diverse natural resources. The results of studies do more often indicate that the actual electrode choice was crucial. Even though cross-comparison between studies is hardly possible, some conclusions on the effect of the electrode material on biocathode processes can still be drawn, from studies that investigated several materials in elsewhere same operational conditions. In their study of the

NanoWeb-RVC electrode, obtained by direct growth of MWCNT on reticulated vitreous carbon, Jourdin *et al.*[5] also investigated unmodified RVC.[5] It was found that microorganisms were not able to form a biofilm on the flat and very smooth surface of the unmodified RVC. Concomitantly, neither acetate nor any other organics were produced throughout the 140 days of operation. MES did not occur on unmodified RVC. On the other hand, NanoWeb-RVC, and in turn EPD-3D, showed high performance toward CO_2 conversion to acetate, with high production rates, selectivity, and product titer, as previously discussed (Figure 5 and Table 1). The MWCNT nanostructure was shown to enhance the microbe–electrode interaction, bacterial attachment, biofilm development and microbial extracellular electron transfer rate.[5,9] The same observation was made by Nie *et al.*[70] who showed that coating a nickel nanowire network on graphite stick enhanced the interfacial interactions between the cathode surface and the microbial biofilm as compared to unmodified graphite stick.[70] This was corroborated by studies using SS electrodes on which no or sporadic microorganisms could attach to Ref. 25. Microorganism attachment was observed on rough or fibrous carbon-based electrodes such as graphite rod, granules, cloth, and felt but never forming a well-developed, continuous biofilm with entangled, multi-layered microorganisms as observed on NanoWeb-RVC and EPD-3D. These suggest that an organized or random nanostructure/nanopattern is favorable for biofilm formation on biocathode materials and for more efficient electron transfer, possibly due to the higher anchoring point availability.[5] A well-developed biofilm was shown to lead to the highest MES performance (Figures 4 and 5(c)) reached to date, with acetate production showed to occur exclusively within the biofilm.[10] Of all the surface modifications investigated by Zhang *et al.*[71] (Figure 3(a)), biofilm coverage/cell density on the electrode surface was unfortunately not completely reported, with only information given on the chitosan-modified surface. A ninefold increase in cell density was measured on the chitosan-modified carbon cloth compared to the unmodified material, which was in line with the 7.6-fold higher acetate production rate recorded. Biofilm coverage should be investigated when possible. Moreover, to date, higher MES performance were obtained by using mixed microbial consortia, than when pure cultures were used, while maintaining high product specificity (Table 1 and Figure 5). Therefore, some of the material strategy studied with pure cultures or less efficient mixed consortia could be worth testing with efficient mixed microbial cultures and optimal operating conditions. Valuable information can still be extracted from those studies, as elaborated above. A combination of electrode modification strategies investigated so far could enhance even further biofilm development and MES performance (*e.g.* positively charged nanostructure on 3D scaffold), but remains to be investigated.

Additionally, it was shown by Ganigue *et al.*[13] that the source of inoculum dramatically affected the product spectrum obtained by MES. As shown in bioanode studies, it is plausible that biocathode material characteristics could also have an effect on microbial enrichment (along with other operational parameters such as applied potential and pH for example). However, it can be observed that mostly different microbial communities were enriched in each study starting from mixed microbial consortium (Table 1). This adds to the complexity of cross-comparison between biocathode studies. It was shown in bioanode studies that different microorganisms are likely to use different EET mechanisms, which requires different electrode material considerations. There is a need to elucidate EET mechanisms in biocathode processes, which remains largely unknown to date.

It has to be stressed here that we investigated in this chapter the effect of electrode material on "biofilm-based" MES processes. Recently, researchers have started investigating bioelectrochemical chemicals production processes solely with microorganisms in suspension (no biofilm) using either soluble mediators or hydrogen as electron carriers between the electrode and the suspended microorganisms. The electrode material requirements are obviously different for those processes and are not covered in this chapter.

12.3.3 Anode electrode materials and modifications can be used for inspiration

Several general considerations can be extracted from the many years of research on anodic processes and biocathode MES processes findings discussed above, as listed in Table 3 and illustrated in Figure 2. These parameters should be assessed when selecting an electrode material for a specific cathodic microbial process. Two reviews on anode electrodes and their modifications provide further information on these considerations.[22,23]

As seen above, the typical electrode materials used in biocathodes are carbon or SS-based (Table 1). The carbon electrodes are shaped as rods, granules, and cloth/felts. The variety of manufacturers and unavailability of the exact compositions of these materials do not allow a direct one-to-one comparison between studies. Therefore, researchers compare electrode modification to their own controls (*i.e.* non-modified electrodes). Carbon electrodes have been more widely investigated as bioanodes of bioelectrochemical systems. Similarly, SS was also more often used as a bioanode, though corrosion risk was present.[86] Therefore, the same kinds of electrode materials were used for both bioanode and biocathode studies, showing biocompatibility for both processes. For example, Sleutels *et al.*[87] used exactly the same graphite felt material to study the oxidation of acetate than Janssen *et al.*[88] for the production of caproate and caprylate from acetate. Current densities between both systems varied from 2.3 A m^{-2} as bioanode to -1.8 A m^{-2} as biocathode. Similarly, Rozendal *et al.*[89] reversed an acetate oxidizing graphite felt based bioanode into a hydrogen producing biocathode, which shows the potential to use the same biocatalysts for both electrode reactions. Performance of bioanodes and biocathodes can be compared based on current densities or production *vs.* removal rates. However, using such performance indicators, one cannot simply state that biocathodes perform worse than bioanodes or *vice versa.* First of all, for microorganisms, the bioanode acts as a terminal electron acceptor, while for biocathodes the electrode acts as an electron donor. This suggests that different electron transfer mechanisms are present, which can already be the basis on explaining any differences. For *G. sulfurreducens*, a well-studied electrochemically active biofilm-forming microorganism, it was shown that deletion of genes for outer-surface proteins previously shown to be essential for optimal electron transfer to electrodes had no impact on electron transfer from electrodes.

Deletion of a gene encoding a putative monoheme *c*-type cytochrome completely inhibited electron transfer from electrodes, but had no impact on electron transfer to electrodes. These results suggested that the mechanisms for electron transfer from electrodes do differ from the mechanisms for electron transfer to electrodes.[90] However, there are also similarities within the electron transfer routes while hydrogenases are both important for oxidation and reduction reactions when hydrogen is involved.[91] Jourdin *et al.*[10] showed that biologically induced hydrogen production drives MES of acetate. Copper was detected at the electrode after biofilm removal, which could come from biological copper synthesis or be the remainings of a copper-containing hydrogenase.

12.3.4 In-depth (bio)electrochemical analysis is needed to quantify the effect of the electrode

With the plurality of known and potential electron transfer mechanisms (and differences between microorganisms) and the effect of electrode materials and designs on the performance, a direct comparison is not always straightforward.[92] A further mechanistic understanding on electron transfer mechanisms and more in-depth (bio)electrochemical analysis (*e.g.* on internal resistances) on the effect of electrode materials, designs, and modifications of cathodes *vs.* anodes is needed to reveal further quantified explanations on performance enhancements.[23] Techniques readily used in studies on BES like microscopy techniques, Raman spectroscopy, electrochemical impedance spectrometry, proteomic techniques, etc. will allow a better understanding of the observed effects.

12.4 Outlook

Biocathodes remain the key element for the production of chemicals via biocatalysts and electricity supply. How are electrons transferred? What is limiting MES? How can higher performances be reached? How to implement the technology? These are questions which researchers and technologists raise during their work. MES is definitively out of its infancy, though there is still more to be

Table 3. Criteria to take into consideration when selecting and designing new electrode materials for cathodic microbial processes and what they impact on.

Electrode material criteria	Impact(s)
Biocompatibility	Improved microorganism adherence
Specific and electroactive surface area (accessible to microorganisms)	Higher biofilm coverage/increased performance
Adequate electrode surface to volume ratio	Maximum performance per reactor volume
Macrostructure	Enhanced mass transfer of substrates and products to and from the surface of the electrode/biofilm, clogging/biofouling
Surface chemistry (surface charge, H-bond, van der Walls bond, polarity, local electrical field, medium composition, molecule adsorption, pH (protonation), ionic strength (double layer thickness), electrode polarization, etc.)	Improved microorganism adherence and electron exchange with bacteria
Micro- and nanotopography (structure, roughness, porosity, etc.)	Microorganism adherence, biofilm coverage, enhanced electron exchange with bacteria, clogging, diffusion, internal resistances
Stability	Chemical stability, *e.g.* corrosion resistance
Toxicity	No impact on microorganisms
Hydrophilicity/hydrophobicity	Interactions to microorganisms/microorganism adherence
Bacterial strains (morphologies, nature, cellular appendages, surface charge density, polarizability, etc.)	Microorganism adherence/enhanced electron exchange with electrode
Ionic conductance (electrolyte)	Microorganisms, internal resistances
Mechanical strength	Practical implementation
Electronic conductance	
Longevity	
Scalability	
Cost of fabrication	
Maintenance	
Environmental impact (manufacturing footprint, recyclability)	
Good processability	

clarified. The rapid developments achieved resulted into a production process reaching acetate production rate of ca. 66 kg $m^{-3}_{electrode}$ day^{-1}, which is similar to that achieved in established industrial fermentation processes, and electron recoveries of 100%.[9] The right electrode surface, as well as the macrostructure design (*e.g.* 3D), operating conditions, supplied catholyte, and the used inoculum all showed to be essential to achieve these results. To date, the highest rates have been achieved by Jourdin *et al.*[5,9,10,14] while the lowest energy requirement (*i.e.* lowest cathode potentials) was reported by Nevin *et al.*[15,28] Lowering energy input and enabling production of more valuable compounds will further support the implementation of the technology. Genetically modified organisms may be needed to establish the later; or an integration with fermentation technology such as chain elongation could be usefull.[93] More in-depth CAPEX/OPEX analysis will be required to reveal the implementation readiness of the technology. Further considerations on feedstock choice and supply and product recovery and purification must be made. As mentioned in Table 3, all kind of considerations must be made with regards to electrodes' scale-up. For a smooth implementation, readily available electrode materials can be simply integrated in existing electrolysis cells. Certainly the side-equipment can be optimized to design effective current collectors or power supplies. Demonstration of MES in real, industrial conditions will further show the potential impact for implementation.

Acknowledgments

Financial support from the Dutch Technology Foundation STW and the company Paques (project no. STW-Paques 12999) is gratefully acknowledged.

List of Abbreviations

BE	biphasic esterification
BES	bioelectrochemical system
CNT	carbon nanotube
CVD	chemical vapor deposition

DET	direct electron transfer
EET	extracellular electron transfer
EPD	electrophoretic deposition
ME	membrane electrolysis
MES	microbial electrosynthesis
MWCNT	multi-walled carbon nanotube
NaBES	sodium 2-bromoethanesulfonate
psa	projected surface area
RVC	reticulated vitreous carbon
SEM	scanning electron microscopy
SHE	standard hydrogen electrode
tsa	total surface area

References

1. S. Chu and A. Majumdar (2012). Opportunities and challenges for a sustainable energy future. *Nature* **488**(7411), 294–303.
2. K. Rabaey and R.A. Rozendal (2010). Microbial electrosynthesis — revisiting the electrical route for microbial production. *Nature Reviews Microbiology* **8**(10), 706–716.
3. D.R. Lovley and K.P. Nevin (2013). Electrobiocommodities: Powering microbial production of fuels and commodity chemicals from carbon dioxide with electricity. *Current Opinion in Biotechnology* **24**(3), 385–390.
4. K. Rabaey, P. Girguis and L.K. Nielsen (2011). Metabolic and practical considerations on microbial electrosynthesis. *Current Opinion in Biotechnology* **22**(3), 371–377.
5. L. Jourdin, S. Freguia, B.C. Donose, J. Chen, G.G. Wallace, J. Keller and V. Flexer (2014). A novel carbon nanotube modified scaffold as an efficient biocathode material for improved microbial electrosynthesis. *Journal of Materials Chemistry A* **2**(32), 13093–13102.
6. L. Jourdin (2015). *Microbial Electrosynthesis from Carbon Dioxide: Performance Enhancement and Elucidation of Mechanisms.* The University of Queensland.
7. H.V.M. Hamelers, A. Heijne, T.H.J.A. Sleutels, A.W. Jeremiasse, D.P. Strik and C.J.N. Buisman (2010). New applications and performance of bioelectrochemical systems. *Applied Microbiology and Biotechnology* **85**(6), 1673–1685.
8. N.S. Lewis and D.G. Nocera (2006). Powering the planet: Chemical challenges in solar energy utilization. *Proceedings of the National Academy of Sciences of the United States of America* **103**(43), 15729–15735.
9. L. Jourdin, T. Grieger, J. Monetti, V. Flexer, S. Freguia, Y. Lu, J. Chen, M. Romano, G.G. Wallace and J. Keller (2015). High acetic acid production rate obtained by microbial electrosynthesis from carbon dioxide. *Environmental Science & Technology* **49**(22), 13566–13574.

10. L. Jourdin, Y. Lu, V. Flexer, J. Keller and S. Freguia (2016). Biologically-induced hydrogen production drives high rate/high efficiency microbial electrosynthesis of acetate from carbon dioxide. *ChemElectroChem* **3**(4), 581–591.
11. M. Mikkelsen, M. Jørgensen and F.C. Krebs (2010). The teraton challenge. A review of fixation and transformation of carbon dioxide. *Energy and Environmental Science* **3**(1), 43–81.
12. G. Centi and S. Perathoner (2009). Opportunities and prospects in the chemical recycling of carbon dioxide to fuels. *Catalysis Today* **148**(3–4), 191–205.
13. R. Ganigue, S. Puig, P. Batlle-Vilanova, M.D. Balaguer and J. Colprim (2015). Microbial electrosynthesis of butyrate from carbon dioxide. *Chemical Communications* **51**(15), 3235–3238.
14. L. Jourdin, S. Freguia, V. Flexer and J. Keller (2015). Bringing high-rate, CO2-based microbial electrosynthesis closer to practical implementation through improved design and operating conditions. *Environmental Science & Technology* **50**(4), 1982–1989.
15. K.P. Nevin, T.L. Woodard, A.E. Franks, Z.M. Summers and D.R. Lovley (2010). Microbial electrosynthesis: Feeding microbes electricity to convert carbon dioxide and water to multicarbon extracellular organic compounds. *mBio* **1**(2), 1–4.
16. C.W. Marshall, D.E. Ross, E.B. Fichot, R.S. Norman and H.D. May (2013). Long-term operation of microbial electrosynthesis systems improves acetate production by autotrophic microbiomes. *Environmental Science and Technology* **47**(11), 6023–6029.
17. C. Liu, J.J. Gallagher, K.K. Sakimoto, E.M. Nichols, C.J. Chang, M.C.Y. Chang and P. Yang (2015). Nanowire–bacteria hybrids for unassisted solar carbon dioxide fixation to value-added chemicals. *Nano Letters* **15**, 3634–3639.
18. M.T. Agler, B.A. Wrenn, S.H. Zinder and L.T. Angenent (2011). Waste to bioproduct conversion with undefined mixed cultures: The carboxylate platform. *Trends in Biotechnology* **29**(2), 70–78.
19. M. Demler and D. Weuster-Botz (2011). Reaction engineering analysis of hydrogenotrophic production of acetic acid by Acetobacterium woodii. *Biotechnology and Bioengineering* **108**(2), 470–474.
20. S. Gildemyn, K. Verbeeck, R. Slabbinck, S.J. Andersen, A. Prévoteau and K. Rabaey (2015). Integrated production, extraction, and concentration of acetic acid from CO_2 through microbial electrosynthesis. *Environmental Science & Technology Letters* **2**(11), 325–328.
21. P. Hu, S. Chakraborty, A. Kumar, B. Woolston, H. Liu, D. Emerson and G. Stephanopoulos (2016). Integrated bioprocess for conversion of gaseous substrates to liquids. *Proceedings of the National Academy of Sciences* **113**(14), 3773–3778.
22. X. Xie, C. Criddle and Y. Cui (2015). Design and fabrication of bioelectrodes for microbial bioelectrochemical systems. *Energy & Environmental Science* **8**(12), 3418–3441.

23. K. Guo, A. Prévoteau, S.A. Patil and K. Rabaey (2015). Engineering electrodes for microbial electrocatalysis. *Current Opinion in Biotechnology* **33**, 149–156.
24. P.-L. Tremblay and T. Zhang (2015). Electrifying microbes for the production of chemicals. *Frontiers in microbiology* **6**, 13–24.
25. E. Blanchet, F. Duquenne, Y. Rafrafi, L. Etcheverry, B. Erable and A. Bergel (2015). Importance of the hydrogen route in up-scaling electrosynthesis for microbial CO_2 reduction. *Energy & Environmental Science* **8**(12), 3731–3744.
26. J.S. Deutzmann, M. Sahin and A.M. Spormann (2015). Extracellular enzymes facilitate electron uptake in biocorrosion and bioelectrosynthesis. *mBio* **6**(2).
27. T. Krieg, A. Sydow, U. Schröder, J. Schrader and D. Holtmann (2014). Reactor concepts for bioelectrochemical syntheses and energy conversion. *Trends in Biotechnology* **32**(12), 645–655.
28. K.P. Nevin, S.A. Hensley, A.E. Franks, Z.M. Summers, J. Ou, T.L. Woodard, O.L. Snoeyenbos-West and D.R. Lovley (2011). Electrosynthesis of organic compounds from carbon dioxide is catalyzed by a diversity of acetogenic microorganisms. *Applied and Environmental Microbiology* **77**(9), 2882–2886.
29. K.B. Gregory, D.R. Bond and D.R. Lovley (2004). Graphite electrodes as electron donors for anaerobic respiration. *Environmental Microbiology* **6**(6), 596–604.
30. K.B. Gregory and D.R. Lovley (2005). Remediation and recovery of uranium from contaminated subsurface environments with electrodes. *Environmental Science & Technology* **39**(22), 8943–8947.
31. S.M. Strycharz, T.L. Woodard, J.P. Johnson, K.P. Nevin, R.A. Sanford, F.E. Löffler and D.R. Lovley (2008). Graphite electrode as a sole electron donor for reductive dechlorination of tetrachlorethene by geobacter lovleyi. *Applied and Environmental Microbiology* **74**(19), 5943–5947.
32. C. Dumas, R. Basseguy and A. Bergel (2008). Microbial electrocatalysis with Geobacter sulfurreducens biofilm on stainless steel cathodes. *Electrochimica Acta* **53**(5), 2494–2500.
33. S.M. Strycharz, S.M. Gannon, A.R. Boles, A.E. Franks, K.P. Nevin and D.R. Lovley (2010). Reductive dechlorination of 2-chlorophenol by anaeromyxobacter dehalogenans with an electrode serving as the electron donor. *Environmental Microbiology Reports* **2**(2), 289–294.
34. J.C. Thrash and J.D. Coates (2008). Review: Direct and indirect electrical stimulation of microbial metabolism. *Environmental Science and Technology* **42**(11), 3921–3931.
35. T. Reda, C.M. Plugge, N.J. Abram and J. Hirst (2008). Reversible interconversion of carbon dioxide and formate by an electroactive enzyme. *Proceedings of the National Academy of Sciences of the United States of America* **105**(31), 10654–10658.
36. C.W. Marshall, D.E. Ross, E.B. Fichot, R.S. Norman and H.D. May (2012). Electrosynthesis of commodity chemicals by an autotrophic microbial community. *Applied and Environmental Microbiology* **78**(23), 8412–8420.
37. Y. Jiang, M. Su, Y. Zhang, G. Zhan, Y. Tao and D. Li (2013). Bioelectrochemical systems for simultaneously production of methane and acetate

from carbon dioxide at relatively high rate. *International Journal of Hydrogen Energy* **38**(8), 3497–3502.
38. S. Min, Y. Jiang and D. Li (2013). Production of acetate from carbon dioxide in bioelectrochemical systems based on autotrophic mixed culture. *Journal of Microbiology and Biotechnology* **23**(8), 1140–1146.
39. N. Xafenias and V. Mapelli (2014). Performance and bacterial enrichment of bioelectrochemical systems during methane and acetate production. *International Journal of Hydrogen Energy* **39**, 21864–21875.
40. Y. Gong, A. Ebrahim, A.M. Feist, M. Embree, T. Zhang, D. Lovley and K. Zengler (2013). Sulfide-driven microbial electrosynthesis. *Environmental Science & Technology* **47**(1), 568–573.
41. E.V. LaBelle, C.W. Marshall, J.A. Gilbert and H.D. May (2014). Influence of acidic pH on hydrogen and acetate production by an electrosynthetic microbiome. *PLoS ONE* **9**(10), e109935.
42. L. Huang, L. Jiang, Q. Wang, X. Quan, J. Yang and L. Chen (2014). Cobalt recovery with simultaneous methane and acetate production in biocathode microbial electrolysis cells. *Chemical Engineering Journal* **253**, 281–290.
43. P. Batlle-Vilanova, S. Puig, R. Gonzalez-Olmos, M.D. Balaguer and J. Colprim (2015). Continuous acetate production through microbial electrosynthesis from CO_2 with microbial mixed culture. *Journal of Chemical Technology & Biotechnology* **91**(4), 921–927.
44. C.G.S. Giddings, K. Nevin, T. Woodward, D.R. Lovley and C.S. Butler (2015). Simplifying microbial electrosynthesis reactor design. *Frontiers in Microbiology* **6**, 1–6.
45. S.A. Patil, J.B.A. Arends, I. Vanwonterghem, J. van Meerbergen, K. Guo, G.W. Tyson and K. Rabaey (2015). Selective enrichment establishes a stable performing community for microbial electrosynthesis of acetate from CO_2. *Environmental Science & Technology* **49**(14), 8833–8843.
46. S. Bajracharya, A. ter Heijne, X.D. Benetton, K. Vanbroekhoven, C.J. Buisman, D.P. Strik and D. Pant (2015). Carbon dioxide reduction by mixed and pure cultures in microbial electrosynthesis using an assembly of graphite felt and stainless steel as a cathode. *Bioresource Technology* **195**, 14–24.
47. G. Mohanakrishna, J.S. Seelam, K. Vanbroekhoven and D. Pant (2015). An enriched electroactive homoacetogenic biocathode for the microbial electrosynthesis of acetate through carbon dioxide reduction. *Faraday Discussions* **183**, 445–462.
48. Y. Kuramochi, Q. Fu, H. Kobayashi, M. Ikarashi, T. Wakayama, H. Kawaguchi, J. Vilcaez, H. Maeda and K. Sato (2013). Electromethanogenic CO_2 conversion by subsurface-reservoir microorganisms. *Energy Procedia* **37**, 7014–7020.
49. M.C.A.A. Van Eerten-Jansen, A.T. Heijne, C.J.N. Buisman and H.V.M. Hamelers (2011). Microbial electrolysis cells for production of methane from CO_2: Long-term performance and perspectives. *International Journal of Energy Research* **36**(6), 809–819.
50. M. Villano, F. Aulenta, C. Ciucci, T. Ferri, A. Giuliano and M. Majone (2010). Bioelectrochemical reduction of CO_2 to CH_4 via direct and indirect

extracellular electron transfer by a hydrogenophilic methanogenic culture. *Bioresource Technology* **101**(9), 3085–3090.
51. P. Clauwaert and W. Verstraete (2009). Methanogenesis in membraneless microbial electrolysis cells. *Applied Microbiology and Biotechnology* **82**(5), 829–836.
52. S. Cheng, D. Xing, D.F. Call and B.E. Logan (2009). Direct biological conversion of electrical current into methane by electromethanogenesis. *Environmental Science and Technology* **43**(10), 3953–3958.
53. D.H. Park, M. Laivenieks, M.V. Guettler, M.K. Jain and J.G. Zeikus (1999). Microbial utilization of electrically reduced neutral red as the sole electron donor for growth and metabolite production. *Applied and Environmental Microbiology* **65**(7), 2912–2917.
54. R.F. Hickey, J. Vanderwielen and M.S. Switzenbaum (1987). The effects of organic toxicants on methane production and hydrogen gas levels during the anaerobic digestion of waste activated sludge. *Water Research* **21**(11), 1417–1427.
55. L. Nollet, D. Demeyer and W. Verstraete (1997). Effect of 2-bromoethanesulfonic acid and Peptostreptococcus productus ATCC 35244 addition on stimulation of reductive acetogenesis in the ruminal ecosystem by selective inhibition of methanogenesis. *Applied and Environmental Microbiology* **63**(1), 194–200.
56. R.S. Oremland and S. Polcin (1982). Methanogenesis and sulfate reduction: Competitive and noncompetitive substrates in estuarine sediments. *Applied and Environmental Microbiology* **44**(6), 1270–1276.
57. J.C.M. Scholten, R. Conrad and A.J.M. Stams (2000). Effect of 2-bromoethane sulfonate, molybdate and chloroform on acetate consumption by methanogenic and sulfate-reducing populations in freshwater sediment. *FEMS Microbiology Ecology* **32**(1), 35–42.
58. R. Sparling and L. Daniels (1987). The specificity of growth inhibition of methanogenic bacteria by bromoethanesulfonate. *Canadian Journal of Microbiology* **33**(12), 1132–1136.
59. A. Chidthaisong and R. Conrad (2000). Specificity of chloroform, 2-bromoethanesulfonate and fluoroacetate to inhibit methanogenesis and other anaerobic processes in anoxic rice field soil. *Soil Biology and Biochemistry* **32**(7), 977–988.
60. M.J. Alperin and W.S. Reeburgh (1985). Inhibition experiments on anaerobic methane oxidation. *Applied and Environmental Microbiology* **50**(4), 940–945.
61. S.H. Zinder, T. Anguish and S.C. Cardwell (1984). Selective inhibition by 2-bromoethanesulfonate of methanogenesis from acetate in a thermophilic anaerobic digestor. *Applied and Environmental Microbiology* **47**(6), 1343–1345.
62. X. Zhu, M. Siegert, M.D. Yates and B.E. Logan (2015). Alamethicin suppresses methanogenesis and promotes acetogenesis in bioelectrochemical systems. *Applied and Environmental Microbiology* **81**(11), 3863–3868.

63. I. Bechthold, K. Bretz, S. Kabasci, R. Kopitzky and A. Springer (2008). Succinic acid: A new platform chemical for biobased polymers from renewable resources. *Chemical Engineering & Technology* **31**(5), 647–654.
64. S.J. Andersen, T. Hennebel, S. Gildemyn, M. Coma, J. Desloover, J. Berton, J. Tsukamoto, C. Stevens and K. Rabaey (2014). Electrolytic membrane extraction enables production of fine chemicals from biorefinery sidestreams. *Environmental Science & Technology* **48**(12), 7135–7142.
65. F. Ozadali, B. Glatz and C. Glatz (1996). Fed-batch fermentation with and without on-line extraction for propionic and acetic acid production by Propionibacterium acidipropionici. *Applied Microbiology and Biotechnology* **44**(6), 710–716.
66. L. Soussan, J. Riess, B. Erable, M.-L. Delia and A. Bergel (2013). Electrochemical reduction of CO_2 catalysed by Geobacter sulfurreducens grown on polarized stainless steel cathodes. *Electrochemistry Communications* **28**, 27–30.
67. M. Sharma, J.L. Varanasi, P. Jain, P. Dureja, B. Lal, X. Dominguez-Benetton, D. Pant and P.M. Sarma (2014). Influence of headspace composition on product diversity by sulphate reducing bacteria biocathode. *Bioresource Technology* **165**, 365–371.
68. M. Rothballer, M. Picot, T. Sieper, J.B.A. Arends, M. Schmid, A. Hartmann, N. Boon, C.J.N. Buisman, F. Barrière and D.P.B.T.B. Strik (2015). Monophyletic group of unclassified γ-Proteobacteria dominates in mixed culture biofilm of high-performing oxygen reducing biocathode. *Bioelectrochemistry* **106**, 167–176.
69. M. Picot, L. Lapinsonnière, M. Rothballer and F. Barrière (2011). Graphite anode surface modification with controlled reduction of specific aryl diazonium salts for improved microbial fuel cells power output. *Biosensors and Bioelectronics* **28**(1), 181–188.
70. H. Nie, T. Zhang, M. Cui, H. Lu, D.R. Lovley and T.P. Russell (2013). Improved cathode for high efficient microbial-catalyzed reduction in microbial electrosynthesis cells. *Physical Chemistry Chemical Physics* **15**(34), 14290–14294.
71. T. Zhang, H. Nie, T.S. Bain, H. Lu, M. Cui, O.L. Snoeyenbos-West, A.E. Franks, K.P. Nevin, T.P. Russell and D.R. Lovley (2013). Improved cathode materials for microbial electrosynthesis. *Energy & Environmental Science* **6**(1), 217–224.
72. A.E. Franks, K.P. Nevin, H. Jia, M. Izallalen, T.L. Woodard and D.R. Lovley (2009). Novel strategy for three-dimensional real-time imaging of microbial fuel cell communities: Monitoring the inhibitory effects of proton accumulation within the anode biofilm. *Energy & Environmental Science* **2**(1), 113–119.
73. U. Schroder (2007). Anodic electron transfer mechanisms in microbial fuel cells and their energy efficiency. *Physical Chemistry Chemical Physics* **9**(21), 2619–2629.
74. S. Chen, G. He, Q. Liu, F. Harnisch, Y. Zhou, Y. Chen, M. Hanif, S. Wang, X. Peng, H. Hou and U. Schroder (2012). Layered corrugated electrode

macrostructures boost microbial bioelectrocatalysis. *Energy & Environmental Science* **5**(12), 9769–9772.
75. S. Chen, H. Hou, F. Harnisch, S.A. Patil, A.A. Carmona-Martinez, S. Agarwal, Y. Zhang, S. Sinha-Ray, A.L. Yarin, A. Greiner and U. Schroder (2011). Electrospun and solution blown three-dimensional carbon fiber nonwovens for application as electrodes in microbial fuel cells. *Energy & Environmental Science* **4**(4), 1417–1421.
76. X. Xie, M. Ye, L. Hu, N. Liu, J.R. McDonough, W. Chen, H.N. Alshareef, C.S. Criddle and Y. Cui (2012). Carbon nanotube-coated macroporous sponge for microbial fuel cell electrodes. *Energy & Environmental Science* **5**(1), 5265–5270.
77. S. Chen, Q. Liu, G. He, Y. Zhou, M. Hanif, X. Peng, S. Wang and H. Hou (2012). Reticulated carbon foam derived from a sponge-like natural product as a high-performance anode in microbial fuel cells. *Journal of Materials Chemistry* **22**(35), 18609–18613.
78. Z. He, J. Liu, Y. Qiao, C.M. Li and T.T.Y. Tan (2012). Architecture engineering of hierarchically porous chitosan/vacuum-stripped graphene scaffold as bioanode for high performance microbial fuel cell. *Nano Letters* **12**(9), 4738–4741.
79. K. Katuri, M.L. Ferrer, M.C. Gutierrez, R. Jimenez, F. del Monte and D. Leech (2011). Three-dimensional microchanelled electrodes in flow-through configuration for bioanode formation and current generation. *Energy & Environmental Science* **4**(10), 4201–4210.
80. X. Xie, L. Hu, M. Pasta, G.F. Wells, D. Kong, C.S. Criddle and Y. Cui (2010). Three-dimensional carbon nanotube-textile anode for high-performance microbial fuel cells. *Nano Letters* **11**(1), 291–296.
81. A. Walcarius, S.D. Minteer, J. Wang, Y. Lin and A. Merkoci (2013). Nanomaterials for bio-functionalized electrodes: Recent trends. *Journal of Materials Chemistry B* **1**(38), 4878–4908.
82. S.A. Patil, S. Gildemyn, D. Pant, K. Zengler, B.E. Logan and K. Rabaey (2015). A logical data representation framework for electricity-driven bioproduction processes. *Biotechnology Advances* **33**(6), 736–744.
83. S.D. Molenaar, A.R. Mol, T.H.J.A. Sleutels, A. ter Heijne and C.J.N. Buisman (2016). Microbial rechargeable battery: Energy storage and recovery through acetate. *Environmental Science & Technology Letters* **3**(4), 144–149.
84. P.-L. Tremblay, D. Höglund, A. Koza, I. Bonde and T. Zhang (2015). Adaptation of the autotrophic acetogen Sporomusa ovata to methanol accelerates the conversion of CO_2 to organic products. *Science Reports* **5**, 16168.
85. M.C.A.A. Van Eerten-Jansen, A.B. Veldhoen, C.M. Plugge, A.J.M. Stams, C.J.N. Buisman and A. Ter Heijne (2013). Microbial community analysis of a methane-producing biocathode in a bioelectrochemical system. *Archaea.*
86. P. Ledezma, B.C. Donose, S. Freguia and J. Keller (2015). Oxidised stainless steel: A very effective electrode material for microbial fuel cell bioanodes but at high risk of corrosion. *Electrochimica Acta* **158**, 356–360.

87. T.H.J.A. Sleutels, H.V.M. Hamelers, R.A. Rozendal and C.J.N. Buisman (2009). Ion transport resistance in Microbial Electrolysis Cells with anion and cation exchange membranes. *International Journal of Hydrogen Energy* **34**(9), 3612–3620.
88. M.C. Van Eerten-Jansen, A. Ter Heijne, T.I. Grootscholten, K.J. Steinbusch, T.H. Sleutels, H.V. Hamelers and C.J. Buisman (2013). Bioelectrochemical production of caproate and caprylate from acetate by mixed cultures. *ACS Sustainable Chemistry & Engineering* **1**(5), 513–518.
89. R.A. Rozendal, A.W. Jeremiasse, H.V.M. Hamelers and C.J.N. Buisman (2008). Hydrogen production with a microbial biocathode. *Environmental Science and Technology* **42**(2), 629–634.
90. S.M. Strycharz, R.H. Glaven, M.V. Coppi, S.M. Gannon, L.A. Perpetua, A. Liu, K.P. Nevin and D.R. Lovley (2011). Gene expression and deletion analysis of mechanisms for electron transfer from electrodes to Geobacter sulfurreducens. *Bioelectrochemistry* **80**(2), 142–150.
91. F.A. Armstrong and J. Hirst (2011). Reversibility and efficiency in electrocatalytic energy conversion and lessons from enzymes. *Proceedings of the National Academy of Sciences* **108**(34), 14049–14054.
92. J. Philips, K. Verbeeck, K. Rabaey and J.B.A. Arends (2015). Electron transfer mechanisms in biofilms. In K. Schott and E.H. Yu, (eds.), *Microbial Electrochemical and Fuel Cells: Fundamentals and Applications*, Elsevier Woodhead Publishing, Cambridge, UK, pp. 67–113.
93. L.T. Angenent, H. Richter, W. Buckel, C.M. Spirito, K.J.J. Steinbusch, C.M. Plugge, D.P.B.T.B. Strik, T.I.M. Grootscholten, C.J.N. Buisman and H.V.M. Hamelers (2016). Chain elongation with reactor microbiomes: Open-culture biotechnology to produce biochemicals. *Environmental Science & Technology* **50**(6), 2796–2810.

Chapter 13

Non-Carbonaceous Electrodes for Microbial Electrochemical Systems

Hernán Romeo*, Diego Massazza†, Rodrigo Parra‡, and Juan Pablo Busalmen†,§

**División Polímeros Nanoestructurados, INTEMA (CONICET-UNMdP), B7608FDQ, Mar del Plata, Argentina*

†Laboratorio de Bioelectroquímica, División Ingeniería de Interfases y Bioprocesos, INTEMA (CONICET-UNMdP), B7608FDQ, Mar del Plata, Argenitna

‡División Cerámicos, INTEMA (CONICET-UNMdP), B7608FDQ, Mar del Plata, Argentina

§jbusalme@fi.mdp.edu.ar

13.1 Introduction

Over the past decade, there has been a substantial expectation on microbial fuel cells (MFCs) as a promising biotechnology to achieve sustainable wastewater treatment. While several particular drawbacks related to MFCs (such as the energy losses associated with the anode, poor control on mass transport of protons to the cathodic compartment, among others) have been taken into consideration to improve their performance, the cascade of global approaches, taking this biotechnology "as a whole", has not been stopped.[1] This way of coming near the subject has been beneficial to rapidly progress on the path of microbially assisted electrochemical systems; however, it has sometimes overlooked some particular and fundamental details when analyzing each part/component of a MFC. Agreeing on this point, we can recognize that the task of pushing forward the MFC technology needs addressing the problem from different and innovative areas. One of those areas is materials science. Electrodes, membranes, current collectors, wires, and resistors, used as irreplaceable pieces in the MFC puzzle, are all under the umbrella of this attractive and powerful discipline. Interestingly, one of the guiding principles that materials science obeys nowadays to address novel functional materials and complex devices is the one that states "learning from Nature".[2] When this guiding principle is thoroughly considered, unforeseen paths open up to create a new generation of synthetic and outstanding hierarchical materials. However, and paradoxically, nature seems to have been left aside during the scientific research of new materials focusing on the improvements needed on the area of MFCs. Among the range of critical components that a MFC needs to work out properly, the electrodes (especially anodes) have found a particular and privileged place on the researchers' interest, as they play the role of energy transducers when colonized by electrically active microbes. When we trace back the evolution of bioanodes used in bioelectrochemical systems (BESs), a spate of numerous and well-conducted studies appear with a distinctive feature: no matter the bioanode configuration (rod, plate, disk, granular, mesh, paper, cloth, felt), the discussion has almost exclusively pointed toward carbonaceous materials.[3–5] This seems to have pushed into the background what nature tells us about the deep interaction between electrogenic bacteria and redox-active minerals in subsurface sediments. This intimate relationship powers the bioenergetics of these microorganisms when solid-phase compounds, such as iron (III), manganese (IV), and uranium (VI) oxides, serve as terminal electron acceptors. It seems that, gazing through a different perspective, the biological reduction of insoluble metal oxides has been right there just under our noses for so long waiting for us to fully exploit the concept. However, the overwhelming difference in attention paid to carbonaceous bioanodes for improving MFC performance compared to metal oxide ones is striking, as well as the lack of basic studies addressing the ultimate interaction between bacteria and oxide surfaces. Nevertheless, and against all odds, non-carbonaceous materials have started coming out the shadows to be positioned as real alternatives to carbon-based electrodes in BESs, more specifically as critical pieces in the development of the MFC biotechnology. We strongly believe that under the "non-carbonaceous" concept, we can find "uncharted" or less explored materials, which can effectively meet the essential criteria into the fascinating microworld that involves channeling the microbial extracellular electron transfer process toward solid electron acceptors. Within this scenario, we must be completely fair, addressing the subject by considering the different classes of non-carbonaceous materials that have been studied so far as bioanodes in BESs, namely (i) metallic electrodes and (ii) metal oxide-based electrodes.

While addressing these topics, we will focus the discussion on the goal of the development of functional electrodes, pointing to the ideal features a bioelectrode should have. Following this idea across the chapter, micro and macroscale characteristics will be taken into consideration just to take the road uphill, from the simplest and more obvious bioelectrode features to the more complex scenario of 3D electroactive bioscaffolds. Along this winding road, a first division into porous and non-porous electrodes will help to get a deep insight into fundamental bioelectrode characteristics. The analysis of total porosity, pore size and shape, surface conductivity, roughness features, and biocompatibility will ensure identifying both weak and strong points facing new challenges in the near future.

Hence, this chapter aims at putting on the table the range of non-carbonaceous electrodes (metallic (Section 13.3) and oxide-based (Section 13.4)) that have been used as ultimate solid electron acceptors into the area of BESs. A particular stress will be made on highlighting the first attempts on using metal oxide-based bioanodes to mediate the growth of microbial communities, reserving a special section (Section 13.4.1.2.1) to introduce electrically conducting titanium oxide-based porous scaffolds, which have recently been employed as anodes demonstrating to outcompete typical carbonaceous electrodes.

Being well aware of the challenge we are taking up, we expect this chapter to attract the attention of those interested in opening up the game toward versatile and high-performance electrode materials other than carbon-based ones, in the fascinating and increasingly growing area of functional electrodes for microbial electrochemical systems.

13.2 Fundamental Considerations on Microbial Anodes

Before pointing directly on the different non-carbonaceous bioanodes developed to date, it will be useful to have in mind the requirements that have been shown to be of paramount importance for the optimum microbial anode performance, no matter its nature. Taking these fundamental considerations into account, the analysis and conclusions that can be drawn will help in visualizing a pattern of different electrode families. In this line of reasoning, an ideal bioanode should exhibit the following characteristics:

(i) High electrical conductivity
(ii) High surface area
(iii) Good biocompatibility
(iv) High surface roughness
(v) 3D structure
(vi) Open porosity
(vii) Structured porosity
(viii) High structural and chemical stability
(ix) Upscaling potentiality
(x) Low cost (material and processing)

Of course, it is almost impossible to meet all these criteria in just one piece of electrode; however, these tips should be considered as a compass to guide us in continuously improving the microbial bioanodes. As a starting point, it has been remarked that the goal of the bioanode development relies on reaching high volumetric current

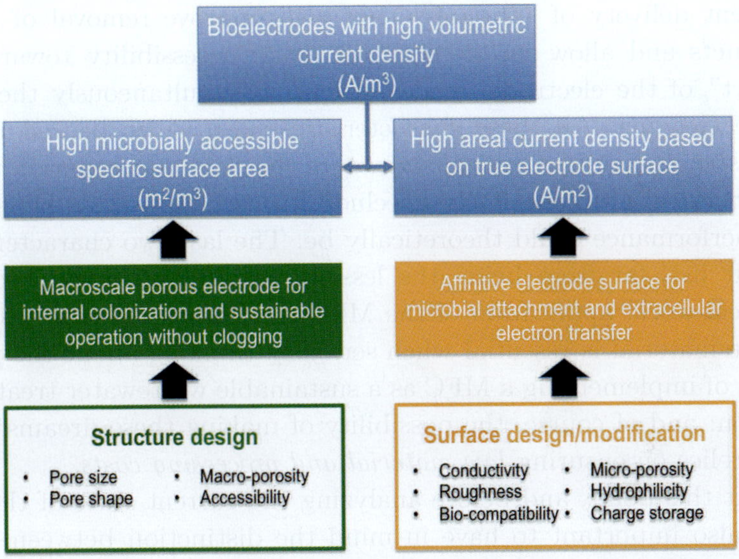

Scheme 1. Goal and strategy for developing high-performance bioanodes (from Ref. 6).

densities (A/m^3),[6] which can only be accomplished by combining high specific surface areas (m^2/m^3) and high current densities per unit area (A/m^2) (Scheme 1).

This deceptively simple task can only be achieved by focusing on the base of the pyramid: the structure design parameters and the surface design features. When doing so, we clearly see the necessity of (inevitably) considering all the points highlighted in our list (from (i) to (x)).

High electrical conductivity combined with a *high surface area* will ensure a high efficiency in electron transfer, which is critical when trying not to have activation losses associated with the electron transfer from the donor (acetate, for instance) to the anode, through the microbial respiratory machinery. *Biocompatibility* and *surface roughness* are other fundamental parameters to be considered if affinitive bioanode surface for microbial adhesion is sought.

In addition, if we pursue a quality leap in microbial anode performance, we inevitably have to face *porous 3D scaffolds* with *open and structured* (*ordered*) *porosity*. These features facilitate not only the efficient delivery of substrates, but also improve removal of waste products and allow for an easy microbial accessibility toward the "heart" of the electrode, while preventing simultaneously the electrode clogging. All these characteristics have to be backed up by *high structural and chemical stability* of the electrode material, which otherwise would completely preclude its use, no matter how good the performance could theoretically be. The last two characteristics of our list, far from being the less important, are critical toward the real-world application of the MFC technology: *up-scaling* of the anode material is essential when seriously thinking about the possibility of implementing a MFC as a sustainable wastewater treatment station; and of course, the possibility of making these dreams come true relies on ensuring *low material and processing costs.*

At this point, and before analyzing the current state of the art, it is also important to have in mind the distinction between those bioanodes used for fundamental bioelectrochemical research (which is the study of the underlying electron transfer mechanisms from bacteria to the solid anode) and those studies related to the improvement of the bioanodes themselves (either in bioelectrochemical reactors or directly in MFC configurations). This distinction will help in understanding the particular choice of anode materials and configurations that can be found in the literature. We will see that those microbial anodes employed in fundamental studies exhibit few of the considerations depicted at the beginning of this section (especially high electrical conductivity, biocompatibility, and chemical stability), making them sometimes not to find place in the wider field of MFCs. On the other hand, we will realize that the most promising bioanodes make use and appeal to most of the fundamental points highlighted. These advanced bioanodes, both from chemical composition and design points of view, are only found in recent publications, which demonstrates that the increase in bioanode complexity (with 3D open structures, highly conductive materials, and high porosity) is the result of the effort of several research groups during the last years.

The fundamental features exposed in this section will be considered when gathering the up-to-date information on metallic (Section 13.3) and oxide-based (Section 13.4) microbial anodes.

13.3 Metallic Microbial Anodes

Metallic electrodes have been mainly employed as anodes to support microbial growth because of their undisputed higher electrical conductivities compared to other materials. This has opened the door to metal-based anodes as feasible alternatives to carbon-based ones. By comparing the electrical conductivity of some metals to that of carbonaceous electrodes (*e.g.* Cu 5.8×10^7 S/m, Au 4.6×10^7 S/m, Ag 6.3×10^7 S/m *vs.* polycrystalline graphite 1×10^5 S/m), the choice of metallic supports to interface redox-active proteins as well as microorganisms is clearly understood.

However, despite the variety of metals that nature offers us, only a limited number appears to be suitable as bioanodes. When gathering the literature on the subject, two distinctive electrode families come up (Scheme 2): those bioanodes based on electrochemically noble (or seminoble) metals (Au, Pt, and Ag); and other group of electrodes based on electrochemically active metals (Ti, Cu, Fe, Ni, and stainless steel).

With the group of active metals, much attention should be paid to what concerns the operative potential window, just to avoid electrode oxidation. This limits the effective operative conditions of a BES, and can be seen as a possible "weak point" in the MFC technology. However, the researchers have managed to use these corrosion-prone

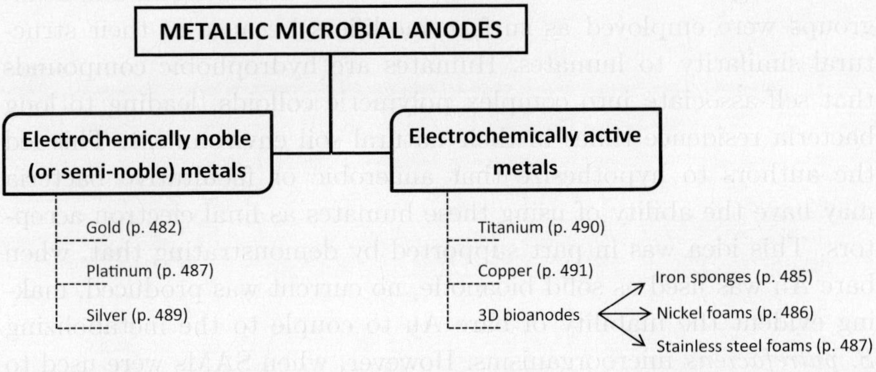

Scheme 2. Metal-based bioanode families.

metals under controlled experimental conditions. In what follows, these two metallic electrode families will be considered.

13.3.1 Noble and seminoble metals as bioanodes (Au, Pt, and Ag)

Gold, platinum, and silver have been used as microbial supports because of their excellent electrical conductivities and electrochemical stability. In addition, as most of them have been studied in a plate-like configuration, an extra advantage must be recognized: a clearly defined exposed surface area. All these features lead to a high electrochemical reversibility, the reason why they have been the preferred option when fundamental electron transfer studies have been conducted. However, their excessive cost precludes (and it will preclude) their use in large technical systems, leaving them out of the MFC framework (at least with regard to pilot- or industrial-scale MFCs).

13.3.1.1 *Gold (Au) bioanodes*

One of the pioneering works on the use of noble metals to explore the electronic coupling between electrogenic bacteria (*Shewanella putrefaciens*) and Au electron acceptors dates back to 2006.[7] In this study, different flat Au electrodes coated with various alkanethiol self-assembled monolayers (SAMs) were used to support microbial growth. Alkanethiols with different chain lengths and headgroups were employed as surface modifiers because of their structural similarity to humates. Humates are hydrophobic compounds that self-associate into complex polymeric colloids, leading to long bacteria residence times in their natural soil environment. This led the authors to hypothesize that anaerobic or facultative bacteria may have the ability of using these humates as final electron acceptors. This idea was in part supported by demonstrating that, when bare Au was used as solid bioanode, no current was produced, making evident the inability of bare Au to couple to the metabolizing *S. putrefaciens* microorganisms. However, when SAMs were used to modify the anode surface, not only electrical current was detected, but also a clear correlation in current density with the chain length and head-groups of the SAMs was evidenced. Relative to graphite and bare Au, the coated electrodes demonstrated to achieve more reproducible experimental conditions and, particularly, certain head-groups (carboxylic acid-terminated monolayers) showed enhanced electronic coupling to bacteria. This was attributed to the possibility of these groups to interact via strong hydrogen bonding with peptide bonds in proteins present in the microorganisms (such as outer membrane cytochromes (OMCs)). This first study triggered the idea that bare Au was possibly an unsuitable anode material for microbial development, surface modification being a suitable path toward improving the biocompatibility features.

The impact of this study was almost immediate, and bare Au was again considered as an anode material for other fundamental investigation.[8] Ultraflat Au was evaluated as an electron acceptor, but this time employing *Geobacter sulfurreducens* as anode-respiring bacteria under a flow-through ministack MFC configuration. In this opportunity, *G. sulfurreducens* effectively produced current over time when inoculated in the anodic chamber of the MFC, reaching stabilized current densities of about 0.7 A/m^2 in a period of 6–10 days. These low densities per unit area were consistent with heterogeneous biofilms developed on Au, as revealed by scanning electron microscopy (SEM) analysis. In addition, the non-homogeneous biofilms showed neither an extracellular polysaccharide binding matrix (as typically observed for biofilms grown on graphite anodes) nor good adherence to Au, as they were readily detached from the anode surface indicating a poor interaction to the metallic substrate (Figure 1(a)). When analyzing the amount of current produced per gram of cell protein, the results surprisingly showed that *G. sulfurreducens* is nearly as effective in transferring electrons to the Au surface as to graphite. These results break with the idea that microorganisms may not use Au anodes as final electron acceptors, and demonstrate that this noble metal can in fact be used as an effective electron receptor. This would expand the range of materials to be considered for microbe/electrode electron transfer reactions.

(a)

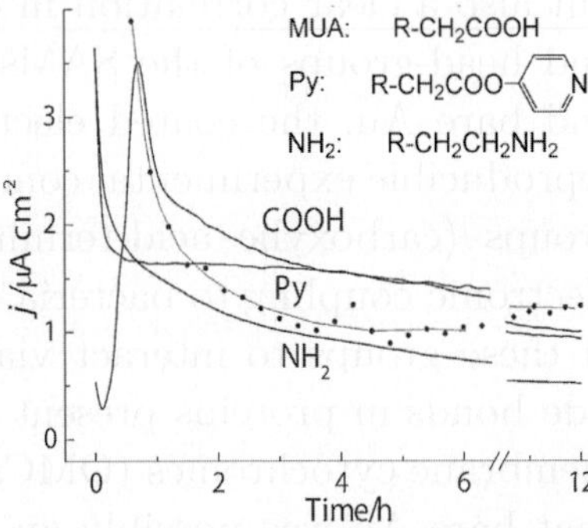

(b)

Figure 1. (a) SEM image of *G. sulfurreducens* biofilms peeled off from a flat Au surface (from Ref. 8); and (b) Chronoamperometric response of *G. sulfurreducens* on different SAM-modified Au plates (from Ref. 9).

Continuing with the bioelectrochemical approaches on flat Au anodes, Kuzume *et al.* grew *G. sulfurreducens* on different SAM-modified Au supports to monitor the structural changes of the OMCs during the EET process.[9] Variations in the steady-state current densities were attributed to the difference in electronic coupling between the terminal anchoring groups of the SAMs and bacterial cells (Figure 1(b)). Contrary to that reported by Crittenden *et al.*,[7] bare Au exhibited the highest stabilized current densities (in the order of 0.01–0.02 A/m^2), comparable to those obtained on mercaptoundecanoic acid (MUA)-modified Au. The authors stressed that although the alkyl chains of the SAMs may act as a tunneling barrier attenuating the electronic coupling, this effect must be compensated by a good electronic communication between the carboxylic terminus and the microorganisms. The study also revealed good biocompatibility of the modified Au surfaces, preventing cell protein denaturation. While it is true that bare Au allows for an effective EET process from *G. sulfurreducens*, carboxyl-anchoring groups make Au more biocompatible, facilitating the heterogeneous electron transfer at the microorganism/electrode interface.

The studies on Au bioanodes have been mainly focused on understanding the underlying electron transfer mechanisms, with only

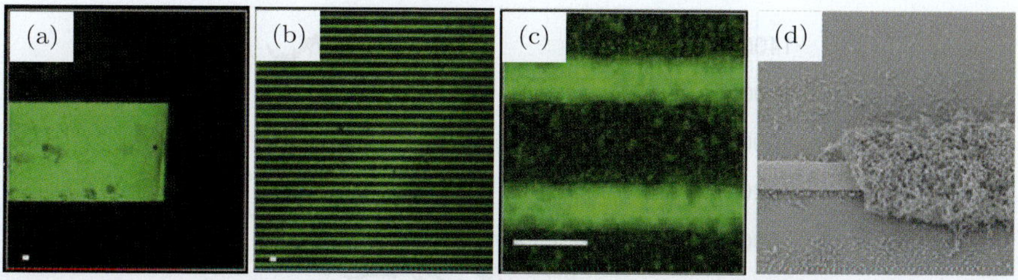

Figure 2. Confocal laser microscopy images of colonized (a) flat Au electrode, (b) patterned Au electrode ((c) 60× magnification of (b)). SEM image (1000×) of a single line colonized by a *G. sulfurreducens* biofilm (d) (from Ref. 10).

scarce experimental approaches setting on the road of improving the metallic anode performance in itself. Among these few works, a study emerged in 2010 proposed altering the geometry of bare Au electrodes to improve the diffusion of organic matter into *G. sulfurreducens* biofilms.[10] For this purpose, two anode designs exposing the same surface areas were compared, one consisting of a rectangular flat Au plate (Figure 2(a)) and other consisting of an array of flat lines (10 μm wide) separated by non-conductive material (Figure 2(b)). The patterned Au electrode was shown to stabilize at a current density fourfold higher than that of its non-patterned counterpart (16 A/m^2 *vs.* 4 A/m^2). However, instead of being a matter of improved substrate diffusion, the differences found were explained by biofilm conductivity, allowing the extension of cells beyond the electrode lines in the grooved bioanode (Figures 2(c) and 2(d)).

While the electrode design was shown to be critical to increase the overall current density, it did not alter the typical current–voltage response characteristic of the employed biofilms. This study evidenced that a wider spacing of biofilms grown on bare Au improves the overall sensitivity toward electron donors, simultaneously allowing cells to better utilize each unit of surface area (while maintaining their typical electrochemical behavior).

Only in the last five years, the researchers have started to play with the metallic electrode structures (from the material point of view) to improve their performance. In this respect, a recent study was focused on single-crystal bare Au microbial anodes to thoroughly

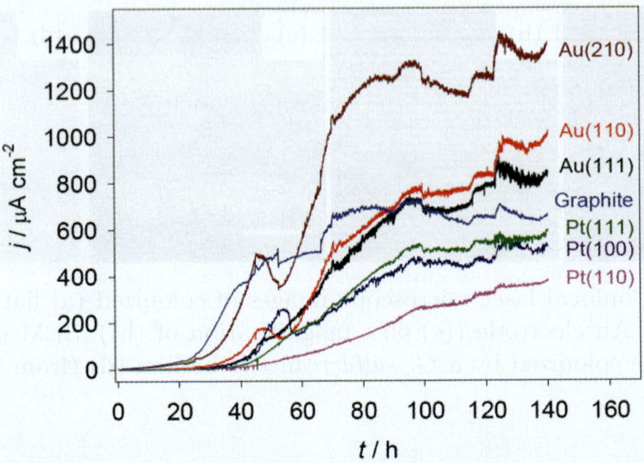

Figure 3. Current density (μA cm^{-2}) over time generated by *G. sulfurreducents* biofilms on different electrode surfaces polarized at 0.2 V *vs.* Ag/AgCl (from Ref. 11).

analyze the influence of the surface atomic arrangement on both the electron transfer process and biofilm formation.[11] Different crystallographic orientations were tested (Au (110), Au (111), and Au (210)), demonstrating that *G. sulfurreducens* interacts with all these surfaces, but showing attachment kinetics and current densities that depend on the metallic face exposed to the culture medium. The highest current densities were achieved with Au (210) (14.4 A/m^2), which was shown to outcompete typical graphite rods used as control bioanode materials (Figure 3).

The results were related to the low atomic density of Au (210), leading to the lowest surface zero charge potential among the tested crystallographic orientations. This conferred the (210) face the most positive charge (at the working potential), facilitating the interaction with negatively charged bacteria. The study evidenced that working on crystal geometry represents a critical parameter for the efficient bacteria–electrode interaction, ensuring a suitable path toward improved microbial devices to optimize energy production.

In summary, only the latest reports have shown a marked interest on the metallic anode performance in itself, which concurs (and it has not to be taken as a mere coincidence) with the highest current densities achieved to date by using Au microbial anodes (14–16 A/m^2).

13.3.1.2 *Platinum (Pt) bioanodes*

When considering the use of other noble metals, one of the most expensive materials appears also on the game. Platinum electrodes have been tested to support the growth of microbial communities, both for fundamental studies and also pointing to MFC improvements. In the recent study by Maestro *et al.*, the influence of the crystallographic structure of single-crystal Pt bioelectrodes (Pt (100), Pt (110), Pt (111), and Pt (210)) was also analyzed and compared to Au.[11] Under constant potential conditions, Pt bioanodes displayed the lowest performance, no matter the surface atomic arrangement exposed to the culture medium (Figure 3). Low and unstable current densities (about 6 A/m^2) were recorded on Pt (111), suggesting that bacterial adhesion (during the first microbial growing stages) could be hampered by negative effects, delaying ultimate electrode colonization. The low efficiency for biofilm growth on Pt found a possible explanation on the superior surface catalytic reactivity of this metal, which may have induced irreversible oxidation of cellular structures. In conclusion, compared to Au electrodes, the up-to-date results have shown that Pt exhibits a poor performance in current production. Beyond these findings, Pt bioanodes were demonstrated to be suitable metallic surfaces for respiratory purposes of *G. sulfurreducens* (depending strikingly on surface crystallographic orientation). This warrants the future use of this metal for fundamental studies related to the underlying EET process to anode surfaces.

If we search for studies dealing particularly with the improvements of MFCs, we can also (and surprisingly because of its cost) find some attempts on the use of Pt anodes. Platinum was used as an ultramicroelectrode (UME) to increase the current densities provided by electroactive biofilms (from garden compost) in electrochemical reactors and under controlled constant potential conditions (−0.2 V *vs.* SCE).[12] Although UMEs (defined as electrodes having at least one dimension equal to or smaller than 25 μm) have been discovered in about 1980, they had never been considered before as microbial supports. The pioneering study by Pocaznoi *et al.*[12] was directly focused on enhancing mass transfer into the biofilm, leading

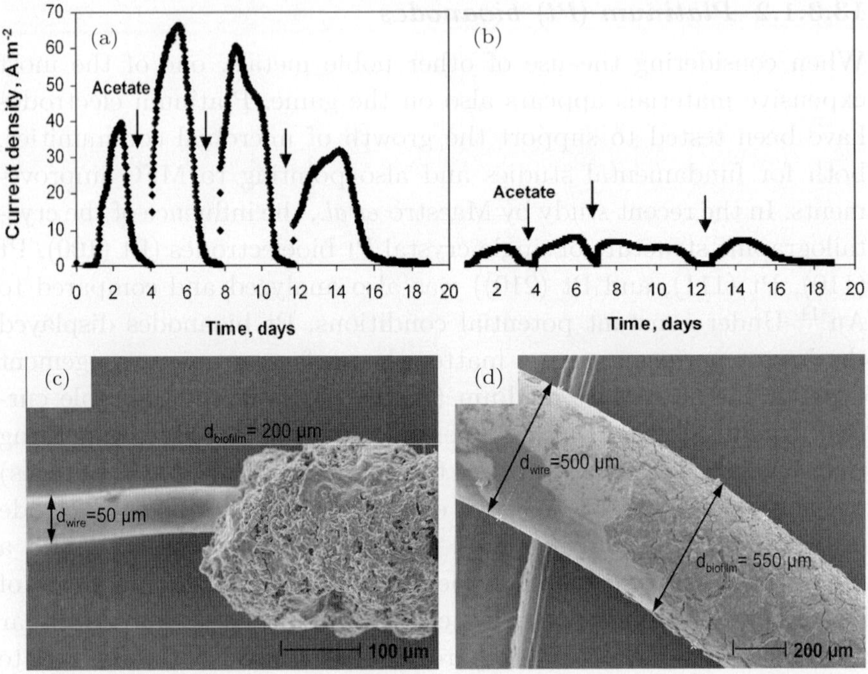

Figure 4. Current density (A/m^2) over time obtained with Pt electrodes used as microbial anodes: (a) 25 μm-diameter Pt wire (66 A/m^2), and (b) 500 μm-diameter Pt wire (7 A/m^2). The SEM images show biofilms formed on Pt electrodes of diverse diameter: (c) 50 μm, and (d) 500 μm (from Ref. 12).

to significant improvements in current densities when using 25 μm diameter Pt wires (66 A/m^2) compared to Pt electrodes (also wires) of larger diameter (7 A/m^2 for both 500 and 1,000 μm diameters) (Figure 4).

While it is true that Pt could hardly be considered as an alternative bioanode in large-size MFCs, Pocaznoi *et al.*[12] suggested the possible use of UMEs in mL- or μL-scale systems, opening the door to miniature MFCs on the view of their various and versatile applications. Among these, power sources for ultrasmall electronics[13] and implantable medical devices[14,15] can be highlighted. Also, micro-MFCs coupled to microfluidic systems have been recently designed for high-throughput identification of electroactive microorganisms.[16] All these cutting-edge applications should be considered in the future, and we recognize (to this respect) the pioneering work by Pocaznoi *et al.*[12] as a promising path to be followed.

13.3.1.3 *Silver (Ag) bioanodes*

Silver electrodes have been used as microbial interfaces on *in situ* fundamental spectro-bioelectrochemical investigations. Particularly, roughened substrates were considered as surface-enhanced Raman spectroscopy (SERS) platforms to grow electroactive microbial biofilms.[17,18] Under controlled potential conditions, *G. sulfurreducens* was grown on Ag rings for a detailed spectroscopic characterization of the OMCs involved in the EET mechanisms.[17] The combination of SERS and cyclic voltammetry (CV)-SERR spectroscopy allowed the authors to probe selectively the heme groups of the membrane-bound cytochromes in the proximity of the electrode surface, thereby providing fundamental protein structural information. Electroactive biofilms grown on roughened Ag produced maximum current densities of 6 A/m^2, typical values achieved for non-porous plate-like bioanodes under turnover conditions. However, these values were only sustained for a few hours, with final stable current densities of 2.7 A/m^2. From the material point of view, and contrary to what was generally believed, Ag electrodes (and the inevitable traces of Ag (I) ions) did not provide any toxic effects on *G. sulfurreducens.* This confirmed other observations on the growing of this strain in the presence of nanoscale Ag precipitates[19,20] and massive Ag plates.[21]

The study by Baudler *et al.*[21] pointed, for the first time, to the use of a palette of metal substrates as alternative bioanodes to carbon-based supports. Particularly, Ag was demonstrated to be innocuous toward electrode-respiring bacteria, providing, in fact, one of the highest current densities (11 A/m^2, polarized at 0.2 V *vs.* Ag/AgCl) compared to the other studied metals (Cu, Ni, Co, Ti, and stainless steel). According to optical images, Ag anodes resulted to be suitable platforms for the formation of uniform and optically dense electrochemically active biofilms. Focusing on the final application, the electrochemical stability window of Ag was determined. When polarized toward positive potentials, Ag electrodes oxidized at relatively low values, with an onset corrosion potential of 0.3 V *vs.* Ag/AgCl. In conclusion, care should be taken during the start-up of a BES just to prevent corrosion effects of the Ag metallic anode.

These restrictions when setting up metallic electrodes in BESs limit their ultimate operative conditions, making the materialization of the MFC technology a challenge yet to come. However, corrosion effects have been reported not only for seminoble metals, but also for other metallic electrode families. We will see that corrosion imprints are even more evident when non-noble metallic substrates are used as microbial supports.

13.3.2 Corrosion-prone metals as bioanodes (Ti, Cu, Fe, Ni, and stainless steel)

The often lower costs of non-noble supports, while preserving excellent electrical conductivities, have made of them increasingly growing alternatives as anodes in BESs. However, their tendency to corrosion has to be particularly considered when setting them in the anodic compartment of a MFC. Unlike noble metals, different potentials are typically set depending on each metal considered.

In addition to typical non-porous (flat) metallic electrodes, several studies dealing with non-noble metals have also centered the attention on 3D porous scaffolds with very high specific surface areas (meshes, foams, felts, sponges). This illustrates that when focusing on the electrode performance (from the material point of view) the range of fundamental characteristics highlighted in Section 13.2 begins, little by little, to gain importance. This change of paradigm is the result of what researchers have learned over the previous years, clearly reshaping our understanding on functional electrodes for microbial electrochemical purposes.

13.3.2.1 *Titanium (Ti) bioanodes*

One of the few studies dealing with Ti as an anode material for microbial purposes dates back to 2008.[22] Non-porous Ti plates (both bare and Pt-coated) were used to evaluate their suitability as bioanodes in MFCs, comparing their performance with flat graphite controls.

The effect of electrode surface roughness on current production was particularly addressed, attempting to improve surface-related parameters toward better microbial adhesion properties. The current densities achieved for both Pt-coated Ti (4 A/m^2) and flat/roughened graphite (3 A/m^2 and 4.5 A/m^2, respectively) were considerable higher than those obtained on bare flat Ti (about 0.1 A/m^2). The lack of current production on the uncoated electrode was explained by a possible anodic passivation of its surface (mediated by formation of Ti oxides). This led to explicitly sentence bare Ti as an unsuitable material toward MFC applications. Nevertheless, as we will see in the following sections (Section 13.4.1.2.1), Ti oxide-based anode materials can play a leading role in the area of functional bioelectrodes. For this to be true, specific atomic arrangements of the Ti oxides have to be achieved, leading to new and interesting material properties. For the anodes evaluated in the work by ter Heijne *et al.*,[22] neither the surface roughness features nor the mass transfer processes were evidenced to be the only variables explaining the differences in current densities under polarization conditions. This suggested that other specific surface properties could also exert important effects on bacterial attachment (yet to be demonstrated).

13.3.2.2 *Copper (Cu) bioanodes*

Based on its exceptional electrical properties, Cu could not be ignored as an option for microbial scaffolds. In 2007, Kargi and Eker used Cu wires as anode materials for electricity generation and simultaneous wastewater treatment in MFCs.[23] This study focused on increasing current and power densities by augmenting the exposed surface area of the electrodes, while treating (with a mixed culture of anaerobic microorganisms) a synthetic wastewater with different chemical oxygen demand (COD) contents (1,000 to 6,000 mg/L). The increase in Cu surface area (from 2.512 to 20.1 cm^2) effectively led to higher current and power densities (0.06 A/m^2 and 0.67 mW/m^2, with the highest exposed surface area). COD contents decreased over time in both the aerobic and anaerobic compartments, with faster depletion in the aerobic chamber. More than 80% COD removal was achieved under aerobic conditions, when the higher influent strength was used. However, corrosion imprints of the Cu bioanodes (if any) were not mentioned, remaining unclear whether they did not take place or were not studied.

As to the possible metal corrosion during the operative performance of Cu bioanodes, Zhu and Logan conducted a study to dispel any doubts about it.[24] They demonstrated that Cu anode corrosion effectively occurs and affects power generation in MFCs. MFCs with Cu-mesh bioanodes initially showed high current generation (similar to abiotic controls used), this fact being related to galvanic corrosion after connection to the external circuit (1,000 Ω). After reaching a maximum current, a rapid decrease followed toward a stabilized steady-state value. No matter the subsequent inoculations of the anodic chamber with microbial suspensions, the obtained current was shown not to increase above 0.003 mA. Corrosion effects were also demonstrated by quantification of produced soluble Cu(II), which inhibited both microorganisms and current generation in the reactors. Contrary to what had been reported up to the publication date, the authors concluded that Cu electrodes should not be used in MFCs as bioanodes, recommending specifically the use of non-corrosive metals as suitable alternatives to Cu-based microbial electrodes.

Other later approaches have also considered Cu as a possible anode option, but paying special attention on polarization conditions. This is the case of Cu plates (and foils) tested by Baudler *et al.*[21] as high performance and economically viable anode materials. Electrochemically active biofilms were grown at −0.2 V *vs.* Ag/AgCl, avoiding metal corrosion and having no impact on the normal electrocatalytic performance of the biofilm. In fact, microbial films evidenced a uniform and optically dense aspect on Cu surfaces. By analyzing the effective operative potential window for Cu, two oxidative processes were observed when the metal was polarized toward positive potentials, evidencing the lowest onset dissolution potential (about 0.5 V *vs.* Ag/AgCl) compared to the other tested materials. However, by ensuring a suitable working potential, Cu was capable of delivering current densities as high as 15 A/m^2, representing one of the highest performances achieved to date to what concerns flat non-porous metallic microbial anodes.

13.3.2.3 *Three-dimensional non-noble metallic bioanodes (Fe, Ni, and stainless steel)*

Following the natural evolution of the concept of advanced functional bioanodes, real interest in 3D electrode platforms can be evidenced from 2012 onward.

In the context of MFCs, and recognizing the lack of intensive real-world applications, a study by Mapelli *et al.*[25] addressed the subject of overcoming typical bioelectrochemical limitations by implementing cast Fe-based metallic sponges, aiming at increasing the exposed electrode surface area and thus the current density at the anodic compartment of a MFC. Cast Fe was selected because of its good biocompatibility features and acceptable electrical conductivity, although the high melting point of the metal can carry several problems during the manufacturing process. Different anode porosities (between 10 and 30 pores per inch) were tested, offering open structures suitable for culture medium transport and bacterial colonization (Figure 5). Iron sponges were set in MFC-like configurations, reaching voltage stability after 48 h operation. Contrary to this, by using a control bioanode (massive Fe with the same dimensions as those of the sponges) the voltage output was never stable. The difference was attributed to a faster and more stable biofilm formation (*S. oneidensis*) on the foamed support. While a maximum voltage of 0.103 V was achieved by using the cast Fe sponges, this value was demonstrated not to be sustained for so long in time, being also rather lower compared to typical MFC voltage outputs. After testing, both bioanodes showed extended corrosion effects (with Fe and P oxides identified by X-ray emission spectrometry). This effect was shown to be responsible for the power loss at the latest growing stages, the extended oxidation patterns compromising the effective electrical conductivity of the macroporous platforms.

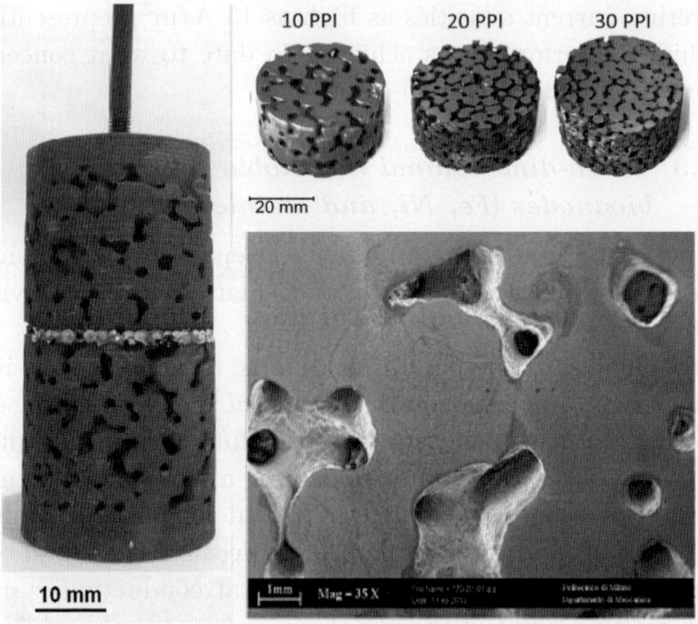

Figure 5. Picture of the Fe-based sponge used as microbial electrode and details of its porosity features (10, 20, and 30 pores per inch [PPI]). The SEM image shows a closer view of the cast porous scaffold (featured for a porosity of 20 PPI) (from Ref. 25).

Although a valid attempt has been made in this study to change the paradigm toward a more complex 3D scenario, corrosion phenomena found on cast Fe foams could seriously impair their performance in future BESs.

Based on the same concept, Ni-coated sponges were also tested as novel and inexpensive anode materials.[26] Open 3D macroporous structures (200–300 μm diameter pores) were used to interface microbial cells (Figure 6(a)). Compared to typical 2D carbon paper, Ni sponges not only offered higher specific surface areas to facilitate microbial attachment, but also supported efficient substrate transfer while exposing highly conductive pathways.

Unlike cast Fe foams,[25] Ni-based platforms demonstrated excellent electrochemical activity and power output stability during electricity generation in MFCs (996 mW/m^2 projected surface area).

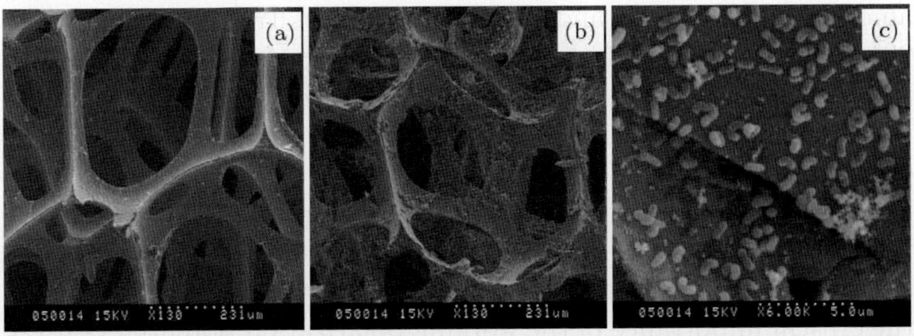

Figure 6. SEM images of Ni-coated sponges before (a) and after (b)–(c) microbial colonization from Ref. 26.

No corrosion effects were evidenced for the porous scaffolds; however, maximum current densities determined on CV tests (10 A/m^2) were of the same order of magnitude as those reported for typical flat electrodes. SEM analysis revealed poor microbial coverage (individual bacteria not forming biofilms), especially on inner electrode surfaces (Figures 6(b) and 6(c)). This is contrary to what authors stated as to the formation of densely adhered biofilms deep inside the 3D structure. The poor surface coverage could explain the low current densities obtained, as well as the absence of electrode pore clogging. Even though the 3D metallic substrate was presented as a suitable material for microorganisms to maintain their electrochemical activity, its surface modification would critically improve the microbial adhesion properties and thus performance, allowing reaching high current densities as those reported for currently available high-performance 3D microbial anodes.[6]

Turning the page, we finally face stainless steel bioanodes. Stainless steel (SS) platforms have shown an interesting progress during the past years and, having their own history (which dates back to 2007 and reaches our days), exhibit a meticulous conceptual evolution. This has made of SS an interesting alternative toward BESs, especially considered (because of its low cost) for large-scale MFCs. Moreover, its high-performance mechanical properties, long-term resistance, and commercial availability under different compositions and morphologies have made of SS materials a common option among microbiologists and material scientists. True, it was not until 2014 that real SS 3D bioanode alternatives came to light; however, the progress on the subject is worth being reviewed (from the pioneering studies on marine MFCs to the latest advances on structural complexity).

The first reports focused on the use of non-porous SS plates to support long-term growth of different microbial communities.[27,28] These studies have been fundamental to light the way forward, revealing the critical importance of the working polarization potential (−0.1 V *vs.* SCE) on both maximum achieved current densities (2–4 A/m^2) and bacterial surface coverage, showing the absence of galvanic oxidation. In particular, low average surface roughness of SS bioanodes (about 0.29 μm)[27] has been shown to be detrimental for the electrode performance, with cell adhesion during the first hours controlling the further electron transfer process to a large extent. Electrochemical efficiencies of SS biohybrid systems have also been suggested to be improved by optimizing surface-related parameters of the metallic anode, with higher current densities achieved (8.2 A/m^2 projected surface area) by growing natural biofilms on SS grids (Figure 7).[28]

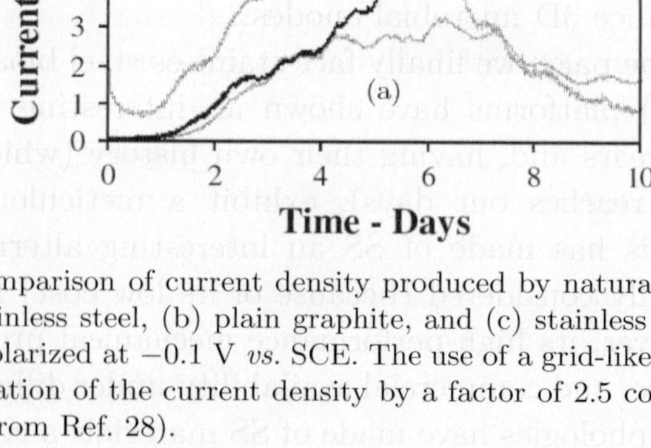

Figure 7. Comparison of current density produced by natural biofilms grown on (a) plain stainless steel, (b) plain graphite, and (c) stainless steel grid. Each electrode was polarized at −0.1 V *vs.* SCE. The use of a grid-like architecture led to the multiplication of the current density by a factor of 2.5 compared to plain stainless steel (from Ref. 28).

Among the plate-like bioanodes, undoubtedly those reported by Pocaznoi *et al.*[12] have been the ones shown to achieve the highest current densities. Under well-controlled electrochemical conditions (–0.2 V *vs.* SCE), and using a three-electrode electrochemical setup, SS plates were shown to achieve current densities as high as 20.6 A/m^2 (projected surface area).[12] This remarkably high areal current density was even increased when the working potential was set at 0.1 V *vs.* SCE (under CV tests), the SS bioanode providing up to 35 A/m^2 showing to outcompete flat graphite plates (11 A/m^2). SS platforms with surface micro and macrostructuring features were also shown not to evidence differences in performance when compared to their flat counterpart. Accordingly, the results revealed that surface topography of SS seems to affect neither the current provided nor biofilm thickness (50–60 μm). Despite this, the high current densities achieved by SS electrodes led to a great interest on this material as a promising microbial anode, with values twice as high as those obtained with typical graphite supports and even higher than those obtained on SS grid electrodes.[28] These facts were, in part, explained from a microbiological point of view, based on the authors' experience for optimally exploiting the soil inoculum used to perform the electrochemical experiments. Other explanation, this time from the material point of view, was focused on the control of the oxide layer that covers the SS surface, the anodic potential used (–0.2 V *vs.* SCE) being low enough to avoid the formation of detrimental n-type semiconducting layers on the anode surface.

All these previous studies have paved the field to profit the characteristics of 3D scaffolds toward advanced microbial anodes. To this respect, SS foams and felts have been lately considered yielding interesting results. The use of foams constitutes the first report regarding true 3D SS microbial platforms (Figure 8(a)).[29] Implemented at low applied potentials (−0.2 V *vs.* SCE), macroporous SS bioanodes (200–800 μm pore diameter) were used to go beyond the highest current densities achieved on plain SS (20.6 A/m^2).[12] Values of about 60–80 A/m^2 (projected surface area) were consequently reached (Figure 8(b)). The open structure of the metallic foam allowed for an extended bacterial colonization, facilitating microbial penetration

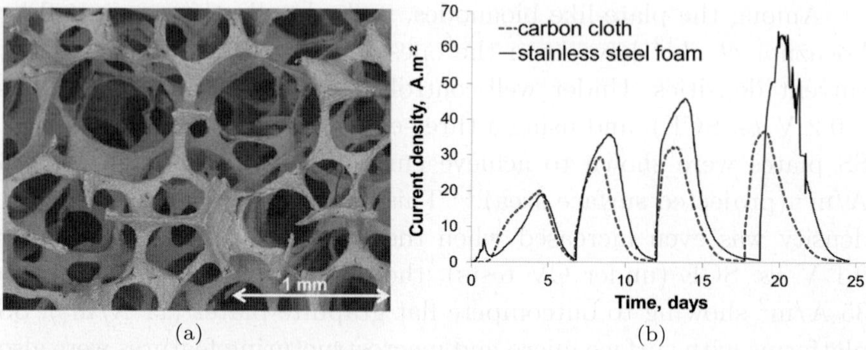

Figure 8. (a) SEM image revealing the open macroporous structure of the SS foam used as bioanode; and (b) Chronoamperometric evolution for the SS foam, compared to control carbon cloth under polarization at −0.2 V *vs.* SCE (from Ref. 29).

into the 3D structure. Importantly, the large mean pore size of the foam allowed avoiding pore clogging after colonization (20 μm biofilm thickness) while facilitating mass transfer to the inner parts of the electrode. However, beyond the clear advantages exhibited by the SS sponges, they showed less electrochemical efficiency compared to carbon cloth, with a larger overpotential required for current increase under CV tests. Voltages higher than 0 V (*vs.* SCE) were not applied to the sponges to avoid disruption of the protective passive layer on the metallic surface.

Clearly, 3D SS platforms have shown to be promising alternatives toward a new generation of microbial anodes. This has led other authors to take over the task of improving even more the performance of SS as 3D scaffolds.[30,31] However, in contrast to what has been reported by Ketep *et al.*,[29] poor biocompatibility of SS has been claimed as a restriction to its direct use as anodes in BESs.[30] In addition, Guo *et al.*[30] highlighted that the high cost of the SS foams, compared to carbon foams (2.5 times higher) and carbon felts (26 times higher), makes their ultimate up-scaling very difficult. This economic restriction led to consider surface modifications of inexpensive SS felts as an attractive strategy to employ them as biointerfaces. *In situ* generation of iron oxide nanoparticles (IONPs) on SS felts (by flame oxidation) was thus proposed, using a Bunsen

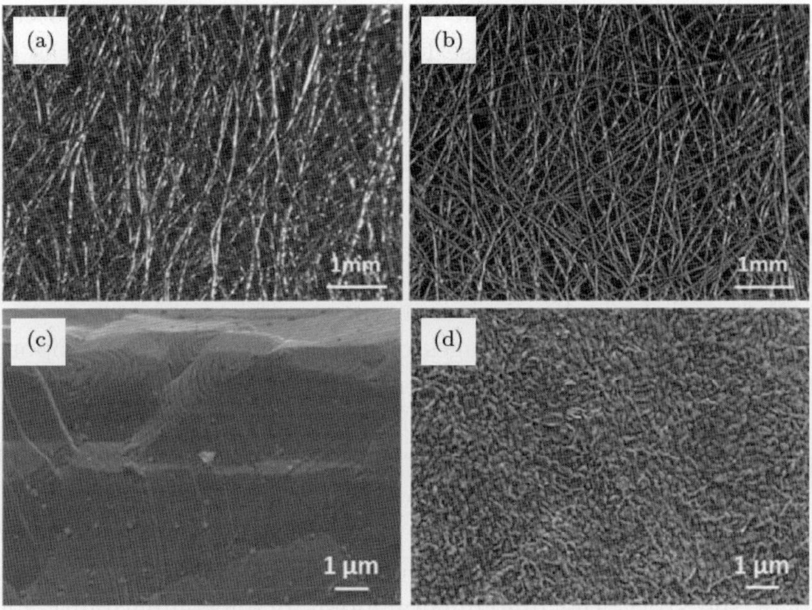

Figure 9. Representative light microscopy images ((a) and (b)) and SEM images ((c) and (d)) of untreated (left column) and flame oxidized (right column) SS felts. Oxidized SS electrodes were uniformly covered by (100–400 nm) Fe_2O_3 nanoparticles (from Ref. 30).

burner[30] or a muffle furnace.[31] The modification with a homogeneous coverage of 100–400 nm IONPs (α-Fe_2O_3) was shown to dramatically enhance the biocompatibility of SS toward *Geobacter*-rich microorganisms, when the bioanodes were used in bioelectrochemical reactors (Figure 9).

Chronoamperometric tests on modified SS felts revealed maximum areal current densities of 19.2 A/m^2 (projected surface area), corresponding to volumetric current densities of 27.4 kA/m^3.[30] However, the values were stabilized to 15 A/m^2 and 21 kA/m^3 when the reactor was switched to continuous mode operation. Electrocatalytically active biofilms grown on IONP-decorated SS felts exhibited the typical sigmoidal response under turnover conditions. The excellent biocompatibility revealed was attributed to the increase in surface roughness, as well as to the stimulating bacterial adhesion effects imparted by oxide layers. The determined onset corrosion potential was shown not to compromise the integrity of the SS felts, as long as the operational anode potential remains to be set at −0.2 V *vs.* Ag/AgCl. Flame oxidized-SS felts account for one of the highest volumetric current densities reported to date for acetate-fed bioelectrochemical reactors. However, due to different reactor designs, culture medium compositions, microbial inoculum, and operational conditions (temperature, convection, potential, etc.), it may be not appropriate to make an exhaustive comparison on the subject.

In summary, the advantages on using metallic 3D macroporous supports as microbial anodes seem to be conclusive toward a new generation of bioelectrodes. In this path, and taking into consideration properties as high electrical conductivity, high accessible surface area, open porosity, structural stability, low cost, and upscaling potentiality; SS felts seem to offer a real alternative towards BESs, especially for MFC technologies. However, far from being the end of the story, the fundamental point of biocompatibility highlighted in the studies by Guo *et al.*[30,31] opens surprisingly the door to other kind of materials that have been practically ignored when considering attractive alternatives as bioanodes. We will see in the next section that *metal oxides* may also offer new horizons in the field of non-carbonaceous microbial electrodes for electrochemical systems.

13.4 Oxide-based Microbial Anodes

Bacterial oxide respiration in anoxic subsurfaces represents a major driver for natural organic carbon oxidation, also influencing a number of biogeochemical processes such as nitrogen, sulfur, and phosphorus cycles.[32] During these natural events, microbes exchange electrons with mineral oxides via different mechanisms mediated, mainly, by a variety of bacterial electron transport chains.[33] Studies on the subject have revealed that natural bacteria/oxide processes are critical on the biological energy gain; but thinking from the oxide point of view, this “marriage” represents an extraordinary opportunity toward the development of advanced bioelectrodes for BESs. To this respect, the possibility of interfering efficiently (by using oxide interfaces) at different redox "stations" along the electron transfer path would directly impact not only on the microbial energy metabolism itself, but also on the energy we could harvest from the electrochemical system. Surprisingly, instead of taking nature as an inspiration source to learn from, the proposal of novel microbial electrode materials (mainly anodes) has almost exclusively relied on carbonaceous supports.

Compared to the literature on carbonaceous and metallic electrodes, oxide bioanode studies are mere scattered islands in the bioelectrochemical ocean. While metallic electrodes have been mainly employed because of their exceptional electrical properties, metal oxides have found (not much) application on the microbial field because of their excellent biocompatibility features. In this section, we will demonstrate that, although lacking of a clear distinction into different electrode families (as is the case with metallic bioanodes), the use of metal oxides to interface microorganisms evidences an evolution that can be usefully displayed on a timeline profile.

13.4.1 Timeline: From subsurface interactions to electrode performance

On the road of bacteria–oxide interactions, we can identify two main patterns over time: the use of metal oxides/hydroxides for fundamental studies, and their use as building blocks for microbial anodes. We will make use of a timeline to introduce the state of the art on the evolution of the subject (Scheme 3). For the purpose of clarity in the graphical representation, we depict the used iron oxide/hydroxide materials on the upper part of the timeline, leaving the bottom part reserved for other metal oxides that have also been considered as microbial anodes to date.

Although both bacterial leaching mechanisms on iron oxide-containing minerals and biologically induced mineralization processes on different bacterial strains have been largely studied over the past years,[34–38] we will only focus on relatively recent investigations (just not to overextend in the subject). Assuming this, the first part of our timeline begins in 2001 and it extends until 2005,

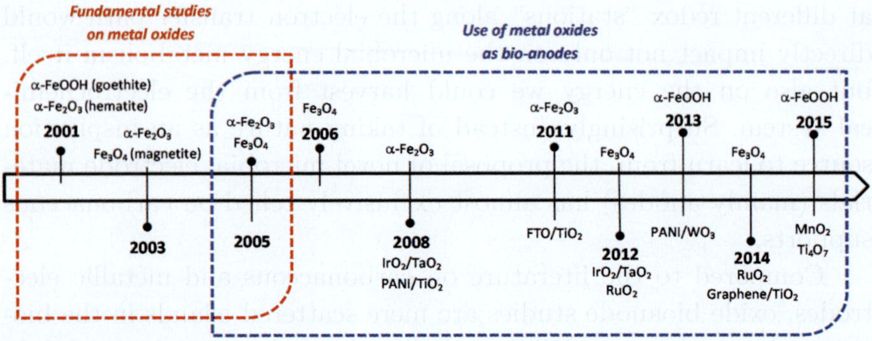

Scheme 3. Timeline evolution on bacteria/metal oxide interactions.

covering only studies on the intimate relationship between bacteria (*Shewanella* strain) and iron oxide/hydroxide surfaces (red dashed box in Scheme 3). The second part (blue dashed box) begins in 2005 with the first attempts on using iron oxides in the bioanodes, reaching our days and exhibiting the most promising oxide-based bioelectrodes reported to date (Ti_4O_7). A distinctive feature of this latter stage in the evolution line is that, except for one study,[39] all the proposed oxides have been considered as just modifiers (not self-standing devices) of the anode system.

13.4.1.1 *Fundamental studies on metal oxides*

Iron is one of the most abundant elements in the Earth crust and, in many subsurface environments, the reduction of its most oxidized form is the predominant redox process. The leading role of the Fe(III)–Fe(II) cycle as a major energy flux in oxygen-depleted environments is unquestionable and, in this respect, the microbial reduction of iron oxyhydroxides/oxides represents the quintessential example. Therefore, it is not surprising that iron oxides have been considered for initial fundamental researches.

Studies on the affinity of bacteria/iron oxide surfaces were conducted by Glasauer *et al.* in 2001.[40] The investigation pointed to the sorption capability of different non-biogenic hydroxyoxides (ferrihydrite, goethite) and oxides (hematite) on *S. putrefaciens* under laboratory conditions, to determine the magnitude and mechanisms of the fundamental interaction. Specifically, the study was conducted to reveal the capability of *Shewanella* to allow the sorption of preformed iron-containing nanominerals commonly found in natural sediments and soils, leading to shed some light on the typical processes involved in the iron cycling.

Interestingly, the interaction between the minerals and cells resulted in the formation of aggregates, which led to increased biomass densities and provided better sedimentation conditions compared to the results obtained for the minerals in the absence of microorganisms. Fundamental conclusions supported the fact that, in addition to natural biomineralization processes, adsorption of preformed minerals on *S. putrefaciens* could also take place in anaerobic environments. This not only represents an alternative route for iron cycling in nature, but also extends the knowledge on the biogeochemical phenomena of the studied bacteria.

The same year, detailed studies on bacterial recognition of mineral surfaces were conducted by Lower *et al.* This work was focused on measuring the infinitesimal forces between *S. oneidensis* and goethite by making use of biological force microscopy, as a means of understanding the bacteria/mineral interface.[41] Under anaerobic conditions and supplemented with lactate as the energy source, *S. oneidensis* exhibited an increased affinity for goethite (by two to five times) compared to the aerobic conditions. Interestingly, the microorganisms were shown to respond to the iron-containing mineral by rapidly developing strong adhesion energies at the interface. The results were interpreted by indicating that after recognition of the mineral surface as terminal electron acceptor, *S. oneidensis* actively mobilizes and/or produces reductase proteins that specifically interact with the oxyhydroxide surface. This study represents one of the first examples on the use of nanomechanical measurements toward biochemical and mineralogical mechanisms that operate under environmentally relevant conditions.

Following the same line of studies, Neal *et al.* continued with the understanding of the intimate relationship of *S. oneidensis* and iron oxide minerals, but digging deeper into the effect of different mineral surface facets upon the biological adhesion properties and electron transfer rates.[42] Cell accumulation studies were conducted on three iron oxide single-crystal faces: hematite (001) and magnetite (100) and (111). Clear differences in bacterial accumulation were observed among the tested minerals, with significantly more cells accumulated at hematite (001) compared to magnetite faces. In addition, modeling of the electron transfer process to the three oxide surfaces predicted that, over the range of existing distances, the highest transfer rates occur at the hematite (001) surface, mirroring the results obtained from accumulation tests. The study presents clear evidences that mineral surface (atomic and electronic) structure definitively influences the oxide-associated cell activity, at least to what *Shewanella* strain concerns. Two years later the same authors, making use of atomic force microscopy (AFM), demonstrated that significantly different adhesive interactions effectively occur when *S. oneidensis* interfaces different crystal arrangements.[43] The results showed that hematite (001) face presents a more optimal surface for bacterial activity than magnetite (100) and (111) facets, suggesting that electrostatic interactions are potential determinants for cell adhesion. The order found in adhesive force (hematite (001) > magnetite (100) > magnetite (111)) resulted to be consistent with their previous observations of maximum cell accumulation at the (001) face of hematite.[42]

The understanding of the fundamental bacteria/oxide interactions has been considered as a cornerstone from both biological and mineralogical points of view; however, the early studies have not considered the possibility of linking the intimate biotic–abiotic relationship to interfere efficiently in the microbial electron transport chain toward energy harvesting devices. While the first examples to this respect began to appear in 2005, it was not until 2008 that metal oxides (other than iron-based ones) flourished with the clear purpose of offering new alternatives as bioanodes in the MFC technology.

13.4.1.2 *Use of metal oxides as bioanodes*

According to the fundamental studies, the first considerations on metal oxides to interface microorganisms in MFCs relied on iron-containing minerals.[44] This study represented the first attempt of modifying typical carbon paper anodes by coating them with ferric oxides, given their known capability to facilitate direct electric contact to OMCs of electricity-producing microorganisms. Particularly, ferric oxide-modified electrodes were employed to enrich iron-reducing bacteria in the culture medium, pointing to more effective inoculation techniques to accelerate MFC startup stages. Results from two-chamber systems revealed that ferric oxide-containing anodes were selectively colonized by acetate-fed microorganisms (from anaerobic sludge) outcompeting methanogenic bacteria for acetate. In addition, voltages obtained from ferric oxide electrodes were consistently higher than those obtained when only carbon paper was used as the bioanode. When normalized to the projected electrode surface area, power densities of 30 mW/m^2 and Coulombic efficiencies as high as 80% were determined when the iron oxide was evaluated (almost four-time higher power densities and two-time higher Coulombic efficiencies than unmodified carbon paper). The results suggested that iron oxide-coated microbial scaffolds could be used to improve typical inoculating techniques, increasing not only the effectiveness of microbial enrichment in the culture medium but also showing to be promising for shortening the MFC startup stage.

The investigation by Kim *et al.*[44] lighted the way forward on the subject, reason why other iron oxides have been also taken into consideration as anode modifiers in the MFC area. Nanocrystalline magnetite,[45] nanocolloids of hematite,[46] layer-by-layer hematite nanorod-based composite films,[47] magnetite/activated carbon/SS rolled meshes,[48] semiconducting nanogoethite,[49] among others[50,51] have all been used as bioelectrodes to increase both the power and electricity outputs in MFCs. However, despite the efforts made (with obvious improvements in anode performance respect to the unmodified electrodes), the results reported to date have only shown a slight increase in the relevant parameters: 0.809 W/m^2, 2.65 A/m^2 when magnetite/activated carbon/SS composites were used as the anode system[48]; and 0.830 W/m^2, 3.2 A/m^2 when magnetite/carbon nanotubes were tested.[50] This situation stimulated the researchers to consider other kinds of oxides as alternatives to iron-containing ones. The task began in 2008 at the hands of Dumas *et al.*, staging for the first-time dimensionally stable anodes (DSAs) to interface *G. sulfurreducens* microorganisms in electrochemical reactors.[27] Dimensionally stable anodes represent a family of electrically conducting scaffolds typically made of a titanium base covered by iridium and tantalum oxides (although diverse oxides can be found in other formulations). While, undoubtedly, DSAs constitute one of the greatest technological breakthroughs of the history of electrochemistry, they had never been considered before as anodes for microbial electrochemical systems. This emphasized the attempt by Dumas *et al.* as a spotlight for the studies to come. The choice of non-porous sheets of DSAs toward MFC applications was made due to their upscaling potentiality,[27] which highlights the intention of the authors in testing new materials feasible for pilot/industrial scale BESs. Under constant polarization at 0.2 V *vs.* Ag/AgCl, DSAs achieved output current densities of about 5 A/m^2, with *G. sulfurreducens* biofilm being fully responsible for the electrocatalytic oxidation of acetate. However, the obtained current values were demonstrated to be lower than those achieved for typical graphite control plates (8 A/m^2). Recognizing this drawback, the authors suggested a potential improvement in performance by working on the electrode surface roughness. Likewise, they raised the issue of comparing DSA electrodes made up of different oxide coatings, with the objective of shortening the initial microbial growth phase. Even though the DSAs had initially shown promising characteristics, it was not until 2012 that they were again considered to interface electrochemically active biofilms.[52] In this opportunity, DSA electrodes embedded in garden compost and polarized at 0.5 V *vs.* SCE gave even lower current densities (1.4 A/m^2) compared to the study by Dumas *et al.*[27] Nevertheless, the differences found in each study could be related to the diverse experimental scenarios under which the oxide-coated anodes were tested.

Other oxides have also been considered to power the anodic chamber of MFCs or in bioelectrochemical reactors. In the list, we can find from typical semiconducting TiO_2[53–55] to other less common options as RuO_2,[56,57] WO_3,[58] MnO_2,[59] and Ti_4O_7.[39]

Specifically, because of its semiconducting behavior, TiO_2 has always been used as a support of electrically conducting materials, which turned out to be the real microbial interfaces. For this reason (and being out of the scope of this chapter), we will not make a detailed analysis of this kind of bioanodes (the case is similar for WO_3). We will simply comment that polyaniline (PANI)/mesoporous TiO_2 composites,[53] fluorine-doped tin oxide (FTO)/TiO_2 hybrids,[54] PANI-functionalized mesoporous WO_3[58], and graphene/TiO_2 architectures[55] have all been taken into consideration to improve the electron transfer efficiency of the basal oxide, pointing to the final improvement of the MFC (or photo-MFC) performance.

The year 2012 embraced for the first time the use of noble metal oxides (RuO_2) to interface electrogenic microorganisms, owing not only to their fast and reversible redox behavior, but also because of their metallic conductivity.[56] In addition, the promising features of RuO_2 as electrode material used in supercapacitors, its high chemical stability as well as good biocompatibility when contacted with bacteria have made of this oxide an alternative microbial platform to promote the biological electron transfer. As in the already mentioned cases, RuO_2 was only employed as an anode modifier (this time on carbon felt electrodes, which were the true current collectors) (Figure 10).

MFCs equipped with RuO_2-modified anodes reached maximum power densities of 3.080 W/m^2 and maximum current densities of about 13 A/m^2 (both expressed as projected area), evidencing

Figure 10. SEM images of the (a) unmodified carbon felt, (b) RuO_2-coated electrode, and (c) magnification of (b) (from Ref. 56).

a substantial improvement in electron transfer in the presence of lactate-fed *S. decolorationis S12* strain. Power and current densities were demonstrated to scale with the amount of oxide deposited on the carbon-based support, with power values listed today as one of the highest ones for dual-chamber MFCs.

The first study by Lv *et al.*[56] put into evidence the beneficial electrochemical interactions associated to the RuO_2 interface, as well as the ability of the oxide to decrease the anode resistance due to the fast and reversible redox reaction between Ru(IV) and Ru(II). Particularly, on the ability of RuO_2 to store electrons, the authors made a later try on its use as an efficient charge accumulator toward integrated microbial pseudocapacitors.[57] The results supported the role of RuO_2 to increase the anode capacitance respect to bare carbon felt controls, leading to more energy stored in open circuit conditions and higher power dissipated under discharging tests. The study evidenced the relevance on the use of RuO_2 to integrate pseudocapacitive surfaces in the anodic chamber of a MFC. This way, by integrating electrochemically active *S. oneidensis* microorganisms into biocapacitors under anaerobic conditions, interesting energy harvesting biodevices powered by microbial organic oxidation can be envisioned. However, expensive metal oxides (like RuO_2) could hardly be considered for future large-scale bioelectrochemical applications.

Coming near the end of the story on metal oxides, it would seem that, considering the obtained results and compared to metallic bioanodes (Section 13.3), the research on oxide-based platforms would not be entirely justified. However, we have to recognize that unlike metallic electrodes, the field on metal oxides as microbial anodes is still in its infancy, with (from our perspective) many improvements yet to come. In this respect, the year 2015 gave birth to two interesting studies contributing to widen the current knowledge on the subject.[39,59] Compared to previous studies, enhanced power and current generation was achieved with MnO_2-coated carbon felt anodes (3.6 W/m^2; 18.7 A/m^2) when used under a flow-through configuration in a two-chamber MFC, thus surpassing the highest values reported to the publication date.[59] Larger amounts of MnO_2 on the felt anode led to higher capacitance values, which contributed to the stabilization of both the anode performance and power output. Manganese dioxide was also demonstrated to electrochemically interact with mixed bacteria in the acetate-fed anodic compartment of the MFC. This work opened the game toward bioanodes operated under a flow-through mode, which evidenced to boost considerably the global system performance.

At this point in the evolution line, we could raise the following question: *Is it possible counting on a material that meets the essential criteria of high electrical conductivity (such as the case of metals), good biocompatibility (such as oxides), and self-standing capability?* Surprisingly, one of the possible answers leads us back to titanium oxide-based materials, especially to the well-known family of Magnéli phases. On the basis of these ceramic structures, electrically conducting 3D porous scaffolds have recently found application as bioanodes, demonstrating to outcompete the best performances achieved by both carbonaceous and metallic bioanodes in the current state of the art.

13.4.1.2.1 *Magnéli phases as functional microbial anodes*

Magnéli phases consist in non-stoichiometric suboxides of titanium with the general formula Ti_nO_{2n-1}, where n is an integer ranging between 4 and 10.[60,61] These compounds, each of them with a distinctive XRD pattern, have been deeply studied by Magnéli *et al.* in a series of communications published in 1957.[62] As the formula suggests ($Ti_nO_{2n-1} \equiv Ti_2O_3 + (n-2)\ TiO_2$), the crystal structure of Magnéli oxides can be viewed as rutile-type slabs of "infinite" extension and different thickness, separated by shear planes with a corundum-like atomic arrangement. This makes the homologous series of Magnéli phases to show a variation in d band occupation (because of the contribution of the *3d* electron of Ti(III) ions), rendering the most interesting feature of these black/bluish titanium oxide-based phases: high electrical conductivity at room temperature (Figure 11).

With conductivities even higher than those of graphite, high chemical stability and the possibility to be cast in different morphologies and sizes, they have been shown to be appealing

	Resistivity / μΩ.cm
Copper	1.7
Aluminum	2.7
Lead	20.6
Titanium	45
Ti_4O_7	630 ←
Carbon (graphite)	1375

(a)

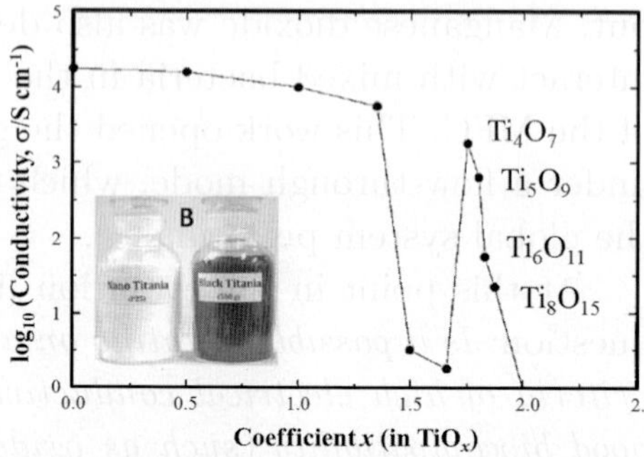

(b)

Figure 11. (a) Room temperature resistivity of Ti_4O_7 compared to those of other typical electrode materials, and (b) phase diagram of the Ti–O system depicting the electrical conductivity of different Magnéli phases. The inset picture shows the typical blue/black aspect of the titanium suboxides compared to common P25 TiO_2 (white powder) (from Ref. 60).

materials for applications in diverse areas such as electrochemistry, electroceramics, resistive switching, catalysis, separation science, and tissue stimulation.[63–68] In fact, electrodes based on Magnéli materials have been commercialized since 2000 as electrodes for batteries, cathodic protection, and water purification under the "Ebonex" trade name (www.ebonex.co.uk).

Basic studies on Magnéli phases have been focused on crystalline structure and physicochemical properties, with a special stress on technological applications and devices.[60] Particularly, from the crystallographic point of view, it has been determined that Magnéli phases are characterized by high concentrations of oxygen vacancies, the resulting atomic structure arranging into reduced ordered lattices. These materials exhibit 2D chains of TiO_6 octahedra joined by the edges and corners forming rutile-type layers, separated (every nth layer, the n corresponding to the chemical formula) by shear planes with a corundum-like atomic feature.[60,61,69] The shear planes represent lattice distortions created because of the accumulation of oxygen vacancies (Figure 12), which, from a bonding point of view (*3d*-Ti/*3d*-Ti metallic bond), leads to an extended electronic conduction along the material at room temperature.

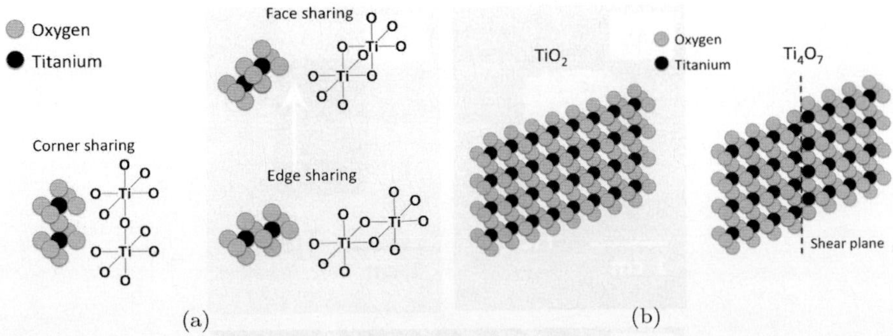

Figure 12. (a) Different orientations of TiO_6 octahedra in Magnéli phase materials (face-, corner- and edge-sharing configurations), and (b) Edge-sharing TiO_6 octahedra in rutile TiO_2, compared to Ti_4O_7 showing a shear plane (dashed line) formed by face-sharing TiO_6 octahedra (from Ref. 61).

Considering the particular structural aspects of Magnéli phases (face-sharing TiO_6 octahedra in shear planes) and those found in crystalline iron(III) oxides (FeO_6 octahedra with a rhombohedral corundum-like structure),[70] a bridge from subsurface environments to Magnéli bioanode applications might be suggested. In this respect, the study by Massazza *et al.* has represented a breakthrough in the area.[39] With the objective of introducing new microbial anode alternatives to boost electrode performance in BESs, the authors proposed (for the first time) the use of Magnéli oxides (Ti_4O_7) to interface electrochemically active microorganisms. Unlike all the previous investigations on metal oxides, this study focused on improving the bioanode performance by taking simultaneously into consideration most of the fundamental points highlighted in section 13.2. From the surface design point of view, Ti_4O_7 platforms were shown to offer high electrical conductivity, excellent biocompatibility toward *G. sulfurreducens* and high surface roughness. Likewise, working on the structure design, 3D electrodes characterized by an open and structured macroporosity with high surface area were introduced as bioscaffolds. The use of an ice-templating technique (so-called ISISA — ice segregation-induced self-assembly) allowed preparing 3D self-standing architectures with macropores (10–15 μm) strongly oriented from one electrode end to the other (Figure 13).

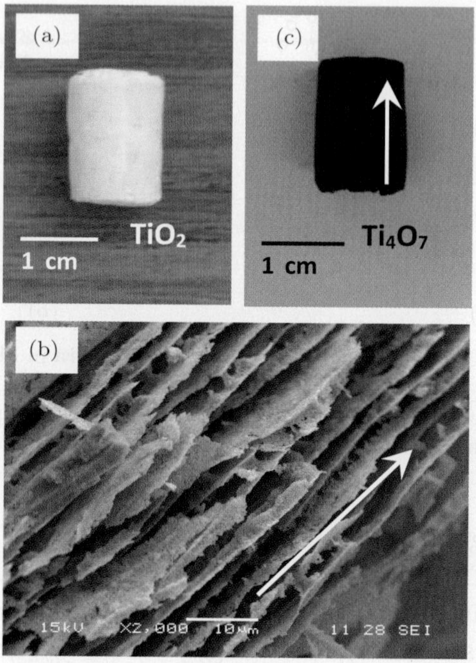

Figure 13. (a) 3D TiO_2 porous scaffold, precursor of Ti_4O_7 electrodes, (b) SEM image of the internal architecture of TiO_2 scaffolds, and (c) Ti_4O_7 electrode obtained by reduction of TiO_2 supports. The arrows in (b) and (c) indicate the orientation of the macroporosity (10–15 μm) generated by ISISA. Aligned pores extended from one electrode end to the other, as evidenced by liquid penetration experiments (from Ref. 39).

All combined, these features allowed the new bioanodes (termed ITTC — *ice templated titanium-based ceramics*) to provide direct paths for acetate supply (along the oriented pores) under a flow-through configuration, while supporting an extended growth of *G. sulfurreducens* microorganisms in anaerobic electrochemical reactors (Figure 14).

Normalized to the projected surface area, the ITTC electrodes provided the highest current densities among those reported for metal oxide-based bioanodes to date (about 129 A/m^2), equating to volumetric current densities of 9.5–10 kA/m^3, unprecedented for self-standing oxide microbial platforms (Figure 15(a)).

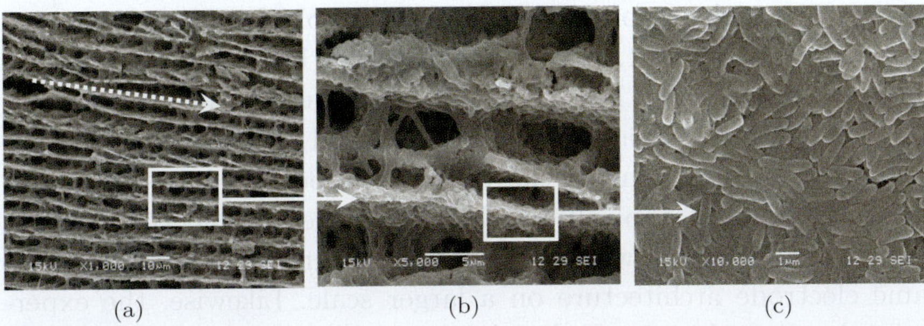

Figure 14. (a) SEM of the internal structure of ITTC after microbial colonization (dotted arrow indicates porosity orientation), (b) closer view of the biofilm adhered to the aligned porosity, and (c) magnification of the biofilm (from Ref. 39).

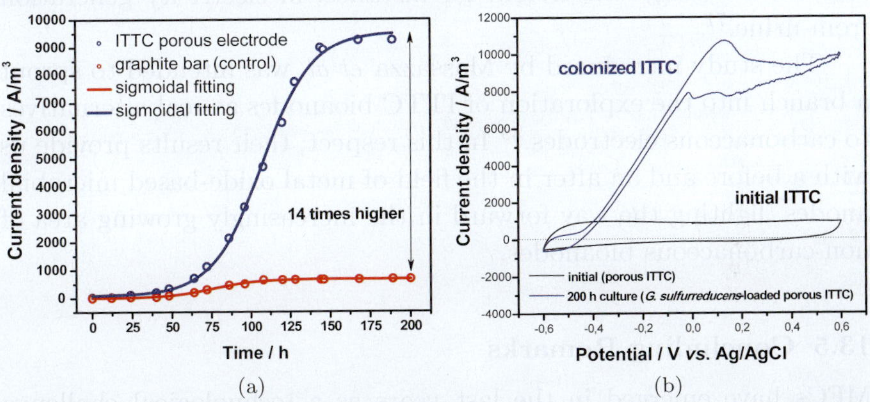

Figure 15. (a) Volumetric current density (A/m^3) over time illustrating growth and electrocatalytic activity of *G. sulfurreducens* on macroporous ITTC, compared to a control graphite bar, and (b) Cyclic voltammograms on ITTC (before and after biofilm formation) (from Ref. 39).

Voltammetry results confirmed not only the biocompatibility of the ITTC electrodes towards the bacterial strain used, but also that they can effectively act as electron acceptors with high efficiency (Figure 15(b)). It is worth mentioning that the new oxide bioanodes remained unaltered in the culture medium during the experimental runs, revealing excellent structural integrity intrinsic to their ceramic nature. The high performance of the ITTC platforms corresponded to power densities of 1.9 kW/m^3 (calculated by assuming a voltage output of 0.2 V under load conditions), surpassing those claimed as the target for sustainable bioelectrochemical technologies. Considering a future upscaling of the ITTC bioanodes facing a large-size bioelectrochemical application, the authors recognized the inherent limitations associated with the processing route (ISISA) to bring the same electrode architecture on a larger scale. Likewise, the experimental approach toward electrodes of bigger size still remains to be seen whether it can produce mechanically stable functional supports. In any cases, the advances revealed by Massazza *et al.* could be directly focused, in a near future, to miniature MFCs (at mL or μL scale). This kind of devices has already proved to be notably efficient in energy conversion; for instance, in electricity generation from urine.[71]

The study introduced by Massazza *et al.* was intended to sprout a branch into the exploration of ITTC bioanodes as real alternatives to carbonaceous electrodes.[39] In this respect, their results provide us with a before and an after in the field of metal oxide-based microbial anodes, lighting the way forward in the increasingly growing area of non-carbonaceous bioanodes.

13.5 Concluding Remarks

MFCs have emerged in the last years as a technological challenge toward biomass energy recovery within a sustainable landscape. The significant bloom in publications on the subject (since 2002–2004 onward) has evidenced an increasing interest in the scientific community. Among the range of components shown to be critical for an optimum performance of the MFC architecture, the electrodes (particularly the anodes) have found a leading role as energy transducers when colonized by electrogenic microbes. However, tracing back the evolution of the bioanodes used in MFCs a clear pattern appears: independently of the configurations proposed, the discussion has almost exclusively been directed to carbonaceous supports.

In this chapter, we have tried to demonstrate that alternative options for microbial anodes (other than carbon-based ones) can be envisioned. With the aim of bringing about a detailed compilation across the current state-of-the-art on *non-carbonaceous bioelectrodes*, we went through the two main families on the subject (metallic and oxide-based electrodes), which should be considered as interesting options in the MFC area facing the future. We have strongly highlighted the requirements for the optimum microbial anode performance, stressing that only the combination of most of these critical features can lead to the production of really competitive microbial anodes in the bioelectrochemical field.

Metallic bioanodes have been mainly employed to support microbial growth because of their higher electrical conductivities compared to other materials. In this respect, and taking into consideration properties as low resistivity, 3D macroporous architectures with high accessible surface area, low-cost and upscaling potentiality, SS felts (and foams, if electrode costs are not considered) seem to have emerged as real alternatives toward BESs.

In the particular case of *metal oxide-based microbial anodes*, we attempted to make a parallel between bacterial oxide respiration in natural anoxic subsurfaces and those processes taking place on the electrode surfaces (with the anode material acting as the final electron acceptor). We consider that this relationship should be taken as an extraordinary opportunity toward the development of advanced bioelectrodes for BESs. In this respect, Ti_4O_7 bioanodes have demonstrated to be reliable and outstanding options to really boost anode performance facing the future.

As a brief summary, Figure 16 depicts a parallel among several of the non-carbonaceous bioanodes compiled in this chapter.

A continuous evolution in electrode complexity toward 3D devices has been evidenced in the last years (from 2012 to 2013 onward . . .). This trend has been accompanied by increasingly higher performances, although some exceptions can also be mentioned (*e.g.* Ni-coated sponges). The evolution on the subject has also shown the appearance of most of the metal oxide bioanodes, leading to a

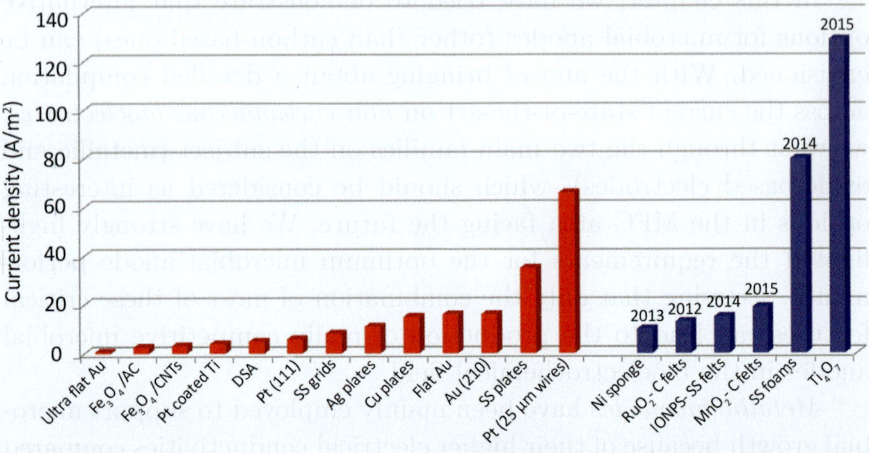

Figure 16. Current density values of different metallic and oxide-based microbial anodes reported to date. Red bars depict 2D platforms. Blue bars indicate 3D scaffolds (in this case, projected surface areas were considered). References: AC (activated carbon), CNTs (carbon nanotubes), DSA (dimensionally stable anodes), SS (stainless steel), IONPs (iron oxide nanoparticles).

research area that is nowadays just opening its eyes toward novel microbial applications.

We would not like concluding without highlighting that, although the performance of a variety of non-carbonaceous materials has been put on the table, we are really aware that due to different reactor designs, culture medium compositions, microbial inoculum sources, and operational conditions, exhaustive conclusions cannot be drawn. Beyond this, we strongly embrace all those promising results on the area, being sure that collective efforts will finally lead to the goal of functional electrodes for microbial electrochemical systems.

The authors of this chapter apologize to all those colleagues whose studies have been unintentionally overlooked when gathering the current literature on the addressed topics.

List of Abbreviations

MFCs microbial fuel cells
BESs bioelectrochemical systems

SAMs	self-assembled monolayers
OMCs	outer membrane cytochromes
SEM	scanning electron microscopy
EET	extracellular electron transfer
UME	ultramicroelectrode
SCE	saturated calomel electrode
SERS	surface-enhanced Raman spectroscopy
COD	chemical oxygen demand
CV	cyclic voltammetry
SS	stainless steel
IONPs	iron oxide nanoparticles
AFM	atomic force microscopy
DSAs	dimensionally stable anodes
PANI	polyaniline
FTO	fluorine-doped tin oxide
XRD	X-ray diffraction
ISISA	ice-segregation induced self-assembly
ITTC	ice-templated titanium-based ceramics
AC	activated carbon
CNTs	carbon nanotubes

References

1. K. Rabaey, L. Angenent, U. Schröder and J. Keller (eds.) (2010). *Bioelectrochemical Systems: From Extracellular Electron Transfer to Biotechnological Application.* IWA Publishing.
2. B.-L. Su, C. Sanchez and X.-Y. Yang (eds.) (2012). *Hierarchically Structured Porous Materials: From Nanoscience to Catalysis, Separation, Optics, Energy and Life Science.* Wiley-VCH Publishing.
3. L. Peng, S.-J. You and J.-Y. Wang (2010). Carbon nanotubes as electrode modifier promoting direct electron transfer from *Shewanella oneidensis. Biosensors and Bioelectronics* **25**(5), 1248–1251.
4. P. Aelterman, S. Freguia, J. Keller, W. Verstraete and K. Rabaey (2008). The anode potential regulates bacterial activity in microbial fuel cells. *Applied Microbiology and Biotechnology* **78**(3), 409–418.
5. V. Flexer, J. Chen, B. Donose, P. Sherrell, G. Wallace and J. Kellera (2013). The nanostructure of three-dimensional scaffolds enhances the current density of microbial bioelectrochemical systems. *Energy & Environmental Science* **6**(4), 1291–1298.

6. X. Xie, C. Criddle and Y. Cui (2015). Design and fabrication of bioelectrodes for microbial bioelectrochemical systems. *Energy & Environmental Science* **8**(12), 3418–3441.
7. S.R. Crittenden, C.J. Sund and J.J. Sumner (2006). Mediating electron transfer from bacteria to a gold electrode via a self-assembled monolayer. *Langmuir* **22**(23), 9473–9476.
8. H. Richter, K. McCarthy, K.P. Nevin, J.P. Johnson, V.M. Rotello and D.R. Lovley (2008). Electricity generation by *Geobacter sulfurreducens* attached to gold electrodes. *Langmuir* **24**(8), 4376–4379.
9. A. Kuzume, U. Zhumaev, J. Li, Y. Fu, M. Füeg, A. Esteve-Nuñez and T. Wandlowski (2013). An in-situ surface electrochemistry approach toward whole-cell studies: Charge transfer between *Geobacter sulfurreducens* and electrified metal/electrolyte interfaces through linker molecules. *Electrochimica Acta* **112**, 933–942.
10. Y. Liu, H. Kim, R. Franklin and D.R. Bond (2010). Gold line array electrodes increase substrate affinity and current density of electricity-producing *G. sulfurreducens* biofilms. *Energy & Environmental Science* **3**(11), 1782–1788.
11. B. Maestro, J.M. Ortiz, G. Schrott, J.P. Busalmen, V. Climent and J.M. Feliu (2014). Crystallographic orientation and electrode nature are key factors for electric current generation by *Geobacter sulfurreducens. Bioelectrochemistry* **98**, 11–9.
12. D. Pocaznoi, B. Erable, M.-L. Delia and A. Bergel (2012). Ultra microelectrodes increase the current density provided by electroactive biofilms by improving their electron transport ability. *Energy & Environmental Science* **5**(1), 5287–5296.
13. F. Qian, M. Baum, Q. Gu and D.E. Morse (2009). A 1.5 microL microbial fuel cell for on-chip bioelectricity generation. *Lab Chip* **9**(21), 3076–81.
14. H.Y. Wang, A. Bernarda, C.Y. Huang, D.J. Lee and J.S. Chang (2011). Micro-sized microbial fuel cell: A mini-review. *Bioresour Technol* **102**(1), 235–243.
15. C. Siu and M. Chiao (2008). A microfabricated PDMS microbial fuel cell. *Journal of Microelectromechanical Systems* **17**(6), 1329–1341.
16. H. Hou, L. Li, P. de Figueiredo and A. Han (2011). Air-cathode microbial fuel cell array: A device for identifying and characterizing electrochemically active microbes. *Biosens Bioelectron* **26**(5), 2680–2684.
17. D. Millo, F. Harnisch, S.A. Patil, H.K. Ly, U. Schröder and P. Hildebrandt (2011). In situ spectroelectrochemical investigation of electrocatalytic microbial biofilms by surface-enhanced resonance Raman spectroscopy. *Angewandte Chemie International Edition* **50**(11), 2625–2627.
18. H. Ly, F. Harnisch, S. Hong, U. Schröder, P. Hildebrandt and D. Millo (2013). Unraveling the interfacial electron transfer dynamics of electroactive microbial biofilms using surface-enhanced Raman spectroscopy. *ChemSusChem* **6**(3), 487–492.
19. N. Law, S. Ansari, F.R. Livens, J.C. Renshaw and J.R. Lloyd (2008). Formation of nanoscale elemental silver particles via enzymatic reduction by

Geobacter sulfurreducens. *Applied Environmental Microbiology* **74**(22), 7090–7093.

20. S. Patil, F. Harnisch and U. Schroder (2010). Toxicity response of electroactive microbial biofilms — a decisive feature for potential biosensor and power source applications. *Chemphyschem* **11**(13), 2834–2837.
21. A. Baudler, I. Schmidt, M. Langner, A. Greiner and U. Schröder (2015). Does it have to be carbon? Metal anodes in microbial fuel cells and related bioelectrochemical systems. *Energy & Environmental Science* **8**, 2048–2055.
22. A. ter Heijne, H.V.M. Hamelers, M. Saakes and C.J.N. Buisman (2008). Performance of non-porous graphite and titanium-based anodes in microbial fuel cells. *Electrochimica Acta* **53**(18), 5697–5703.
23. F. Kargi and S. Eker (2007). Electricity generation with simultaneous wastewater treatment by a microbial fuel cell (MFC) with Cu and Cu–Au electrodes. *Journal of Chemical Technology and Biotechnology* **82**(7), 658–662.
24. X. Zhu and B.E. Logan (2013). Copper anode corrosion affects power generation in microbial fuel cells. *Journal of Chemical Technology and Biotechnology* **89**(3), 471–474.
25. C. Mapelli, V. Mapelli, L. Olsson, D. Mombelli, A. Gruttadauria and S. Barella (2013). Viability study of the use of cast iron open cell foam as microbial fuel cell electrodes. *Advanced Engineering Materials* **15**(3), 112–117.
26. X. Liu, X. Du, X. Wang, N. Li, P. Xu and Y. Ding (2013). Improved microbial fuel cell performance by encapsulating microbial cells with a nickel-coated sponge. *Biosens Bioelectron* **41**, 848–851.
27. C. Dumas, R. Basseguy and A. Bergel (2008). DSA to grow electrochemically active biofilms of *Geobacter sulfurreducens*. *Electrochimica Acta* **53**(7), 3200–3209.
28. B. Erable and A. Bergel (2009). First air-tolerant effective stainless steel microbial anode obtained from a natural marine biofilm. *Bioresource Technology* **100**(13), 3302–3307.
29. S. Ketep, A. Bergel, A. Calmeta and B. Erablea (2014). Stainless steel foam increases the current produced by microbial bioanodes in bioelectrochemical systems. *Energy & Environmental Science* **7**(5), 1633–1637.
30. K. Guo, B.C. Donose, A.H. Soeriyadi, A. Prevoteau, S.A. Patil, S. Freguia, J.J. Gooding and K. Rabaey (2014). Flame oxidation of stainless steel felt enhances anodic biofilm formation and current output in bioelectrochemical systems. *Environmental Science and Technology* **48**(12), 7151–7156.
31. K. Guo, A. Soeriyadi, H. Feng, A. Prévoteau, S. Patil, J. Gooding and K. Rabaey (2015). Heat-treated stainless steel felt as scalable anode material for bioelectrochemical systems. *Bioresource Technology* **195**, 46–50.
32. D.J. Richardson, M.J. Edwards, G.F. White, N. Baiden, R.S. Hartshorne, J. Fredrickson, L. Shi, J. Zachara, A.J. Gates, J.N. Butt and T.A. Clarke (2012). Exploring the biochemistry at the extracellular redox frontier of bacterial mineral Fe(III) respiration. *Biochemical Society Transaction* **40**(3), 493–500.

33. F. Kracke, I. Vassilev and J.O. Kromer (2015). Microbial electron transport and energy conservation — the foundation for optimizing bioelectrochemical systems. *Frontier in Microbiology* **6**, 1–18.
34. D.R. Lovley, S.J. Giovannoni, D.C. White, J.E. Champine, E.J.P. Phillips, Y.A. Gorby and S. Goodwin (1993). *Geobacter metallireducens* gen. nov. sp. nov., a microorganism capable of coupling the complete oxidation of organic compounds to the reduction of iron and other metals. *Archives of Microbiology* **159**(4), 336–344.
35. F. Caccavo, Jr., D.J. Lonergan, D.R. Lovley, M. Davis, J.F. Stolz and M.J. McInerney (1994). *Geobacter sulfurreducens* sp. nov., a hydrogen- and acetate-oxidizing dissimilatory metal-reducing microorganism. *Applied and Environmental Microbiology* **60**(10), 3752–3759.
36. H.A. Lowenstam (1981). Minerals formed by organisms. *Science* **211**(4487), 1126–1131.
37. M. Grantham and P. Dove (1996). Investigation of bacterial-mineral interactions using Fluid Tapping Mode^{TM} atomic force microscopy. *Geochimica et Cosmochimica Acta* **60**(13), 2473–2480.
38. H. Dong, J. Fredrickson, D. Kennedy, J. Zachara, R. Kukkadapu and T. Onstott (2000). Mineral transformations associated with the microbial reduction of magnetite. *Chemical Geology* **169**(3–4), 299–318.
39. D. Massazza, J. Busalmen, R. Parra and H.E. Romeo (2015). New ceramic electrodes allow reaching the target current density in bioelectrochemical systems. *Energy & Environmental Science* **8**(9), 2707–2712.
40. S. Glasauer, S. Langley and T.J. Beveridge (2001). Sorption of Fe (hydr)oxides to the surface of *Shewanella putrefaciens*: Cell-bound fine-grained minerals are not always formed de novo. *Applied and Environmental Microbiology* **67**(12), 5544–5550.
41. S.K. Lower, M.F. Hochella, Jr. and T.J. Beveridge (2001). Bacterial recognition of mineral surfaces: Nanoscale interactions between *Shewanella* and alpha –FeOOH. *Science* **292**(5520), 1360–1363.
42. A.L. Neal, K.M. Rosso, G.G. Geesey, Y.A. Gorby and B.J. Little (2003). Surface structure effects on direct reduction of iron oxides by *Shewanella oneidensis*. *Geochimica Et Cosmochimica Acta* **67**(23), 4489–4503.
43. A. Neal, T. Bank, M. Hochella and K. Rosso (2005). Cell adhesion of *Shewanella oneidensis* to iron oxide minerals: Effect of different single crystal faces. *Geochemical Transactions* **6**(4), 77–84.
44. J.R. Kim, B. Min and B.E. Logan (2005). Evaluation of procedures to acclimate a microbial fuel cell for electricity production. *Applied Microbiology and Biotechnology* **68**(1), 23–30.
45. D.A. Lowy, L.M. Tender, J.G. Zeikus, D.H. Park and D.R. Lovley (2006). Harvesting energy from the marine sediment-water interface II. Kinetic activity of anode materials. *Biosensors and Bioelectronics* **21**(11), 2058–2063.
46. R. Nakamura, F. Kai, A. Okamoto, G.J. Newton and K. Hashimoto (2009). Self-constructed electrically conductive bacterial networks. *Angewandte Chemie Internation Edition English* **48**(3), 508–511.

47. J. Ji, Y. Jia, W. Wu, L. Bai, L. Ge and Z. Gu (2011). A layer-by-layer self-assembled Fe_2O_3 nanorod-based composite multilayer film on ITO anode in microbial fuel cell. *Colloids and Surfaces A: Physicochemical and Engineering Aspects* **390**(1–3), 56–61.
48. X. Peng, H. Yu, X. Wang, Q. Zhou, S. Zhang, L. Geng, J. Sun and Z. Cai (2012). Enhanced performance and capacitance behavior of anode by rolling Fe_3O_4 into activated carbon in microbial fuel cells. *Bioresources and Technology* **121**, 450–453.
49. X. Peng, H. Yu, X. Wang, N. Gao, L. Geng and L. Ai (2013). Enhanced anode performance of microbial fuel cells by adding nanosemiconductor goethite. *Journal of Power Sources* **223**, 94–99.
50. I.H. Park, M. Christy, P. Kim and K.S. Nahm (2014). Enhanced electrical contact of microbes using Fe_3O_4/CNT nanocomposite anode in mediatorless microbial fuel cell. *Biosen Bioelectron* **58**, 75–80.
51. L. Wang, L. Su, H. Chen, T. Yin, Z. Lin, X. Lin, C. Yuan and D. Fu (2015). Carbon paper electrode modified by goethite nanowhiskers promotes bacterial extracellular electron transfer. *Materials Letters* **141**, 311–314.
52. O. Nercessian, S. Parot, M.L. Delia, A. Bergel and W. Achouak (2012). Harvesting electricity with *Geobacter bremensis* isolated from compost. *PLoS One* **7**(3), 1–8.
53. Y. Qiao, S.-J. Bao, C.M. Li, X.-Q. Cui, Z.-S. Lu and J. Guo (2008). Nanostructured polyaniline/titanium dioxide composite anode for microbial fuel cells. *ACS Nano* **2**(1), 113–119.
54. R. Thorne, H. Hu, K. Schneider, P. Bombelli, A. Fisher, L. Peter, A. Dentc and P. Cameron (2011). Porous ceramic anode materials for photo-microbial fuel cells. *Journal of Materials Chemistry* **21**(44), 18055–18060.
55. C. Zhao, W.-J. Wang, D. Sun, X. Wang, J.-R. Zhang and J.-J. Zhu (2014). Nanostructured graphene/TiO_2 hybrids as high-performance anodes for microbial fuel cells. *Chemistry A European Journal* **20**(23), 7091–7097.
56. Z. Lv, D. Xie, X. Yue, C. Feng and C. Wei (2012). Ruthenium oxide-coated carbon felt electrode: A highly active anode for microbial fuel cell applications. *Journal of Power Sources* **210**, 26–31.
57. Z. Lv, D. Xie, F. Li, Y. Hu, C. Wei and C. Feng (2014). Microbial fuel cell as a biocapacitor by using pseudo-capacitive anode materials. *Journal of Power Sources* **246**, 642–649.
58. Y. Wang, B. Li, L. Zeng, D. Cui, X. Xiang and W. Li (2013). Polyaniline/mesoporous tungsten trioxide composite as anode electrocatalyst for high-performance microbial fuel cells. *Biosensors and Bioelectronics* **41**, 582–588.
59. C. Zhang, P. Liang, Y. Jiang and X. Huang (2015). Enhanced power generation of microbial fuel cell using manganese dioxide-coated anode in flow-through mode. *Journal of Power Sources* **273**, 580–583.
60. J. Smith, F. Walsh and R. Clarke (1998). Electrodes based on Magnéli phase titanium oxides: The properties and applications of Ebonex® materials. *Journal of Applied Electrochemistry* **28**(10), 1021–1033.

61. F. Walsh and R. Wills (2010). The continuing development of Magnéli phase titanium sub-oxides and Ebonex® electrodes. *Electrochimica Acta* **55**(22), 6342–6351.
62. S. Andersson, B. Collen, U. kuylenstierna and A. Magneli (1957). Phase analysis studies on the Titanium-Oxygen system. *Acta chemica Scandinavica* **11**, 1641–1652.
63. Z. Luo, S. Sang, Q. Wu and S. Liu (2013). A conductive additive for Zn electrodes in secondary Ni/Zn batteries: The Magneli phase titanium sub-oxides conductive ceramic Ti_nO_{2n-1}. *ECS Electrochemistry Letters* **2**(2), A21–A24.
64. A. Gusev, E. Avvakumov, A. Medvedev and A. Masliy (2007). Ceramic electrodes based on Magneli phases of titanium oxides. *Science of Sintering* **39**(1), 51–57.
65. D.H. Kwon, K.M. Kim, J.H. Jang, J.M. Jeon, M.H. Lee, G.H. Kim, X.S. Li, G.S. Park, B. Lee, S. Han, M. Kim and C.S. Hwang (2010). Atomic structure of conducting nanofilaments in TiO_2 resistive switching memory. *Nat Nanotechnology* **5**(2), 148–153.
66. J.P. Strachan, M.D. Pickett, J.J. Yang, S. Aloni, A.L. David Kilcoyne, G. Medeiros-Ribeiro and R. Stanley Williams (2010). Direct identification of the conducting channels in a functioning memristive device. *Advanced Materials* **22**(32), 3573–3577.
67. A. Kitada, G. Hasegawa, Y. Kobayashi, K. Kanamori, K. Nakanishi and H. Kageyama (2012). Selective preparation of macroporous monoliths of conductive titanium oxides $Ti_nO_{(2n-1)}$ (n = 2, 3, 4, 6). *Journal of American Chemical Society* **134**(26), 10894–10898.
68. M. Canillas, E. Chinarro, M. Carballo-Vila, J. Jurado and B. Moreno (2013). Physico-chemical properties of the Ti_5O_9 Magneli phase with potential application as a neural stimulation electrode. *Journal of Materials Chemistry B* **1**(46), 6459–6468.
69. X. Zhong, I. Rungger, P. Zapol and O. Heinonen (2015). Electronic and magnetic properties of Ti_4O_7 predicted by self-interaction-corrected density functional theory. *Physical Review B* **91**(11), 115143.
70. E. Bykova, L. Dubrovinsky, N. Dubrovinskaia, M. Bykov, C. McCammon, S.V. Ovsyannikov, H.P. Liermann, I. Kupenko, A.I. Chumakov, R. Ruffer, M. Hanfland and V. Prakapenka (2016). Structural complexity of simple Fe_2O_3 at high pressures and temperatures. *Nature Communications* **7**, 10661.
71. I. Ieropoulos, J. Greenman and C. Melhuish (2013). Miniature microbial fuel cells and stacks for urine utilisation. *International Journal of Hydrogen Energy* **38**(1), 492–496.

Section IV

Imaging and Characterization of Bioelectrodes

Chapter 14

Imaging and Characterization of Microbial Electrodes

Yang Lu*,‡ and Bogdan C. Donose†,§
*Advanced Water Management Centre,
The University of Queensland,
Brisbane, QLD 4072, Australia
†School of Chemical Engineering,
The University of Queensland,
Brisbane, QLD 4072, Australia
‡y.lu@awmc.uq.edu.au
§b.donose@uq.edu.au

14.1 Introduction

The imaging of microbial electrodes (ME) requires complex instrumentation, and an entire family of equipment and techniques based on the fluorescence and electron microscopy have been developed over the last few decades. These methods are commonly used by researchers in the field, but the majority of these users need a *pit-stop* document aimed at highlighting state of the art approaches. In this chapter, we are outlining what is currently possible, hoping to aid readers in selecting the most meaningful approach to solve their research puzzles. Each section discusses instrumentation and technology in relationship to application in research. Two of the most common trends are covered: fluorescence imaging (FI) and scanning electron microscopy (SEM).

14.2 Fluorescence Imaging

Biofilms are an important component of the ME and therefore, physical parameters such as surface coverage, biovolume, and biofilm roughness, as well as microbiological parameters such as identification, distribution, and cellular processes are crucial for researchers. Beyond optical imaging, FI is able to provide the aforementioned information.

Fluorescence is a property of a fluorescent molecule releasing the energy absorbed as light of longer wavelength. Fluorescing molecules, including dyes, probes, or proteins, are used as reporter and designed according to the particular application for FI, providing particular experimental observations from cell status, identification, quantification, and distribution to cellular processes. FI enables the differentiation of targets from others, due to target-specific staining or binding, and offers quick *in situ* access, as no pre-extraction is needed. For all these reasons, FI has emerged as a popular method in biological research. In the study of biofilm in ME, three techniques are the most popular: Fluorescent *in situ* hybridization (FISH), LIVE/DEAD staining (LDS), and fluorescent protein-based (FPB) imaging. In the following parts, we will introduce these techniques with examples of their applications in ME research and suggest specific precautions.

14.2.1 Fluorescent *in situ* hybridization

FISH became available in 1989 and is commonly applied in medicine and diagnostic.[1] Application of FISH in microbial studies commonly targets 16S rRNA sequences in microorganism of phylum level to species level.[2] 16S rRNA gene is the most common biomarker due to its properties, including its presence in all living microorganisms (except viruses) and presence of conserved and variable regions. Oligonucleotide probes, typically 18–20 nucleotides long, are designed to target the specific complementary region of 16S rRNA. The fluorescent dye attached to probes then enables detection in fluorescent microscopy. Probes representing different taxonomy levels can be applied to the same microorganism and produce overlaid color. Thus, relative abundances of specific (genus or species) population

Table 1. Common applied FISH probes in ME studies.

Target population(s)	FISH probe-fluorochrome	Probe sequences (5′–3′)	References
Bacteria	*EUB338*	GCT GCC TCC CGT AGG AGT	5
	EUB338II	GCA GCC ACC CGT AGG TGT	6
	EUB338III	GCT GCC ACC CGT AGG TGT	6
Gamma-proteobacteria	*GAM42a*	GCCTTCCCACATCGTTT	7
Geobacter	*GEO1a*	CTC ACG CAC TTC GGG ACC AG	8
	GEO1b	CTC ACG CAC TTC GGG ACC AA	8

within general (phylum or class) community can be are analyzed. The overlaid color also provides opportunity for multiple probes in a single hybridization. For example, Lu *et al.*[3] utilized five probes in a single image for the detection of fermenters, acetoclastic, and hydrogentrophic methanogens.

FISH may not be suitable for high-resolution microbial community characterization, as prior knowledge being required, either metabolic function of interest from literature such as methanogenesis in anaerobic digestion, or microbial profiling from another molecular analysis such as 16S rRNA amplicon sequencing. FISH detects *in situ*, properties such as distribution of microorganisms retained. Moreover, gene copy number does not affect FISH quantification, thus it is usually used to support analysis of PCR-based techniques. While communities of ME are relatively less diverse, FISH became a fast and accurate method to identify, quantify, and localize major functional groups. Common FISH probes used in ME studies are listed in Table 1. FISH was generally conducted with the protocol reported by Amann.[4] 4′,6-Diamidino-2-phenylindole (DAPI) that strongly binds to A-T rich regions of DNA is sometimes used as counterstain for FISH to visualize the whole DNA contents in the sample.

14.2.2 Application of FISH in ME studies

FISH is the most commonly used FI technique in ME studies with variable purposes including the following.

Identifying microorganisms to indicate biofilm development. Villano *et al.*[9] showed successful enrichment by FISH of trichloroethene dechlorinating *Desulfitobacterium* and *cis*-dichloroethene dechlorinating *Dehalococcoides* with H_2 as the sole electron donor, respectively. The presence of general bacteria and specific Betaproteobacteria on the biocathode surface were illustrated in successful development and long-term operation of an autotrophic biocathode using CO_2 as the sole carbon source.[10] Picot[11] compared the attachment of bacteria, specifically *Geobacter,* on modified graphite anode and confirmed the development of an efficient electroactive biofilm on anode reduced by aryl amine groups or positively charged benzylphosphonium groups. Archaea were shown by FISH in the mixed biomass of the upflow MFC[12] due to the excess substrate. Attachments of biofilm to electrode with surface modification were also tested with FISH on 3D base.[13] Biofilm showed different coverage and thickness according to surface charge modification. Researchers also have applied FISH[14] to measure the thickness of homogeneous biofilm developed from domestic wastewater and further confirmed the dominance of *Geobacter spp.* (Figure 1).

Monitoring abundance link to performance. In a bioelectrochemical system fed with glycerol, microbial community was first elucidated by 16S rRNA amplicon sequencing and further monitored by FISH. The decrease of *Citrobacter* followed by reduction of 1,3-PDO production, showing the correlation of *Citrobacter* to glycerol digestion.[15] With FISH examination, *Geobacteraceae* remained constant when MFCs were converted to MECs.[16] However, MECs fed with diary manure wastewater failed to generate current with little *Geobacteraceae*, suggesting the important role of *Geobacteraceae* in the performance of MECs. In the study of cathodic nitrate reduction, FISH was used in conjugation to phylochip with 16S rRNA (*c*DNA) to characterize active microbes in cathode biofilm, showing

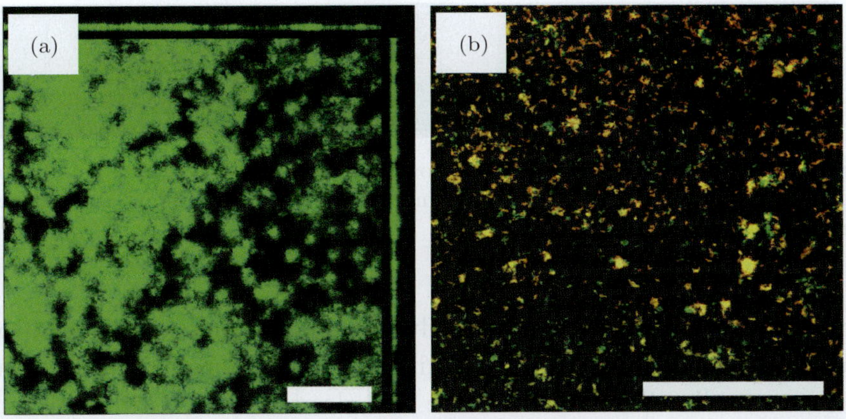

Figure 1. FISH showing a homogenous coverage of bacteria (green) on the electrode (a) with top and lateral inserts representing the orthogonal slices of the biofilm for thickness measurement. (b) Further analysis on the scratched off biofilms detected *Geobacter* (yellow: as an overlay of green and red). (Reproduced from Virdis *et al.*[14]). Scale bars: 100 μm.

a more diverse microbial community with high abundance of *Proteobacteria, Firmicutes*, and *Chloroflexi*. Furthermore, with reactor performance in terms of current production and rates of nitrogen removal, they concluded the richness and evenness corresponds to cathode performance.[17]

Accessing spatial distribution to study function. In the co-culture study of *Clostridium cellulolyticum* and *Geobacter sulfurreducens*, investigating power generation in MFCs from cellulose, Ren *et al.*[18] demonstrated a function-related distribution with cellulose fermentation and electrons transfer happening in suspension by *Clostridium*, and on electrode by *Geobacter*, respectively by FISH and quantitative PCR analysis. Virdis *et al.*[19] also identified a two-layered biofilm (Figure 2) consisting of putative nitrifying bacteria in the outer layer and denitrifying bacteria in the inner layer during simultaneous nitrification and denitrification at a biocathode.

14.2.3 LIVE/DEAD staining

The principle of LDS is that cell membrane of a dead cell is compromised and this enables the penetration of specific stains, such as

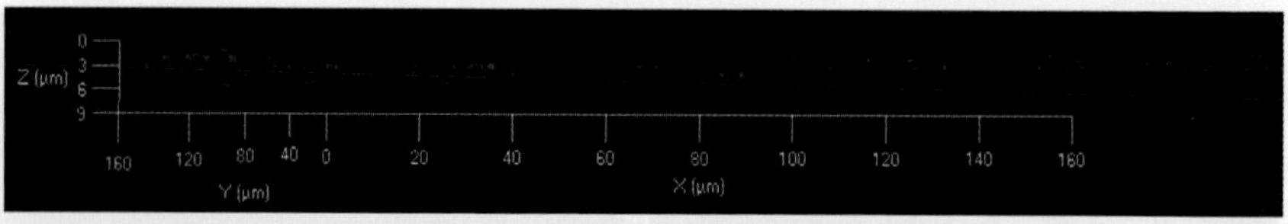

Figure 2. Z-stack confocal images showing the distribution of nitrifying bacteria (green) and denitrifying bacteria (blue) on the cathodic biofilm. (Reproduced from Virdis *et al.*[19]).

propidium iodide, emitting red fluorescence. Other stains, such as acridine orange and SYTO9, emitting green fluorescence, are permeant to all cells no matter live and dead. When both stains are applied, dead cells are differentiated with red fluorescence due to quenching from live cells with only green fluorescence. Fresh samples are required for LDS to minimize viability lost after be removed from the system. However, the reproducibility of the results using this technique is arguable.

LDS was commonly used in ME studies to provide measurements of biofilm including coverage[20] and thickness[21] or identification of the most active part in the biofilm. The structure of dead inner core covered by live outer-layer was commonly observed in microbial fuel cells (Figure 3) with variable settings, suggesting electron transfer are not limited by the presence of dead layer in between metabolically active cells and the electrode surface.[22–24] Electrons can still be transferred long-range via either the conductive pili[25] or the dead inner-layer as conductive matrix.[24] By monitoring the anode biofilm at different levels of current production with LDS, it was reported that *Geobacter sulfurreducens* remains viable regardless to the biofilm thickness and current increased.[26]

14.2.4 Fluorescent protein-based imaging

Proteins with self-fluorescence, green fluorescent protein (GFP) emitting stable green fluorescence, were initially extracted from *Aequorea victoria* jellyfish in the late 1960s and 1970s. GFP fluorescence is species-independent and the gene (*gfp*) encodes GFP is relatively small and non-toxic. Successful expression of GFP has been reported

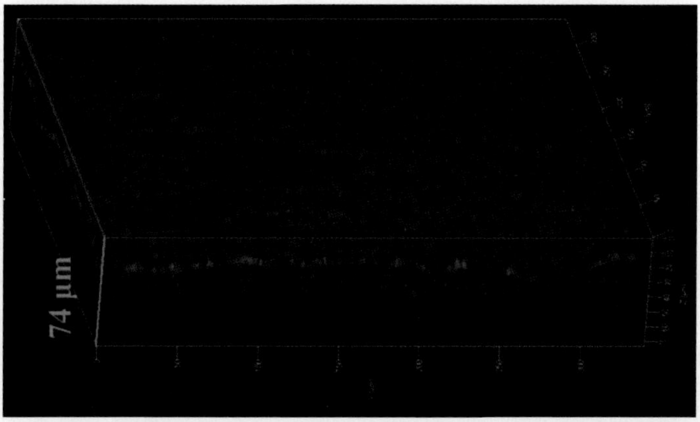

Figure 3. LIVE/DEAD staining shows live outer-layer (green) and dead inner core distribution of anodic biofilm with the measurements of thickness. (Reproduced from Sun *et al.*[24])

in many different types of hosts. The localization and quantification can then be accessed by measuring the green fluorescence emitted by GFP. Compared to FISH, fluorescent protein-based analysis enables real time, non-invasive observation. This is especially important when monitoring the on-line impact of environmental factors to the microbial community. Furthermore, by inserting a gene promoter upstream of the *gfp* gene, promoter activity can be monitored in the same manner.

FPB analysis is mainly used in studies with mono-cultures due to the limitation of genetic modification. Introducing *gfp* gene to single species can be easily performed with assistance from antibiotics. Antibiotics select the successful modified microorganisms as the antibiotic resistance gene is inserted along with the *gfp* gene. In a mixed culture, however, it requires selective insertion of *gfp* gene to target microorganism and non-selective insertion of antibiotic resistance gene to all microorganisms, which is not practically achievable. Oxygen is required for chromophore maturation by most fluorescent proteins, limiting their usage in anaerobic bacteria.

However, recent discoveries facilitate possible application of fluorescent proteins under anaerobic conditions. Fluorescent protein gfp*mut*3 was reported detectable at 0.1 ppm dissolved oxygen.[27]

Codon-modified fluorescent protein expressed in anaerobic bacteria can acquire fluorescence in fixed cells with subsequent exposure to air.[28,29] Followed by the discovery of flavin-mononucleotide-based anaerobic fluorescent protein,[30] more and more anaerobic fluorescent proteins became available.[31–33]

Fluorescent protein-based analysis is limited in the field of ME but extremely useful for tracking the real-time response in pure cultures. Anode biofilm development under different external resistance was recorded by detecting the fluorescent from gene-modified *Shewanella oneidensis* strain MR-1 p519nGFP (constitutively expressed green fluorescent protein) to investigate the link between biofilm structure development and material properties.[34] A comprehensive study[25] was conducted to monitor the real-time change of essential current-producing genes in *Geobacter sulfurreducens* by labeling the bacteria with mCherry and labeling the genes with designed GeoFR. Based on a uniform gene expression profile (Figure 4), they conclude that all cells within the biofilm, including the ones at a great distance to anode, are metabolic active and contribute to current production.

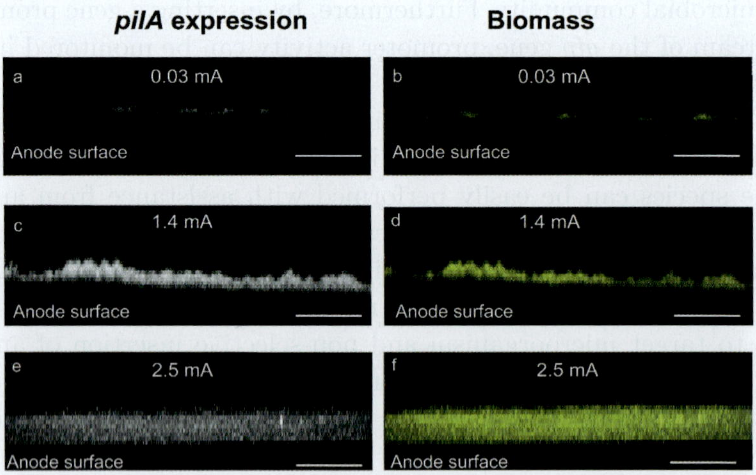

Figure 4. Expressions of GeoFR representing *pilA* promoter (white) and mCherry representing biomass (yellow) are uniform throughout the anode biofilms. (Reproduced from Franks *et al.*[25]). Scale bars: 75 μm.

Precautions and tips

Sample collection. Biofilm samples are normally detached from surface in a sterilized environment and centrifuged to collect cell pellets. Washing with PBS before removal of biofilm may be necessary to avoid contamination from planktonic liquid. Planktonic cells are collected with centrifuge or filtration. Clear images can be obtained from detached biofilms and is better for classification and quantification purpose as the influence of extracellular polymeric substances can be minimized. However, spatial distribution can only be accessed if the biofilm is fixed as a whole. Spatial distribution analyses were reported on ME using flat surface electrode due to handling difficulty with rod shape electrodes. Another caution necessary to reduce disruption is to avoid adding liquid directly to electrode surface during sample preparation to minimize shearing.

Sample visualization. Wide-field epifluorescence microscopy or confocal microscopy are commonly used for sample visualization. Confocal microscopy is able to eliminate out-of-focus noise thus suitable for thick fluorescently labeled biofilms. The ability of confocal microscopy to collect serial optical sections also provides 3D structure of biofilm. Although thickness can be obtained from confocal microscopy, shadow effects due to absorption of light by biofilm material overlying the focal plane should be considered in thickness analysis. An alternative approach for thickness measurement of thicker biofilm is to perform cryosectioning.

Cryosectioning refers to sectioning the biofilm to thinner layers in either vertical or horizontal axis. Embedding with OCT should be performed directly on biofilm and frozen at −20°C. Vertical cryosectioning may be directly performed on biofilm attached to a soft material such as carbon sheet. With harder materials, frozen biofilm may be detached with a sterile razor blade before vertical cryosection or horizontal cryosection should be used instead. Biofilm sections can be attached to glass slides and fluorescent imaging can be performed on the thinner layers to minimize background interference.

Image analysis then provides information including physical biofilm properties, microbial abundance, and distribution. COMSTAT2[35] was successfully used to extract a number of quantitative parameters from biomass such as fractal dimensions, thickness, and surface area. Digital image analysis in microbial ecology (DAIME)[36] and PHLIP toolbox[37] performs additional biovolume fraction to calculate percentage of specific targets from multi-channel imaging with automated image processing.

Following the recent development of molecular sequencing technologies, microbial communities can be commonly studied in detail, from identification and quantification to function. However, fluorescent imaging provides not only additional evidence, but also a powerful way to spatially track viable microorganisms and active gene expressions *in situ* and even *in vivo*. These are of utmost importance for the study of MES in which biofilms are the major point of interest.

14.3 Scanning Electron Microscopy

The publication of one of the earliest known works describing the concept of a SEM by Max Knoll in 1935[38] has drafted the sketch of what had to became among the most popular and influential scientific tools in the present day. Accelerated further developments helped morphing it into a versatile instrument[39–41] serving research fields such as material science, biology, medicine, and nanotechnology, just to name some of the rapidly emerging prominent ones.

ME for microbial bioelectrochemical systems have been created and their investigations started at a point in time[42–44] when SEM was already a mature technique and the benefits stemmed without further delay since sample preparation methods were readily available and operation modes have been exhaustively tested by those exploring the wealthy mines of biological samples.

Depending on the pressure/vacuum level in the sample chamber, SEM techniques can be typically dividend in two: conventional, which requires vacuum and environmental, where samples can be imaged in any pressure/gas conditions.

Table 2. Examples of typical use of SEM for ME imaging.

Organisms	Conventional references	Environmental
Shewanella oneidensis	48	49
Geobacter sulfurreducens	50,51	
Escherichia coli	52	
Pseudomonas aeruginosa	53,54	
Rhodopseudomonas palustris	55	
Alcaligenes species/ Paracoccus species		56
Mixed	11,57,58	59

In conventional SEM, a fine probe of electrons with energies typically in the range of few tens of kV is focused on the specimen following a pattern of parallel lines. The image is collected through processing of various signals generated upon the impact of the incident electrons with the samples surface. These signals are mostly secondary electrons, backscattered high-energy electrons and characteristic X-rays, which may allow elemental analysis.

The environmental mode (ESEM) allows imaging any samples in practically any gaseous environment without the need of conductive coatings and has as its core technological advancement the use of a gaseous detection device (GDD).[45–47] One would expect that the implications for MEs imaging would be tremendous since analyzing wet samples represents a great advantage. As it would be later shown, environmental imaging in MEs is not yet very broadly used, despite obvious advantages.

Table 2 includes a list of most relevant works in pure and/or mixed culture MES employing conventional and/or environmental SEM.

As can be observed from Table 2, the prevalent method for imaging MEs is the conventional mode, but considering the advantages of ESEM, such as minimal preparation requirements and the maintenance of samples state close to their natural/experiment conditions, it is expected that the environmental mode would gain momentum.

14.3.1 Sample preparation

A typical schedule for ME preparations includes *chemical fixation* with diluted solution of glutaraldehyde followed by deionized water postfix wash,[58] *dehydration*, in high vacuum desiccators or gradually increasing ethanol solutions, *mounting* with conductive tape on stubs, followed by conductive layer coating (carbon,[10] platinum, iridium).

As opposed to SEM, ESEM requires only sample fixation (*e.g.* 4 h immersion in buffered glutaraldehyde followed by washing) and deposition on a stub.[49]

Examples of conventional SEM

Examples provided in Figure 5 allow observing biofilm morphological features, such as shape, density, and dimension of the microorganisms (a)–(e), presence of flagella, filaments (d, e) or compact biofilm (f).

Examples of environmental SEM

In addition to the morphological features revealed by conventional SEM, environmental SEM highlights the presence of extracellular polymeric substance (Figure 6(b)) as well as individual organisms (Figure 7) in a state closer to their native configuration.

The conventional SEM technique is now the dominating approach in imaging ME due to its relatively easy sample preparation and high quality of the obtained micrographs. Nevertheless, researchers need to be aware of inducing potential artifacts upon fixation and dehydration, such as the collapse of biofilm scaffolds. Additionally, the high vacuum inside the SEM chamber has the potential to fracture the biofilm layer. All these could lead, for example, to inaccurate biofilm thickness measurements. A potential solution to maintaining the biofilms intact is using the environmental SEM approach, although it comes with more challenges in terms of optimizing the measurement conditions, such as stage temperature and pressure.

Since ESEM avoids the pitfalls of the conventional mode, *i.e.* dehydration and vacuum, it is expected to attract more adepts among

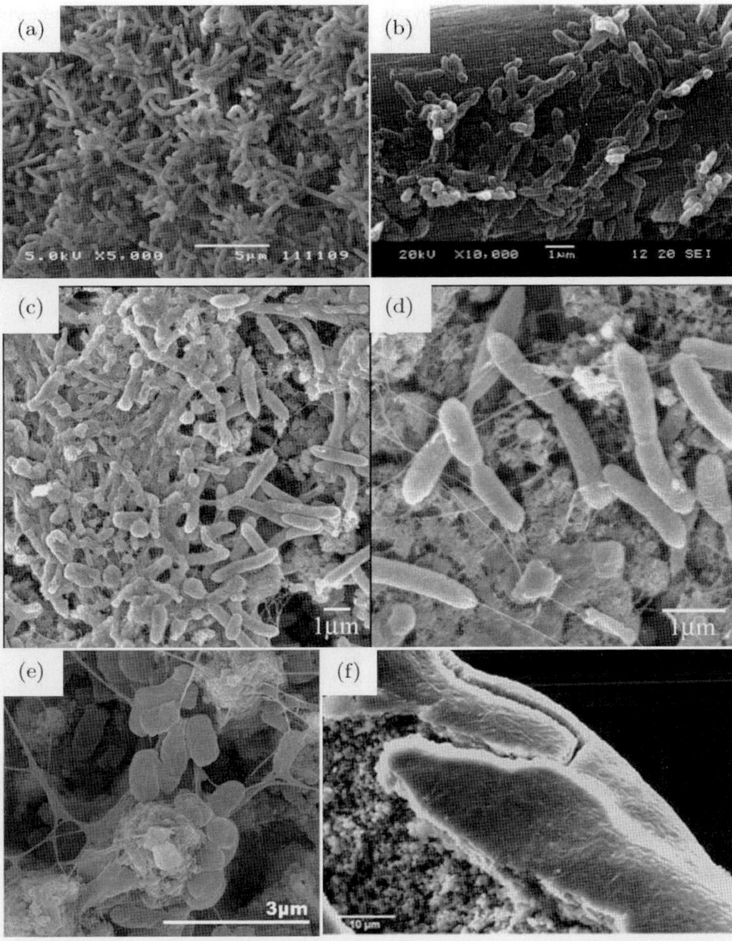

Figure 5. SEM micrographs of ME based on pure and mixed cultures: *G. sulfurreducens* (a[51]), *P. aeruginosa* (b[54]), *E. coli* (c, d[52]) and mixed cultures (e[57], f[58]).

microscopists and bioelectrochemists in their attempts to imaging samples in a state as close as possible to their operational conditions.

Continuous developments in sample preparation together with technological advancements on all fronts of electron microscopy warrant a strong position for SEM as a powerful and reliable ally for researchers in their quest to understand and utilize microbial electrodes.

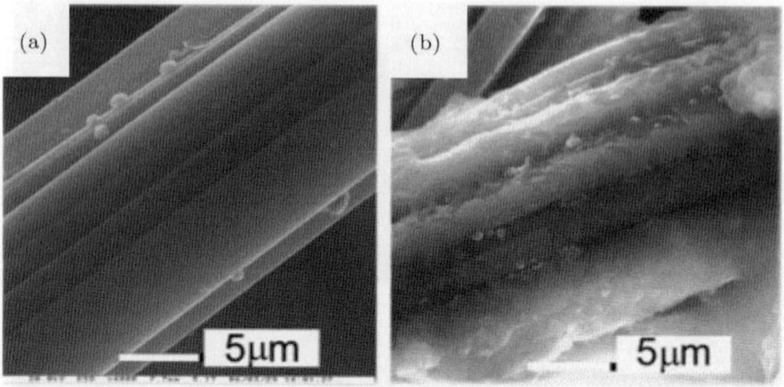

Figure 6. ESEM micrographs of graphite felt exposed to *Shewanella oneidensis* DSP10 (a) and biofilm coated felt (b[49]).

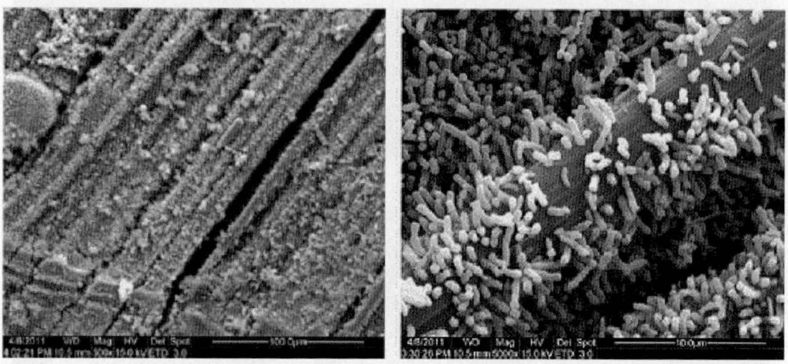

Figure 7. ESEM polyaniline modified carbon cloth mixed culture biofilm (left) and zoom-in (right).[59]

List of Abbreviations

ME	microbial electrodes
FI	fluorescence imaging
SEM	scanning electron microscopy
FISH	fluorescent *in situ* hybridization
rRNA	ribosomal ribonucleic acid
DAPI	4′,6-diamidino-2-phenylindole
MFC	microbial fuel cells
MEC	microbial electrolysis cells
PCR	polymeric chain reaction

GFP	green fluorescent protein
LDS	live/dead staining
ESEM	environmental scanning electron microscope
GDD	gaseous detection device

References

1. E.F. Delong, G.S. Wickham and N.R. Pace (1989). Phylogenetic stains — ribosomal rna-based probes for the identification of single cells. *Science* **243**, 1360–1363.
2. R.I. Amann, W. Ludwig and K.H. Schleifer (1995). Phylogenetic identification and *in-situ* detection of individual microbial-cells without cultivation. *Microbiological Reviews* **59**, 143–169.
3. Y. Lu, F. Slater, R. Bello-Mendoza, and D.J. Batstone (2013). Shearing of biofilms enables selective layer based microbial sampling and analysis. *Biotechnology and Bioengineering* **110**, 2600–2605.
4. R. Amann, B.M. Fuchs and S. Behrens (2001). The identification of microorganisms by fluorescence *in situ* hybridisation. *Current Opinion in Biotechnology* **12**, 231–236.
5. R.I. Amann, L. Krumholz and D.A. Stahl (1990). Fluorescent-oligonucleotide probing of whole cells for determinative, phylogenetic, and environmental-studies in microbiology. *Journal of Bacteriology* **172**, 762–770.
6. H. Daims, A. Bruhl, R. Amann, K.H. Schleifer and M. Wagner (1999). The domain-specific probe EUB338 is insufficient for the detection of all Bacteria: Development and evaluation of a more comprehensive probe set. *Systematic and Applied Microbiology* **22**, 434–444.
7. W. Manz, R. Amann, W. Ludwig, M. Wagner and K.H. Schleifer (1992). Phylogenetic oligodeoxynucleotide probes for the major subclasses of proteobacteria — problems and solutions. *Systematic and Applied Microbiology* **15**, 593–600.
8. S. Demaneche, H. Sanguin, J. Pote, E. Navarro, D. Bernillon, P. Mavingui, W. Wildi, T.M. Vogel and P. Simonet (2008). Antibiotic-resistant. soil bacteria in transgenic plant fields. *Proceedings of the National Academy of Sciences of the United States of America* **105**, 3957–3962.
9. M. Villano, L. De Bonis, S. Rossetti, F. Aulenta and M. Majone (2011). Bioelectrochemical hydrogen production with hydrogenophilic dechlorinating bacteria as electrocatalytic agents. *Bioresource Technology* **102**, 3193–3199.
10. L. Jourdin, S. Freguia, B.C. Donose and J. Keller (2015). Autotrophic hydrogen-producing biofilm growth sustained by a cathode as the sole electron and energy source. *Bioelectrochemistry* **102**, 56–63.
11. M. Picot, L. Lapinsonniere, M. Rothballer and F. Barriere (2011). Graphite anode surface modification with controlled reduction of specific aryl diazonium salts for improved microbial fuel cells power output. *Biosensors & Bioelectronics* **28**, 181–188.

12. Z. He, S.D. Minteer and L.T. Angenent (2005). Electricity generation from artificial wastewater using an upflow microbial fuel cell. *Environmental Science & Technology* **39**, 5262–5267.
13. K. Guo, S. Freguia, P.G. Dennis, X. Chen, B.C. Donose, J. Keller, J.J. Gooding and K. Rabaey (2013). Effects of surface charge and hydrophobicity on anodic biofilm formation, community composition, and current generation in bioelectrochemical systems. *Environmental Science & Technology* **47**, 7563–7570.
14. B. Virdis, D. Millo, B.C. Donose, Y. Lu, D.J. Batstone and J.O. Kromer (2016). Analysis of electron transfer dynamics in mixed community electroactive microbial biofilms. *Rsc Advances* **6**, 3650–3660.
15. M. Zhou, S. Freguia, P.G. Dennis, J. Keller and K. Rabaey (2015). Development of bioelectrocatalytic activity stimulates mixed-culture reduction of glycerol in a bioelectrochemical system. *Microbial Biotechnology* **8**, 483–489.
16. P.D. Kiely, R. Cusick, D.F. Call, P.A. Selembo, J.M. Regan and B.E. Logan (2011). Anode microbial communities produced by changing from microbial fuel cell to microbial electrolysis cell operation using two different wastewaters. *Bioresource Technology* **102**, 388–394.
17. K.C. Wrighton, B. Virdis, P. Clauwaert, S.T. Read, R.A. Daly, N. Boon, Y. Piceno, G.L. Andersen, J.D. Coates and K. Rabaey (2010). Bacterial community structure corresponds to performance during cathodic nitrate reduction. *Isme Journal* **4**, 1443–1455.
18. Z. Ren, L.M. Steinberg and J.M. Regan (2008). Electricity production and microbial biofilm characterization in cellulose-fed microbial fuel cells. *Water Science and Technology* **58**, 617–622.
19. B. Virdis, S.T. Read, K. Rabaey, R.A. Rozendal, Z.G. Yuan and J. Keller (2011). Biofilm stratification during simultaneous nitrification and denitrification (SND) at a biocathode. *Bioresource Technology* **102**, 334–341.
20. Y.C. Yong, Y.Y. Yu, X.H. Zhang and H. Song (2014). Highly active bidirectional electron transfer by a self-assembled electroactive reduced-graphene-oxide-hybridized biofilm. *Angewandte Chemie-International Edition* **53**, 4480–4483.
21. S.M. Strycharz-Glaven, R.H. Glaven, Z. Wang, J. Zhou, G.J. Vora and L.M. Tender (2013). Electrochemical investigation of a microbial solar cell reveals a nonphotosynthetic biocathode catalyst. *Applied and Environmental Microbiology* **79**, 3933–3942.
22. S.T. Read, P. Dutta, P.L. Bond, J. Keller and K. Rabaey (2010). Initial development and structure of biofilms on microbial fuel cell anodes. *Bmc Microbiology*, 10.
23. E.G. Di Domenico, G. Petroni, D. Mancini, A. Geri, L. Di Palma and F. Ascenzioni (2015). Development of electroactive and anaerobic ammonium-oxidizing (anammox) biofilms from digestate in microbial fuel cells. *Biomed Research International*, DOI: 10.1155/2015/ 351014.
24. D. Sun, S.A. Cheng, A.J. Wang, F.J. Li, B.E. Logan and K.F. Cen (2015). Temporal-spatial changes in viabilities and electrochemical properties of anode biofilms. *Environmental Science & Technology* **49**, 5227–5235.

25. A.E. Franks, R.H. Glaven and D.R. Lovley (2012). Real-time spatial gene expression analysis within current-producing biofilms. *Chemsuschem* **5**, 1092–1098.
26. G. Reguera, K.P. Nevin, J.S. Nicoll, S.F. Covalla, T.L. Woodard and D.R. Lovley (2006). Biofilm and nanowire production leads to increased current in Geobacter sulfurreducens fuel cells. *Applied and Environmental Microbiology* **72**, 7345–7348.
27. M.C. Hansen, R.J. Palmer, C. Udsen, D.C. White and S. Molin (2001). Assessment of GFP fluorescence in cells of Streptococcus gordonii under conditions of low pH and low oxygen concentration. *Microbiology-Sgm* **147**, 1383–1391.
28. E.M. Ransom, K.B. Williams, D.S. Weiss and C.D. Ellermeier (2014). Identification and characterization of a gene cluster required for proper rod shape, cell division, and pathogenesis in clostridium difficile. *Journal of Bacteriology* **196**, 2290–2300.
29. E.M. Ransom, C.D. Ellermeier and D.S. Weiss (2015). Use of mCherry red fluorescent protein for studies of protein localization and gene expression in clostridium difficile. *Applied and Environmental Microbiology*, **81**, 1652–1660.
30. T. Drepper, T. Eggert, F. Circolone, A. Heck, U. Krauss, J.K. Guterl, M. Wendorff, A. Losi, W. Gartner and K.E. Jaeger (2007). Reporter proteins for *in vivo* fluorescence without oxygen. *Nature Biotechnology* **25**, 443 445.
31. T. Drepper, R. Huber, A. Heck, F. Circolone, A.K. Hillmer, J. Buchs and K.E. Jaeger (2010). Flavin mononucleotide-based fluorescent reporter proteins outperform green fluorescent protein-like proteins as quantitative *in vivo* real-time reporters. *Applied and Environmental Microbiology* **76**, 5990–5994.
32. J.M. Landete, S. Langa, C. Revilla, A. Margolles, M. Medina and J.L. Arques (2015). Use of anaerobic green fluorescent protein versus green fluorescent protein as reporter in lactic acid bacteria. *Applied Microbiology and Biotechnology* **99**, 6865–6877.
33. J.M. Landete, A. Peiroten, E. Rodriguez, A. Margolles, M. Medina and J.L. Arques (2014). Anaerobic green fluorescent protein as a marker of Bifidobacterium strains. *International Journal of Food Microbiology* **175**, 6–13.
34. J.S. McLean, G. Wanger, Y.A. Gorby, M. Wainstein, J. McQuaid, S.I. Ishii, O. Bretschger, H. Beyenal and K.H. Nealson (2010). Quantification of electron transfer rates to a solid phase electron acceptor through the stages of biofilm formation from single cells to multicellular communities. *Environmental Science & Technology* **44**, 2721–2727.
35. M. Vorregaard (2008). Comstat2 — a modern 3D image analysis environment for biofilms, Technical University of Denmark: Kongens Lyngby, Denmark.
36. H. Daims, S. Lucker and M. Wagner (2006). Daime, a novel image analysis program for microbial ecology and biofilm research. *Environmental Microbiology* **8**, 200–213.
37. L.N. Mueller, J.F.C. de Brouwer, J.S. Almeida, L.J. Stal and J.B. Xavier (2006). Analysis of a marine phototrophic biofilm by confocal laser scanning microscopy using the new image quantification software PHLIP. *BMC Ecology* **6**, 1–15.

38. M. Knoll (1935). Charge potential and secondary emissions of electron irradiated bodies. *Physikalische Zeitschrift* **36**, 861–869.
39. J. Goldstein, D.E. Newbury, D.C. Joy, C.E. Lyman, P. Echlin, L. Lifshin, L. Sawyer and J.R. Michael (2003). Scanning Electron Microscopy and X-ray Microanalysis, New York, 3rd edn.
40. C.W. Oatley (1982). The early history of the scanning electron-microscope. *Journal of Applied Physics* **53**, R1–R13.
41. E. Ruska (1986). The emergence of the electron-microscope — connection between realization and 1st patent application, documents of an invention. *Journal of Ultrastructure and Molecular Structure Research* **95**, 3–28.
42. D.R. Bond, D.E. Holmes, L.M. Tender and D.R. Lovley (2002). Electrode-reducing microorganisms that harvest energy from marine sediments. *Science* **295**, 483–485.
43. S.K. Chaudhuri and D.R. Lovley (2003). Electricity generation by direct oxidation of glucose in mediatorless microbial fuel cells. *Nature Biotechnology* **21**, 1229–1232.
44. B.E. Logan and K. Rabaey (2012). Conversion of wastes into bioelectricity and chemicals by using microbial electrochemical technologies. *Science* **337**, 686–690.
45. G.D. Danilatos (1990). Equations of charge-distribution in the environmental scanning electron-microscope (Esem). *Scanning Microscopy* **4**, 799–823.
46. G.D. Danilatos (1990). Mechanisms of detection and imaging in the Esem. *Journal of Microscopy-Oxford* **160**, 9–19.
47. G.D. Danilatos (1990). Design and construction of an environmental Sem. 4. *Scanning* **12**, 23–27.
48. L. Peng, S.J. You and J.Y. Wang (2010). Electrode potential regulates cytochrome accumulation on Shewanella oneidensis cell surface and the consequence to bioelectrocatalytic current generation. *Biosensors & Bioelectronics* **25**, 2530–2533.
49. J.C. Biffinger, J. Pietron, R. Ray, B. Little and B.R. Ringeisen (2007). A biofilm enhanced miniature microbial fuel cell using Shewanella oneidensis DSP10 and oxygen reduction cathodes. *Biosensors & Bioelectronics* **22**, 1672–1679.
50. D.R. Bond and D.R. Lovley (2003). Electricity production by Geobacter sulfurreducens attached to electrodes. *Applied and Environmental Microbiology* **69**, 1548–1555.
51. H. Richter, K. McCarthy, K.P. Nevin, J.P. Johnson, V.M. Rotello and D.R. Lovley (2008). Electricity generation by Geobacter sulfurreducens attached to gold electrodes. *Langmuir* **24**, 4376–4379.
52. Y. Qiao, S.-J. Bao, C.M. Li, X.-Q. Cui, Z.-S. Lu and J. Guo (2008). Nanostructured polyaniline/titanium dioxide composite anode for microbial fuel cells. *ACS Nano* **2**, 113–119.
53. J. Jass, J.W. Costerton and H.M. Lappinscott (1995). The effect of electrical currents and tobramycin on pseudomonas-aeroginosa biofilms. *Journal of Industrial Microbiology* **15**, 234–242.

54. T. Song, Y. Xu, Y. Ye, Y. Chen and S. Shen (2009). Electricity generation from terephthalic acid using a microbial fuel cell. *Journal of Chemical Technology & Biotechnology* **84**, 356–360.
55. D.F. Xing, Y. Zuo, S.A. Cheng, J.M. Regan and B.E. Logan (2008). Electricity generation by Rhodopseudomonas palustris DX-1. *Environmental Science & Technology* **42**, 4146–4151.
56. P.K. Dutta, K. Rabaey, Z. Yuan, R.A. Rozendal and J. Keller (2010). Electrochemical sulfide removal and recovery from paper mill anaerobic treatment effluent. *Water Research* **44**, 2563–2571.
57. J.E. Mink, J.P. Rojas, B.E. Logan and M.M. Hussain (2012). Vertically grown multiwalled carbon nanotube anode and nickel silicide integrated high performance microsized (1.25 mu L) microbial fuel cell. *Nano Letters* **12**, 791–795.
58. L. Jourdin, S. Freguia, B.C. Donose, J. Chen, G.G. Wallace, J. Keller and V. Flexer (2014). A novel carbon nanotube modified scaffold as an efficient biocathode material for improved microbial electrosynthesis. *Journal of Materials Chemistry A* **2**, 13093–13102.
59. B. Lai, X.H. Tang, H.R. Li, Z.W. Du, X.W. Liu and Q. Zhang (2011). Power production enhancement with a polyaniline modified anode in microbial fuel cells. *Biosensors & Bioelectronics* **28**, 373–377.

Chapter 15

Spectroscopic Methods for Characterizing Redox Chemistry at Metalloprotein-Modified Electrodes

Philip A. Ash and Kylie A. Vincent*
Department of Chemistry, University of Oxford,
Inorganic Chemistry Laboratory,
South Parks Road, Oxford, OX1 3QR
**kylie.vincent@chem.ox.ac.uk*

15.1 Combining Bioelectrochemistry with Spectroscopic Methods — Challenges and Opportunities

A range of spectroscopic methods are available that can be valuable in obtaining information about redox chemistry occurring at metal sites in proteins immobilized on electrodes. Ideally, the current response at the electrode can be correlated with spectroscopic changes at metal sites in the protein to help understand redox events observed electrochemically. However, spectroscopic characterization of bioelectrodes is often difficult because the amount of biomolecule on the electrode may be low, and it is often challenging to combine the requirements of spectroscopic sampling with an optimal electrochemical cell geometry. Further challenges arise when examining catalytic processes at functional metalloenzyme electrodes due to the requirement for effective mass transport of the substrate to the enzyme, and product away from the enzyme, in a spectroelectrochemical cell where diffusion may be constrained. In this section, we address some of the spectroscopic approaches that can be used to examine metalloprotein electrodes, the limitations and challenges in these approaches, and the opportunities for further development in these areas. We focus on studies that examine functional electrodes on which a metalloenzyme is engaging in catalysis.

15.2 UV–Visible Spectroscopy

Many important metalloenzymes contain one or more heme centers, and UV–Vis spectroscopy is ideal for monitoring redox or coordination changes at the heme sites. Several groups have demonstrated the possibility of carrying out *in situ* UV–Vis spectroelectrochemistry on heme proteins immobilized on optically transparent metal oxide electrodes.[1–3] Combined electrochemical and UV–Vis spectroscopic experiments on hemoglobin adsorbed on a TiO_2 electrode were used by Durrant *et al.* to demonstrate a combined spectroscopic and electrochemical method for sensing NO as it binds to the heme sites of the protein.[4] Butt *et al.* demonstrated a UV–Vis spectroelectrochemical approach for studying the pentaheme nitrite reductase NrfA from *Escherichia coli* adsorbed onto high surface area mesoporous SnO_2.[5,6] Figure 1(a) shows the arrangement of heme centers in NrfA, and Figure 1(b) shows electrochemical and UV–Vis spectroscopic data obtained simultaneously.[5] The absorbance at 552 nm reports on the redox state of low spin heme centers, and its evolution with potential fits well to a sum of four Nerstian $n = 1$ contributions from the four low spin electron relay heme centers (red data sets). The absorbance at 442 nm reports on the redox state of high spin heme and fits well to a single $n = 1$ contribution, consistent with the high spin active site heme center (blue data sets). Redox potentials for each of the heme sites extracted from the spectroscopic data correlate well with those obtained from fitting the current response from the simultaneous electrochemical measurement. Thus, these experiments provided new information on the behavior of individual redox sites within this multi-heme enzyme. The high surface area offered by the SnO_2 electrode was significant here, as individual non-turnover

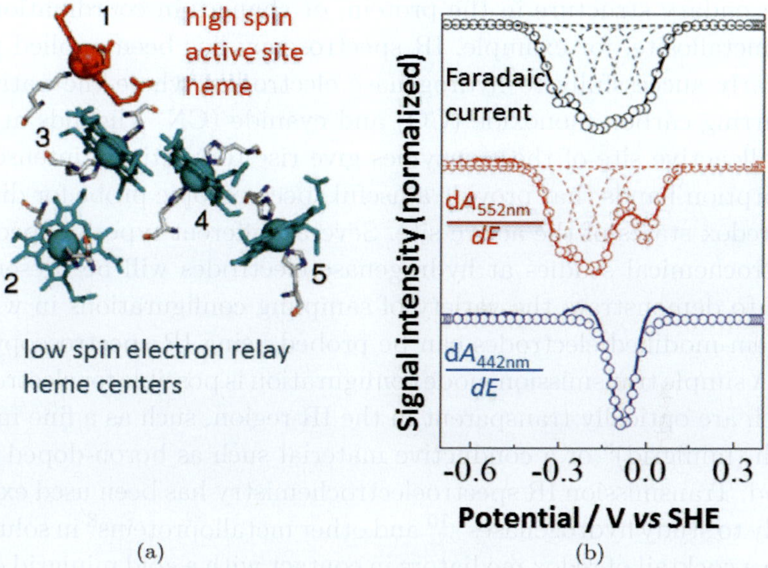

Figure 1. UV–Vis spectroelectrochemistry at a SnO_2 electrode modified with *E. coli* multi-heme nitrite reductase NrfA reveals signals from individual heme centers in the enzyme; (a) arrangement of heme centers in NrfA; (b) Faradaic current and corresponding spectral response with applied potential (solid lines), fitted peaks (dashed lines), and the sum of the fitted peaks (circles). Adapted with permission from Ref. 5. Copyright (2008) American Chemical Society.

signals are not resolved in the current response when the same enzyme is studied at a conventional gold or graphite electrode.[5]

Other optically transparent metal oxides such as indium-doped tin oxide (ITO) may also be suitable for UV–Vis spectroscopic studies of metalloprotein electrodes.[7] Characterization of metalloproteins on metal oxide electrodes offers the possibility of making a direct connection to photoelectrocatalytic applications of metalloenzymes on these materials.

15.3 Infrared Spectroelectrochemistry

Infrared (IR) spectroscopy can provide information on chemical bonds in biomolecules on an electrode. When applied to metalloprotein electrodes, it can be used to provide information on the nature of covalent linkers between an electrode and a biomolecule, the retention of secondary structure in the protein, or changes in coordination at the metallosite, for example. IR spectroscopy has been applied particularly successfully to hydrogenase electrodes, where the natively occurring carbon monoxide (CO) and cyanide (CN^-) ligands at the metalloactive site of these enzymes give rise to relatively intense IR absorption bands that provide a useful spectroscopic probe for different redox states of the active site. Several different types of spectroelectrochemical studies at hydrogenase electrodes will be presented here to demonstrate the variety of sampling configurations in which protein-modified electrodes can be probed using IR spectroscopy.

A simple transmission mode configuration is possible for electrodes which are optically transparent in the IR region, such as a fine metal mesh (minigrid)[8] or a conductive material such as boron-doped diamond. Transmission IR spectroelectrochemistry has been used extensively to study hydrogenases[9,10] and other metalloproteins[8] in solution with a cocktail of redox mediators in contact with a gold minigrid electrode, as well as for hydrogenase trapped in a polymer electrode.[11] Rüdiger *et al.* have described an approach to assembly of hydrogenase electrodes for H_2/O_2 enzyme fuel cells that involves trapping the enzyme in a viologen-functionalized redox polymer (Figure 2(a)).[11] The viologen groups act as redox mediators in a potential range close to their midpoint potential in the polymer, –0.3 V *vs.* SHE, thus ensuring that the hydrogenase is not exposed to oxidizing potentials at which it would normally be inactivated. As shown by comparison of cyclic voltammograms at the viologen polymer electrode to those in which the enzyme exchanges electrons directly at a graphite electrode, the redox polymer appears to confer protection for the enzyme against oxidative inactivation (Figure 2(b)). Rüdiger *et al.* performed *in situ* IR spectroscopic studies to confirm this hypothesis by coating the enzyme–polymer mixture onto an optically transparent gold minigrid working electrode, stepping the potential at this electrode slowly from –0.6 to +0.2 V *vs.* SHE, and allowing the hydrogenase–redox polymer system to equilibrate at each potential before recording spectra.[11] With these measurements, they were able to show that there is no detectable population of the oxidized inactive hydrogenase states known as Ni–A and Ni–B in the redox polymer at any of the applied

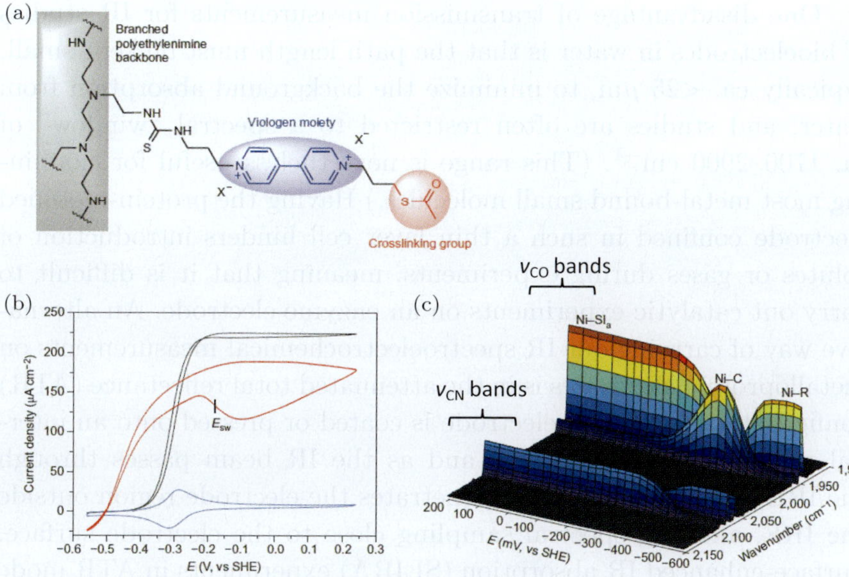

Figure 2. Entrapment of *Desulfvibrio vulgaris* Miyazaki F NiFe hydrogenase in a viologen-modified redox polymer prevents access to inactive oxidized states by limiting the potential experienced by the enzyme. (a) Structure of the viologen-containing polymer; (b) cyclic voltammograms at 20 mV s^{-1} at a rotating disk electrode, showing the response of hydrogenase in the redox hydrogel, under H_2 (black) and inhibitory CO (blue), and (for comparison) immobilized hydrogenase directly exchanging electrons with a highly ordered pyrolytic graphite electrode (red); (c) IR spectra showing that the only states of the hydrogenase active site formed in the polymer are the catalytically active states known as Ni–SI_a, Ni–C, and Ni–R states. Reprinted by permission from Macmillan Publishers Ltd: Nature Chemistry, Ref. 11, Copyright 2014.

potentials under N_2 (Figure 2(c)). The authors also propose a mechanism for protection against O_2 whereby the viologen groups close to the surface of the polymer layer reduce O_2 to water, preventing O_2 penetration and attack on hydrogenase molecules deeper in the polymer matrix. Since the transmission IR spectra report on the entire hydrogenase sample in the hydrogel sample, it was not possible to obtain depth-dependent information to test this further hypothesis. A similar finding on protection of a hydrogenase from O_2 in a 3D electrode was made by Xu and Armstrong in studying a porous carbon electrode loaded with hydrogenase.[12]

One disadvantage of transmission measurements for IR studies of bioelectrodes in water is that the path length must be kept small, typically ca. <25 μm, to minimize the background absorption from water, and studies are often restricted to a spectral "window" of ca. 1700–2900 cm^{-1}. (This range is nevertheless useful for examining most metal-bound small molecules.) Having the protein-modified electrode confined in such a thin layer cell hinders introduction of solutes or gases during experiments, meaning that it is difficult to carry out catalytic experiments on an enzyme electrode. An alternative way of carrying out IR spectroelectrochemical measurements on metalloprotein electrodes is in the attenuated total reflectance (ATR) configuration. Here, the electrode is coated or pressed onto an internal reflection element (IRE), and as the IR beam passes through the IRE, an evanescent wave penetrates the electrode region outside the IRE, allowing spectral sampling close to the electrode surface. Surface-enhanced IR absorption (SEIRA) experiments in ATR mode typically employ a very thin gold layer on the IRE as the bioelectrode. Biomolecules can be attached to the gold via self-assembled thiolate monolayers selected to suit the surface properties of the protein of interest. Nanostructure in the thin gold electrode layer gives rise to a surface enhancement effect, which increases the IR signal intensity from species within a few nanometers of the gold surface, and hence ATR-SEIRA is a near-field spectroscopic technique. This permits (sub)monolayers of protein on gold to be probed, as well as the electrode–protein interface. SEIRA spectroscopy has been applied to a number of electrode-immobilized metalloproteins to examine, for example, small molecule ligands at metal sites such as the CO and CN^- in hydrogenases,[13] structural changes in proteins revealed by changes in the amide vibrational bands,[14] or orientation changes in immobilized proteins.[15]

ATR-IR spectroscopy has also been used to study enzymes immobilized by direct adsorption at a carbon electrode in an approach termed protein film IR electrochemistry (PFIRE).[16,17] In this case, the IR signal from the protein is increased by utilizing a very high surface area carbon electrode assembled from carbon black particles.

This has been applied to study NiFe hydrogenases under catalytic turnover controlled at the electrode (Figure 3).[16] An ATR-IR flow cell was designed to provide efficient mass transport for supply of substrate (H_2) to the hydrogenase under rapid catalytic turnover, and cyclic voltammograms recorded under 1 bar H_2 closely resemble those of hydrogenase on a rotating disk electrode (Figure 3(a)). Spectra recorded in the H_2 oxidation region under both argon (non-turnover conditions) and H_2 (turnover conditions) (Figure 3(c)) reveal important differences. In Figure 3(c), only the CO stretching region of the spectrum is shown, and since the enzyme has a single CO ligand, each band represents a different redox or coordination state of the active site within the enzyme. Under the inert atmosphere, the hydrogenase exists entirely in the oxidized inactive state known as Ni–B at this potential. In contrast, the range of states present during turnover under H_2 must represent states involved in catalysis. The states termed Ni–R, Ni–C, and Ni–SI are well-established as catalytic intermediates for NiFe hydrogenases, but importantly, the presence of the low wavenumber pair of bands assigned to the "Ni–L" state(s), (see Figure 3(d)) suggests that a Ni(I) intermediate may also be involved.[16] These studies, therefore, provided insight into the mechanism of H_2 oxidation by NiFe hydrogenases.

15.4 Surface Enhanced Resonance Raman Spectroscopy

Resonance Raman spectroscopy can be very sensitive to structural features of metalloproteins immobilized on a nanostructured silver electrode because resonance enhancement by laser excitation close to an electronic transition in the protein is combined with signal enhancement due to plasmonic effects arising at the electrode surface. This approach has been used most widely for heme proteins, where UV–Vis excitation in the Soret band or Q-band regions enhances vibrational bands mainly associated with C–C and C–N stretching modes of the porphyrin ring in the 1,700–1,300 cm^{-1} region.[18] These are very sensitive to the redox state, spin state, and coordination

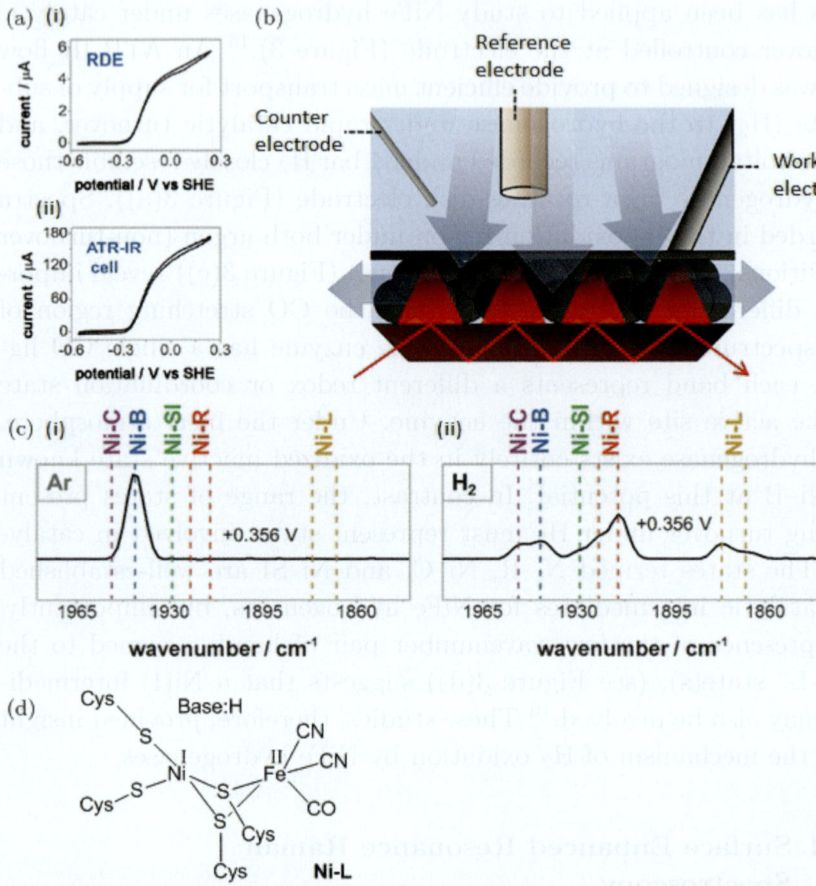

Figure 3. Protein film IR electrochemistry permits study of metalloenzymes on a carbon electrode under catalytic turnover conditions: experiments on NiFe hydrogenase I from *E. coli* adsorbed on a carbon black BP2000 electrode. (a) Comparison of cyclic voltammograms recorded (i) on a pyrolytic graphite rotating disk electrode; and (ii) on carbon black in the ATR-IR cell, both at 10 mV s^{-1}, pH 6.0 and under 1 bar H_2. (b) Schematic representation of the ATR-IR cell showing the carbon black particle working electrode, the solution compartment containing the counter and reference electrodes, and the flow of solution through the cell. (c) ATR-IR spectra recorded under Ar and H_2 (as indicated) at +0.356 V showing the v_{CO} bands from different states of the hydrogenase active site under non-turnover and turnover conditions, respectively. (d) Comparison of spectra under non-turnover and turnover conditions provide strong evidence for a Ni(I) state known as Ni–L in the catalytic cycle; two forms of Ni–L are observed in the spectra, likely to be due to different protonation states close to the metallocenter. Panels (a)–(c) are reproduced from Ref. 16 with permission, John Wiley and Sons, copyright 2015.

state of the heme and are therefore useful in identifying any changes that accompany immobilization of a heme protein, and it is also possible to extract information on heme protein orientation on the electrode.[18]

As an example, resonance Raman spectroscopy was used by Silveira *et al.* to examine a cytochrome cd_1 nitrite reductase and its physiological redox partner, cytochrome c_{552}, to investigate structural changes around the heme centers that often seem to impede electroactive protein immobilization.[19] Cytochrome cd_1 nitrite reductase was immobilized onto a mixed self-assembled monolayer on the silver surface. The 413 nm laser excites in the Soret band region and is focused on a small spot on the electrode. Rotation of the electrode during experiments minimizes the length of light exposure to a given part of the immobilized protein sample. As shown in Figure 4, spectra of cytochrome cd_1 were recorded both in solution (resonance Raman, RR spectra) and with the enzyme immobilized on a self-assembled monolayer on the silver electrode (surface enhanced resonance Raman, SERR spectra). For both the oxidized and the reduced enzyme, there is a fairly good correspondence between band positions of characteristic heme porphyrin vibrational modes known as ν_4, ν_3, ν_2, and ν_{10} for the RR spectra of the enzyme in solution and SERR spectra of the immobilized enzyme, showing that the native structure is retained upon immobilization. The authors note that the absence of band broadening in SERR spectra also indicates a uniform orientation of the immobilized enzyme. Disadvantages of the resonance Raman approach arise from the need for intense laser irradiation of the sample, which can lead to photodamage or photoreduction, or simply prevent examination of photosensitive states.

15.5 Fluorescence Methods

Fluorescence spectroelectrochemical approaches have been applied to several type 1/blue copper containing metalloproteins immobilized at gold electrodes to monitor, indirectly, the redox state of a metal site within the protein in response to electrode potential or catalytic turnover.[20,21] This approach relies upon the observation that Förster

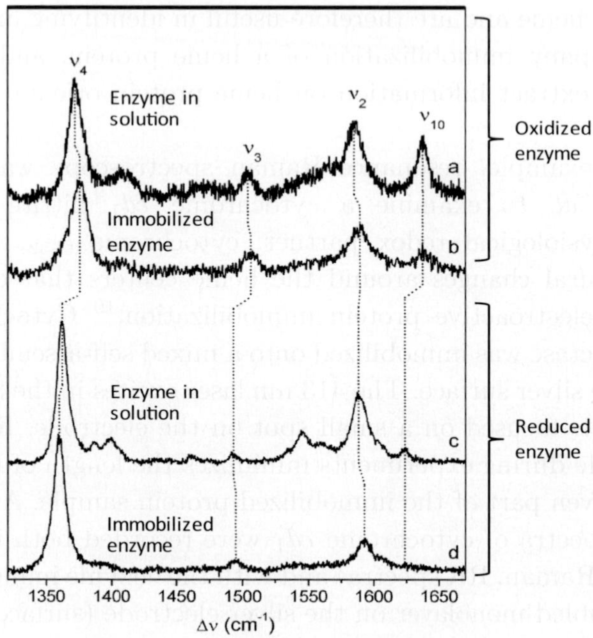

Figure 4. Comparison of cytochrome cd_1 in solution and immobilized on a silver electrode surface modified with a mixed self-assembled monolayer. (a) Oxidized ferric enzyme; and (c) sodium ascorbate reduced ferrous enzyme: solution resonance Raman (RR) spectra; (b) +300 mV *vs.* SHE; and (d) −300 mV *vs.* SHE; surface enhanced resonance Raman (SERR) spectra of the enzyme immobilized on a silver electrode surface. The spectra were recorded with 413 nm excitation with laser power and accumulation time of 2–3 mW and 40 s (RR) or 1.5–2.5 mW and 30 s (SERR). Reproduced (with figure labels added) with permission from Ref. 19, "SERR Spectroelectrochemical Study of Cytochrome cd_1 Nitrite Reductase Co-immobilized with physiological Redox Partner Cytochrome c_{552} on Biocompatible Metal Electrodes". Copyright: 2015 Silveira *et al.* This is an open access article distributed under the terms of the Creative Commons Attribution License.

resonance energy transfer (FRET) lowers the fluorescence intensity of a dye introduced at a specific location in a protein when a nearby copper site is in the oxidized state, Cu(II), and that FRET is lowered (dye fluorescence increases) when the metal is in the reduced Cu(I) state. These approaches require modification of the protein with a fluorescent dye, but have the advantage of being incredibly sensitive, even down to the single molecule level.[22]

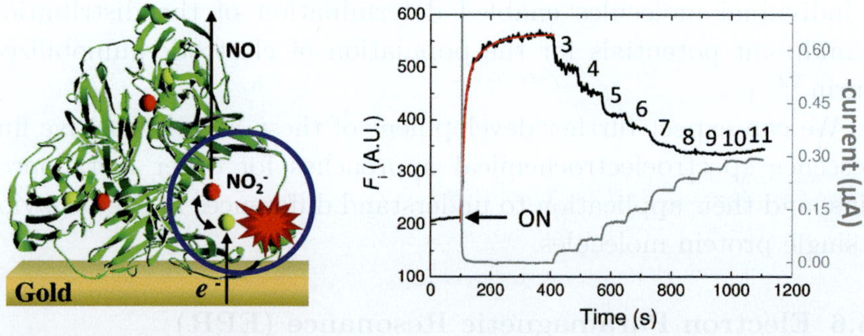

(a) (b)

Figure 5. Fluorescence spectroelectrochemistry applied to a copper nitrite reductase modified with dye. (a) The dye, "ATTO 565" (red star) is attached at a cysteine site close to one of the type 1/"blue" copper centers (yellow sphere). The catalytic type 2 copper site is shown as a red sphere. (b) The potential is stepped to −200 mV *vs.* Ag/AgCl at 100 s ('ON'), which reduces the type 1 Cu center, switching on fluorescence. Injection of nitrite initiates catalysis, and subsequent nitrite injections cause stepwise increases in the magnitude of the catalytic current. FRET-induced loss of fluorescence intensity indicates increasing oxidation of the type 1 Cu centers accompanying these steps. Adapted with permission from Ref. 21. Copyright (2011) American Chemical Society.

Jeuken *et al.* used a fluorescence approach to probe heterogeneity in the rate of interfacial electron transfer to the type 1 copper centers of a homotrimeric copper nitrite reductase immobilized on a self-assembled monolayer (SAM) on a gold electrode, and to assess the oxidation state of the type 1 copper during enzyme catalysis (Figure 5). Fitting of kinetic data indicated two interfacial electron transfer rates, $k_{\mathrm{ET}\,1/3}$ and $k_{\mathrm{ET}\,2/3}$, consistent with one-third of the monomers in the trimeric enzyme being closer to the electrode than the other two-thirds.[21]

Akkilic *et al.* have demonstrated that a similar concept can be used for detection of electron transfer to single protein molecules on SAM-modified gold electrodes.[22] They prepared a sparse coverage of azurin, modified with a fluorescent dye called Cy-5, on a SAM-modified gold electrode, and were then able to observe electron transfer to individual azurin molecules based on quenching of dye fluorescence when the type 1 Cu center of azurin became oxidized. Analysis of fluorescence changes with electrode potential for a series of individual molecules enabled determination of the distribution in midpoint potentials for the population of electrode-immobilized azurin.[22]

We can expect further development of these highly sensitive fluorescence spectroelectrochemical approaches for other metalloproteins, and their application to understand differences in the behavior of single protein molecules.

15.6 Electron Paramagnetic Resonance (EPR)

EPR spectroscopy is applied widely in potentiometric titrations of metalloprotein solutions, providing midpoint potentials for transitions between redox states and enabling characterization of specific paramagnetic states of metalloproteins. The sampling demands of the technique, and the fact that many metalloprotein studies are carried out at low temperature, make it difficult to apply EPR spectroscopy directly to metalloprotein electrodes. However, Polliotto *et al.* have reported EPR spectra of a FeFe hydrogenase enzyme immobilized on TiO_2 engaging in light-triggered electron uptake from the metal oxide surface[23]; although this work does not involve an electrode, it does make important links to related work directed towards developing hydrogenase-modified TiO_2 electrodes in photoelectrochemical devices. This work thus shows the feasibility of conducting EPR experiments on immobilized metalloenzymes in direct electron transfer contact with a surface.

15.7 X-ray Absorption Spectroscopy

X-ray absorption spectroscopy methods are likely to be particularly difficult to apply to electrode-immobilized metalloproteins due to the risk of X-ray induced photoreduction of static samples, and the fact that experiments are therefore typically conducted on frozen samples. Best *et al.* have demonstrated redox control of cytochrome *c* in solution at a reticulated vitreous carbon working electrode in an EXAFS flow cell (Figure 6(a)),[24] which provides an indication that it may be possible to develop X-ray absorption spectroscopy approaches for protein electrodes in the future. Driven by two syringe pumps, the cytochrome *c* solution flows back-and-forth through the sampling

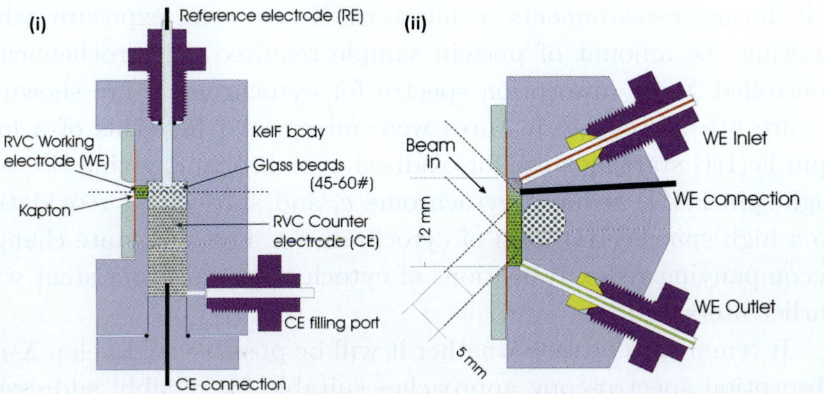

(a)

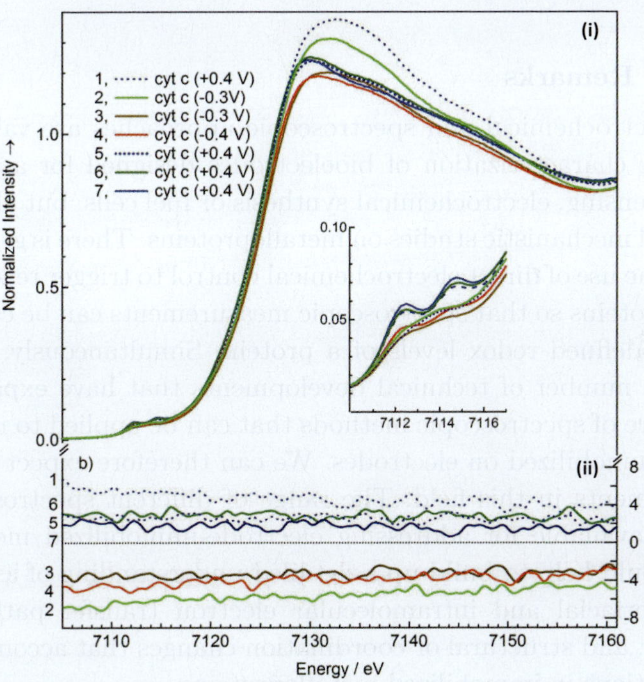

(b)

Figure 6. Scope for metalloprotein X-ray absorption spectroelectrochemistry. (a) Vertical (i) and horizontal (ii) sections through the working electrode compartment of a fluorescence X-ray absorption spectroelectrochemical cell. Solution flow through the cell during measurement is controlled by syringe pumps. (b) Results obtained for cytochrome *c* (5 mM) with 4,4′-bipyridine as redox mediator. Reproduced with permission from Ref. 24. Copyright (2016) International Union of Crystallography.

cell during measurements to minimize X-ray beam exposure, while limiting the amount of protein sample required. Electrochemically controlled X-ray absorption spectra for cytochrome *c* are shown in Figure 6(b). Pre-edge features were interpreted in terms of a low-spin Fe(III) starting state for oxidized cytochrome *c*, giving rise to a high-spin Fe(II) reduced cytochrome *c*, and subsequent reoxidation to a high-spin Fe(III) form of cytochrome *c*. The spin state changes accompanying redox transitions of cytochrome *c* are consistent with earlier literature.

It remains to be seen whether it will be possible to develop X-ray absorption spectroscopy approaches suitable for reliably addressing electrode-immobilized metalloprotein samples.

15.8 Final Remarks

Coupled electrochemical and spectroscopic approaches are valuable both for the characterization of bioelectrodes designed for applications in biosensing, electrochemical synthesis or fuel cells, but also for fundamental mechanistic studies on metalloproteins. There is growing interest in the use of direct electrochemical control to trigger reactions of metalloproteins so that spectroscopic measurements can be carried out on well-defined redox levels of a protein. Simultaneously, there have been a number of technical developments that have expanded the repertoire of spectroscopic methods that can be applied to metalloproteins immobilized on electrodes. We can therefore expect ongoing developments in this field. The range of different spectroscopic approaches available for addressing electrode-immobilized metalloproteins should help to build up a detailed understanding of aspects such as interfacial and intramolecular electron transfer pathways and kinetics, and structural or coordination changes that accompany redox transitions in immobilized metalloproteins.

List of Abbreviations

ATR	attenuated total reflectance
EPR	electron paramagnetic resonance
EXAFS	extended X-ray absorption fine structure

FRET	Förster resonance energy transfer
IRE	internal reflection element
ITO	indium-doped tin oxide
PFIRE	protein film infrared electrochemistry
SAM	self-assembled monolayer
SEIRA	surface-enhanced infrared absorption
SERR	surface-enhanced resonance Raman
SHE	standard hydrogen electrode

References

1. E. Topoglidis, Y. Astuti, F. Duriaux, M. Grätzel and J. R. Durrant (2003), *Langmuir, Direct electrochemistry and nitric oxide interaction of heme proteins adsorbed on nanocrystalline tin oxide electrodes*, **19**, 6894.
2. G. L. Kemp, S. J. Marritt, X. Li, J. R. Durrant, M. R. Cheesman and J. N. Butt (2009), *Biochemical Society Transactions, Opportunities for mesoporous nanocrystalline SnO_2 electrodes in kinetic and catalytic analyses of redox proteins*, **37**, 368.
3. C. Renault, V. Balland, E. Martinez-Ferrero, L. Nicole, C. Sanchez and B. Limoges (2009), *Chemical Communications, Highly ordered transparent mesoporous TiO_2 thin films: an attractive matrix for efficient immobilization and spectroelectrochemical characterization of cytochrome c*, 7494.
4. E. Topoglidis, C. J. Campbell, A. E. G. Cass and J. R. Durrant (2006), *Electroanalysis, Nitric oxide biosensors based on the immobilization of hemoglobin on mesoporous titania electrodes*, **18**, 882.
5. S. J. Marritt, G. L. Kemp, L. Xiaoe, J. R. Durrant, M. R. Cheesman and J. N. Butt (2008), *Journal of the American Chemical Society, Spectroelectrochemical characterization of a pentaheme cytochrome in solution and as electrocatalytically active films on nanocrystalline metal-oxide electrodes*, **130**, 8588.
6. C. W. J. Lockwood, B. Burlat, M. R. Cheesman, M. Kern, J. Simon, T. A. Clarke, D. J. Richardson and J. N. Butt (2015), *Journal of the American Chemical Society, Resolution of key roles for the distal pocket histidine in cytochrome c nitrite reductases*, **137**, 3059.
7. I. Ashur and A. K. Jones (2012), *Electrochimica Acta, Immobilization of azurin with retention of its native electrochemical properties at alkylsilane self-assembled monolayer modified indium tin oxide*, **85**, 169.
8. D. Moss, E. Nabedryk, J. Breton and W. Mäntele (1990), *European Journal of Biochemistry, Redox-linked conformational changes in proteins detected by a combination of infrared spectroscopy and protein electrochemistry*, **187**, 565.
9. B. Bleijlevens, F. A. van Broekhuizen, A. L. De Lacey, W. Roseboom, V. M. Fernandez and S. P. J. Albracht (2004), *Journal of Biological Inorganic Chemistry, The activation of the [NiFe]-hydrogenase from Allochromatium vinosum. An infrared spectro-electrochemical study*, **9**, 743.

10. A. L. DeLacey, C. Stadler, V. M. Fernandez, E. C. Hatchikian, H.-J. Fan, S. Li and M. B. Hall (2002), *Journal of Biological Inorganic Chemistry, IR spectroelectrochemical study of the binding of carbon monoxide to the active site of Desulfovibrio fructosovorans Ni-Fe hydrogenase*, **7**, 318.
11. N. Plumeré, O. Rüdiger, A. A. Oughli, R. Williams, J. Vivekananthan, S. Pöller, W. Schuhmann and W. Lubitz (2014), *Nature Chemistry, A redox hydrogel protects hydrogenase from high-potential deactivation and oxygen damage*, **6**, 822.
12. L. Xu and F. A. Armstrong (2015), *RSC Advances, Pushing the limits for enzyme-based membrane-less hydrogen fuel cells - achieving useful power and stability*, **5**, 3649.
13. D. Millo, P. Hildebrandt, M.-E. Pandelia, W. Lubitz and I. Zebger (2011), *Angewandte Chemie International Edition, SEIRA spectroscopy of the electrochemical activation of an immobilized [NiFe] hydrogenase under turnover and non-turnover conditions*, **50**, 2632.
14. A. Schwaighofer, C. Steininger, David M. Hildenbrandt, J. Srajer, C. Nowak, W. Knoll and Renate L. Naumann (2013), *Biophysical Journal, Time-resolved surface-enhanced IR-absorption spectroscopy of direct electron transfer to cytochrome c oxidase from R. sphaeroides*, **105**, 2706.
15. P. Kielb, M. Sezer, S. Katz, F. Lopez, C. Schulz, L. Gorton, R. Ludwig, U. Wollenberger, I. Zebger and I. M. Weidinger (2015), *ChemPhysChem, Spectroscopic observation of calcium-induced reorientation of cellobiose dehydrogenase immobilized on electrodes and its effect on electrocatalytic activity*, **16**, 1960.
16. R. Hidalgo, P. A. Ash, A. J. Healy and K. A. Vincent (2015), *Angewandte Chemie International Edition, Infrared spectroscopy during electrocatalytic turnover reveals the Ni-L active site state during* H_2 *oxidation by a NiFe hydrogenase*, **54**, 7110.
17. P. A. Ash, J. Liu, N. Coutard, N. Heidary, M. Horch, I. Gudim, T. Simler, I. Zebger, O. Lenz and K. A. Vincent (2015), *Journal of Physical Chemistry B, Electrochemical and infrared spectroscopic studies provide insight into reactions of the NiFe regulatory hydrogenase from Ralstonia eutropha with* O_2 *and CO*, **119**, 13807.
18. D. H. Murgida and P. Hildebrandt (2008), *Chemical Society Reviews, Disentangling interfacial redox processes of proteins by SERR spectroscopy*, **37**, 937.
19. C. M. Silveira, P. O. Quintas, I. Moura, J. J. G. Moura, P. Hildebrandt, M. G. Almeida and S. Todorovic (2015), *PLOS ONE, SERR spectroelectrochemical study of cytochrome* cd_1 *nitrite reductase co-immobilized with physiological redox partner cytochrome* c_{552} *on biocompatible metal electrodes*, DOI:10.1371/journal.pone.0129940.
20. J. M. Salverda, A. V. Patil, G. Mizzon, S. Kuznetsova, G. Zauner, N. Akkilic, G. W. Canters, J. J. Davis, H. A. Heering and T. J. Aartsma (2010), *Angewandte Chemie International Edition, Fluorescent cyclic voltammetry*

of immobilized azurin: direct observation of thermodynamic and kinetic heterogeneity, **49**, 5776.
21. Ł. Krzemiński, L. Ndamba, G. W. Canters, T. J. Aartsma, S. D. Evans and L. J. C. Jeuken (2011), *Journal of the American Chemical Society, Spectroelectrochemical investigation of intramolecular and interfacial electron-transfer rates reveals differences between nitrite reductase at rest and during turnover*, **133**, 15085.
22. N. Akkilic, M. Kamran, R. Stan and N. J. Sanghamitra (2015), *Biosensors and Bioelcctronics, Voltage-controlled fluorescence switching of a single redox protein*, **67**, 747.
23. V. Polliotto, S. Morra, S. Livraghi, F. Valetti, G. Gilardi and E. Giamello (2016), *International Journal of Hydrogen Energy, Electron transfer and* H_2 *evolution in hybrid systems based on [FeFe]-hydrogenase anchored on modified* TiO_2, **41**, 10547.
24. S. P. Best, A. Levina, C. Glover, B. Johannessen, P. Kappen and P. A. Lay (2016), *Journal of Synchrotron Radiation, XAS spectroelectrochemistry: reliable measurement of X-ray absorption spectra from redox manipulated solutions at room temperature*, **23**, 743.

Chapter 16

Spectroelectrochemistry of Microbial Biofilms

Diego Millo* and Bernardino Virdis[†,‡]

*Biomolecular Spectroscopy/LaserLaB Amsterdam, Vrije Universiteit Amsterdam, De Boelelaan 1081, 1081 HV Amsterdam, The Netherlands

[†]The University of Queensland, Advanced Water Management Centre (AWMC) Level 4, Gehrmann Building (60), Brisbane, QLD 4072, Australia

[‡]b.virdis@awmc.uq.edu.au

16.1 Introduction and Motivation

Since the first observation that electric energy can be liberated from the microbial disintegration of organic compounds in 1911,[1] electrochemical techniques have always had played a prominent role in the study of BESs (see Chapter 2), becoming increasingly invaluable tools to grow and characterize the functioning of electrochemically active biofilms. However, despite their importance, electrochemical techniques do not provide a comprehensive understanding of living microbial biofilms. In general, the electrochemical signals of biofilms arise from the interplay between many different processes, often difficult to disentangle, such as Faradic and non-Faradic processes. The former comprise electron transfer (ET) processes, such as the oxidation and reduction of redox mediators wiring the cell metabolism to the electrode; the latter do not involve redox reactions and consist of capacitive and charge transport processes such as the charging of the electrical double layer at the biofilm–electrode interface, as well as the charge transport through the biofilm. The non-Faradic processes involve all charged molecules moving inside the biofilm under the effect of an electric field (migration) and/or under a concentration gradient (diffusion). In the case of electroactive microbial biofilms, the metabolic substrate (often the charged molecules acetate and lactate) and the metabolic product H_3O^+ migrate and diffuse through the biofilm, generating two opposite gradients inside the biofilm (see Figure 1).[2]

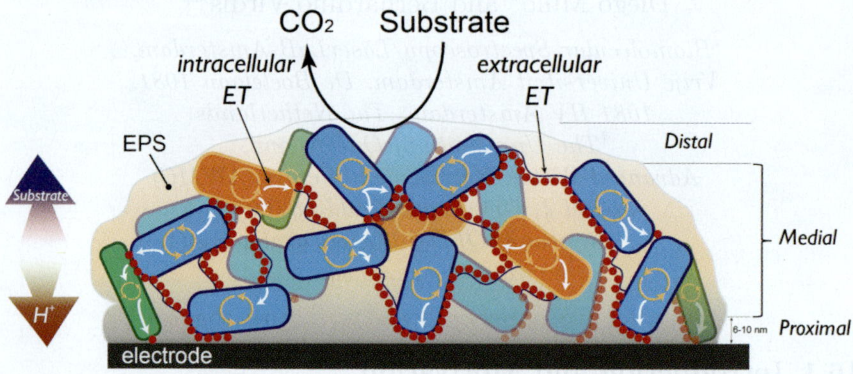

Figure 1. Schematic representation of catalytic substrate oxidation by Mc electroactive biofilms according to the redox conduction[3] and metallic-like conduction[4] ET mechanisms. Microbial cells are embedded into a self-produced extracellular polymeric matrix attached to the electrode. The yellow arrows represent the intracellular metabolism oxidizing the substrate and delivering electrons to intracellular reducing equivalents (*e.g.* NADH). The white arrows represent the ET from this internal redox poll to the ET chain by redox reactions involving membrane-bound cytochromes gating ET to the extracellular environment. Outside of the microbial cells, electrons transfer from cells occupying medial and distal biofilm locations (see Section 16.2.1) to the electrode is mediated by cytochromes associated with the cell surface, aligned along pili and/or conductive nanowires (the blue wire), or dispersed within the EPS matrix (herein collectively referred to as OMCs and indicated as red dots in the figure). Electrocatalytic oxidation generates two opposite concentration gradients of substrate and products, including protons, within the biofilm. For clarity purposes, the presence of alternative ET mechanisms such as the one promoted by soluble redox mediators is not shown.

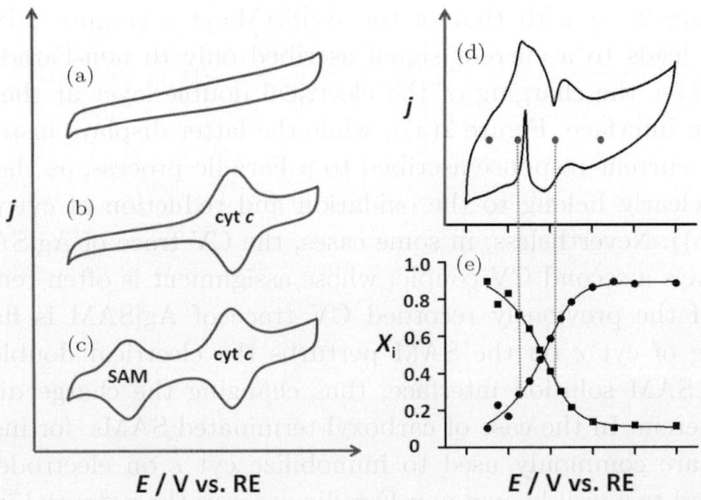

Figure 2. Left panel: Graphical representations of CV traces presented in Millo *et al.*[5] Trace (a) represents the CV signal of Ag|SAM before cyt *c* immobilization (*i.e.* a proper blank). Traces (b) and (c) represent the CV traces of Ag|SAM|cyt *c* without and with the interference of a non-Faradic process due to the potential-induced protonation of the carboxyl-terminated SAM, respectively. Right panel: (d) CV trace of a biofilm as reported in Millo *et al.*[7] (e) Molar fraction X_i of the oxidized (circles) and reduced (squares) heme groups of the OMCs obtained from the fitting of SERR spectra operating under stationary conditions. The dotted lines connect the inflections points of the Nernstian fit in E (*i.e.* the formal potential $E^{0\prime}$ of the OMCs) to the corresponding midpoint potential $E_{1/2}$ of the redox couples in D. In all panels, the potential E increases from left to right.

In the case of a CV experiment, where the potential E of the electrode carrying the biofilm is varied linearly with time, the resulting current, which is plotted *vs.* the applied potential, is affected to some extent by Faradic and non-Faradic processes, although their contributions to the CV signal are not easy to assign. To understand the difficulty of the assignment, we compare in Figure 2 the CV trace of a Mc biofilm grown on a rough Ag electrode (Figure 2(d)) with that of a well-known protein system, cyt *c* immobilized on a rough Ag electrode chemically modified with an appropriate self-assembled monolayer (SAM) (Figures 2(b) and 2(c)). In the case of cyt *c*, the assignment of the CV peaks is relatively straightforward. In fact, one can compare the signal of the Ag|SAM (*i.e.* the blank in Figure 2(a)) with that of the Ag|SAM|cyt *c* (Figure 2(b)): the former leads to a current signal ascribed only to non-Faradic processes (*i.e.* the charging of the electrical double layer at the SAM-solution interface, Figure 2(a)), while the latter displays instead the typical current response ascribed to a Faradic process, as the redox peaks clearly belong to the oxidation and reduction of cyt *c* (Figure 2(b)). Nevertheless, in some cases, the CV trace of Ag|SAM|cyt *c* displays a second CV couple, whose assignment is often tentative. Even if the previously recorded CV trace of Ag|SAM is flat, the binding of cyt *c* on the SAM perturbs the electrical double layer at the SAM–solution interface, thus changing the charge distribution therein. In the case of carboxyl-terminated SAMs, for instance, which are commonly used to immobilize cyt *c* on electrodes, this may lead to a well-known non-Faradic process: the potential-induced protonation and deprotonation of the carboxylic group of the SAM exposed to the solution. This non-Faradic process gives two voltammetric peaks having often comparable intensity and identical current *vs.* scan rate dependency as those of cyt *c* oxidation and reduction (Figure 2(c)).[5,6] To complicate the situation even further, the non-Faradic process, which is not always evident on the blank signal (Ag|SAM, Figure 2(a)), occurs in the same potential window where some non-native cyt *c* species are electrochemically active. In this scenario, without the support of ancillary experimental approaches, such as the spectroelectrochemical techniques outlined in this chapter, the assignment of the second redox couple in Figure 2(c) is not always conclusive.

If discriminating between a Faradic and a non-Faradic process is difficult in the case of a very well-known system, which can be built up and characterized layer by layer (*i.e.* Ag, Ag|SAM, and Ag|SAM|cyt *c*), it is clearly much more difficult in the case of a microbial biofilm, where more complex CV traces arise (see Figure 2(d)), often featuring multiple peaks, the presence of an unusual cleft such as in the case of biofilms dominated by *Geobacteraceae* (the origin of which is still unknown), and in general, a more complex background signal. Most importantly, contrary to the Ag|SAM|cyt *c*, in the case of biofilms, we do not have appropriate blank helping us to distinguish between Faradic and non-Faradic processes. In fact, it is impossible to selectively suppress the redox active species without affecting the matrix of the biofilm comprising a complex architecture of cells and metabolites. For this reason, the assignment of the CV peaks in the case of biofilms is often tentative and it is limited to the redox active species. A common way to assign the CV peaks to a specific redox species is by comparing the midpoint potential of the CV trace of the biofilm with that of the isolated redox species. Even if this method works in the case of soluble redox mediator that can be added to the biofilm without affecting its integrity,[8] it is not always applicable. The reason is that the very same molecule may act differently when isolated and when embedded in a living biofilm. Even if this does not seem the case of soluble redox mediators, it is certainly the case of cytochromes interacting directly with the electrode surface. In the case of cyt *c*, the interaction with a biocompatible electrode surface (Ag|SAM) does not change the structure of the protein (which remains native), but it lowers considerably the midpoint potential of protein (by ca. 70 mV) with respect to the same protein in solution. Should protein denaturation occur upon binding, the potential shift of the bound protein may be even more pronounced.[9] For this reason, comparing the midpoint potential of isolated and surface-bound proteins does not always lead to a conclusive assignment. This situation, which is well known to occur in the case of isolated proteins and enzymes, was recently observed also for the OMCs of *Dace* by Alves *et al.*[10] who showed how an isolated OMC in direct contact with a rough Ag electrode lost its native structure and became CV-silent, while the same protein embedded together with many others in a microbial biofilm grown on the same electrode displayed a well-defined CV trace with a much higher midpoint potential. The cited study was carried out with a combination of surface-enhanced resonance Raman (SERR) spectroscopy (see below) and CV. This approach proved that the CV-silent isolated protein was immobilized on the electrode and was still redox-active, albeit it could not be detected by CV. Interestingly, if the authors would have based their interpretation on the missing CV trace only, they would have erroneously concluded that the isolated protein does not bind the electrode, or it does not promote the ET. This example shows the limitations of electrochemical techniques and the importance of using ancillary approaches to probe what inevitably escapes the electrochemical analyses.

In the case of electrochemical impedance spectroscopy (EIS), which is an electrochemical technique providing much information on the electrical and electrochemical properties of coatings attached on electrodes (*e.g.* SAMs and biofilms), it is possible to distinguish between different Faradic and non-Faradic processes such as electron transfer mechanisms, and mass transfer phenomena. Even if EIS gained much consensus when applied to the study of SAMs and more complex architectures such as tethered bilayer lipid membrane, data interpretation is still controversial in the case of biofilms.[11] The reason is that EIS data are commonly analyzed by fitting to an equivalent electrical circuit model, which is a theoretical circuit constituted by common electrical elements such as resistors and capacitors, which should retain all the electrical characteristics of the biofilm under consideration. Clearly, the design of a proper equivalent circuit requires much information on the architecture of the biofilm, as well as an in-depth knowledge of the physical and biochemical processes occurring therein.[12] Even if much progress has been done in this direction, substantial information on the composition and the electrochemical processes occurring at the biofilm–electrode interface is still largely missing, and thus, the design on any equivalent circuit of a biofilm is based to some extent on assumptions, which will affect the interpretation of the results accordingly.

Disentangling the different contributions of an electrochemical signal is not only a problem in the attempt to discriminate between Faradic and non-Faradic processes and/or ascribe a redox couple to the correct redox mediator, but it is also an issue when it comes to distinguish between events occurring in different locations of the biofilm. For instance, it is not possible to know *a priori* if a CV trace of a biofilm is dominated by the relative small amount of OMCs at the biofilm–electrode interface or the more abundant OMCs embedded in the bulk biofilm promoting the long-range ET. Some help in this sense can come from the use of ultramicroelectrodes, which pierce the biofilm and allow monitoring the electrochemical properties (*e.g.* pH and redox potential) of a small volume of the biofilm.[13] However, this is a disruptive technique, which perturbs the sample while measuring it and thus its applicability may not be representative of the situation in intact biofilms. Some indirect information on the location of the redox species inside the biofilm can come from the analysis of plots depicting the peak current of the voltammograms *vs.* the scan rate. In fact, in a CV trace governed by surface-confined redox species, the peak current varies linearly with the scan rate; on the contrary, if the redox process is governed by diffusion, a linear trend will be observed when the peak current is plotted *vs.* the square root of the scan rate. This approach may help to discriminate between surface-confined (*i.e.* attached on the electrode) and diffusive redox species (*note*: In this case, since it is not clear if the electron carriers are extracellular cytochromes which are actually able to diffuse through the biofilm, it is generally considered that the electrons diffuse through the biofilm), thus helping identifying the chemical composition at the biofilm–electrode interface. However, the applicability of this method for differentiating between surface-confined and diffusive redox species was recently called into question due to the difficulty to safely discriminate between linear and nonlinear trends.[14–16] Accordingly, also the applicability of the so-called Laviron method (*i.e.* an easy way to extract the heterogeneous ET rate constant from CV measurements in the case of surface-confined redox species[17]) is questionable. In fact, lacking sound evidence supporting the presence of surface-confined redox species (a consequence of the impossibility to distinguish between a linear and a nonlinear trend in the current *vs.* scan rate plot quantitatively), which is the essential prerequisite for its applicability, this method cannot be safely applied.

However, along with the abovementioned limitations, the main drawback of electrochemical analyses is the lack of structural insights. In fact, electrochemical techniques do not provide direct information on the structure of the species involved in an electrochemical process. These limitations may greatly impair our ability to understand the functioning of electroactive biofilms. Fortunately, they can be overcome to some extent by combining electrochemical techniques with various spectroscopic methods to be performed on the same sample. This means that the same biofilm studied by electrochemical means is subjected to a spectroscopic measurement performed under electrochemical control. Since the most common electroanalytical techniques used to study microbial biofilms operate under potential control (*i.e.* the applied potential is either constant or varied in a controlled way in time and the recorded current is plotted *vs.* the applied potential or the time), this means that the spectroscopic measurements are also performed under potential control.

This chapter aims to show why spectroscopic techniques are nowadays essential to complement electrochemical analyses on biofilms, not only by explaining the intrinsic limitations of electrochemical approaches (see above), but also by providing much information on the scientific achievements obtained thanks to spectroelectrochemical analyses, and indicating possible future developments in the field of biofilms (see below).

16.2 Spectroscopic Techniques

Spectroscopic techniques applied to the study of biofilms provide structural information at two different levels: chemical structures of constituting molecules, as well as their location inside the biofilm. While the former capability has been extensively used to probe the structure of species promoting the ET (*i.e.* OMCs and soluble redox mediators), the possibility to locate the position of these molecules within the biofilms has been only recently explored. Nevertheless, it allows an unprecedented level of insight into the spatial variability of the ET processes, as it makes it possible to study ET at defined biofilm locations and hence, also at defined distances from the electrode. In this respect, we hereby introduce the definition of three different biofilm locations called proximal, medial and distal, each one characterized by an increasing distance from the electrode (Figure 1). An operative definition of these locations should rely on the analytical technique of choice. In this respect, we define as proximal, the location of the biofilm closest to the electrode within the detection limit of the analytical technique. For instance, if the analytical technique of choice allows detecting the OMC within 10 nm from the electrode surface, we define as proximal the location of the biofilm comprised between the electrode plane and 10 nm from the electrode. Similarly, the distal region consists of the biofilm location comprised between the outmost biofilm layer (*i.e.* the farthermost from the electrode surface) and the ultimate underlying layer that is accessible by the analytical technique. Consequently, the medial location is the one in between the distal and the proximal ones, and is often called "bulk." This definition, which we will adopt in the rest of this chapter, aims to identify the location of the biofilm accessible to a specific technique and avoid confusions about the terminology present in literature, which is often based on the observer's point of view and not — as in this case — on the electrode surface. In view of this definition, a molecule inside a biofilm may be present in the proximal, medial, and/or distal location of the biofilm. In the case of the OMCs, which are present inside the entire biofilm, much research effort aimed to map the redox states of the OMCs in different biofilm locations. In fact, according to the redox conduction model predicted by Strycharz *et al.* a fast ET at the biofilm–electrode interface coupled with a slow long-range ET would have caused a gradient of OMCs redox states with the most oxidized OMCs in close contact with the electrode.[3] To test this hypothesis, different groups adopted different techniques, based on UV–Vis or CRR spectroscopy (see below).[18–21]

Along with the identification of a proximal, medial and distal biofilm location, one must also distinguish between extracellular and the intracellular locations. In fact, biofilms are heterogeneous coatings consisting of individual bacterial cells embedded inside a complex network of ET pathways and EPS substances constituting the biofilm architecture. Extracellular locations comprise nanowires, surface-exposed proteins and EPS (*i.e.* everything inside the biofilm and outside the cell membrane), while intracellular locations comprise everything placed inside the microbial cell. Combining these two different types of insights (proximal/medial/distal and extracellular/intracellular) should give the most comprehensive description of the functioning of a biofilm. In this respect, even if different spectroscopic techniques allow focussing on different proximal/medial/distal biofilm locations, the extracellular/intracellular dimension of a biofilm is still largely unexplored, mainly due to the limited accessibility of the cell to the envisioned spectroscopic techniques. Noteworthy, electrochemical techniques cannot provide such a detailed spatial resolution. For this reason, researchers must be aware about the different spatial resolutions of the electrochemical and the spectroscopic techniques of choice when comparing results obtained with different techniques. This aspect, which is not always discussed in the literature, is indeed one of the most critical parameters affecting the interpretation of the data.

The spectroscopic techniques used to study biofilms in combination with electrochemical techniques are variations of three different approaches: UV–Vis, Raman, and IR spectroscopy. All these approaches are non-disruptive and, contrary to many fluorescence techniques, do not require staining or other biofilm manipulations, thus being able to study intact living biofilms operating under physiologically relevant conditions, that is, *in vivo*. In UV–Vis spectroscopy, a biofilm is grown onto an optically transparent electrode (*e.g.* ITO) and this system is placed between a light source and a detector (see Figure 3(a)). To reach the detector, the electromagnetic radiation travels through the biofilm and the electrode, and is absorbed accordingly. UV–Vis probes the electronic transitions of molecules embedded inside the biofilm. Due to the overwhelming presence of cytochromes in electroactive biofilms, UV–Vis spectra probe mainly the heme groups of the OMCs present in the biofilm, as shown by the strong absorption bands around 410–420 and 520–550 nm.[22,23] Unfortunately, this technique has a very poor spatial resolution (the diameter of the incident beam is in the cm range) and cannot distinguish between proximal/medial/distal locations of the biofilm.

Raman and IR are vibrational spectroscopic techniques providing fingerprint spectra of molecules. Each band of these spectra corresponds to a specific vibrational mode of the molecule under consideration. As such, both techniques are able to probe subtle structural changes occurring in the molecule of interest. Raman and IR spectroscopy rely on different mechanisms (light scattering *vs.*

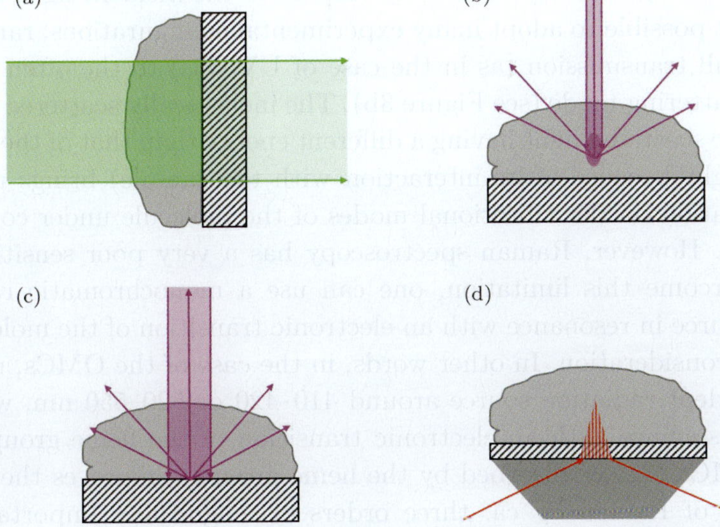

Figure 3. Schematic representation of the most frequent spectroelectrochemical configurations adopted to study electroactive biofilms (in grey). (a) In a UV–Vis setup, the UV–Vis radiation passing through the biofilm and the optically-transparent ITO electrode (striped), is absorbed mainly by the biofilm. (b) In a RR and a CRR setup, the monochromatic radiation is focused on a portion of the biofilm grown onto the electrode (striped). The scattered light (upward arrows) is collected by the spectrometer (not shown). (c) In a SERR setup, the monochromatic radiation is focused onto the nanostructured Ag electrode (striped). The scattered light (upward arrows) is collected by the spectrometer (not shown). (d) In a SEIRA setup operating in the ATR configuration, the IR beam (red arrow on the left side) is focused on one side of the optical element (in grey) and is enhanced by the nanostructured Au layer (striped), also serving as the working electrode. The surface-enhanced IR beam is absorbed by the surface-confined portion of the biofilm, and sent to the detector (not shown) through the optical element. Figure adapted from Ref. 26.

light absorption) and have different selection rules (*i.e.* Raman-active vibrations are not necessarily IR-active and vice versa). For this reason, despite some similarities (if a vibrational mode is both Raman- and IR-active, it will have the same frequency in both spectra), they are considered as complementary techniques.

In Raman spectroscopy, a monochromatic radiation source is focused on the sample — usually through a microscope objective — and the scattered light is collected by a detector. The possibility to collect the light at any angle with respect to the incident light beam made it possible to adopt many experimental configurations, ranging from full transmission (as in the case of UV–Vis) to the often used backscattering mode (see Figure 3b). The inelastically scattered light (*i.e.* the scattered light having a different energy than that of the incident light because of its interaction with the sample) brings much information on the vibrational modes of the molecule under consideration. However, Raman spectroscopy has a very poor sensitivity. To overcome this limitation, one can use a monochromatic radiation source in resonance with an electronic transition of the molecule under consideration. In other words, in the case of the OMCs, using an incident radiation source around 410–420 or 520–550 nm, which is in resonance with an electronic transition of the heme groups of the OMCs (*i.e.* is absorbed by the heme groups), improves the sensitivity of Raman by ca. three orders of magnitude. Importantly, operating under resonance Raman (RR) conditions improves also the specificity of the spectroscopic approach because the spectrum of the heme will be enhanced above that of other non-resonant molecules. For instance, the Raman spectrum of cyt *c* obtained with the 633 nm excitation wavelength is very crowded, displaying the contributions of the heme group and the polypeptide matrix simultaneously.[24] On the contrary, the RR spectrum of cyt *c* obtained with the 413, the 514, and the 532 nm excitation wavelength is much more intense (thus lowering the detection limit from the mM to the μM range), and it also provides selectively the vibrational modes of the heme group, while those of the protein matrix are negligible.[25] To date, the possibility to selectively enhance the spectrum of the heme group(s) above all the other molecules present in the biofilm, in combination with much information available in literature on RR of heme species, provided unprecedented structural insight on the redox-, coordination-, and spin-state of the OMCs in living biofilms.[26]

The spatial resolution of RR is dictated by the diameter of the laser beam, which depends much on the optics of the setup (see below). Usually, the spatial resolution on the XY plane is in the order of few μm, while on the Z-axis is much larger.[27–29] To improve the spatial resolution on the Z-axis, some authors performed RR using a confocal microscope, possibly equipped with a motorized optical stage, a technique called CRR spectroscopy. In this way, one can obtain RR spectra of relatively small areas of the biofilm (see Figure 3b). Plotting the intensity of one or more spectral bands as a function of the position of the stage allows obtaining detailed maps of the molecule(s) associated to that band in the entire biofilm. This methodology was used to map the distribution of the redox states of the OMCs inside a biofilm under different metabolic conditions.[30] Importantly, the absolute intensities of the CRR spectra are dependent on the location where they are sampled. In fact, the intensity of CRR spectra obtained for biofilm locations readily accessible to the incident light beam will be more intense than spectra obtained for hidden biofilm locations, where the incident and the scatter light must travel across the biofilm, a process which causes unavoidable losses of light and thus lower the overall intensity of the spectrum. For this reason, the redox state of the OMCs inside the biofilm cannot be quantified accurately by analyzing the absolute intensity of one band, which is both dependent on the oxidation state and the biofilm location, but it should be indeed inferred from the relative intensities of at least two oxidation marker bands of the heme groups.[31–33] To make the proximal location more accessible to CRR spectroscopy, one possibility is to grow a biofilm on an optically transparent electrode and focus the laser beam trough the electrode. Another way to study the proximal area, albeit with a much lower spatial resolution on the Z-axis, is operating under surface-enhanced conditions. SER spectroscopy is a special Raman application exploiting the surface enhancement provided by a rough metal surface. In SER spectroscopy, a nanostructured metal surface, which acts also as an electrode in many spectroelectrochemical applications, amplifies the intensity of the incident light locally, so that the intensity of the Raman spectrum of a molecule immobilized in the close vicinity of the electrode surface (<7 nm) is greatly enhanced (see Figure 3c). The enhancement is as high as 10^6, and, in combination with RR, it increases even further up to 10^8–10^{12}, thus allowing to study samples in the pM concentration range.[34] As stated above, this approach probes the proximal layer of a biofilm with a spatial resolution of few nm. In fact, the surface enhancement is distance dependent and scales down rapidly, so that above 10 nm from the surface, no SER signal can be detected. This approach may be either performed by growing biofilms on SER-active substrates such as nanostructured Ag and Au electrodes, or by growing a biofilm on a different electrode material and then add (or synthesize *in situ*) Ag or Au nanoparticles providing the surface enhancement. Both approaches may provide complementary insight into the structure of the molecules forming the biofilm.

Another surface enhancement approach is SEIRA, which was the first spectroelectrochemical approach applied to the study of biofilms.[35] As in the case of SER spectroscopy, also SEIRA exploits the surface-enhancement of a rough metal surface. However, due to the different underlying mechanism of Raman and IR, the experimental configuration is different than in the case of SER spectroscopy.[26] In a SEIRA setup operating in the ATR mode, a thin Au layer is deposited onto an IR-transparent optical element, typically a Si prism. The IR radiation, which is focused onto the Au surface from the side of the optical element, is amplified by the nanoscopically rough Au surface (see Figure 3d). The surface enhancement of SEIRA is modest in comparison to that of SER spectroscopy (only 10^2–10^3), but it is sufficient to detect a surface-confined molecule in the submonolayer concentration.[34,36] Even if the surface enhancement of SEIRA is lower than that of SE(R)R, its penetration depth into the solution is larger, and may probe molecules placed at 10–12 nm from the electrode. Contrary to SERR, SEIRA cannot operate in the resonance mode, and thus it is a much less selective technique. Moreover, the IR beam is very large (typically in the cm range) and thus SEIRA has a very bad spatial resolution on the XY plane. However, being outstanding sensitive to subtle changes occurring at level of single chemical bonds, SEIRA is a very powerful tool to study the chemical composition of the proximal location of a biofilm. In fact, even if it is difficult to interpret the very crowded SEIRA spectra, which is the reason why this technique has not exploited its full potential in the field of biofilms yet, the ability to detect simultaneously H_2O, CO_2, proteins, EPS, soluble (*e.g.* redox mediators and metabolic substrates) and insoluble organic molecules (*e.g.* the phospholipids comprising the cellular membranes), makes SEIRA a powerful tool to study the biofilm–electrode interface with unprecedented precision at the molecular level.

Noteworthy, all the above spectroelectrochemical approaches can be performed in the stationary or in the time-resolved mode. In other words, each spectroscopic technique may be used to record a spectrum of a biofilm under steady state conditions (*e.g.* a biofilm kept at constant applied potential), or monitor the changes occurring after perturbing the system. The perturbation may be caused by substrate or O_2 injection into the spectroelectrochemical cell, or simply by applying a potential step to the electrode carrying the biofilm. The latter approach was sometimes used to quantify the rates of the ET promoted by OMCs. In fact, in the absence of metabolic substrate, applying an oxidizing potential to a fully reduced biofilm triggers the oxidation of the OMCs (and similarly, applying a reducing potential to a fully oxidized biofilm triggers the reduction of OMCs). The change of the oxidation states of the OMCs resulting from the application of the potential step can be monitored by recording spectra at different times after the perturbation. The time resolution achievable may vary enormously since it depends on the time needed to obtain a spectrum of sufficient quality to provide the requested information. In the case of SE(R)R and RR, which are the most sensitive and thus faster techniques, most setups allow obtaining good quality spectra in ca. 0.1–0.5 s. This time resolution may be sufficient to detect the complete oxidation and/or reduction of the OMCs inside a biofilm, a process that takes seconds to minutes to occur. However, the time resolution can be increased significantly, as in the case of time-resolved-SER and time-resolved-SEIRA, where dedicated experimental setups were specifically designed to probe events in the sub-ms time frame.[34]

16.3 Equipment for Spectroelectrochemical Analyses

Spectroelectrochemical measurements require the use of specifically designed cells, often home-made, which need to (i) meet the different requirements of each spectroscopic technique (*e.g.* UV–Vis require an optically transparent electrode), (ii) focus on the region of interest of the biofilm (*e.g.* proximal, medial, distal, see above), and (iii) allow for multiple and sequential spectroelectrochemical measurements, for instance to observe biofilm growth over time, while reducing the exposure of the biofilm to air when the electrode is transferred from the growth vessel to the spectroelectrochemical cell. In fact, all spectroelectrochemical cells must (iv) avoid atmospheric O_2 to come in contact with the biofilm, since the prolonged exposure suppresses the electrocatalytic activity by inducing a rapid oxidation of the OMCs, which hinders their capacity to interact with the electrode.[37] Finally, spectroelectrochemical setups must (v) maximize the amount of light reaching the sample, so as to increase the intensity of the spectroscopic signal, minimizing at the same time the effects of photodegradation of the sample. Prolonged exposure of biological samples to light may in fact damage the sample irreversibly.[21,34] [*Note*: In some cases, photodegradation is evident from large burns on the electrode sample. However, in the majority of cases, the signs of photodegradation are less obvious, and include protein denaturation, and photoreduction or photooxidation of redox-active molecules such as the OMCs. The latter signs may be difficult to interpret in the case of a biofilm, where little is known about the native conformations and the redox states of the OMCs in the biofilm.] This effect is more evident for techniques using excitation sources in resonance with the sample, such as RR, CRR, SERR, CSERR and UV–Vis and high-power laser sources focused on the sample through a microscopic objective. In the case of surface-enhanced techniques, photodegradation may be prevented using rotating electrodes or movable spectroelectrochemical cells. Appropriate control experiments using the specific analytical setup should be always performed to make sure the exposure to the excitation light is sufficiently low to avoid photodegradation effects during the whole duration of the measurement. To avoid O_2 contaminations in the spectroelectrochemical cell, one may either purge the cell with N_2, or, if the cell geometry and the setup configuration do not allow adopting this strategy, equip the cell with an external flow system so to introduce previously purged O_2-free solutions.[38] Alternatively, one can always consider the opportunity to place the entire spectroelectrochemical setup, consisting of the cell and the spectrometer, inside a glove box.[39] This is just to show that there is no one standard solution: each spectroelectrochemical cell is unique because it is designed according to the unique characteristics of the setup (optics and geometry) and the scope of the analyses.

16.4 Achievements in Spectroelectrochemistry of Microbial Biofilms

To date, the main focus of spectroelectrochemical studies on electroactive biofilms was the elucidation of the ET mechanism. In fact, substantial information available in literature on the spectroelectrochemical properties of isolated heme proteins (albeit not for the OMCs) in solution and on electrode surfaces, provided the bases to detect the OMCs embedded in a living biofilm under electrochemical control. All spectroelectrochemical studies clearly proved that the OMCs play a central role in ET (Table 1). In fact, upon changing the applied potential, different researchers observed a corresponding change in the oxidation state of the OMCs detected by spectroscopic means. This phenomenon has been observed by different research groups for different microbial cultures, and different electrode materials, and it is thus accepted by the scientific community, although the exact mechanism and the contribution of alternative and/or competing ET pathways (*e.g.* the individual contributions of OMCs and conductive nanowires to the observed current densities of *Gs* biofilms) is still under debate. The experimental approach that brought to this result shows the main advantage of spectroelectrochemical over electrochemical analyses. In fact, basing on the study of the CV traces of electroactive biofilms under different metabolic conditions, Fricke *et al.* noticed that the electrocatalytic oxidation of the metabolic substrate acetate was promoted by two of the four redox couples that are detectable under non-turnover CV trace[48] (*i.e.* in the absence of the metabolic substrate acetate) (see Figure 2(d)). The authors did not assign these two redox couples to the OMCs but stated — safely and correctly — that "[t]he shape of the voltammogram appears typical for bacteria, whose

Table 1. Summary of achievements in spectroelectrochemistry of microbial biofilms. Electrochemical techniques — CA (chronoamperometry): whereby a fixed potential is applied to the electrode carrying the biofilm and the current is recorded over time, allowing sufficient time (usually >10 min) for the biofilm to equilibrate to each potential, at which point the current is recorded and plotted; cyclic voltammetry (CV): the potential is swept linearly with respect of time within a defined potential range and at a certain scan rate; Potential step: the potential is stepped from an initial value to a final value to monitor the subsequent relaxation transition between the two conditions; linear sweep voltammetry (LSV): similar to CV, the potential is swept linearly within a defined potential range but on one direction only. With (*) are indicated electrochemical techniques applied in conjunction with spectra acquisition.

Spectroscopy	Electro-chemistry	Biofilm	Electrode material	Hierarchy	Remarks	Reference
SERR	CA* and CV	Mc	Roughened Ag	Proximal	A	7
	CA, CV, Potential step*				B	40
	CA*, CV	*Dace*			C	10
CRR	CA* and CV	Mc	Carbon rod	Medial, distal	D	30
	CA*	*Gs*	ITO		E	3, 19
	CA, CV		Au		F	20
	CA*, CV*	Mc	Glassy carbon	Medial, distal	G	41
	CA*, CV, Potential step*		ITO		H	21
CSERR	CA	*So*	Ag/AgCl ink	Proximal to the nanopar-ticles	I	42
SEIRA	CA*, CV	*Gs*	Au	Proximal	J	35
					K	43
SERR and SEIRA	CA*, CV				L	44
					M	45

(*Continued*)

Table 1. (*Continued*)

Spectroscopy	Electro-chemistry	Biofilm	Electrode material	Hierarchy	Remarks	Reference
UV–Vis	CA, CV	*Sl*	ITO	Proximal	N	46
	CA*, CV				O	47
	CA*, CV*	*Gs*		Proximal, medial, distal	P	22
	CA*, CV*, LSV*				Q	23
	CA*, CV*				R	18

Notes:

[A]First *in situ* spectroscopic characterization of OMCs embedded in electrochemically active biofilms.

[B]Measurements in the time-resolved mode provided direct information on the dynamics of the ET process at the biofilm–electrode interface, for which the heterogeneous ET rate constant was determined to be 0.03 s^{-1}.

[C]*Dace* biofilms were found able to control the interactions of outer membrane protein OmcB in direct contact to the electrode in order to minimize protein denaturation, which was shown to occur when the isolated protein was directly bound to the electrode surface.

[D]Biofilm development was monitored over time by multiple, sequential measurements performed *ex situ* at different growth stages, without impacting the metabolic or catalytic activity of the biofilm. The study showed the suitability of CRR spectroscopy to resolve the spatial redox electrochemistry of electroactive biofilms *in vivo*.

[E]*In situ* redox profiling of biofilms revealing the presence of a concentration gradient of oxidized (and reduced) OMCs spanning several tens of micrometers from the electrode surface, in line with the redox conduction model suggesting that ET occurs via electron hopping between OMCs.

[F]*Ex situ* measurements of internal cytochrome redox gradient. The presence of oxidized OMCs even when a low electrochemical potential was applied suggested that the majority of OMCs have redox potential too negative to be directly involved in the coupling of acetate oxidation and electrode reduction.

[G]*Ex situ* measurements of redox changes of OMCs at a defined distance from the electrode during potential sweeps revealed the presence of a significant time delay between the oxidation of OMCs at the electrode surface and the resulting oxidation of distal OMCs as probed by CRRM. This delay was used to estimate the diffusivity of the electrons through the biofilm.

[H]*In situ* measurements of redox changes of OMCs at a defined distance from the electrode during the application of potential steps allowed direct comparison of ET process at the biofilm/electrode interface with the long-range ET in the bulk biofilm.

[I]Plasmonic resonance was achieved by *in situ* precipitation of Ag nanoparticles by *So*, allowing the mapping of OMCs, redox mediating molecules flavins as well as other components of the extracellular polymeric substance (polysaccharides) and phosphate.

(*Continued*)

Table 1. *(Continued)*

[J]First reported spectroelectrochemical investigation (using ATR-SEIRA) of interfacial interactions between a polarized electrode surface and suspended *Gs* cells deposited on the Au electrode.

[K]ATR-SEIRA was used to characterize two different redox pairs on *Gs* surface expressed in response to different poised potentials. The results suggested the expression of high potential OMCs responsible for alternative ET pathways, in line with what shown by voltammetry.

[L]Combined *in situ* ATR-SEIRA and SERR spectroscopy (on Ag nanoparticles) investigation of conformational changes of OMCs in *Gs* during heterogeneous (interfacial) ET between biofilm and a variety of ω-functionalized SAMs on Au and Ag.

[M]Combined *in situ* ATR-SEIRA and SERR spectroscopy (on Ag nanoparticles) approach applied to monitor the electrochemical turnover of OMCs on active *Gs* directly in contact with the electrode.

[N]UV–Vis evanescent wave (EW) spectra of hemes chemically marked with NO was used to corroborate electrochemical observations that identified surface-confined OMCs as responsible for linking *Sl* to the electrode.

[O]Evanescent wave (EW) UV–Vis spectra suggests a large positive shift of the midpoint potential of surface-confined OMCs embedded in the biofilm if compared to purified proteins, possibly ascribed to the continuous supply of electrons from the central metabolism of *Sl*.

[P]First spectroelectrochemical report showing the fully reversible redox conversion of OMCs in viable, fully grown *Gs*.

[Q]Potential-dependent changes linked with real-time measurements on spectral signature in fully-grown *Gs* shown a significant lag between the changes in redox state of OMCs in the biofilm and changes of the applied potential during potential sweep, in agreement with voltammetric data showing semi-infinite diffusional kinetics at fast scan rates. Ascribing a feature of the CV trace to the periplasmic cytochrome PpcA, this is the first report attempting to distinguish between extracellular and intracellular biofilm locations (see Section 16.1).

[R]Electrons were observed to accumulate within OMCs as biofilm growth progressed from an individual cells layer on the electrode surface to a multi-layer of cells stacked together. Not all the OMCs pool seemed to lag behind changes in the electrode potential but only those in the potential range of −0.1 to 0.05 V SHE were affected. This behavior was consistent to a multi-step diffusional process limited in the bulk biofilm, resulting in a measurable gradient of reduced OMCs extending from the electrode surface.

electrochemical activity is ascribed to outer membrane cytochromes", and referred to two other papers. In the first one, Park *et al.*[49] wrote that "[i]t is plausible that the electrochemical activities in the isolate EG3 might be due to cell surface cytochrome(s)", and in the second one, Kim *et al.*[50] stated that "[t]he electrochemical activity is believed to be due to the direct ET from the electron carriers located on the cell surface, possibly outer membrane cytochromes, to the electrode". This was the situation before the application of spectroelectrochemical techniques: basing on CV traces only, the assignment of the redox couples was tentative. The first confirmation of the assignment came in 2008,[35] and it was further confirmed from many subsequent analyses.[7,22,23,46,47] The typical spectroelectrochemical approach to safely assign the redox peaks of the CV trace to the OMCs is the redox titration. In this experiment, the potential of the electrode carrying the biofilm is set to a constant value (stationary conditions) until the system reaches the equilibrium (*i.e.* no net current is passing through the external circuit). Then, the spectrum of the system is recorded while the potential is kept at this set value. Repeating this process at different potential values within the potential window where the redox couples of interest are electrochemically active (*i.e.* sufficiently broad to achieve complete oxidation and/or reduction), one will obtain a set of spectra and one for each applied potential. Fitting the spectra allows extracting the relative amount of oxidized and reduced heme species in equilibrium at a specific applied E. As shown in Figure 2(e), this approach allowed to prove that only two of the four redox couples observed in the CV trace correspond to the oxidation state of the heme groups detected by spectroscopic means.[7] Noteworthy, when applied to Ag|SAM|cyt *c*, the same experimental strategy allows to safely assign the redox couples in Figure 2(c) and thus discriminate between the Faradic process and the non-Faradic potential-induced protonation–deprotonation of the SAM (see above).[5] The redox titration of the biofilm was the common starting point of all the envisioned spectroelectrochemical techniques. Then, each research group made a step forward into different directions, according to its preferences and abilities.

16.4.1 Heme structure of the OMCs

The use of SERR and RR spectroscopies provided unprecedented insights into the structure of the heme groups of the OMCs embedded inside the biofilm. This is a crucial information to understand the factors controlling the redox properties of the biofilm at the molecular level. To date, research on Mc and *Dace* biofilms showed that the large majority of the OMCs have a *c*-type heme where the central Fe atom is in the six-coordinated low-spin configuration, having a His-Fe-His acting as axial ligands.[7,10,41] This information was obtained for living biofilms grown on electrodes[7] and it is consistent with precedent spectroscopic and subsequent crystallographic data of isolated OMCs.[51,52]

16.4.2 Kinetics of the ET process

To date, the kinetics of the ET processes taking place inside biofilms has attracted much research effort. Research in this field was conducted with several electrochemical approaches (mainly CV and CA), with very little spectroelectrochemical insights. In their seminal work, Strycharz *et al.*[3] identified five steps contributing to current in *Gs* biofilms grown onto a glassy carbon electrode and developed a model to extract kinetic information from the CV trace of the biofilm under turnover conditions (*i.e.* in the presence of the metabolic substrate acetate). This model could fit the experimental data in the presence of a non-limiting interfacial ET at the biofilm–electrode interface with a heterogeneous ET rate constants k^0 in the order of 10^4 s^{-1}. Noteworthy, this is not a direct measurement of the ET rate, but a parameter used by the authors to fit the experimental CV traces. Direct measurements obtained with the Laviron method (see above) on biofilms of *Gs* gave a k^0 ranging from 0.15 to 0.82 s^{-1}.[53] Similar experiments on *So* gave k^0 ranging from $4.0 \cdot 10^{-6}$ to 3 s^{-1},[54,55] with the fastest ET observed for *Sl* (150–230 s^{-1}).[46] In a subsequent study by Bonanni *et al.*[56] who applied a model similar to that of Strycharz *et al.* albeit using CA instead of CV, the experimentally determined ET rate constant between the surface-confined OMCs and the electrode was 13 s^{-1}, which is almost three orders of magnitude lower than the values predicted by Strycharz *et al.* for *Gs* biofilms. In the first spectrelectrochemical characterization, Ly *et al.*[40] performed a time-resolved-SERR spectroscopy study of a Mc biofilm grown on nanostructured Ag, and found a heterogeneous ET rate constant of 0.03 s^{-1} for the interfacial ET process promoted by the surface-confined OMCs. This value has been then confirmed by subsequent studies[21] and is in line with the theoretical work by Korth *et al.*[57] who proposed an ET rate constant of 1.2 s^{-1} for the interfacial and the long-range ET. Despite a large body of evidence has progressively lowered the very high ET rate constant proposed by Strycharz *et al.*, the large differences in the heterogeneous ET rate constants (spawning almost six orders of magnitude!) are probably due to intrinsic differences between the studied systems (*i.e.* bacterial strain and electrode material). In fact, the possibility to have an OMC in direct contact with the electrode surface, possibly with the plane of the heme perpendicular to the electrode surface, may certainly promote fast ET processes, while an OMC separated from the electrode by a layer of EPS may easily explain a low heterogeneous ET rate constant. To overcome this apparent discrepancy, much information on the chemical composition of the biofilm at the biofilm-electrode interface is needed.

16.4.3 Mechanism of the ET process: OMCs *vs.* conductive nanowires

In the attempt to establish if the ET inside biofilms is dominated by OMCs or conductive nanowires — a controversial topic that has recently monopolized the scientific debate in the field of BESs — many research groups tried to use CRR and UV–Vis to probe the presence of a redox gradient inside the biofilm. In fact, basing on the redox conduction model developed by Strycharz *et al.* if the ET process inside a biofilm is promoted OMCs through a hopping mechanism — and not by the metallic-like electron conduction proposed for the conductive nanowires — a redox gradient should be observed inside the biofilm, with the more reduced OMCs facing the biofilm–solution interface, and the more oxidized proteins located at the biofilm–electrode interface.[3] Even if the existence of this redox gradient is quite evident for *Gs*, it is still not clear yet if the small differences in the relative intensity and position of the bands in the RR spectra can explain changes in the redox state of OMCs ascribed to the long-range ET. For this reason, it is always advisable to perform a proper data fitting (to be previously tested on a known system such as cyt *c*), so to avoid underestimating or overestimating the observed changes of the redox states of the OMCs. Interestingly, the redox gradient was never observed in Mc biofilms dominated by *Geobacteraceae*. To date, it is not clear if this substantial difference is due to a different ET mechanism operating in *Gs* and Mc biofilms, or to the different experimental approach applied in the cited papers. For instance, Lebedev *et al.* performed the experiments on biofilms exposed to air,[20] while Robuschi *et al.* and Virdis *et al.* operated under strict anoxic conditions.[19,21] For this reason, more experiments in this direction and proper quantitative analysis are still warranted to be conclusive on this aspect.

16.4.4 Looking beyond the obvious

Even if OMC detection dominated by far the spectroelectrochemical analyses of biofilms, it is worth mentioning that biofilms comprise also many other molecules, such as soluble redox mediators (see below), EPS, metabolic substrates, and carotenoids.[41,42,58,59] Interestingly, while the first three molecules were largely expected, and thus actively searched, the presence of the last one was rather unexpected. Even if the role of carotenoids in mixed culture biofilms dominated by *Geobacteraceae* is not clear yet, their amount was found to increase with the age of biofilms, thus making it possible to use Raman spectroscopy to identify the age of a biofilm, an important parameter with much impact on the performance of the biofilms. For this reason, we believe that careful spectroscopic analysis beyond the well-known spectral signatures of the expected molecules may greatly help to identify other molecules, and possibly also to understand their role. Therefore, we encourage spectroelectrochemists to broaden their horizons beyond the ET.

16.5 Conclusions and Future Prospects

As shown at the beginning of this chapter, deriving conclusions about the functioning of complex systems such as electroactive biofilms from the data obtained with one technique only can easily lead to over interpretation of the results. The reason is that, lacking substantial information on the architecture of the biofilms, the interpretation is unavoidably based on many different assumptions, including biofilm architecture, mode of electron transfer, distribution of concentration gradients, etc. which, however reasonable, may be wrong or misleading. For this reason, we think that more sound interpretations may be obtained by using complementary experimental approaches to be performed simultaneously on the same sample. In this perspective, we see a bright future for spectroelectrochemistry of biofilms. In fact, even if spectroscopy techniques have been extensively applied to unravel the direct-ET process (*i.e.* the ET promoted mainly by OMCs), there are many other aspects where they can provide novel insights. In the frame of redox-active species, one could try to prove more complex ET mechanisms such as those involving the simultaneous mediated-ET (*i.e.* the ET promoted by soluble redox mediators) and direct-ET such as in *Shewanellae*. This possibility was initially explored by Jain *et al.* for *Sl* biofilms by means of various electrochemical techniques and fluorescence spectroscopy of the cell-free supernatant to track the soluble redox mediators,[60] and, more recently, by Schkolnick *et al.* who observed CRRM spectra of the soluble redox mediator flavin promoting the mediated-ET in a *So* biofilm.[42] These early studies show that the application of spectroelectrochemical methodologies to unravel the interplay between the direct- and the mediated-ET is still in its infancy, and much can be done in this respect by spectroelectrochemical analyses, especially since the list of redox mediators seems far from being complete.[61]

Along with these possibilities, spectroelectrochemical techniques should also push their boundaries to improve their spatial resolution as well as their ability to capture increasingly smaller details of living biofilms. In this sense, the most frequent expectations are the following.

16.5.1 Resolution

Monitor individual proteins embedded inside a biofilm or on a bacterial cell is not possible with conventional confocal techniques. In fact, the theoretical diffraction-limited spatial resolution on the XY plane is defined by the laser wavelength and the numerical aperture (NA) of the objective being used, according to the following equation:

$$\text{Spatial resolution} = 0.61\lambda/\text{NA}$$

where λ is the wavelength of the laser. Thus, for a 532 nm laser with an objective having NAs of 0.6 and 1, the spatial resolution is 541 and 325 nm, respectively. This spatial resolution allows detecting ca. half bacterial cell, but it is far too poor to detect single proteins. Depth (Z) spatial resolution is more complex, and depends on the specifications of the confocal microscope being used, but it is ca. 1–2 μm for a true confocal design.

16.5.2 Heme groups in OMCs

Resolving the individual heme groups of an OMC embedded in a living biofilm is not possible with the envisioned spectroelectrochemical approaches, especially if the OMC under consideration is embedded inside a biofilm. This is indeed possible *in vitro* for isolated heme proteins with six heme groups at maximum,[62] by means of NMR, a technique which is well-established in the case of isolated proteins[63] and still in its infancy with respect to spectroelectrochemical applications on biofilms.[64] However, since many different OMCs are embedded inside the biofilm — and many have more than six heme groups — and the heterogeneity of the biofilm will probably cause a dispersion of redox potentials in a relative small potential window, not much progress is expected in this direction in the immediate future.

16.5.3 OMCs identification

Identifying specific OMCs (*i.e.* distinguish an OmcB from an OmcS or an OmcZ) is impossible in the absence of a unique spectroscopic signature of the heme group. In fact, most OMCs have *c*-type heme groups having the His-Fe-His axial ligation, and thus they all have the very same UV–Vis and RR spectrum. Some discrimination is possible for those OMCs having a different axial ligation, as in the case of *Gs*, whose CRRM spectra were sometimes characterized by the presence of a His-Fe-Met axial ligation,[41] possibly belonging to an OmcF: the only outer surface-exposed cytochrome of *Gs* having that axial ligation.[65] Noteworthy, the opportunity to use mutated bacterial strains to grow biofilms carrying OMCs with non-native axial ligands is certainly a feasible option. However, one should consider that site-specific mutations may vary the $E^{0\prime}$ of the heme groups and thus their involvement in the ET process. In the context of the biofilm, which is a living community where many ET pathways may operate simultaneously, it is likely that the cell metabolism and the microbial community will try to compensate for the inflicted mutations, for instance by preferring a different ET route or ET mechanism over others. Thus, if on the one hand, the site-specific mutation may allow identification of a specific OMC inside the biofilm, on the other hand, it may affect the ET process and the cell metabolism in a way that comparing these results with those of the biofilm with the native OMC may be difficult, if not impossible.

16.5.4 Conductive nanowires

The possibility to track conductive nanowires and monitor their contribution to the ET by spectroscopic means has often been considered and discussed. Unfortunately, conductive nanowires are pretty much elusive both from the electrochemical and the spectroscopic point of view. Moreover, since the nanowires of *So* and *Gs* have been proposed to differ substantially (membrane protrusions *vs.* proteinaceous structures),[66,67] it is likely that different experimental approaches should be used in these cases. In the former, the challenge arises from the difficulty, if not the impossibility, to distinguish the spectroscopic signal of a cellular membrane from that of a nanowire/membrane vesicle; in the latter case, the absence of specific spectroscopic markers for the nanowires (the same amino acids of pilA can be found in many other proteins inside the biofilm, including the OMCs) makes it very difficult to track them. Much insight can come from the specific labeling of nanowires with spectroscopically active compounds, possibly sensitive to the redox state of the environment. This strategy may help to map the nanowires in living biofilms under potential control. Interestingly, also the work on isolated nanowires of *Gs* may be relevant. In fact, integrating photoactive donors and acceptors in a pilA protein at a distance such that tunneling ET cannot occur, and measuring the rates of the ET between them according to the spectroelectrochemical procedure pioneered by Gray and Winkler, would finally provide an unambiguous evidence for the proposed metallic-like conductivity of the nanowires.[68]

16.5.5 Strain identification in Mc biofilms

Identifying the bacterial strain from Raman analysis is indeed possible in principle.[69] The preference for Raman is dictated by the ability to provide fingerprint spectra of the specimen with high spatial resolution, and improve the detection limit enormously by operating the RR or in the SER mode. To date, the most popular strategy consists of performing CRM analysis to probe the EPS of non-electroactive biofilms. However, in the case of electroactive biofilms, many different factors such as the presence of gradients inside the biofilm, different metabolic and redox states of the bacterial cells inside the biofilm, and the presence of different molecules at different growth stages of the biofilm (*e.g.* the carotenoids) can complicate the analysis considerably. For this reason, much work would be needed to construct an appropriate database and develop a method to identify individual bacterial strains inside Mc biofilms.

While the challenges outlined in Sections 16.5.2 and 16.5.3 are very difficult, Section 16.5.1 can be achieved by tip-enhanced RRS under electrochemical control. This technique consists of scanning the specimen with a surface-enhancement active AFM tip. Contrary to the other surface-enhancement-active substrates (*i.e.* rough electrode surface or a nanoparticle), this strategy allows to focus the surface-enhancement on the specimen with unprecedented precision. Most importantly, this strategy allows beating the optical diffraction limit thus reaching very high spatial resolution, which is dictated by the size of the tip (and not by the wavelength and the NA of the objective as in the case of confocal techniques), and is thus capable of studying single proteins in living biofilms. The applicability of

TERR spectroscopy to biological samples, possibly in combination with an appropriate electrochemical technique such as scanning electrochemical microscopy makes it possible to use it for biofilms, albeit probing only the distal location of the biofilm.

In conclusion, we show here how in the past eight years, spectroelectrochemistry of electroactive biofilms has promoted an extraordinary leap forward in the understanding of ET processes in these fascinating biological systems. From providing evidence of the role played by OMCs in the ET, to its use as a tool for the *in vivo* characterization of the biofilm morphology without requiring staining or labeling, and virtually without prior knowledge of the sample; from the study of individual cellular components to the probing of redox process non-accessible to traditional electrochemical techniques alone, spectroelectrochemistry has proven its potential as an inseparable tool available to the bioelectrochemist interested in unravelling complex dynamics and features underlying the microbial catalysis at electrodes. Owing to the unprecedented levels of information that can be obtained by using spectroscopy techniques in combination with electrochemical methods, we feel that the use of spectroelectrochemical approaches will steadily increase in the study of biofilms. For this reason, much interaction and collaboration is needed between electrochemists and spectroscopists, to understand the mutual strengths and limitations of their respective analytical approaches and develop proper spectroelectrochemical methods to tackle the most complex challenges emerging in the research on electroactive microbial biofilms.

List of Abbreviations

ATR	attenuated total reflection
AFM	atomic-force microscopy
BES	bioelectrochemical system
CA	chronoamperometry
CRR	confocal resonance Raman
CSERR	confocal surface-enhanced resonance Raman
CV	cyclic voltammetry
cyt *c*	cytochrome *c*

Dace	*Desulfuromonas acetoxidans*
E	potential of an electrode *vs.* a reference
EIS	electrochemical impedance spectroscopy
EPS	extracellular polysaccharide substances
ET	electron transfer
Gs	*Geobacter sulfurreducens*
IR	infrared
ITO	indium tin oxide
Mc	mixed community
NADH	nicotinamide adenine dinucleotide
OMC	outer membrane cytochrome
RR	resonance Raman
SAM	self-assembled monolayer
SEIRA	surface-enhanced infrared absorption
SER	surface-enhanced Raman
SERR	surface-enhanced resonance Raman
Sl	*Shewanella loihica*
So	*Shewanella oneidensis*
TERR	tip-enhanced resonance Raman
UV–Vis	ultraviolet–visible

References

1. M.C. Potter (1911). Electrical effects accompanying the decomposition of organic compounds. *Proceedings of the Royal Society of London B: Biological Sciences* **84**(571), 260–276.
2. D.R. Bond, S.M. Strycharz, L.M. Tender and C.I. Torres (2012). On electron transport through geobacter biofilms. *ChemSusChem* **5**(6), 1099–1105.
3. S.M. Strycharz, A.P. Malanoski, R.M. Snider, H. Yi, D. Lovley and L.M. Tender (2011). Application of cyclic voltammetry to investigate enhanced catalytic current generation by biofilm-modified anodes of geobacter sulfurreducens strain DL1 *vs.* variant strain KN400. *Energy and Environmental Science* **4**(3), 896–913.
4. N. Malvankar, M. Vargas, K.P. Nevin, A. Franks, C. Leang, B.-C. Kim, K. Inoue, T. Mester, S.F. Covalla, J.P. Johnson *et al.* (2011). Tunable metallic-like conductivity in microbial nanowire networks. *Nature Nanotech* **6**(9), 573–579.
5. D. Millo, A. Bonifacio, A. Ranieri, M. Borsari, C. Gooijer and G. van der Zwan (2007). pH-induced changes in adsorbed cytochrome C. Voltammetric

and surface-enhanced resonance raman characterization performed simultaneously at chemically modified silver electrodes. *Langmuir* **23**(19), 9898–9904.
6. H.S. White, J.D. Peterson and Q. Cui (1998). Voltammetric measurement of interfacial acid/base reactions. *Journal of Physical Chemistry B* **102**, 2930–2934.
7. D. Millo, F. Harnisch, S.A. Patil, H.K. Ly, U. Schröder, P. Hildebrandt (2011). *In situ* spectroelectrochemical investigation of electrocatalytic microbial biofilms by surface-enhanced resonance raman spectroscopy. *Angewandte Chemie International Edition English* **50**(11), 2625–2627.
8. E. Marsili, D.B. Baron, I.D. Shikhare, D. Coursolle, J.A. Gralnick and D.R. Bond (2008). Shewanella secretes flavins that mediate extracellular electron transfer. *Proceedings of Natural Academy of Science USA* **105**(10), 3968–3973.
9. A. Ranieri, D. Millo, G. Di Rocco, G. Battistuzzi, C.A. Bortolotti, M. Borsari and M. Sola (2015). Immobilized cytochrome C bound to cardiolipin exhibits peculiar oxidation state-dependent axial heme ligation and catalytically reduces dioxygen. *Journal of Biology Inorganic Chemistry* **20**(3), 531–540.
10. A. Alves, H.K. Ly, P. Hildebrandt, R.O. Louro and D. Millo (2015). Nature of the surface-exposed cytochrome-electrode interactions in electroactive biofilms of desulfuromonas acetoxidans. *Journal of Physical Chemistry B* **119**(25), 7968–7974.
11. X. Dominguez-Benetton, S. Sevda, K. Vanbroekhoven and D. Pant (2012). The accurate use of impedance analysis for the study of microbial electrochemical systems. *Chemical Society Reviews* **41**(21), 7228.
12. Z. He and F. Mansfeld, (2009). Exploring the use of electrochemical impedance spectroscopy (EIS) in microbial fuel cell studies. *Energy and Environmental Science* **2**(2), 215–219.
13. J.T. Babauta, H.D. Nguyen and H. Beyenal (2011). Redox and pH microenvironments within shewanella oneidensis MR-1 biofilms reveal an electron transfer mechanism. *Environmental Science Technology* **45**(15), 6654–6660.
14. J.N. Roy, S. Babanova, K.E. Garcia, J. Cornejo, L.K. Ista and P. Atanassov (2014). Catalytic biofilm formation by shewanella oneidensis MR-1 and anode characterization by expanded uncertainty. *Electrochimica Acta* **126**, 3–10.
15. R. Rousseau, M.L. Delia and A. Bergel (2014). A theoretical model of transient cyclic voltammetry for electroactive biofilms. *Energy and Environmental Science* **7**(3), 1079–1094.
16. P. Atanassov (2013). *Microbial Fuel Cells with Artificial Biofilms. Topical Meeting of the International Society of Electrochemistry & XXII International Symposium on Bioelectrochemistry and Bioenergetics of the Bioelectrochemical Society*, 17–21 March 2013, Bochum, Germany.
17. E. Laviron (1979). General expression of the linear potential sweep voltammogram in the case of diffusionless electrochemical systems. *Journal of Electroanalytical Chemistry and Interfacial Electrochemistry* **101**(1), 19–28.

18. Y. Liu and D.R. Bond (2012). Long-distance electron transfer by *G. Sulfurreducens* biofilms results in accumulation of reduced C-type cytochromes. *ChemSusChem* **5**(6), 1047–1053.
19. L. Robuschi, J.P. Tomba, G.D. Schrott, P.S. Bonanni, P.M. Desimone and J.P. Busalmen (2013). Spectroscopic slicing to reveal internal redox gradients in electricity-producing biofilms. *Angewandte Chemie International Edition English* **52**(3), 925–928.
20. N. Lebedev, S.M. Strycharz and L.M. Tender (2014). Spatially resolved confocal resonant raman microscopic analysis of anode-grown *geobacter sulfurreducens*biofilms. *Chemphyschem* **15**(2), 320–327.
21. B. Virdis, D. Millo, B.C. Donose, Y. Lu, D.J. Batstone and J.O. Krömer (2016). Analysis of electron transfer dynamics in mixed community electroactive microbial biofilms. *RSC Advances* **6**(5), 3650–3660.
22. A. Jain, G. Gazzola, A. Panzera, M. Zanoni and E. Marsili (2011). Visible spectroelectrochemical characterization of *geobacter sulfurreducens* biofilms on optically transparent indium tin oxide electrode. *Electrochimica Acta* **56**(28), 10776–10785.
23. Y. Liu, H. Kim, R.R. Franklin and D.R. Bond (2011). Linking spectral and electrochemical analysis to monitor C-type cytochrome redox status in living *geobacter sulfurreducens* biofilms. *ChemPhysChem* **12**(12), 2235–2241.
24. G. Niaura, A.K. Gaigalas and V.L. Vilker (1996). Non-resonant SERS study of the adsorption of cytochrome C on a silver electrode. *Journal of Electroanalytical Chemistry* **416**(1–2), 167–178.
25. P. Hildebrandt (1995). Resonance raman spectroscopy of cytochrome C. in A.G. Mauk and R.A. Scott (eds.), *Cytochrome c. A Multidisciplinary Approach*, University Science Books, pp. 285–314.
26. D. Millo (2012). Spectroelectrochemical analyses of electroactive microbial biofilms. *Biochem. Soc. Trans.* **40**(6), 1284–1290.
27. D. Millo (2015). An electrochemical strategy to measure the thickness of electroactive microbial biofilms. *ChemElectroChem* **2**, 288–291.
28. A. Egner and S.W. Hell (2006). Aberrations in confocal and multi-photon fluorescence microscopy induced by refractive index mismatch. in J. Pawley (ed.), *Handbook of Biological Confocal Microscopy*, Springer US: Boston, MA, pp. 404–413.
29. N.J. Everall (2000). Confocal raman microscopy: Why the depth resolution and spatial accuracy can be much worse than you think. *Applied Spectroscopy* **54**(10), 1515–1520.
30. B. Virdis, F. Harnisch, D.J. Batstone, K. Rabaey and B.C. Donose (2012). Non-invasive characterization of electrochemically active microbial biofilms using confocal raman microscopy. *Energy and Environmental Science* **5**, 7017–7024.
31. D.H. Murgida and P. Hildebrandt (2005). Redox and redox-coupled processes of heme proteins and enzymes at electrochemical interfaces. *Physical Chemistry Chemical Physics* **7**(22), 3773–3784.

32. P. Hildebrandt and M. Stockburger (1989). Cytochrome C at charged interfaces. 1. Conformational and redox equilibria at the electrode/electrolyte interface probed by surface-enhanced resonance Raman spectroscopy. *Biochemistry* **28**(16), 6710–6721.
33. S. Oellerich, H. Wackerbarth and P. Hildebrandt (2002). Spectroscopic characterization of nonnative conformational states of cytochrome C. *Journal of Physical Chemistry B* **106**(25), 6566–6580.
34. F. Siebert and P. Hildebrandt (2007). *Vibrational spectroscopy in life science.* Wiley-VCH Verlag GmbH & Co. KGaA, Weinheim, Germany.
35. J.P. Busalmen, A. Esteve-Nunez, A. Berna and J.M. Feliu (2008). C-type cytochromes wire electricity-producing bacteria to electrodes. *Angewandte Chemie International Edition English* **47**(26), 4874–4877.
36. M. Osawa (2006). Surface-enhanced infrared absorption spectroscopy. in J.M. Chalmers (ed.). *Handbook of Vibrational Spectroscopy*, John Wiley & Sons, Ltd: Chichester, UK, pp. 785–799.
37. D. Millo and H.K. Ly (2015). Towards the understanding of the effect of oxygen on the electrocatalytic activity of microbial biofilms: A spectroelectrochemical study. *RSC Advances* **5**(112), 92042–92044.
38. A. Bonifacio, D. Millo, P.H.J. Keizers, R. Boegschoten, J.N.M. Commandeur, N.P.E. Vermeulen, C. Gooijer and G. van der Zwan (2007). Active-site structure, binding and redox activity of the heme–thiolate enzyme CYP2D6 immobilized on coated Ag electrodes: A surface-enhanced resonance Raman scattering study. *Journal of Biological Inorganic Chemistry* **13**(1), 85–96.
39. D. Millo, P. Hildebrandt, M.E. Pandelia, W. Lubitz and I. Zebger (2011). SEIRA spectroscopy of the electrochemical activation of an immobilized [NiFe] hydrogenase under turnover and non-turnover conditions. *Angewandte Chemie International Edition English* **50**(11), 2632–2634.
40. H.K. Ly, F. Harnisch, S.-F. Hong, U. Schröder, P. Hildebrandt, D. Millo (2013). Unraveling the interfacial electron transfer dynamics of electroactive microbial biofilms using surface-enhanced Raman spectroscopy. *ChemSusChem* **6**(3), 487–492.
41. B. Virdis, D. Millo, B.C. Donose and D.J. Batstone (2014). Real-time measurements of the redox states of C-type cytochromes in electroactive biofilms: A confocal resonance Raman microscopy study. *PLoS ONE* **9**(2), e89918.
42. G. Schkolnik, M. Schmidt, M.G. Mazza, F. Harnisch and N. Musat (2015). *In Situ* analysis of a silver nanoparticle-precipitating *Shewanella* biofilm by surface enhanced confocal Raman microscopy. *PLoS ONE* **10**(12), e0145871.
43. J.P. Busalmen, A. Esteve-Nunez, A. Berna and J.M. Feliu (2010). ATR-SEIRAs characterization of surface redox processes in *G. Sulfurreducens*. *Bioelectrochemistry* **78**(1), 25–29.
44. A. Kuzume, U. Zhumaev, J. Li, Y. Fu, M. Füeg, A. Esteve-Nunez, T. Wandlowski (2013). An *in-situ* surface electrochemistry approach toward whole-cell studies: Charge transfer between *geobacter sulfurreducens* and electrified metal/electrolyte interfaces through linker molecules. *Electrochimica Acta* **112**, 933–942.

45. A. Kuzume, U. Zhumaev, J. Li, Y. Fu, M. Füeg, M. Estévez, Z. Borjas, T. Wandlowski and A. Esteve-Nunez (2014). An *in-situ* surface electrochemistry approach towards whole-cell studies: The structure and reactivity of a *Geobacter Sulfurreducens* submonolayer on electrified metal/electrolyte interfaces. *Physical Chemistry Chemical Physics* **16**(40), 22229–22236.
46. A. Okamoto, R. Nakamura, K. Ishii and K. Hashimoto (2009). *In vivo* electrochemistry of C-Type cytochrome-mediated electron-transfer with chemical marking. *Chembiochem* **10**(14), 2329–2332.
47. R. Nakamura, K. Ishii and K. Hashimoto (2009). Electronic absorption spectra and redox properties of C type cytochromes in living microbes. *Angewandte Chemie International Edition English* **121**(9), 1634–1636.
48. K. Fricke, F. Harnisch and U. Schröder (2008). On the use of cyclic voltammetry for the study of anodic electron transfer in microbial fuel cells. *Energy and Environmental Science* **1**(1), 144–147.
49. H.S. Park, B.H. Kim, H.S. Kim, H.J. Kim, G.T. Kim, M. Kim, I.S. Chang, Y.K. Park and H.I. Chang (2001). A novel electrochemically active and Fe(III)-reducing bacterium phylogenetically related to *clostridium butyricum* isolated from a microbial fuel cell. *Anaerobe* **7**(6), 297–306.
50. H. Kim, H. Park, M. Hyun, I. Chang, M. Kim and B. Kim (2002). A mediatorless microbial fuel cell using a metal reducing bacterium, *Shewanella Putrefaciens*. *Enzyme Microb Tech* **30**(2), 145–152.
51. X. Qian, T. Mester, L. Morgado, T. Arakawa, M.L. Sharma, K. Inoue, C. Joseph, C.A. Salgueiro, M.J. Maroney and D. Lovley (2011). Biochemical characterization of purified OmcS, a C-type cytochrome required for insoluble Fe(III) reduction in *Geobacter Sulfurreducens*. *Biochimica et Biophysica Acta (BBA) — Bioenergetics* **1807**(4), 404–412.
52. T.A. Clarke, M.J. Edwards, A.J. Gates, A. Hall, G.F. White, J. Bradley, C.L. Reardon, L. Shi, A.S. Beliaev, M.J. Marshall *et al.* (2011). Structure of a bacterial cell surface decaheme electron conduit. *Proceedings of National Academy of Science USA* **108**(23), 9384–9389.
53. Y. Yuan, S. Zhou, N. Xu and L. Zhuang (2011). Electrochemical characterization of anodic biofilms enriched with glucose and acetate in single-chamber microbial fuel cells. *Colloids Surf B Biointerfaces* **82**(2), 641–646.
54. C.E. Turick, A.S. Beliaev, B.A. Zakrajsek, C.L. Reardon, D.A. Lowy, T.E. Poppy, A. Maloney and A.A. Ekechukwu (2009). The Role of 4-hydroxyphenylpyruvate dioxygenase in enhancement of solid-phase electron transfer by *Shewanella Oneidensis* MR-1. *FEMS Microbiology Ecology* **68**(2), 223–225.
55. D.B. Baron, E. LaBelle, D. Coursolle, J.A. Gralnick and D.R. Bond (2009). Electrochemical measurement of electron transfer kinetics by *Shewanella Oneidensis* MR-1. *Journal of Biological Chemistry* **284**(42), 28865–28873.
56. P.S. Bonanni, G.D. Schrott, L. Robuschi and J.P. Busalmen (2012). Charge accumulation and electron transfer kinetics in *Geobacter Sulfurreducens* biofilms. *Energy and Environmental Science* **5**(3), 6188–6195.

57. B. Korth, L.F.M. Rosa, F. Harnisch and C. Picioreanu (2015). A framework for modeling electroactive microbial biofilms performing direct electron transfer. *Bioelectrochemistry* **106**(Pt A), 194–206.
58. P. Bhosale (2004). Environmental and cultural stimulants in the production of carotenoids from microorganisms. *Applied Microbiology and Biotechnology* **63**(4), 351–361.
59. Y.-T. Zheng, M. Toyofuku, N. Nomura and S. Shigeto (2013). Carotenoid accumulation correlates with aggregation and biofilm development in rhodococcus Sp. SD-74. *Analytical Chemistry* **85**(15), 7295–7301.
60. A. Jain, X. Zhang, G. Pastorella, J.O. Connolly, N. Barry, R. Woolley, S. Krishnamurthy and E. Marsili (2012). Electron transfer mechanism in *Shewanella Loihica* PV-4 biofilms formed at graphite electrode. *Bioelectrochemistry* **87**, 28–32.
61. J. Oram and L.J.C. Jeuken (2016). A re-evaluation of electron-transfer mechanisms in microbial electrochemistry: *Shewanella* releases iron that mediates extracellular electron transfer. *ChemElectroChem* **3**(5), 829–835, http://doi.org/10.1002/celc.201500505.
62. M.N. Alves, A.P. Fernandes, C.A. Salgueiro and C.M. Paquete (2016). Unraveling the electron transfer processes of a nanowire protein from *Geobacter Sulfurreducens. Biochimica et Biophysica Acta (BBA) — Bioenergetics* **1857**(1), 7–13.
63. C.M. Paquete and R.O. Louro (2013). Unveiling the details of electron transfer in multicenter redox proteins. *Accounts of Chemical Research* **47**(1), 56–65.
64. R.S. Renslow, J.T. Babauta, P.D. Majors and H. Beyenal (2013). Diffusion in biofilms respiring on electrodes. *Energy and Environmental Science* **6**(2), 595–607.
65. T.C. Santos, M.A. Silva, L. Morgado, J.M. Dantas and C. Salgueiro (2015). Diving into the redox properties of *Geobacter Sulfurreducens* cytochromes: A model for extracellular electron transfer. *Dalton Transactions* **44**, 9335–9344.
66. S. Pirbadian, S.E. Barchinger, K.M. Leung, H.S. Byun, Y. Jangir, R.A. Bouhenni, S.B. Reed, M.F. Romine, D.A. Saffarini, L. Shi *et al.* (2014). *Shewanella Oneidensis* MR-1 nanowires are outer membrane and periplasmic extensions of the extracellular electron transport components. *Proceedings of Natural Academy of Science USA* **111**(35), 12883–12888.
67. N. Malvankar, M. Vargas, K.P. Nevin, P.-L. Tremblay, K. Evans-Lutterodt, D. Nykypanchuk, E. Martz, M.T. Tuominen, D. Lovley (2015). Structural basis for metallic-like conductivity in microbial nanowires. *mBio* **6**(2), e00084–15.
68. H.B. Gray and J.R. Winkler (2005). Long-range electron transfer. *Proceedings of Natural Academy of Science USA* **102**(10), 3534–3539.
69. S. Pahlow, S. Meisel, D. Cialla-May, K. Weber, P. Rösch, J. Popp (2015). Isolation and identification of bacteria by means of raman spectroscopy. *Advanced Drug Delivery Reviews* **89**, 105–120.

Chapter 17

Scanning Electrochemical Microscopy: A New Tool for Studying Enzymatic Reactions

Dodzi Zigah*,‡ and Olivier Fontaine†
*Institut des Sciences Moléculaires, UMR 5255 CNRS, Université de Bordeaux, Talence, France
†Institut Charles Gerhardt de Montpellier, UMR 5253 CNRS, ENSCM, Université de Montpellier, France
‡dodzi.zigah@u-bordeaux.fr

17.1 Introduction

Developing bioelectrochemical devices, such as glucose biosensors,[1–3] or biofuel cells,[4,5] requires enzyme immobilization and understanding the fundamental functioning of the catalytic activity of immobilized enzymes. After immobilization, the enzymes may have a loss of activity. For this reason, it is important to have a technique that can study these systems. Moreover, with the size reduction of biosensors and biofuel cells, it is important to be able to investigate enzymatic activity at the micro/nanoscale. The term "enzyme electrode" was employed in the 1960s, when glucose oxidase (GOx) was employed to oxidize glucose at the surface of an electrode and to measure the concentration of glucose in biological solutions.[6] After the immobilization of the enzyme on the surface of the electrodes, it is important to obtain local information about the reactivity of the modified electrode. Scanning tunneling microscopy (STM) and atomic force microscopy (AFM) were used but could only provide information regarding surface morphology.[7–11] It is important to have information about the morphology of the immobilized enzyme but it is even more important to have local information about their reactivity. Scanning electrochemical microscopy (SECM) is the perfect tool to make this type of investigation. It was only a few years after the first SECM experiments that attempts were made to use this technique to investigate enzymatic reactions.[12,13] The novelty of this technique lies in the possibility to study the enzymatic activity of the enzymes locally after their immobilization and to obtain electrochemical images of this activity. SECM is a type of scanning probe microscopy method. Like AFM or STM, a tip is used to scan a surface and extract local information. In this chapter, after an introduction to experimental and theoretical basics, we will focus on the use of SECM to extract kinetic parameters of enzymatic reactions. Then we will present the methods employed to obtain electrochemical images. SECM provide information without damaging the immobilized enzymes and can be used in physiological conditions. With this method, the characterization of immobilized enzymes on surfaces of different nature (insulating, conductor, organic, etc.) can be realized. In the literature, excellent reviews and book chapters about the use of SECM to characterize enzyme electrodes can be found.[14–17] In this chapter, we present SECM for non-specialists, to open this powerful technique to a larger community. If the readers want more details, we invite them to refer to these reviews and book chapters.

17.2 Principle of Scanning Electrochemical Microscopy

The main experimental and theoretical bases of SECM were established by Bard *et al.* at the end of the 1980s.[18,19] The particularity of SECM is the use of a tip called an ultramicroelectrode (UME) that can be moved in the three directions of space to probe the surfaces (Figure 1). By moving the UME, it is possible to study different area on the same surface, but it is also possible by scanning the UME in

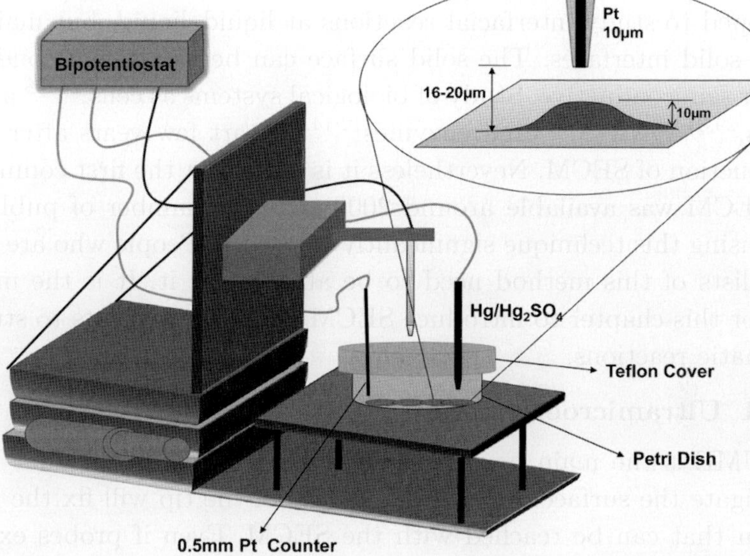

Figure 1. Scheme of the SECM with the UME, the counterelectrode, the reference electrode with a bipotentiostat. Reproduced with permission from Ref. 38.

x, y direction to obtain an electrochemical image. By consequence, the UME allows to locally study the electrochemical properties of surfaces, as for example the local electroenzymatic activity, which is not possible in classical electrochemistry experiments. The definition of UME is ambiguous but mainly electrode whose characteristic length is less than 25 μm are called UME.[20,21] The information about the surface are mainly obtained through the current measure at the tip. SECM can be used with a potentiostat or a bipotenstiostat, and the working electrode is most of the time the UME, there is a counter-electrode and a reference electrode, like in classical electrochemical experiment.[22,23] In the SECM literature, the surface under investigation is usually called "substrate". However, to avoid confusion with the biological term employed for enzymatic reaction, we choose to replace "substrate" by "surface or sample".

The information about the surface are obtained when the current register at the probe is disturbed by the surface. The UME can be moved, in SECM configuration, at the nanometer scale when piezoelectric elements are used to position the tip. This technique is developed to study interfacial reactions at liquid–liquid, but mainly liquid–solid interfaces. The solid surface can be insulating, conductive, or semiconductive. Study of biological systems as cells,[24–28] antibodies,[29,31] DNA,[32–36] and enzymes[12,14,37] start few years after the introduction of SECM. Nevertheless it is only after the first commercial SECM was available around 2000 that the number of publications using this technique significantly increased. People who are not specialists of this method need to be able to use it. It is the main goal for this chapter to introduce SECM for non-specialists to study enzymatic reactions.

17.2.1 Ultramicroelectrodes

The UME is the main part of the SECM. It is the probe that will investigate the surface. Moreover, the size of the tip will fix the resolution that can be reached with the SECM. Even if probes exists with different shapes (hemisphere, band, etc.), the most popular one for SECM experiment is the disk-shaped UME. The disk UME are commercial or can be easily home-made. The tip is a wire embedded in a glass capillary, and only a disk surrounding by the glass in exposed. The characteristic length of the UME is its radius noted "a". The total radius of the UME including the glass part is noted rg. In the SECM literature, it is defined as a dimensionless parameter $\mathrm{RG} = rg/a$. RG can affect the tip current and its positioning. The usual UME has an RG value between 2 and 10. Including their small size, UME has other advantages as rapid response times and minimization of the ohmic drop. By consequence, these electrodes can be used in highly resistive media.[20,39]

Considering an electrochemical oxidation limited by diffusion at the surface of the electrode. That means that the potential applied is high enough to oxidize all the reduce species initially present on its surface. The following reaction takes place at the electrode surface:

$$\mathrm{Red} \rightarrow \mathrm{Ox} + e^{-}. \quad (1)$$

For a large planar electrode, where the diffusion layer is much smaller than the radius of the electrode (Figure 2(a)), the current

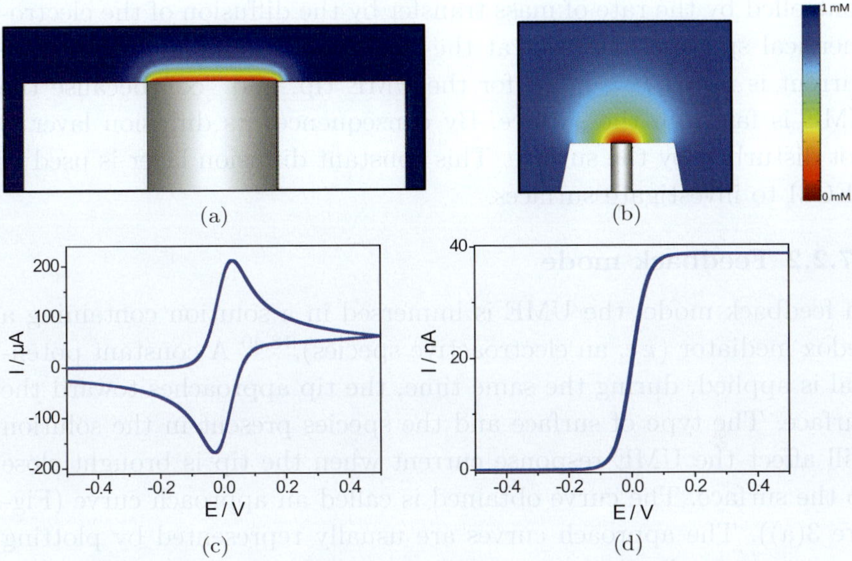

Figure 2. Concentration profile for a reduce species during an oxidation reaction at (a) a macroelectrode, (b) an UME. Shape of the voltammogramm obtained for (c) a macroelectrode, (d) an UME. Concentration profile and theoreticals curves are obtained by simulation with Comsol Multiphysics.

decreases with the time. The current decays proportionally to the inverse square root of time. But for an UME, where the diffusion layer is much larger than the radius of the electrode (Figure 2(b)), the current is independent of the time.[20] The expression of the steady-state current is given by the following equation:

$$I_{SS} = 4\text{nFD}_{\text{red}}c_{\text{red}}a, \tag{2}$$

where n is the number of electron exchange during the electrochemical reaction, F the Faraday constant, D_{red} the diffusion coefficient of the reduce species, c_{red} the concentration and a the radius of the probe. The voltammogram obtained with an UME has a sigmoidal shape because a steady state is reached when the potential applied is higher than the standard potential of the electroactive species (Figure 2(d)). This sigmoidal shape is due to the diffusion of the electroactive species that is not anymore linear but radial. This increases the mass transport to the tip. The steady-state current is controlled by the rate of mass transfer by the diffusion of the electrochemical species that react at the microelectrode. The steady-state current is noted $i_{T,\infty}$, "T" for the UME tip, and "∞" because the UME is far from the surface. By consequence, its diffusion layer is not disturbed by the surface. This constant diffusion layer is used in SECM to investigate surfaces.

17.2.2 Feedback mode

In feedback mode, the UME is immersed in a solution containing a redox mediator (*i.e.* an electroactive species).[18,40] A constant potential is applied, during the same time, the tip approaches toward the surface. The type of surface and the species present in the solution will affect the UME response current when the tip is brought close to the surface. The curve obtained is called an approach curve (Figure 3(a)). The approach curves are usually represented by plotting the current at the tip normalized by the current far from the surface ($I_T = i_T/i_{T,\infty}$) in function of the tip-surface distance normalized by the UME radius ($L = d/a$). There are two limiting cases called

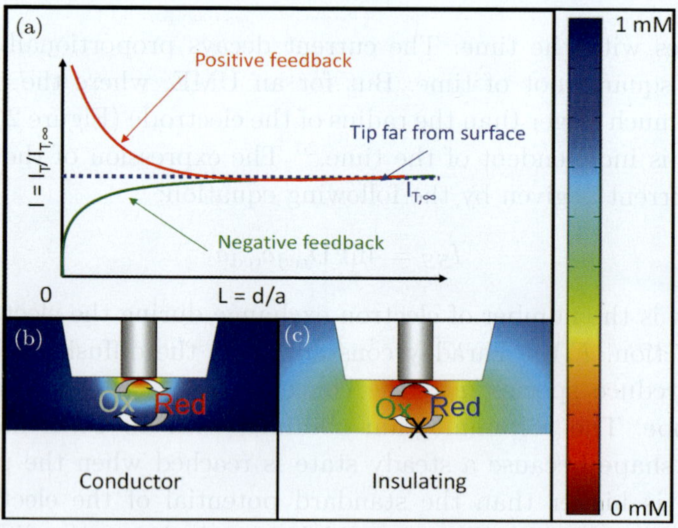

Figure 3. (a) Approach curves for negative and positive feedback. Concentration profile for a reduced species; (b) for a positive feedback; (c) for a negative feedback.

negative feedback and positive feedback. When the UME approaches an insulating surface, as a glass substrate, the species generated at the UME cannot react at the insulator (Figure 3(c)). By reducing the distance between the tip and the insulator the current measure decreases $i_T < i_{T,\infty}$, because the insulator surface blocks the diffusion of the species that should react at the UME. If the surface is a conductor, the rate of the reaction on the surface can be controlled by the potentiostat. That is why the SECM usually function with a bipotentiostat so the potential can be applied on both the tip and the surface under investigation. In the case of positive feedback $i_T > i_{T,\infty}$, when the distance between the tip and the surface decreases the current increases. The species electrogenerated at the UME can react with the surface by consequence an additional flux of the initial species provided by the surface will go to the probe (Figure 3(b)).

The approach curves depend on the rate constant (k_{app}) for electron transfer between the electrogenerated mediator and the surface. For an insulator, $k_{\text{app}} \rightarrow 0$ and for a conductor $k_{\text{app}} \rightarrow \infty$, between these two limit cases, there exists intermediate values of k_{app} (Figure 4).

Moreover, one of the strong advantage of the SECM is the possibility to realize an approach above a conducting surface without applying a potential to the substrate (Figure 5). In the case where the solution contains only one form of the electroactive couple Ox/Red, for example, only the oxidized form. The conductive substrate has a potential following the Nernst equation,

$$E_S = E^0 + \frac{RT}{nF} \ln \frac{[\text{Ox}]}{[\text{Red}]}. \tag{3}$$

In this case, [Red] ≈ 0 and $E_S \gg E^0$. By consequence, all the reduced species reaching the conductor surface will be oxidized. In this case, when the distance between the tip and the surface decreases, the current increases $i_T > i_{T,\infty}$, it is a positive feedback.

17.2.3 Generation collection mode

There are two different generation-collection modes: the tip generation/surface collection (TG/SC) mode and the tip collection/surface

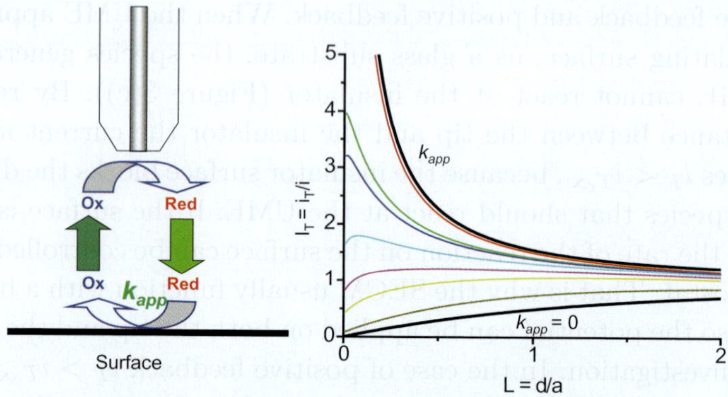

Figure 4. Left, scheme of the reaction between the reduce species electrogenerated at the UME and the surface. The kinetic of this reaction is called k_{app}. Right, approach curves for different value of K_{app} are represented.

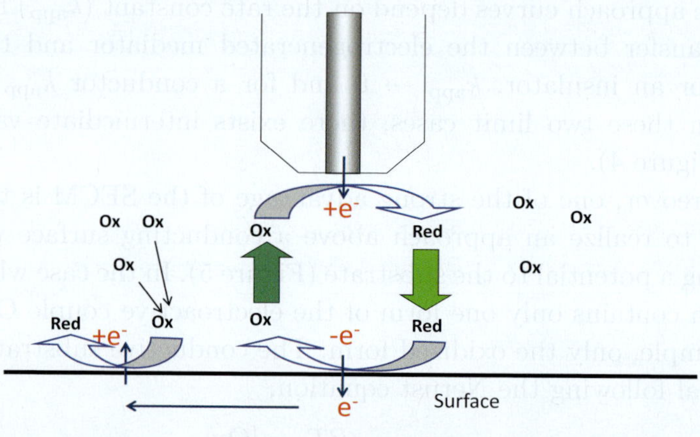

Figure 5. Scheme explaining the positive feedback on open circuit surfaces.

generation mode (TC/SG) (Figure 6). In the TG/SC experiment, the tip electrogenerates a species that diffuse to the surface to be detected. For example, if the oxidation reaction $R \rightarrow O + e^-$ takes place at the tip, $O+e^- \rightarrow R$ will take place on the surface. The rotating ring-disk electrode or the thin layer cells have the same principle. Homogenous chemical reactions are usually studied with this mode by measuring the collection efficiency as a function of d.[23,41,42] The collection efficiency is the ratio between the substrate current and the tip current (i_S/i_T). If there is no homogenous reaction, because

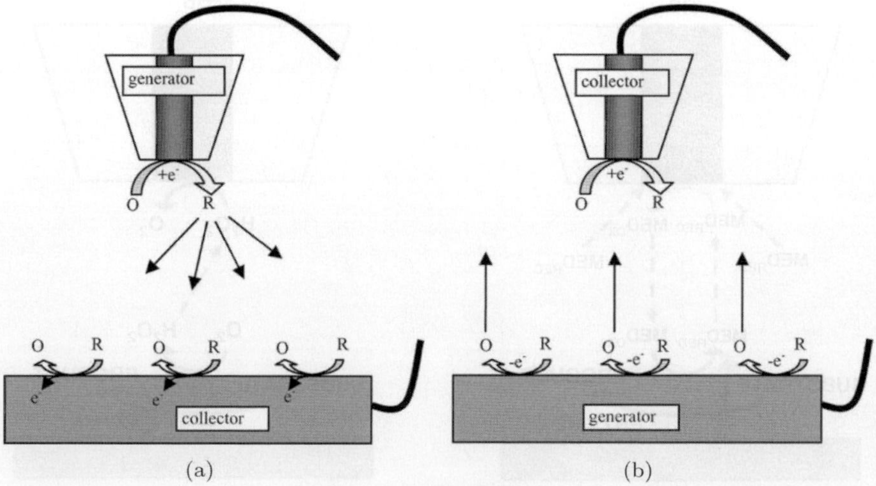

Figure 6. Scheme of (a) Tip generator/substrate collector mode. The electrogenerated species R at the tip diffuses to the substrate to be reoxidized. (b) Substrate generator/tip collector mode. *O* is electrogenerated at the substrate and diffuse to the tip and is consumed at the tip. In these two modes, the tip and the substrate currents are recorded. Reproduced with permission from Ref. 23.

the substrate is usually larger than the tip, the coefficient efficiency is 100% for $d < a$. The other mode SG/TC was the first mode used in SECM historically. It was mainly used to measure concentration profiles over the surface.[43]

Immobilized enzymes can be studied using these two different modes (Figure 7). For a qualitative point of view, by moving the tip above the enzymes, in the good conditions, it is possible to obtain an electrochemical image of the enzyme activity. For a more quantitative measurement, it is better to use approach curves made using the FB mode. In the second part of this chapter, we will discuss about the extraction of kinetic measurement by SECM on an enzymatic reaction by using the FB mode and also by the GC mode.

17.3 Kinetic Determination of Enzymatic Activity by SECM

The scheme in Figure 7 describes the analysis of the immobilized enzyme in FB mode with a mediator in solution and in GC mode

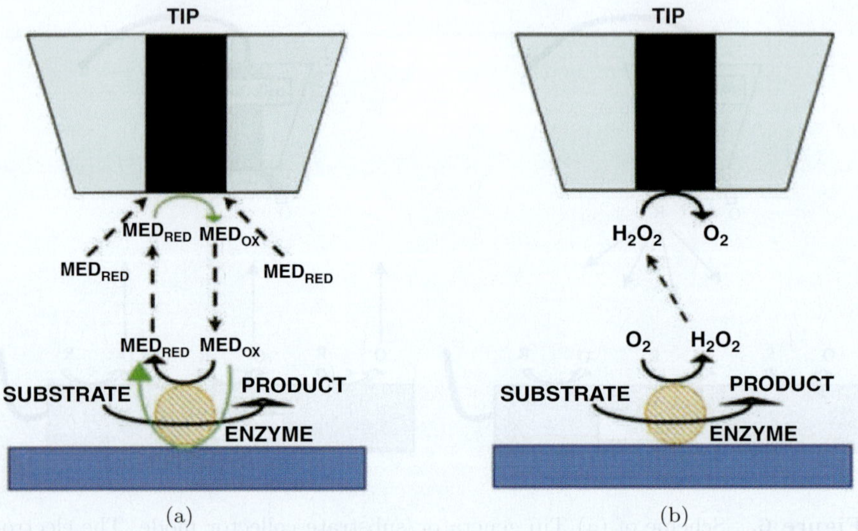

Figure 7. Scheme of the principles of SECM on an immobilized enzyme. (a) In feedback mode, (b) In generation collection mode. Reproduced with permission from Ref. 15. All rights reserved. © Royal Society of Chemistry.

where the dixoygen is consumed by the enzyme and the produced H_2O_2 is detected by the UME.

17.3.1 Kinetic extraction in feedback mode

During the past few years, the enzyme that has been most studied by SECM is an oxydoreductase, the GOx.[37,44–46] This enzyme is the most popular because it is used to monitor the glucose concentration in blood. GOx can be found in the construction of glucose biosensors[1–3] or in biofuel cells.[4,5] That is why study its properties after its immobilization on surfaces is crucial. The immobilization of enzymes can affect their catalytic properties, and this needs to be verified. Furthermore, the miniaturization of this type of devices requires tool like SECM, which can be used to study locally the behavior of the enzyme, regardless of whether the enzyme is immobilized on a conductive or insulating surface.[46–51] GOx catalyzes the oxidation of glucose to gluconolactone; in normal condition, during the catalysis O_2 is reduced in H_2O_2. But most of the SECM experiments are done in inert atmosphere and an electroactive mediator is used to replace O_2. Indeed, even if the couple O_2/H_2O_2 can be considered as the natural electroactive mediator for GOx, because of the low solubility of O_2 in physiological conditions (around 0.2 mM), it is far from being the optimal mediator.[2] The mediators that replace O_2 are most of the time hexacyano-complexes of iron,[13,52,53] ruthenium,[54,55] and cobalt.[56]

The oxidized species is generated at the UME surface, and this species will serve as an electron acceptor instead of oxygen for GOx. The enzymatic reaction starts under the tip because it is where the oxidized species is formed and during the reaction the initial reduced species is regenerated. As in positive feedback, the additional flux of oxidant increases the current. The additional flux needs to be high enough to compensate for the decrease of current due to the hindering of the mediator diffusion as a consequence of the surface proximity.

The first SECM experiment on enzyme has been made to study the GOx covalently attached on insulating surface or trapped within porous membrane.[12] In this work, to be able to measure the apparent kinetics of the surface-catalyzed electron-transfer reaction, theory for the SECM feedback current was developed to model the zero- and the first-order electron transfer kinetics expected for such surface catalysis.

To be able to study the kinetic of the electrocatalyst for the immobilized enzyme, three criteria are required:

(1) The tip must be able to operate under diffusion control (no inhibition because of the enzymatic reaction).
(2) The enzymatic reaction must not be disturbed by other species.
(3) The regeneration rate of the enzymatic reaction has to be high enough to compete with the mass transport from the bulk of the solution.

These criteria are described in this equation

$$k_{\text{cat}}\Gamma_{\text{enzyme}} \geq 10^{-3}\frac{Dc}{a}, \tag{4}$$

where Γ_{enzyme} is the enzyme surface coverage, D the diffusion coefficient, c the concentration, and a the UME radius.

The shape of the approach curves depends upon the rate at which the enzyme catalytically regenerates the reduction species. Quantitatively, the catalytic reaction under substrate saturation (for example, high glucose concentration) is characterized by the rate constant k_{cat} (also called k_2). The rate-determining step is the reaction between Ox and GOx_{red}.

For the case of GOx the simplified reaction mechanism is:

$$GOx_{ox} + \text{glucose} \underset{k_1}{\overset{k_{-1}}{\rightleftarrows}} [GOx_{ox} - \text{glucose}],$$

$$[GOx_{ox} - \text{glucose}] \xrightarrow{k_2} GOx_{red} + \text{gluconolactone},$$

$$GOx_{red} + Ox \xrightarrow{k_2} GOx_{ox} + \text{red}.$$

For large excess of glucose, when the concentration of glucose is higher than the Michaelis–Menten constant, this reaction mechanism leads to the following simplified rate law:

$$\frac{d[\mathrm{Ox}]}{dt} = \frac{2k_2\Gamma_{\mathrm{GOx}}}{\frac{K_{M,\mathrm{Ox}}}{[\mathrm{Ox}]} + 1} \quad \text{with } K_{M,\mathrm{Ox}} = \frac{k_2}{k_3}.$$

Because the mass transport rate depends on the distance between the surface and the tip, the approach curve will give information about the kinetics of the enzymatic reaction but only in the case where the tip reaction is not diffusion-controlled. This pioneering work of Bard *et al.* demonstrates the potential of SECM in feedback mode for quantifying and imaging enzymatic surface reactions, and also the influence of the immobilization method on the SECM response. They studied the effect of the insulating surface morphology on feedback response and differentiated between zero- and first-order reaction kinetics for different redox mediators.[12]

In another work, Matsue *et al.* analyze quantitatively by SECM a monolayer of diaphorase immobilized on glass[57] and gold surfaces.[58–60] Diaphorase is an enzyme that catalyzes the oxidation of a nicotinamide coenzyme (NADH) by a redox mediator. On the glass surface, with ferrocenymethanol as the mediator, without NADH, the oxidation current at the UME decreases when the tip approaches the enzyme layer (negative FB). But in presence of the enzyme, the mediator is regenerated (positive FB), but the regeneration is not limited by diffusion. Then, it was possible to extract the quantitative parameter K_m by fitting the experimental curve with those obtained by digital simulation.

Reaction at the UME:

$$M_{\text{red}} \rightarrow M_{\text{ox}} + e^{-}.$$

Reaction between the mediator and the enzymatic monolayer:

$$2M_{\text{ox}} + Dp_{\text{red}} \rightarrow 2M_{\text{red}} + Dp_{\text{ox}}.$$

Reaction between the enzyme and the substrate:

$$Dp_{\text{ox}} + \text{NADH} \rightarrow Dp_{\text{red}} + \text{NAD}^{+},$$

where M_{ox}, and M_{red} are respectively the oxidized and reduced forms of the mediator, and Dp_{ox} and Dp_{red} are respectively, the oxidized and reduced forms of diaphorase (Figure 8).

It is common to have an enzyme immobilized on a conducting surface, for example to design electrochemical sensors or biofuel cells, but analyzing this type of system with SECM in FB mode

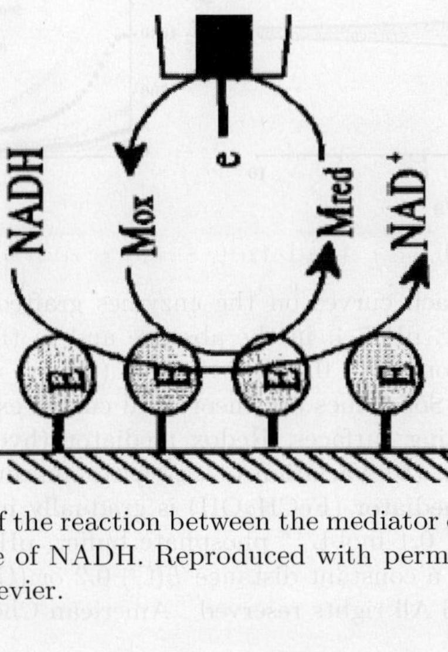

Figure 8. Scheme of the reaction between the mediator catalyzed by diaphorase monolayer in presence of NADH. Reproduced with permission from Ref. 60. All rights reserved © Elsevier.

can be difficult. Indeed, the current at the UME depends on two parameters: the catalytic activity of the enzyme and the reaction of the electroactive species with the conducting surface. That is why, FB mode is limited to very active enzymatic systems with high turnover rates.[12,46,61,62] Pellissier *et al.* were able to use the FB mode on a conducting surface by controlling the penetration of the mediator inside the enzymatic layer. They used the *in situ* generation of the aryldiazonium[66–68] salt to obtain a thick organic layer with amino group that can be further linked to the GOx thanks to a cross-linker. By optimization of the blocking properties of the organic layer to the mediator, enzymatic catalysis can easily be detected. In Figure 9, approaches are made on the glassy carbon electrode modified by the cross-linked GOx layer. The experiment starts in the absence of glucose, and then the successive addition of glucose was made, allowing the approach curve to gradually switch from a negative to a positive feedback until the concentration of glucose is no longer the limiting parameter.

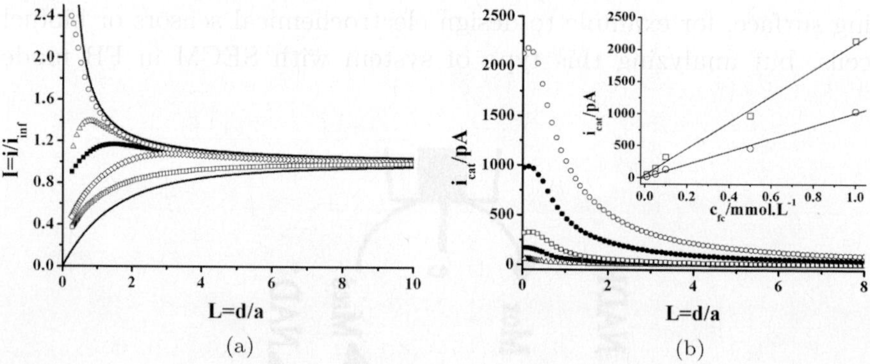

Figure 9. (a) Approach curves on the enzymes grafted on glassy carbon in 0.1 M phosphate buffer, pH 7.4, in the absence and in the presence of glucose at different concentrations. (□) 0, (◊) 6×10^{-4}, (■) 1.3×10^{-3}, (△) 3×10^{-3}, (O) 3×10^{-2} mol L^{-1}. Solid lines are theoretical curves expected for the totally insulating and conducting surfaces. Redox mediator (hydroxymethyl)ferrocene at concentration 1 mM. (b) Variation of i_{cat} with the dimensional distance L. The concentration of mediator ($FcCH_2OH$) is gradually increased by successive addition of mediator in 0.1 mol·L^{-1} phosphate buffer, pH 7.4. The insets show the variations of i_{cat} at a constant distance L(□) 0.2 or (O) 1. Reproduced with permission from Ref. 46 All rights reserved. American Chemical Society.

Extracting the current due to the enzymatic catalytic activity allows to be more quantitative[69]

$$i_{\text{cat}}(L) = (I(L) - I_{\text{ins}}(L)) \cdot i_{T,\infty},$$

where I_{ins} is the normalized current expected when approaching an insulating surface.

The same principle was used on carbon electrode modified by polypyrrole[63] and on gold surface modified by a thioalkane SAMs.[64,65]

17.3.2 Using the GC mode

As explained above, in GC mode, the UME is close to the surface and detects an electroactive species generate by the surface. This mode is similar to other applications of UME where the tip is used to map the concentration profiles of electroactive species generated at a larger electrode.[70] The main interest of the GC mode in the possibility of detect different molecules produced by the enzymatic reaction and its good sensitivity.[71] In the GC mode, the shape of the approach curve depends on the ratio between the enzymatic activity and the reagent bulk concentration. In this case, the current first increases and then at small distances reaches a maximum before decreasing sharply (Figure 10).[37] In opposition to the classic negative FB, the diffusion of the electroactive species to the tip is hindered, limiting the tip current. In this case, it is the reaction on the surface that is limited by the hindered diffusion. But when the tip is far from the surface ($d > 5a$), the tip is a passive sensor and does not perturb the local concentration. In FB mode, the species product at the UME reacts with the surface that allows to confine the enzymatic reaction below the UME.[72] In GC mode, the enzymatic reaction is less dependent on the presence of the UME, and the reaction may occur simultaneously across the whole surface. But in this mode, it is not possible to obtain directly the distance between the tip and the substrate. One possibility to overcome this difficulty is to add an electroactive species that will not react with the enzyme but that can be used in the negative FB mode to know precisely the distance between the tip and the substrate.

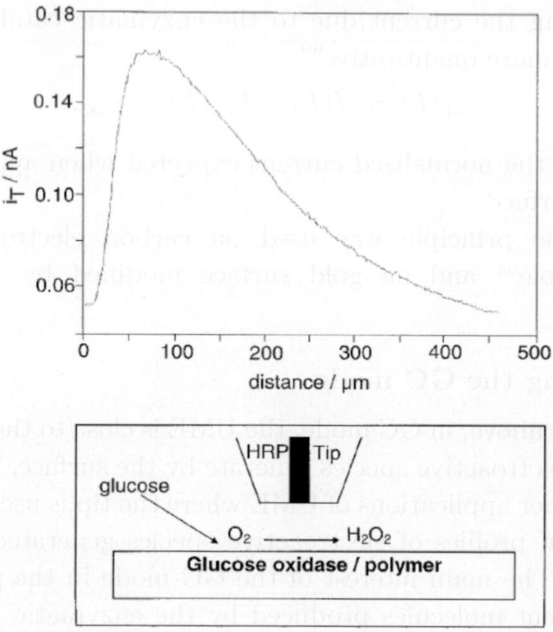

Figure 10. Approach curve using an UME cover with horseradish peroxidase to detect H_2O_2 generated by the glucose oxidase in presence of glucose and oxygen. Reproduced with permission from Ref. 37. All rights reserved. © American Chemical Society.

The GC mode offers a better sensitivity than the FB mode. Indeed, in the FB mode, the contribution of the enzymatic flux to the UME signal must be distinguished from the background current due to the mediator from the bulk of the solution. By consequence, lower activities of immobilized enzyme can be detected. Using the following equation, it is possible to obtain the minimum catalytic rate (k_{cat}):

$$k_{\text{cat}}\Gamma_{\text{enzyme}} \geq \frac{Dc_{\text{lim}}}{r_S}.$$

where c_{lim} is the detection limit of the tip, Γ_{enzyme} is the enzyme surface coverage, D the diffusion coefficient, and r_s the surface radius. We can estimate that enzymes with $k_{\text{cat}} > 1\ \text{s}^{-1}$ may be studied if $\Gamma_{\text{enzyme}} = 10^{-12}\ \text{mol cm}^{-2}$, $D = 5 \times 10^{-6}\ \text{cm}^2\ \text{s}^{-1}$, $r_s = 5 \times 10^{-3}$ cm, and c_{lim} is estimated at $10^{-6}\ \text{mol dm}^{-3}$.

17.4 Imaging Enzymatic Reactions

Imaging a surface with the GC mode leads to poor resolution because spatial resolution is governed by the mass transport of the product (diffusion, natural convection). Then only little information about the mechanism involved in the modified electrode can be extracted. By contrast, the FB mode can have a very good spatial resolution because it will depend on the size of the UME. Also it is possible in this mode to perform detailed mechanistic investigations.

Wittstock *et al.*[44] used the SECM to first prepare enzymatically active spots on self-assembled monolayers on gold. They locally immobilized a periodate-oxidized GOx and used the imaging capabilities of SECM to map the enzymatic activity. In this work, both modes, FB and GC, were used. They obtained better results with the GC mode because with the FB mode interference due to the competition between mediator regeneration by the enzymatic reaction and by the direct electron transfer at the gold surface (Figure 11). They demonstrated that GC imaging is a powerful tool for the analysis of biosensing surfaces because the results obtained are independent of the nature of the support onto which the biomolecules are immobilized. It is also easier to achieve higher sensitivity.

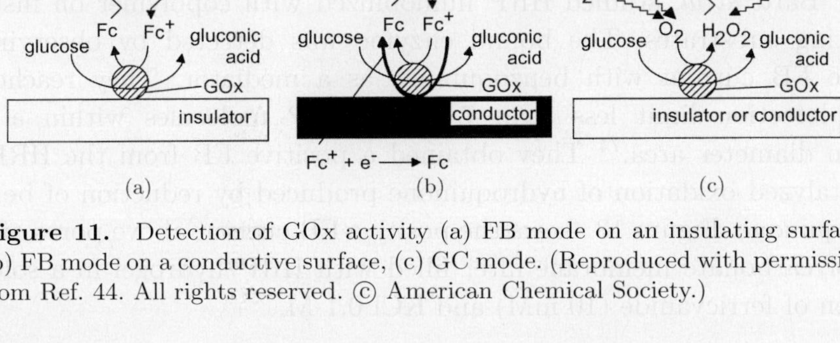

Figure 11. Detection of GOx activity (a) FB mode on an insulating surface, (b) FB mode on a conductive surface, (c) GC mode. (Reproduced with permission from Ref. 44.

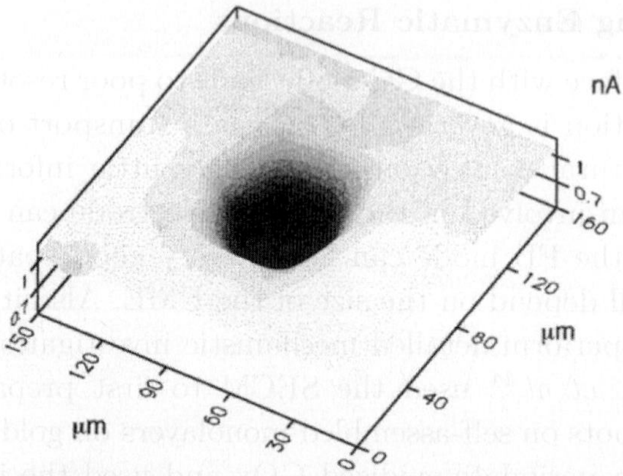

Figure 12. SECM image of a deactivated circle of diaphorase-immobilized surface in a 0.5 mM FMA, 5 mM NADH in 1 M KCl, 0.1 M phosphate buffer solution. Reproduced with permission from Ref. 59. All rights reserved. American Chemical Society.

Matsue *et al.*,[59] as in the Wittstock work, used the SECM to prepare and to characterize the micropatterning of enzyme (Figure 12). But in this work, this is a diaphorase, a Flavin enzyme that had been immobilized. Diaphorase catalyzes the oxidation of a NADH by a redox mediator. It can be used to build NADH sensing devices.[58,73] The SECM images were used to determine the catalytic activity of the immobilized diaphorase. From the image, they extracted the surface concentration of the active diaphorase (Γ_{Dp}). In order to do that, digital simulation with an explicit finite difference method incorporating the heterogeneous enzyme reaction at the surface was used. They calculated theoretically the dependence of the maximum current as a function of Γ_{Dp}.

Bard *et al.* studied HRP immobilized with copolymer on insulating substrates. The bound enzyme was detected by observing the FB current with benzoquinone as a mediator. They reached a detection limit less than 7×10^5 HRP molecules within a 7 μm diameter area.[74] They obtained a positive FB from the HRP-catalyzed oxidation of hydroquinone produced by reduction of benzoquinone. Figure 13 shows the positive FB current above pores of a polycarbonate membrane filter filled with HRP/hydrogel in a solution of ferricyanide (10 mM) and KCl 0.1 M.

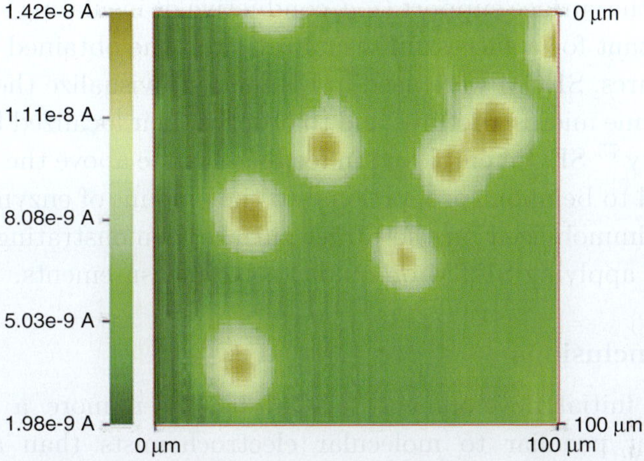

Figure 13. SECM image of the polycarbonate filter membrane supported by an electrode. Reproduced with permission from Ref. 74. All rights reserved. © American Chemical Society.

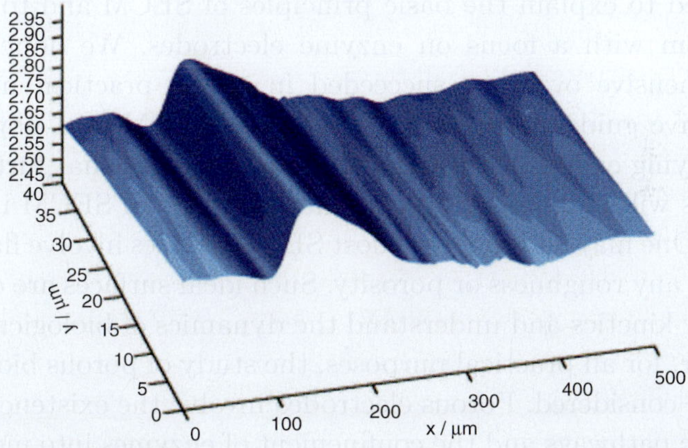

Figure 14. SECM image in FB mode of the microstructure (width 100 μm) of hydrogel containing the enzyme. Reproduced with permission from Ref. 75. All rights reserved. © Elsevier.

Schumann *et al.* developed a micropattern of complex biosensor architectures based on redox hydrogel. These systems are used in amperometric ethanol or glucose biosensors. They were studied by SECM after immobilization on gold, graphite, or glass surface. The choice of the surface support (*e.g.* conductive or insulating) proved to be important for a successful visualization of the obtained biosensor architectures. SECM were used in FB mode to visualize the immobilized enzyme microstructures and to observe their localized biochemical activity.[75] SECM employed in feedback mode above the biosensor was found to be about proportional to the amount of enzyme/redox-hydrogel immobilized on the target surface, demonstrating the possibility of applying SECM on quantitative measurements.

17.5 Conclusions

Since its initial development, SECM has been more a specialist instrument peculiar to molecular electrochemists than a democratized tool accessible to biochemists, materials chemists, and interdisciplinary communities. The obscure formalism employed by electrochemists, together with its recent development, could partially explain the relative inaccessibility of this technique. In this context, we aimed to explain the basic principles of SECM and to clarify its formalism with a focus on enzyme electrodes. We hope that this comprehensive overview succeeded in giving practical insight and instructive guidance regarding the use of SECM as a powerful tool for studying enzymatic reactions. We also believe that further developments will be necessary to extend the scope of SECM in the near future. One may assume that most SECM studies involve flat surfaces without any roughness or porosity. Such ideal surfaces are convenient to study kinetics and understand the dynamics of biological systems. However, for all practical purposes, the study of porous bioelectrodes must be considered. Porous electrodes involve the existence of different ionic pathways and the confinement of enzymes into meso and/or macropores. It is clear that the development of new tools to study porous enzyme electrodes via SECM will be the next step.

List of Abbreviations

AFM	atomic force microscopy
DNA	deoxyribonucleic acid
FB	feedback

GC	generation collection
GOx	glucose oxidase
HRP	horseradish peroxidase
NAD	nicotinamide adenine dinucleotide
NAD^+/NADH	oxidized and reduced state (with two electrons and one proton) of NAD
SAM	self-assembled monolayer
SC	surface collection
SECM	scanning electrochemical microscopy
SG	surface generation
SHE	standard hydrogen electrode
STM	scanning tunneling microscopy
TC	tip collection
TG	tip generation
UME	ultramicroelectrode

References

1. J. Wang (2008). Electrochemical glucose biosensors. *Chemical Reviews* **108**(2), 814–825.
2. A. Heller and B. Feldman (2008). Electrochemical glucose sensors and their applications in diabetes management. *Chem. Rev.* **108**(7), 2482–2505.
3. S. Cosnier (1999). Biomolecule immobilization on electrode surfaces by entrapment or attachment to electrochemically polymerized films: A review. *Biosensors and Bioelectroncis* **14**(5), 443–456.
4. S. Calabrese Barton, J. Gallaway and P. Atanassov (2004). Enzymatic biofuel cells for implantable and microscale devices. *Chemical Reviews* **104**(10), 4867–4886.
5. F. Mao, N. Mano and A. Heller (2003). Long tethers binding redox centers to polymer backbones enhance electron transport in enzyme "wiring" hydrogels. *Journal of American Chemical Society* **125**(16), 4951–4957.
6. S.J. Updike and G.P. Hicks (1967). The enzyme electrode. *Nature* **214**(5092), 986–988.
7. R. Czajka, C. Koopal, M. Feiters, J. Gerritsen, R. Nolte and H. Vankempen (1992). Scanning tunneling microscopy study of polypyrrole films and of glucose-oxidase as used in a 3rd-generation biosensor. *Bioelectrochemistry and Bioenergy* **29**(1), 47–57.
8. Q. Chi, J. Zhang, S. Dong and E. Wang (1994). Direct observation of native and unfolded glucose-oxidase structures by scanning-tunneling-microscopy. *Journal of Chemical Society Faraday Trans.* **90**(14), 2057–2060.

9. C. Lanzellotto, G. Favero, M.L. Antonelli, C. Tortolini, S. Cannistraro, E. Coppari and F. Mazzei (2014). Nanostructured enzymatic biosensor based on fullerene and gold nanoparticles: Preparation, characterization and analytical applications. *Biosensors and Bioelectroncis* **55**, 430–437.
10. S. Han, M. Zhu, Z. Yuan and X. Li (2001). A methylene blue-mediated enzyme electrode for the determination of trace mercury(II), mercury(I), methylmercury, and mercury–glutathione complex. *Biosensors and Bioelectroncis* **16**(1–2), 9–16.
11. M.A. Rahman, N.-H. Kwon, M.-S. Won, E.S. Choe and Y.-B. Shim (2005). Functionalized conducting polymer as an enzyme-immobilizing substrate: An amperometric glutamate microbiosensor for *in vivo* measurements. *Analytical Chemistry* **77**(15), 4854–4860.
12. D.T. Pierce, P.R. Unwin and A.J. Bard (1992). Scanning electrochemical microscopy. Studies of enzyme-mediator kinetics for membrane- and surface-immobilized glucose oxidase. *Analytical Chemistry* **64**(17), 1795–1804.
13. S.A. Jaffari and A.P.F. Turner (1997). Novel hexacyanoferrate(III) modified graphite disc electrodes and their application in enzyme electrodes — Part I. *Biosensors and Bioelectroncis* **12**(1), 1–9.
14. G. Wittstock (2001). Modification and characterization of artificially patterned enzymatically active surfaces by scanning electrochemical microscopy. *Fresenius J. Anal. Chem.* **370**(4), 303–315.
15. E. Casero, L. Vazquez, A. Maria Parra-Alfambra and E. Lorenzo (2010). AFM, SECM and QCM as useful analytical tools in the characterization of enzyme-based bioanalytical platforms. *Analyst* **135**(8), 1878–1903.
16. L. Stoica, S. Neugebauer and W. Schuhmann (2008). Scanning electrochemical microscopy (SECM) as a tool in biosensor research. In R. Renneberg and F. Lisdat (eds.), *Biosensing for the 21st Century*, Springer-Verlag Berlin: Berlin, Vol. 109, pp. 455–492.
17. Benjamin R Horrocks Gunther Wittstock (2012). Biotechnological applications. In *Scanning Electrochemical Microscopy*, Second Edition; CRC Press, pp. 317–378.
18. A.J. Bard, F.R.F. Fan, J. Kwak and O. Lev (1989). Scanning electrochemical microscopy. Introduction and principles. *Analytical Chemistry* **61**(2), 132–138.
19. D. Mandler and A.J. Bard (1989). Scanning electrochemical microscopy: The application of the feedback mode for high resolution copper etching. *Journal of Electrochemical Society* **136**(10), 3143–3144.
20. C. G. Zoski (2002). Ultramicroelectrodes: Design, fabrication, and characterization. *Electroanalysis* **14**(15–16), 1041–1051.
21. C. Amatore, M.R. Deakin and R.M. Wightman (1987). Electrochemical kinetics at microelectrodes. *Journal of Electroanalytical Chemistry and Interfacial Electrochemistry* **225**(1), 49–63.
22. A.J. Bard, G. Denuault, C. Lee, D. Mandler and D.O. Wipf (1990). Scanning electrochemical microscopy — a new technique for the characterization and modification of surfaces. *Accounts of Chemical Research* **23**(11), 357–363.

23. P. Sun, F.O. Laforge and M.V. Mirkin (2007). Scanning electrochemical microscopy in the 21st century. *Physical Chemistry and Chemical Physics* **9**(7), 802–823.
24. T.J. Schroeder, J.A. Jankowski, K.T. Kawagoe, R.M. Wightman, C. Lefrou and C. Amatore (1992). Analysis of diffusional broadening of vesicular packets of catecholamines released from biological cells during exocytosis. *Analytical Chemistry* **64**(24), 3077–3083.
25. C.M. Frey, A. Eifert, H. Schütz, H. Barth, B. Mizaikoff and C. Kranz (2016). Macroscopic and microscopic electrochemical investigation of clostridium botulinum C2IIa embedded in supported lipid membranes. *Electrochimica Acta* **209**, 341–349.
26. I. Morkvenaite-Vilkonciene, A. Ramanaviciene and A. Ramanavicius (2016). 9,10-phenanthrenequinone as a redox mediator for the imaging of yeast cells by scanning electrochemical microscopy. *Sensors and Actuators B: Chemical* **228**, 200–206.
27. H. Shiku, T. Shiraishi, H. Ohya, T. Matsue, H. Abe, H. Hoshi and M. Kobayashi (2001). Oxygen consumption of single bovine embryos probed by scanning electrochemical microscopy. *Analytical Chemistry* **73**(15), 3751–3758.
28. B. Liu, S.A. Rotenberg and M.V. Mirkin (2000). Scanning electrochemical microscopy of living cells: Different redox activities of nonmetastatic and metastatic human breast cells. *Proceedings of the National Academy of Science USA* **97**(18), 9855–9860.
29. G. Wittstock, K. Yu, H.B. Halsall, T.H. Ridgway and W.R. Heineman (1995). Imaging of immobilized antibody layers with scanning electrochemical microscopy. *Analytical Chemistry* **67**(19), 3578–3582.
30. Z.P. Aguilar, W.R. Vandaveer and I. Fritsch (2002). Self-contained microelectrochemical immunoassay for small volumes using mouse IgG as a model system. *Analytical Chemistry* **74**(14), 3321–3329.
31. W. Song, Z. Yan and K. Hu (2012). Electrochemical immunoassay for CD10 antigen using scanning electrochemical microscopy. *Biosensors and Bioelectroncis* **38**(1), 425–429.
32. K. Yamashita, M. Takagi, S. Takenaka, K. Uchida and H. Kondo (2001). Visualization of DNA microarrays by scanning electrochemical microscopy (SECM). *The Analyst* **126**(8), 1210–1211.
33. N. Moradi, A. Noori, M.A. Mehrgardi and M.F. Mousavi (2016). Scanning electrochemical microscopy for electrochemical detection of single-base mismatches by tagging ferrocenecarboxylic acid as a redox probe to DNA. *Electroanalysis* **28**(4), 823–832.
34. B. Chen, Q. Hu, Q. Xiong, F. Zhang and P. He (2016). An ultrasensitive scanning electrochemical microscopy (SECM)-Based DNA biosensing platform amplified with the long self-assembled DNA concatemers. *Electrochimica Acta* **192**, 127–132.
35. J. Wang, F. Song and F. Zhou (2002). Silver-enhanced imaging of DNA hybridization at DNA microarrays with scanning electrochemical microscopy. *Langmuir* **18**(17), 6653–6658.

36. B. Liu, A.J. Bard, C.Z. Li and H.B. Kraatz (2005). Scanning electrochemical microscopy. Studies of self-assembled monolayers of DNA in the absence and presence of metal ions. *Journal of Physical Chemistry B* **109**(11), 5193–5198.
37. B. Horrocks, D. Schmidtke, A. Heller and A. Bard (1993). Scanning electrochemical microscopy. 24. Enzyme ultramicroelectrodes for the measurement of hydrogen-peroxide at surfaces. *Analytical Chemistry* **65**(24), 3605–3614.
38. D. Koley and A.J. Bard (2010). Triton X-100 concentration effects on membrane permeability of a single HeLa cell by scanning electrochemical microscopy (SECM). *Proceedings of the National Academy of Science* **107**(39), 16783–16787.
39. K. Aoki (1993). Theory of ultramicroelectrodes. *Electroanalysis* **5**(8), 627–639.
40. J. Kwak and A.J. Bard (1989). Scanning electrochemical microscopy. Apparatus and two-dimensional scans of conductive and insulating substrates. *Analytical Chemistry* **61**(17), 1794–1799.
41. F. Zhou, P.R. Unwin and A.J. Bard (1992). Scanning electrochemical microscopy. 16. study of second-order homogeneous chemical reactions via the feedback and generation/collection modes. *Journal of Physical Chemistry* **96**(12), 4917–4924.
42. C. Lee, J. Kwak and F.C. Anson (1991). Application of scanning electrochemical microscopy to generation/collection experiments with high collection efficiency. *Analytical Chemistry* **63**(14), 1501–1504.
43. J.V. Macpherson, J.-P.G. Mussy and J.-L. de Delplancke (2002). High-resolution electrochemical, electrical, and structural characterization of a dimensionally stable Ti/TiO2/Pt Electrode. *Journal of Electrochemical Society* **149**(7), B306–B313.
44. G. Wittstock and W. Schuhmann (1997). Formation and imaging of microscopic enzymatically active spots on an alkanethiolate-covered gold electrode by scanning electrochemical microscopy. *Analytical Chemistry* **69**(24), 5059–5066.
45. Y. Takahashi, A.I. Shevchuk, P. Novak, Y. Murakami, H. Shiku, Y.E. Korchev and T. Matsue (2010). Simultaneous noncontact topography and electrochemical imaging by SECM/SICM featuring ion current feedback regulation. *Journal of American Chemical Society* **132**(29), 10118–10126.
46. M. Pellissier, D. Zigah, F. Barriere and P. Hapiot (2008). Optimized preparation and scanning electrochemical microscopy analysis in feedback mode of glucose oxidase layers grafted onto conducting carbon surfaces. *Langmuir* **24**(16), 9089–9095.
47. D. Zigah, M. Pellissier, B. Fabre, F. Barriere and P. Hapiot (2009). Covalent immobilization and SECM analysis in feedback mode of glucose oxidase on a modified oxidized silicon surface. *Journal of Electroanalytical Chemistry* **628**(1–2), 144–147.
48. C.A. Gunawan, E.V. Nam, C.P. Marquis, J.J. Gooding, P. Thordarson and C. Zhao (2016). Scanning electrochemical microscopy of cytochrome c peroxidase through the orientation-controlled immobilisation of cytochrome c. *ChemElectroChem* **3**(7), 1150–1156.

49. L. Stratmann, J. Clausmeyer and W. Schuhmann (2015). Non-destructive patterning of carbon electrodes by using the direct mode of scanning electrochemical microscopy. *ChemPhysChem*, n/a-n/a.
50. N.R. Mohamad, N.H.C. Marzuki, N.A. Buang, F. Huyop and R.A. Wahab (2015). An overview of technologies for immobilization of enzymes and surface analysis techniques for immobilized enzymes. *Biotechnology and Biotechnology Equipment* **29**(2), 205–220.
51. W. Nogala, K. Szot, M. Burchardt, M. Jönsson-Niedziolka, J. Rogalski, G. Wittstock and M. Opallo (2010). Scanning electrochemical microscopy activity mapping of electrodes modified with laccase encapsulated in sol–gel processed matrix. *Bioelectrochemistry* **79**(1), 101–107.
52. S. Zhang, N. Wang, H. Yu, Y. Niu and C. Sun (2005). Covalent attachment of glucose oxidase to an Au electrode modified with gold nanoparticles for use as glucose biosensor. *Bioelectrochemistry* **67**(1), 15–22.
53. S.-H. Lee, H.-Y. Fang and W.-C. Chen (2006). Amperometric glucose biosensor based on screen-printed carbon electrodes mediated with hexacyanoferrate–chitosan oligomers mixture. *Sens Actuators B Chem.* **117**(1), 236–243.
54. R. Pauliukaite, S.B. Hočevar, E.A. Hutton and B. Ogorevc (2008). Novel electrochemical microsensor for hydrogen peroxide based on iron-ruthenium hexacyanoferrate modified carbon fiber electrode. *Electroanalysis* **20**(1), 47–53.
55. A.L. Crumbliss, H.A.O. Hill and D.J. Page (1986). The electrochemistry of hexacyanoruthenate at carbon electrodes and the use of ruthenium compounds as mediators in the glucose/glucose oxidase system. *Journal of Electroanalytical Chemistry Interfacial Electrochemistry* **206**(1), 327–331.
56. M. Florescu and C.M.A. Brett (2004). Development and characterization of cobalt hexacyanoferrate modified carbon electrodes for electrochemical enzyme biosensors. *Analytical Letters* **37**(5), 871–886.
57. H. Shiku, T. Matsue and I. Uchida (1996). Detection of microspotted carcinoembryonic antigen on a glass substrate by scanning electrochemical microscopy. *Analytical Chemistry* **68**(7), 1276–1278.
58. T. Sawaguchi, T. Matsue and I. Uchida (1992). Catalytic capability of diaphorase bound to a self-assembled thiol monolayer at a gold electrode. *Bioelectrochemistry and Bioenergetics* **29**(1), 127–133.
59. H. Shiku, T. Takeda, H. Yamada, T. Matsue and I. Uchida (1995). Microfabrication and characterization of diaphorase-patterned surfaces. *Analytical Chemistry* **67**(2), 312–317.
60. H. Yamada, H. Shiku, T. Matsue and I. Uchida (1994). Microvoltammetric characterization of diaphorase monolayer at glass surfaces. *Bioelectrochemistry and Bioenergetics* **33**(1), 91–93.
61. B. Chen, Q. Hu, Q. Xiong, F. Zhang and P. He (2016). An ultrasensitive scanning electrochemical microscopy (SECM)-based DNA biosensing platform amplified with the long self-assembled DNA concatemers. *Electrochimica Acta* **192**, 127–132.

62. C. Zhao and G. Wittstock (2004). Scanning electrochemical microscopy of quinoprotein glucose dehydrogenase. *Analytical Chemistry* **76**(11), 3145–3154.
63. C. Kranz, G. Wittstock, H. Wohlschläger and W. Schuhmann (1997). Electrochemical microsystem technologies imaging of microstructured biochemically active surfaces by means of scanning electrochemical microscopy. *Electrochimica Acta* **42**(20), 3105–3111.
64. C. Kranz, T. Lötzbeyer, H.-L. Schmidt and W. Schuhmann (1997). Lateral visualization of direct electron transfer between microperoxidase and electrodes by means of scanning electrochemical microscopy. *Biosensors and Bioelectroncis* **12**(4), 257–266.
65. D. Oyamatsu, N. Kanaya, Y. Hirano, M. Nishizawa and T. Matsue (2003). Area-selective immobilization of multi enzymes by using the reductive desorption of self-assembled monolayer. *Electrochemistry* **71**(6), 439–441.
66. M. Delamar, R. Hitmi, J. Pinson and J. M. Saveant (1992). Covalent modification of carbon surfaces by grafting of functionalized aryl radicals produced from electrochemical reduction of diazonium salts. *Journal of American Chemical Society* **114**(14), 5883–5884.
67. C. Bourdillon, M. Delamar, C. Demaille, R. Hitmi, J. Moiroux and J. Pinson (1992). Immobilization of glucose oxidase on a carbon surface derivatized by electrochemical reduction of diazonium salts. *Journal of Electroanalytical Chemistry* **336**(1–2), 113–123.
68. S. Baranton and D. Bélanger (2005). Electrochemical derivatization of carbon surface by reduction of *in situ* generated diazonium cations. *Journal of Physical Chemistry B* **109**(51), 24401–24410.
69. M. Burchardt and G. Wittstock (2008). Kinetic studies of glucose oxidase in polyelectrolyte multilayer films by means of scanning electrochemical microscopy (SECM). *Bioelectrochemistry* **72**(1), 66–76.
70. C. Amatore, K. Knobloch and L. Thouin (2007). Alteration of diffusional transport by migration and natural convection. Theoretical and direct experimental evidences upon monitoring steady-state concentration profiles at planar electrodes. *Journal Electroanalytical Chemistry* **601**(1–2), 17–28.
71. M. Burchardt and G. Wittstock (2013). Micropatterned multienzyme devices with adjustable amounts of immobilized enzymes. *Langmuir* **29**(48), 15090–15099.
72. K. Borgwarth, C. Ricken, D.G. Ebling and J. Heinze (1996). Surface analysis by scanning electrochemical microscopy: Resolution studies and applications to polymer samples. *Analytical Biocatalytical and Chemistry* **356**(3–4), 288–294.
73. H.-C. Chang, A. Ueno, H. Yamada, T. Matsue and I. Uchida (1991). Immobilized-enzyme electrode for nicotinamide adenine dinucleotide (Reduced form)(NADH) sensing and application to the kinetic studies of NADH dependent dehydrogenases. *Analyst* **116**(8), 793–796.

74. J.F. Zhou, C. Campbell, A. Heller and A.J. Bard (2002). Scanning electrochemical microscopy. Imaging of horseradish peroxidase immobilized on insulating substrates. *Analytical Chemistry* **74**(16), 4007–4010.
75. M. Niculescu, S. Gaspar, A. Schulte, E. Csoregi and W. Schuhmann (2004). Visualization of micropatterned complex biosensor sensing chemistries by means of scanning electrochemical microscopy. *Biosensors and Bioelectronics* **19**(10), 1175–1184.

Outlook

Nicolas Brun*,§ and Victoria Flexer†,‡

*Institut Charles Gerhardt de Montpellier, UMR5253 CNRS, ENSCM, Université de Montpellier, France

†Centro de Investigación y Desarrollo en Materiales Avanzados y Almacenamiento de Energía de Jujuy-CIDMEJu (CONICET-Universidad Nacional de Jujuy), Argentina

‡Centro de Investigación y Transferencia Jujuy, CIT-JUJUY, Argentina

§nicolas.brun@enscm.fr

The frontier — or the so-called *white area* — between chemistry and biology was cleverly defined by Nobel Prize laureate Jean-Marie Lehn through a complexity-diversity diagram. While chemistry has been endlessly inspired by the extreme complexity of biology, the diversity of chemistry — limited only by the imagination of the chemist — has been indefinitely supplying biology with novel molecular tools. In particular, biomimetism has been an active field of research in materials science.[1] Chemists have been fascinated by the way biology can synthesize sophisticated materials in thermodynamically unfavorable conditions. In particular, exoskeletons of unicellular organisms, such as diatoms, have been a source of inspiration to develop cooperative and compartmentalized synthetic approaches for the design of innovative hierarchical structures.[2]

Bioelectrochemistry fills another exciting section of this *white area*, where biomimetism stimulates creativity. Bioelectrochemistry is an interdisciplinary field of research where microbiology, synthetic biology, biochemistry, molecular chemistry, electrochemistry, materials science, and engineering overlap and inspire each other. As defined by Philip N. Bartlett, *bioelectrochemistry is the study and application of biological electron transfer processes.*[3] Electron transfer processes at and across biological membranes drive essential reactions in living systems and play a central role in respiration and photosynthesis. Over the last decades, chemists and electrochemists have been studying such bioelectrochemical processes and have been drawing their inspiration from electron transport chains in mitochondria, bacteria and chloroplasts to design and engineer bioelectrochemical systems (BES). The main essence of BES lies in promoting biological electron transfer processes at interfaces and involves wiring biological entities, *i.e.* biofilms, bacteria and/or enzymes, to electrode surfaces. In this context, the design of functional electrodes has been of major importance in such a way that we believed this topic would deserve being addressed in a dedicated book.

Beyond the main scope of this book, the future advances in bioelectrochemistry might rely on our faculty to not only apply biological electron transfer processes, but also mimic and even surpass them. Our multi-disciplinary approach should not be restricted to the direct translation of biological entities from their natural environment to synthetic electrode surfaces. Even though promoting biological electron transfers at interfaces is still challenging, we should rethink our strategy and define new paradigms. Over the last years, considerable progress has been made in probing the molecular structure of enzymes by X-ray crystallographic analyses, magnetic resonance spectroscopy, and molecular dynamic simulations.[4] Thanks to these tools, chemists could not only integrate biological entities into functional electrodes, as described in this book, but could also attempt to reproduce the complexity of biology through the design of active-site synthetic analogs (Figure 1).[5–10] Such analogs have been seen as promising alternatives to fragile and expensive enzymes, with a particular interest for hydrogenases. Interestingly, synthetic active sites could be immobilized onto electrode materials for potential applications in electrolyzers and fuel cells.[11] Artificial metalloenzymes

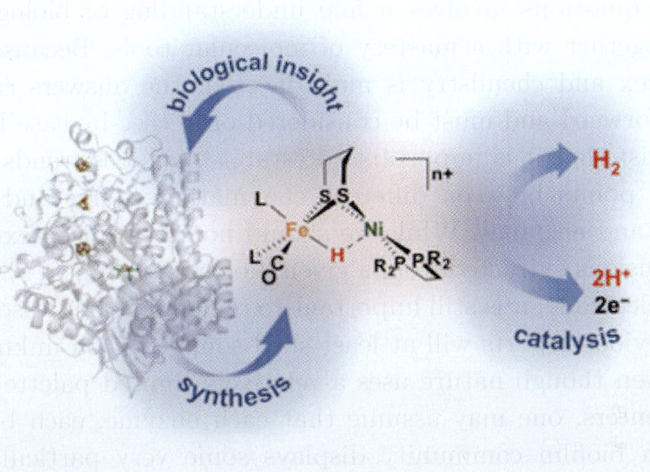

Figure 1. From biological insight to the synthesis of active-site analogs and to the design of bioinspired electrocatalysts. Example of a [NiFe]-hydrogenase. Reproduced with permission from Ref. 9. Copyright 2016, American Chemical Society Publishing.

were also synthesized through the association of active-site synthetic analogs with biological macromolecular hosts, such as proteins.[7,12] These recent developments have led to the emergence of *bioinspired electrochemistry*, bridging the gap between molecular electrochemistry and bioelectrochemistry. To complete the definition of Philip N. Bartlett, while bioelectrochemistry studies and applies biological electron transfer processes, bioinspired electrochemistry studies and mimics (surpasses?) their main molecular components. In a near future, we can presage further breakthroughs in the design of bioinspired electrodes and electrochemical systems.

Besides fundamental aspects regarding both enzymatic and microbial electrochemical systems, we hope that this book endowed you with useful keys to solve general and more specific questions. How to efficiently inject or withdraw electrons to or from biological entities? How to provide enzymes and bacteria with an environment conducive to long-term stability and high electrochemical activity? What electrode material — in terms of chemical nature, structure and morphology — for what application? Answering all of these questions involves a fine understanding of biological processes together with a mastery of molecular tools. Because biology is complex and chemistry is multi-faceted, the answers cannot be straightforward and must be considered on a case-by-case basis. We must insist that it is important to establish general trends and find common points between different enzymatic systems and between bacteria and enzymes. While we should not attempt to extrapolate all conclusions extracted from a given bioelectrochemical system to a new bioelectrode, it is still important to remember that lessons learnt from previous reports will at least shed some light on unknown systems. Even though nature uses a relatively limited palette of redox active centers, one may assume that each enzyme, each bacterium and each biofilm community displays some very particular specificities regarding its own electron transfer mechanism(s), Moreover, the morphology, surface charge density, and surface functionality of the electrode material together with the way the biological entities are immobilized dramatically influence the electron transfer process at the interface. Last but not least, from implantable biosensors to microbial electrosynthesis, bioelectrochemistry finds many practical applications and bioelectrodes can be included in a broad range of devices with radically different needs and constraints.

We hope that this comprehensive overview succeeded in giving practical insight and instructive guidance regarding the design and preparation of functional bioelectrodes. Besides, we hope that it will inspire you to keep on filling the *white area* between chemistry and biology with further inventive contributions and revolutionary innovations.

References

1. C. Sanchez, H. Arribart and M.M.G. Guille (2005). Biomimetism and bioinspiration as tools for the design of innovative materials and systems. *Nature Materials* **4**(4), 277–288.
2. S. Mann, S.L. Burkett, S.A. Davis, C.E. Fowler, N.H. Mendelson, S.D. Sims, D. Walsh and N.T. Whilton, Sol-gel synthesis of organized matter. *Chemistry of Materials* **9**(11), 2300–2310.
3. P.N. Bartlett (2008). *Bioelectrochemistry: Fundamentals, Experimental Techniques, and Applications.* John Wiley & Sons, Ltd.

4. A. Volbeda, P. Amara, C. Darnault, J.M. Mouesca, A. Parkin, M.M. Roessler, F.A. Armstrong and J.C. Fontecilla-Camps (2012). X-ray crystallographic and computational studies of the O-2-tolerant [NiFe]-hydrogenase 1 from Escherichia coli. *Proceedings of the National Academy of Sciences of the United States of America* **109**(14), 5305–5310.
5. C. Tard and C.J. Pickett (2009). Structural and functional analogues of the active sites of the [Fe]-, [NiFe]-, and [FeFe]-hydrogenases. *Chemical Reviews* **109**(6), 2245–2274.
6. G. Berggren, A. Adamska, C. Lambertz, T.R. Simmons, J. Esselborn, M. Atta, S. Gambarelli, J.M. Mouesca, E. Reijerse, W. Lubitz, T. Happe, V. Artero and M. Fontecave (2012). Biomimetic assembly and activation of [FeFe]-hydrogenases. *Nature* **499**(7456), 66–69.
7. T.R. Simmons, G. Berggren, M. Bacchi, M. Fontecave and V. Artero (2014). Mimicking hydrogenases: From biomimetics to artificial enzymes. *Coordination Chemistry Reviews* **270**, 127–150.
8. M. Gomez-Mingot, J.P. Porcher, T.K. Todorova, T. Fogeron, C. Mellot-Draznieks, Y. Li and M. Fontecave (2015). Bioinspired tungsten dithiolene catalysts for hydrogen evolution: A combined electrochemical, photochemical, and computational study. *Journal of Physical Chemistry B* **119**(43), 13524–13533.
9. D. Schilter, J.M. Camara, M.T. Huynh, S. Hammes-Schiffer and T.B. Rauchfuss (2016). Hydrogenase enzymes and their synthetic models: The role of metal hydrides. *Chemical Reviews* **116**(15), 8693–8749.
10. D. Brazzolotto, M. Gennari, N. Queyriaux, T.R. Simmons, J. Pécaut, S. Demeshko, F. Meyer, M. Orio, V. Artero and C. Duboc (2016). Nickel-centred proton reduction catalysis in a model of [NiFe] hydrogenase. *Nature Chemistry* **8**, 1054–1060.
11. A. Le Goff, V. Artero, B. Jousselme, P.D. Tran, N. Guillet, R. Metaye, A. Fihri, S. Palacin and M. Fontecave (2009). From hydrogenases to noble metal-free catalytic nanomaterials for H-2 production and uptake. *Science* **326**(5958), 1384–1387.
12. Y. Lu, N. Yeung, N. Sieracki and N.M. Marshall (2009). Design of functional metalloproteins. *Nature* **460**(7257), 855–862.

Index

A

acetate, 436, 441, 446, 460
acetate production, 444, 460
acetate production rate, 452, 465
Acetobacterium, 140
acetogenesis, 138
acetogenic biocathode, 443
acetogens, 100, 142, 441
Acidiphilium cryptum, 143
Acidithiobacillus ferrooxidans, 114
Acidobacteria, 105
Actinobacteria, 125
Aeruginosa, 537
air cathodes, 97
Alcaligenes, 535
alcohol dehydrogenase, 232, 365
algal microbial solar cells (AMSCs), 102–104, 144
amide, 550
amperometric electrode, 3, 10–11, 35
anion exchange membrane (AEM), 63, 101, 136, 138, 144
anode surface modification, 404
anode surface modification strategies, 408
anodic biofilms, 52, 56
apoenzyme, 179, 354
apparent electron diffusion coefficient, 181
Arcobacter, 140
artificial redox mediator, 8, 15–17, 19, 26–28
ascorbate oxidase, 13
assembly, 345
  drop casting, 345–346
  layer-by-layer, 345, 360, 362
  self-assembled monolayer (SAM), 348
atomic force microscopy (AFM), 136, 144
attenuated total reflectance, 550–552
available biofilm area, 452
azurin, 555–556

B

bacteria, 77, 83, 86–87, 89, 91, 527–532
  Geobacter sulfurreducens, 84–85, 91
  proteobacteria, 92
  Shewanella, 85, 91
bacteria adhesion, 53
bacteria–electrode interaction, 452
bacterial microbial solar cells (*see also* BMSCs), 102–103, 145
bacteroidetes, 135
BEAMRs, 127, 145
benthic unattended generators (BUGs), 98
bilirubin oxidase (*see also* BOD), 13, 26, 385
bio-based economy, 430
bioanodes, 449, 478
biocatalysts, 432, 442, 462, 463

biocathode(s), 102, 122, 129, 134, 141, 384, 429–431, 440, 443, 448, 452, 459, 462, 463
biocathode material, 452, 460–461
biocompatibility, 462
bioelectrochemical systems (BESs), 93, 138, 145, 403, 430, 563, 585, 591
bioelectrochemically assisted microbial reactors (BEAMRs), 127, 145
biofilm, 95, 100, 103, 105–106, 108, 110, 113–115, 117, 119, 121, 133, 136, 141–142, 451, 454, 459–460, 526, 528–534, 536, 538
biofilm coverage, 460
biofilm development, 443–444, 449, 452, 461
biofouling, 136
biofuel, 100, 102–104, 121, 126, 141
biofuel cell, 87, 274, 345, 348, 359, 361
    enzymatic fuel cells, 80–81, 83, 87, 89
    flexible electrode, 348–349
    implantable, 347–348, 350
    microbial fuel cell, 83, 87, 89, 91
biofuel methane, 124
biomimetism, 627
    bioinspired electrochemistry, 629
bioremediation, 95, 121, 141
biosensors, 3, 4, 8–11, 14, 16–19, 21, 25, 35
blue multi-copper oxidases, 355
breakthrough developments, 431
Brevibacillus, 119
Butler–Volmer expression, 278

C

[Cp*Rh(bpy)Cl]$^+$, 230, 241, 245, 252–253, 255, 257
c-type cytochromes (OMCs), 105–106, 109, 111, 129
    cyt *c*, 565–567, 574, 583, 586, 592
    cytochromes, 564, 567, 569, 572, 581–582, 589, 592
Canna indica, 102, 124
carbon aerogel, 388
carbon allotropes, 341
carbon black, 383
carbon cryogel, 388
carbon dioxide, 429
carbon electrode, 90, 341, 381
    carbon materials, 82
    fractured glassy carbon, 78
    highly ordered pyrolytic graphite, 78
carbon fiber brushes, 417
carbon fiber-based materials, 417
carbon gel, 388
carbon nanotube(s), 341, 343, 364, 383, 452
    assembly, 345, 348
    buckypaper, 348, 350
    functionalization, 346, 348, 351, 353–354
    nanocomposite, 364–365, 359
    porosity, 347, 349
carbon paper, 347, 350–351, 385
carbon-based electrodes, 58, 404, 407
carbonized sponge-like electrode, 417
carboxylate platform, 439
carotenoids, 586, 590
casing, 24–26
catalytic, 545, 550, 555
catalytic constants, 19, 32–33
catalytic turnover, 551–553
cathode biofilms, 441
cathode material, 440, 442
cathode potentials, 444
cathodic biofilms, 52, 57, 66
cathodic microbial process, 431, 462
cathodic performance, 416
cation exchange membrane (CEM), 63, 101, 136, 138, 145
cellobiose dehydrogenase, 13
ceruloplasmin, 13
chemical fixation, 536
chemical oxygen demand (COD), 122, 126, 138, 145
chitosan, 346–347, 358

chloroflexi, 529
chronoamperometry
CA, 580–581, 584, 591
Citrobacter, 528
Clostridiaceae, 127
Clostridium, 106, 529
Clostridium ljungdahlii, 100
Clostridium pasteurianum, 113
CO_2, 100, 106, 120, 127, 138, 431–432, 436, 439, 441–442, 447
CO_2 capture, 453
CO_2 conversion, 460
CO_2 into chemicals, 431, 436
CO_2 to acetate, 449, 451
CO_2-based technologies, 431
coating, 418
cofactor, 6, 30
coli, 537
Comamonas, 125
commercial anode materials, 404
commercial electrode, 347, 441, 458
commercially, 458
commercially available carbonaceous electrode, 444, 454–455
commercially available electrodes, 417, 431
commercially available metal materials, 448
comparison, 462
concentration profiles, 275
conducting polymer, 292, 311–312
conductivity, 94, 115, 122, 128, 133, 136–137, 143–144
confocal microscopy, 533
confocal resonance Raman spectroscopy
confocal resonance Raman, 591
CRR, 571, 575, 578, 580, 585, 591
confocal surface-enhanced resonance Raman spectroscopy
confocal surface-enhanced resonance Raman, 591
CSERR, 578, 580, 591
converting and storing electrical energy, 430
coordination, 547, 551, 558
coordination state, 551
coproduct, 14–16, 19
corrosion, 416
Corynebacterium, 143
cosubstrate, 6, 7, 11, 15
coulombic efficiency (*see also* CE), 97–98, 126, 137, 139, 143, 145
coulometric electrode, 3, 10–11
criteria, 429, 431, 454, 464
cross-comparison, 461
cross-comparison of data, 455
cross-linker, 191, 347–348
cryosectioning, 533
CueO, 384
current collectors, 64
current density, 409, 457
cyanobacteria, 102, 127
cyclic voltammetry, 580, 591
CV, 565–569, 579–581, 584, 591
cyclic voltammograms, 195
CymA, 112, 131
cytochrome, 94, 106, 108–110, 112–114, 120, 129, 133, 553–554, 556–557
cytochrome *c* peroxidase, 13
cytochrome P450, 233–235, 246, 249, 251

D

3D, 347, 359, 451, 456
D-fructose dehydrogenase, 384
D-lactate dehydrogenase, 231
D-sorbitol dehydrogenase, 231
dehalococcoides, 528
dehydrogenase, 230
Desulfitobacterium, 118, 528
Desulfitobacterium hafniense, 119
Desulfobulbus, 99
Desulfobulbus propionicus, 114
Desulfovibrio, 120
Desulfovibrio G11, 135
Desulfovibrio paquesii, 135

Desulfovibrio vulgaris, 134
Desulfuromonas, 111, 141
Desulfuromonas acetoxidans, 592
Dace, 567, 580, 583, 592
diaphorase, 229–232, 253, 257
DIET, 105, 119–120, 123, 128, 133, 145
direct electron transfer (DET), 11–13, 17–19, 31, 37–38, 344, 347, 353–356
direct extracellular electron transfer (DEET), 105–106, 110–111, 113–114, 116, 145
dissimilatory metal-reducing bacteria (DMRB), 104–105, 113, 145
DmsA, 131
DmsE, 112
DmsEFA, 112

E

3D electrode material performance, 457
EAB formation impacting factors, 54–55
EABs formation process, 52
EFCs, 8, 23–26, 29–31, 39
effective mass transfer, 443
electric configurations, 408
electric energy, 439
electricigens, 93–94, 100–103, 122–125, 127–128, 132–133, 137, 141–143
electro-fermentation, 436
electroactive biofilm, 408
electroactive microbial biofilms, 564
electroactive microorganisms, 429
electrochemical biosensors, 302, 304, 317, 328
electrochemical impedance spectroscopy (EIS), 568, 592
electrochemical systems (MESs), 93
electrochemically active biofilms, 51
electrochemically active microorganisms (EAMs), 51–52, 54, 65, 93–95, 102, 104, 119, 121, 135, 145
electrode, 462
electrode material, 440, 456, 459, 462, 465
electrode material considerations, 461
electrode modification, 351, 462
electrode volume, 456
electrode-immobilized metalloproteins, 550, 556, 558
electrogen(s), 106, 129, 142
electrogenic, 142
electromethanogenesis, 100
electron diffusion, 276, 280
electron hopping, 274
electron shuttle, 13–14, 16
electron transfer (ET), 10, 12, 16, 19–20, 442–443, 449, 452, 460, 462–463, 563–564, 568–571, 577, 579, 581–582, 584–587, 589–592
electrophoretic deposition, 393
electrotrophs, 93, 100
encapsulation, 386
enterobacter, 140
environmental mode (ESEM), 535–536, 538
enzymatic fuel cells, 3, 22, 381
enzymatic reactions, 600
enzyme activity, 607
enzyme cascades, 30
enzyme electrodes, 273
enzyme engineering, 12, 31
enzyme immobilization, 353, 360, 599
enzyme kinetics, 31–36
enzyme specificity, 3–5, 7, 15, 23, 25
enzyme/mediator coimmobilization, 182
epifluorescence microscopy, 533
EPR spectroscopy, 556
Escherichia coli, 535, 546–547, 552
EXAFS, 556
external diffusion layer, 282
extracellular electron transfer (EET), 94–95, 104–106, 108–119, 121, 123–124, 128, 130–131, 142, 144–145, 408

extracellular electron transfer (EET) mechanisms, 440
extracellular polymeric substances, 533

F

Förster resonance energy transfer, 554
Faradic process, 566, 583
Faradic, 563, 565, 567–568
fermentation, 116, 127
ferrocene, 8, 16, 28, 353
ferrocyanide, 8
finite difference method, 276
Firmicutes, 105, 135, 529
first-generation biosensors, 19
flavin adenine dinucleotide, 6–7, 387
flavoenzyme, 229, 232–233, 245, 251, 253, 257
flow cell, 551, 556
fluorescence, 555
fluorescence imaging (FI), 525–526, 528
fluorescent *in situ* hybridization (FISH), 526–529, 531
fluorescent dye, 554–555
fluorescent protein-based (FPB), 526, 530–532
foam, 347, 359
fructose dehydrogenase, 13, 349
fuel cells, 548, 558
fuel efficiency, 30
fuels or commodities, 430
fully integrated systems, 21–22

G

gas-diffusion layer, 385
gaseous detection device (GDD), 535
GatDH, 246
Geoalkalibacter subterraneus, 114
Geobacter, 105–106, 109, 111, 114–115, 117, 121, 124, 128, 140, 527–530, 532, 535
Geobacter metallireducens, 100, 120, 128
Geobacter sulfurreducens, 100, 105–106, 108–110, 113–115, 119–120, 123, 128–130, 135, 142, 537, 592
Gs, 579–582, 584–585, 588–589, 592
Geobacteraceae, 109, 111, 566, 586
Geothrix, 117
Geothrix fermentans, 114
gluconolactone
glucose, 79
glucose, 78, 80–81
electrochemical glucose sensors, 90
gluconolactone, 78
glucose biosensors, 87
glucose dehydrogenase, 13, 27, 231, 353
glucose oxidase (GOx), 7–8, 14–16, 26–27, 354, 384, 599, 609
Glyceria maxima, 125
graphene, 357
exfoliation method, 358
porosity, 359
graphene oxide, 358–359
porosity, 360
reduction method, 360
green fluorescent protein (GFP), 530

H

H_2, 442, 459
H_2 production, 444
Helicobacter pylori, 118
heme(s), 106, 113, 115, 130–131, 546–547, 551, 553
hemoglobin, 546
heterotrophic, 103
heterotrophs, 139
hexaheme, 109
hierarchical pore, 386
hierarchical pore structure, 347, 382
high surface area, 546
higher market value, 447
holoenzyme, 179
homoacetogens, 127–128, 138

horseradish peroxidase, 13
hydrogel-based, 313–314
hydrogel(s), 292–294, 296, 298, 300, 302–306, 308, 310–314, 316–321, 324–327, 385
hydrogen, 100–101, 104, 116, 119, 121, 123, 126–127, 130–131, 133–136, 138, 141, 442–443, 457, 462
hydrogenase, 345, 356, 548–549, 551–552, 556, 628
synthetic active sites, 628

I

ideal BES electrode, 404
ImcH, 129
immobilized enzyme, 600
impact of electrode surface on microbial electrocatalysis, 404–405
implantable biofuel cells, 23–25, 35, 347–348, 350
implementation of the technology, 465
incomplete data generation or analysis, 431
infrared spectroscopy
IR, 576, 592
IR spectroscopy, 572
integrated enzyme electrodes, 180
integrated systems, 184
ion exchange membrane, 60, 62
isoelectric point, 194

K

K_{MS}, 7
k, 7
k_{cat}, 7
Ketjen black, 383
Koutecky–Levich curves, 286
Koutecky–Levich plot, 282–285

L

L-glutamate dehydrogenase, 231
laccase, 13, 27, 275, 346–347, 349, 355–356, 364–365, 390
Lachnospiraceae, 127
lactoperoxidase, 13
large-scale applicability, 448
large-scale application, 456
large-scale chemical production, 446
layer-by-layer electrostatic self-assembly, 191
linear sweep voltammetry
LSV, 580–581
LipDH, 247
lipoamide dehydrogenase, 230
LIVE/DEAD staining (*see also* LDS), 526, 529–531
lock and key model, 4
low spin, 546

M

3D macroporous electrode materials, 454
M. barkeri, 120
M. harundinacea, 120, 128
MacA, 106, 130
macropores, 382
Magnéli phases, 509
magnesium oxide (MgO)-templated mesoporous carbon, 390
mass transfer, 452, 454
mass transfer limitation, 444, 456
mass transport, 545, 551
material modification, 451
mediated electron transfers, 353, 382
mediator, 276–277
concentration, 286
mediator-associated extracellular electron transfer, 105, 145
membrane, 24, 26, 178
MES(s), 94–95, 102–104, 121–123, 126, 128–139, 141–145, 430–433, 436–437, 439, 442, 447–448, 450–451, 453–460, 463, 465, 535
MES of acetate, 442
MES of acetate from CO_2, 452
MES of organics from CO_2, 453
MES performance, 440, 444, 448, 454, 460–461
mesopores, 382

mesoporous carbon, 394
mesoporous silica, 386
metal nanoparticles, 363
  gold nanoparticle, 364
  palladium nanoparticles, 364
  platinum nanoparticles, 364
  porous aerogels, 364
metal-based electrode materials, 404, 407
metallic electrodes, 60
metallic microbial anodes, 481
metalloenzyme, 545
metalloprotein, 546–547, 550–551, 553, 556–558
methane, 99–100, 104, 120, 126, 138, 141
methane production, 446
Methanobacillus omelianskii, 120
Methanobacterium ruminantium, 120
methanogens, 118, 120, 125, 126, 128, 141
Methanosaeta, 128
Methanosaeta (*Methanotrix*) *harundinacea*, 120
Methanosarcina, 128
Methanosarcina barkeri, 120, 128
Methanothermobacter thermautotrophicus, 120
Michaelis–Menten enzyme kinetics, 275
Michaelis–Menten mechanism, 19, 22
microbe–electrode interaction, 449, 460
microbial bioelectrochemcial systems, 534
microbial desalination cells (MDCs), 94–95, 101, 145
microbial electrocatalysis, 403
microbial electrochemical system, 145
microbial electrochemical technologies, 93
microbial electrodes (*see also* MEs), 100, 132, 145, 525–528, 530, 532, 534–536
microbial electrolysis cells (MECs), 94–95, 99–100, 127, 135, 138, 141, 145
microbial electrosynthesis, 94, 145, 429
microbial fuel cells (*see also* MFCs), 94–95, 97–100, 102, 104, 119, 121, 124, 126, 128, 132–133, 136, 137–140, 142–143, 145, 530
microbial solar cells (*see also* MSCs), 102, 124, 145
microorganisms, 432
miniaturized devices, 22, 24–25
mixed cultures, 442
mixed microbial consortia, 436, 444–445, 447, 457, 461
mixed microbial consortium, 452
mixed microbial cultures, 461
molecular wires, 14
monoheme, 129
MtrA, 112, 131
MtrABC, 112
MtrB, 112, 131
MtrC, 112, 117
MtrC (OmcB), 131
MtrD, 112, 131
MtrDEF, 112
MtrE, 131
MtrF, 112, 131
multi-copper oxidases, 13
multi-carbon chemicals, 429
multi-walled carbon nanotubes (MWCNTs), 343, 452

N

Nafion, 136–138
nanocarbon(s), 342, 361
  carbon nanodots, 362
  graphene quantum dots, 361
  herringbone carbon nanofibers, 362
  nanodiamonds, 362
  single-walled carbon nanohorns, 361
nanowire, 589

natural redox mediator, 7, 13, 15, 17
new chemical producing platform technology, 430
nicotinamide adenine dinucleotide cofactor (NAD), 79, 83, 365
  NAD-dependent dehydrogenase, 218, 230, 232–234, 237, 241–243, 245–247, 249–250, 252–255
nitrite reductase, 546–547, 553–555
non-Faradic process(es), 563–568
non-carbonaceous electrodes, 477

O

Ochrobactrum, 125
Ochrobactrum anthropi, 121
off-peak energy storage, 430
ohmic, 122, 134–135, 138
OMC(s), 85, 106, 108–114, 118, 120, 123, 129–130, 145, 565, 567–568, 570–572, 574, 577–579, 581–585, 587–589, 591
  c-type outer membrane cytochrome, 84
  OmcF, 85, 91
OmcA, 112, 131
OmcB, 109–110, 129–130
OmcC, 109, 111, 129–130
OmcE, 110, 129–130
OmcF, 110, 129–130
OmcG, 111
OmcH, 111
OmcT, 111, 129
OmcZ, 108, 110, 129–130
OmpB, 130
on-site production of organics, 439
open circuit potential, 31
optically transparent electrodes
  boron-doped diamond, 548
  indium-doped tin oxide, 547
  metal minigrid, 548
  SnO_2, 547
  TiO_2, 546
optically transparent metal oxide electrodes, 546
optimizing, 440
orientation, 550, 553
Os compounds, 28
osmium-based complexes, 8, 186, 349
outer membrane *c*-type cytochrome, 145
outer membrane cytochrome, 592
  OMC(s), 564, 592
oxide-based microbial anodes, 500
  cytochrome *c* oxidase, 84
  NADH-dehydrogenase, 84
  succinate dehydrogenase, 84
  ubiquinol-cytochrome *c* reductase, 84
oxidoreductases, 6–7, 77, 79, 81–82, 89
  glucose oxidase, 79–82, 90, 354
  laccase, 79, 81–82, 91, 346–347, 349, 355–356, 364–365
oxygen, 79, 81–84, 86–87
  dioxygen, 78–79, 81, 83
oxygen-reducing, 274

P

P. putida, 132
Paracoccus, 535
path length, 550
Pelobacter, 109, 111, 120
Pelotomaculum thermopropionicum, 120
performance, 431, 442–443, 451, 453, 456, 458, 460, 463
performance enhancement, 436, 452
performance improvements, 440
performance indicator, 433, 437, 455, 458, 462
pH, 442, 444
phenazine, 116, 118–119, 132
photoelectrocatalytic, 547
photoelectrochemistry, 556
photoheterotrophs, 127
photosynthesis, 102–103
photosynthetic microbial electrochemical cell (*see also* PMEC), 127, 145

phototrophic, 102, 115
phototrophs, 103
Pichia guilliermondii, 118
pilA, 115, 120
pili-associated extracellular electron transfer (*see also* PEET), 105, 115, 145
ping–pong mechanism, 6–7, 13, 14, 32
plant microbial solar cells (*see also* PMSCs), 102, 124, 145
platform, 436
platinum electrode, 79
  platinum, 78, 81
polymer backbones, 188
porous electrode model, 274
porphyrin, 551, 553
potentiometric electrode, 3
power density, 24, 26, 29, 31, 409
power management system (*see also* PMS), 99, 145
PpcA, 106, 109, 130
PpcA *c*-type cytochromes, 129
PpcB, 106
PpcD, 106
PpcE, 106
practical application, 440, 449
practical implementation, 459
preparation routes, 296
probe, 526, 535
product separation/recovery, 446
product specificity, 436, 461
product specificity and production rates, 448
production of acetate, 443
production of chemicals, 430, 463
production of fuels and chemicals, 431
production rate, 436, 449, 457, 460
productivity, 443
prospective applications, 67
prosthetic group, 6–7, 11–12, 17, 37, 201
protein film IR electrochemistry, 550, 552
proteobacteria, 105, 135, 527–529
proton exchange membrane (PEM), 94–95, 97–98, 122, 127, 135, 137, 142, 144–145
Pseudomonas, 115, 118, 125
Pseudomonas aeruginosa, 116, 118, 123, 132, 535
purposely built anode materials, 404
purposely built electrodes, 417–418, 431, 444–445, 451, 453, 455, 457–459
Pyrobaculum, 121
pyrolytic graphite, 549, 552

Q

Q-band, 551
quartz crystal microbalance, 196
quinones, 8, 28

R

Raman spectroscopy, 573, 586
  Raman, 572, 575, 590
reactor configurations, 65
redox conduction, 564, 571, 581, 585
redox hydrogel, 184, 273, 288
  dimensionless description, 278
  engineering and design, 275, 286
  enzyme–redox hydrogel film, 274
  film thickness, 282
  monolayer equivalents, 273
  partition coefficient, 278
  thin-film system, 280
redox mediator, 173, 178, 180, 353, 548, 557, 563–564, 567–568, 576, 586–587
redox polymer, 349, 384, 548–549
redox potential, 288
redox state, 546, 548, 551, 553, 556
reductase, 84
reduction, 431, 442
resonance Raman spectroscopy
  RR, 574–575, 577–578, 583, 585, 588, 590, 592
  SERR, 578
reticulated vitreous carbon (RVC, *see also* RVE), 347, 452

Rhodoferax ferrireducens, 114
Rhodopseudomonas palustris, 535
rotating disk electrode (RDE), 282, 549, 551–552
Ru complexes, 8
Ru compounds, 28

S

scanning electrochemical microscopy (SECM), 600, 605, 609
scanning electron microscopy (SEM), 525, 534–537
screen printer, 386
second generation biosensor, 19–20
second substrate, 6–7, 13–15, 18
sediment microbial fuel cell (SMFCs), 94, 98–99, 145
SEET, 113, 116, 118, 145
selectivity, 439, 442–443, 460
self-assembled monolayer, 348, 550, 553–555
sensing, 274
separators, 60, 63
sequencing, 527–528, 534
Shewanella, 105–106, 111, 117, 121
Shewanella loihica, 592
  Sl, 581, 584, 587, 592
  So, 589
Shewanella oneidensis, 105, 111–112, 116–118, 123, 131, 535, 538, 592
  So, 580, 587, 592
Shewanella putrefaciens, 119
shuttle-mediated extracellular electron transfer (SEET), 105, 145
simple transmission, 548
single molecule, 554
single-chambered microbial fuel cells (SCMFCs), 94, 97–98, 126, 133, 138–140, 145
Sl, 582
small tetraheme cytochrome *c* (STC), 112, 131, 145
soluble mediators, 174
soret band, 551, 553
spectroelectrochemistry, 546–548, 555, 557, 563, 579, 587, 591
  spectroelectrochemical, 566, 575, 577
spectroscopy, 547
  electron paramagnetic resonance, 556
  fluorescence, 553–555
  infrared, 547–550, 552
  Raman spectroscopy, 551, 553–554
  spectroscopic, 570, 572, 574, 577, 579, 581, 583, 586, 588–589
  surface enhanced resonance Raman spectroscopy, 551, 553–554
  surface-enhanced IR absorption, 550
  UV–Vis, 546–547
  X-ray absorption, 557
  X-ray absorption spectroscopy, 556, 558
spin, 558
spin state, 551, 558
Sporolactobacillaceae, 127
Sporomusa ovata, 100
stainless steel anodes, 407
Streptococcaceae, 127
substrate, 277
sulfate-reducing bacteria (*see also* SRB), 128, 145
surface area, 18, 29, 37–40
surface functionalization, 351–353, 409
surface modification strategies, 416
surface-enhanced infrared absorption, 592
  SEIRA, 576–577, 580, 592
surface-enhanced Raman, 592
  SER, 575–577, 590, 592
  surface-enhanced resonance Raman (SERR), 565, 567, 576, 580, 583–584, 592

surface-modified electrode, 431, 458
  chitosan-modified surface, 346, 460
  modifying the electrode, 440
  positive surface charge, 449
  surface modifications, 351–353, 440, 449–450, 455, 460
  surface treatments, 449
syntrophic, 120, 123–124, 133, 144

T

TERR spectroscopy, 591
tetraheme, 112
Thermincola, 106
Thermincola ferriacetica, 113
Thermincola potens, 113
Thermoanaerobacter, 106, 113
Thermoanaerobacteriacea, 127
Thiele modulus, 279
thin film coating, 409
third generation biosensors, 19
Ti_4O_7, 506
tip-enhanced resonance Raman, 592
  TERR, 592
total petroleum hydrocarbon (TPH), 125, 145
transmission mode, 549, 550
two-chambered MFC (2CMFC), 95, 124, 135, 144
two-substrate enzymes, 7

U

ultraviolet–visible spectroscopy
  UV–Vis, 571–572, 574, 577, 581–582, 585, 588, 592
UME, 605, 609

V

valuable organic, 439
viologens, 8, 28

W

wastewater, 98–99, 101, 104, 121, 124, 126–127, 136–139, 142
wired enzyme, 202

X

X-ray, 535